CLIMATE CHANGE: GLOBAL RISKS, CHALLENGES AND DECISIONS

This book provides an up-to-date synthesis of knowledge relevant to climate change, from the fundamental science of the climate system to the approaches and actions needed to deal with the challenge.

This broad synthesis is unique in that the topics dealt with range from the basic science documenting the need for policy action to the technologies, economic instruments and political strategies that can be employed in response to climate change. Ethical and cultural issues constraining the societal response to climate change are also discussed. As scientific evidence and understanding accumulate, it becomes ever more convincing that the global climate system is moving beyond the patterns of natural variability within which human civilisations have developed and thrived. The good news is that many of the tools and approaches necessary to deal effectively with climate change already exist. The challenge of the twenty-first century is to integrate these instruments into the development trajectories of contemporary societies. This book provides a handbook for those who want to understand and contribute to meeting this challenge.

The book covers a very wide range of disciplines: core biophysical sciences involved with climate change (geosciences, atmospheric sciences, ocean sciences, and ecology/biology) as well as economics, political science, health sciences, institutions and governance, sociology, ethics and philosophy, and engineering. As such it will be invaluable for a wide range of researchers and professionals wanting a cutting-edge synthesis of climate change issues, and for advanced student courses on climate change.

The book was written by a team of authors led by Katherine Richardson, Will Steffen and Diana Liverman. Additional authors are Terry Barker, Frank Jotzo, Daniel M. Kammen, Rik Leemans, Timothy M. Lenton, Mohan Monasinghe, Balgis Osman-Elasha, Hans Joachim Schellnhuber, Nicholas Stern, Coleen Vogel and Ole Wæver.

KATHERINE RICHARDSON is Vice-Dean at the Faculty of Science at the University of Copenhagen and Professor in Biological Oceanography. She has been active both as a member and chairman of several national and international research committees and advisory bodies including the scientific steering committee of the International Geosphere-Biosphere Programme. She is Chairman of the Danish Government's Commission on Climate Change Policy. She was also chairman of the Scientific Steering Committee for the international scientific congress Climate Change: Global Risks, Challenges and Decisions. The focus of her research is carbon cycling in the ocean and how changing climate conditions influence biodiversity in the ocean and the ability of biological processes in the ocean to remove CO_2 from the atmosphere. Richardson has authored over 75 scientific publications and a large number of popular scientific works, including *Our Threatened Oceans* (2009, Haus Publishing; with Stefan Rahmstorf).

WILL STEFFEN is Executive Director of the Climate Change Institute at the Australian National University (ANU), Canberra, and is also Science Adviser, Department of Climate Change and Energy Efficiency Australian Government. From 1998 to mid-2004, he served as Executive Director of the International Geosphere-Biosphere Programme, based in Stockholm, Sweden. His research interests span a broad range within the fields of climate change and Earth System science, with an emphasis on incorporation of human processes in Earth System modelling and analysis; and on sustainability, climate change and the Earth System.

Both Will Steffen and Katherine Richardson were authors on the book *Global Change and the Earth System: A Planet Under Pressure* (2004, Springer).

DIANA LIVERMAN holds appointments at the University of Arizona (where she directs the Institute of the Environment) and Oxford University (working with the Environmental Change Institute). Her main research interests include climate impacts, vulnerability and adaptation, and climate policy, especially the role of the developing world and non-state actors in both mitigation and adaptation. She has written numerous books and articles on the environment, climate and development and advised government, business and NGOs on climate issues. Currently she chairs the scientific advisory committee of the International Global Environmental Change and Food Security Program, co-chairs the US National Academies panel on Informing America's Climate Choices and edits the Annual Review of Environment and Resources.

CLIMATE CHANGE: GLOBAL RISKS, CHALLENGES AND DECISIONS

KATHERINE RICHARDSON

University of Copenhagen, Denmark

WILL STEFFEN

Australian National University, Canberra

DIANA LIVERMAN

University of Arizona and University of Oxford

and

Terry Barker, Frank Jotzo, Daniel M. Kammen, Rik Leemans, Timothy M. Lenton, Mohan Munasinghe, Balgis Osman-Elasha, Hans Joachim Schellnhuber, Nicholas Stern, Coleen Vogel, Ole Wæver

With contributions to chapters by

Myles R. Allen, Giles Atkinson, Marilyn Averill, Jonathan Bamber, Paul M. Barker, Jørgen Bendtsen, Pam Berry, Roberto Bertollini, Nathaniel L. Bindoff, Edward Blandford, Sarah G. Bonham, Niel H. A. Bowerman, Maxwell Boykoff, Ronald D. Brunner, Gregory Buckman, Diarmid Campbell-Lendrum, Josep G. Canadell, Benjamin Cashore, Lynda Chambers, Nakul Chettrin, John A. Church, Kerry H. Cook, Paul Crutzen, Dorthe Dahl-Jensen, Peter Dann, Simon Dietz, Catia M. Domingues, Harry Dowsett, S. S. Drijfhout, Jeff R. Dunn, Hallie Eakin, Thomas Elmqvist, Matthew England, Polly Ericksen, Kirsten Findell, Jean-Pierre Gattuso, Mette Kildegaard Graversen, Nicolas Gruber, Stephen J. Hall, Christian Pilegaard Hansen, Alan M. Haywood, Kieran P. Helm, Jennifer Helgeson, Cameron Hepburn, Daniel J. Hill, Ove Hoegh-Guldberg, Larry Horowitz, John Ingram, Arne Jacobson, Chris D. Jones, Peter Kanowski, Sylvia I. Karlsson-Vinkhuyzen, Lance Kim, Brigitte Knopf, Niels Elers Koch, Katrine Krogh Andersen, Paul Leadly, Hiram Levy II, Valerie N. Livina, Jason Lowe, Jens Friis Lund, Daniel J. Lunt, Amanda H. Lynch, Ariel Macaspac Penetrante, Omar Masera, Constance Mcdermott, Warwick J. McKibbin, Anthony J. McMichael, Anders Melin, Kevin J. Noone, Jørgen E. Olesen, Jisung Park, Donald Perovich, Per F. Peterson, Jonathan Pickering, Stefan Rahmstorf, V. Ramaswamy, Michael R. Raupach, Leanne Renwick, Johan Rockström, Dominic Roser, Minik Rosing, Håkon Sælen, Ulrich Salzmann, Marko Scholze, Thomas Schneider Von Deimling, M. Daniel Schwarzkopf, Frances Seymour, Eklabya Sharma, Drew Shindell, Pete Smith, David A. Stainforth, Konrad Steffen, Martin Stendel, Hanne Strager, Carol Turley, Chris Turney, Paul J. Valdes, S. L. Weber, Neil J. White, Susan E. Wijffels, Mark Williams, Peter J. Wood, Jan Zalasiewicz, Robert J. Zomer

CAMBRIDGE
UNIVERSITY PRESS

CAMBRIDGE UNIVERSITY PRESS
Cambridge, New York, Melbourne, Madrid, Cape Town,
Singapore, São Paulo, Delhi, Tokyo, Mexico City

Cambridge University Press
The Edinburgh Building, Cambridge CB2 8RU, UK

Published in the United States of America by Cambridge University Press, New York

www.cambridge.org
Information on this title: www.cambridge.org/9780521198363

First published 2011

Printed in the United Kingdom at the University Press, Cambridge

A catalogue record for this publication is available from the British Library

Library of Congress Cataloguing in Publication data
Richardson, Katherine, 1954-
Climate Change: Global Risks, Challenges and Decisions / Katherine Richardson,
Will Steffen, Diana Liverman; additional authors, Terry Barker [and ten others];
with contributions to chapters by Myles R. Allen [and many others].
p. cm
Includes bibliographical references and index.
ISBN 978-0-521-19836-3
1. Climatic changes. 2. Climatic changes – Government policy. I. Title.
QC903.R48   2011
363.738′74–dc22   2010042731

ISBN 978-0-521-19836-3 Hardback

To the memory of climate scientist Steve Schneider (1945–2010),
a committed climate change communicator and important mentor to many
whose work is represented in this book.

Contents

Colour plate section between pages 298 and 299.

Writing team

Professor Katherine Richardson (lead author)
Center for Macroecology, Evolution and Climate
Faculty of Science
University of Copenhagen
Tagensvej 16 DK-2200 Copenhagen
Denmark

Professor Will Steffen (lead author)
ANU Climate Change Institute
Coombs Extension
The Australian National University
Canberra
ACT 0200
Australia

Professor Diana Liverman (lead author)
Institute of the Environment
The University of Arizona
PO Box 210158b
Tucson
Arizona 85721
USA
and
Institute of the Environment
Oxford University
OXI 3QY
UK

Dr Terry Barker
Cambridge Centre for Climate Change Mitigation Research
Department of Land Economy
University of Cambridge
19 Silver Street
Cambridge
CB3 9EP
UK

Dr Frank Jotzo
Research Fellow
Research School of Pacific and Asian Studies
The Australian National University
Coombs Building
Canberra
ACT 0200
Australia

Professor Daniel M. Kammen
Renewable and Appropriate Energy Laboratory (RAEL)
University of California, Berkeley
4152 Etcheverry Hall
Berkeley, CA 94720–1731
USA

Professor Rik Leemans
Environmental Systems Analysis Group
Wageningen University
Droevendaalsesteeg 4
PO Box 47
6700AA WAGENINGEN
The Netherlands

Professor Timothy M. Lenton
School of Environmental Sciences
University of East Anglia
Norwich NR4 7TJ
UK

Professor Mohan Munasinghe
Munasinghe Institute for Development (MIND)
10/1, De Fonseka Place
Colombo 5
Sri Lanka

Dr Balgis Osman-Elasha
Climate Change Unit
Higher Council for Environment and Natural Resources
HCENR – Gamaa Street- Khartoum /Sudan
Khartoum, 10488
Sudan

Professor Hans Joachim Schellnhuber CBE
Potsdam Institute for Climate Impact Research (PIK)
P.O. Box 60 12 03
14412 Potsdam
Germany

Professor Nicholas Stern
Suntory and Toyota International Centres for Economics and Related Disciplines
London School of Economics and Political Science
Houghton Street
London WC2A 2AE
UK

Professor Coleen Vogel
School of Geography, Archaeology and Environmental Studies
University of the Witwatersrand
1 Jan Smuts Avenue
Private Bag 3 Wits
2050 Johannesburg
South Africa

Professor Ole Wæver
Center for Advanced Security Theory
Department of Political Science
University of Copenhagen
Øster Farimagsgade 5
1353 Copenhagen K
 Denmark

Contributors: expert boxes

Myles R. Allen, Oxford University
Giles Atkinson, London School of Economics
Marilyn Averill, University of Colorado at Boulder
Jonathan Bamber, Bristol University
Paul M. Barker, CSIRO
Jørgen Bendtsen, VitusLab Denmark

Pam Berry, Oxford University
Roberto Bertollini, World Health Organization
Nathaniel L. Bindoff, University of Tasmania
Edward Blandford, University of California, Berkeley
Sarah G. Bonham, Leeds University
Niel H. A. Bowerman, Oxford University
Maxwell Boykoff, University of Colorado
Ronald D. Brunner, University of Colorado
Gregory Buckman, Australian National University
Diarmid Campbell-Lendrum, World Health Organization
Josep G. Canadell, CSIRO
Benjamin Cashore, Yale University
Lynda Chambers, Bureau of Meteorology, Australia
Nakul Chettrin, International Centre for Integrated Mountain Development
John A. Church, CSIRO
Kerry H. Cook, University of Texas at Austin
Paul Crutzen, Max-Planck-Institut für Chemie
Dorthe Dahl-Jensen, University of Copenhagen
Peter Dann, Phillip Island Nature Parks, Australia
Simon Dietz, London School of Economics
Catia M. Domingues, CSIRO
Harry Dowsett, U.S. Geological Survey
S. S. Drijfhout, Koninklijk Nederlands Meteorologisch Instituut
Jeff R. Dunn, CSIRO
Hallie Eakin, Arizona State University
Thomas Elmqvist, Stockholm University
Matthew England, University of New South Wales
Polly Ericksen, Livestock Research Institute, Nairobi
Kirsten Findell, NOAA
Jean-Pierre Gattuso, l'Observatoire Océanologique de Villefranche-sur-Mer
Mette Kildegaard Graversen, Fødevareøkonomisk Institut
Nicolas Gruber, ETH Zurich
Stephen J. Hall, Consultative Group on International Agricultural Research
Christian Pilegaard Hansen, Copenhagen University
Alan M. Haywood, Leeds University
Kieran P. Helm, University of Tasmania
Jennifer Helgeson, London School of Economics
Cameron Hepburn, Oxford University
Daniel J. Hill, British Geological Survey
Ove Hoegh-Guldberg, University of Queensland
Larry Horowitz, NOAA
John Ingram, Oxford University

Arne Jacobson, Humboldt State University
Chris D. Jones, Met Office UK
Peter Kanowski, Australian National University
Sylvia I. Karlsson-Vinkhuyzen, Finland Future Research Centre, Turku School of Economics
Lance Kim, University of California, Berkeley
Brigitte Knopf, Potsdam Institute for Climate Impact Research
Niels Elers Koch, University of Copenhagen
Katrine Krogh Andersen, Danish Meterological Institute
Paul Leadly, Université Paris-Sud 11
Hiram Levy II, NOAA
Valerie N. Livina, University of East Anglia
Jason Lowe, Met Office UK
Jens Friis Lund, University of Copenhagen
Daniel J. Lunt, Bristol University
Amanda H. Lynch, Monash University
Ariel Macaspac Penetrante, International Institute for Applied Systems Analysis
Omar Masera, Universidad Nacional Autónoma de México
Constance McDermott, Oxford University
Warwick J. McKibbin, Australian National University
Anthony J. McMichael, Australian National University
Anders Melin, Lund University
Kevin J. Noone, ITM Stockholms Universitet
Jørgen E. Olesen, University of Aarhus
Jisung Park, Oxford University
Donald Perovich, US Army Corps of Engineers
Per F. Peterson, University of California, Berkeley
Jonathan Pickering, Australian National University
Stefan Rahmstorf, Potsdam Institute for Climate Impact Research
V. Ramaswamy, NOAA
Michael R. Raupach, CSIRO
Leanne Renwick, Phillip Island Nature Parks
Johan Rockström, Stockholm Environment Institute
Dominic Roser, University of Zurich
Minik Rosing, University of Copenhagen
Håkon Sælen, Centre for International Climate and Environmental Research
Ulrich Salzmann, British Antarctic Survey
Marko Scholze, Bristol University
Thomas Schneider von Deimling, Potsdam Institute of Climate Impact Research
M. Daniel Schwarzkopf, NOAA
Frances Seymour, Consultative Group on International Agricultural Research
Eklabya Sharma, International Centre for Integrated Mountain Development
Drew Shindell, NASA

Pete Smith, University of Aberdeen

David A. Stainforth, London School of Economics

Konrad Steffen, University of Colorado

Martin Stendel, Danish Meterological Institute

Hanne Strager, University of Copenhagen

Carol Turley, Plymouth Marine Laboratory

Chris Turney, University of New Southwales

Paul J. Valdes, Bristol University

S. L. Weber, Koninklijk Nederlands Meteorologisch Instituut

Neil J. White, CSIRO

Susan E. Wijffels, CSIRO

Mark Williams, University of Leicester

Peter J. Wood, Australian National University

Jan Zalasiewicz, University of Leicester

Robert J. Zomer, International Centre for Integrated Mountain Development

Foreword

An important pinnacle was reached in the journey towards addressing one of the greatest global challenges of our time at the UNFCCC climate change conference (COP15) in Copenhagen in December 2009.

For the first time since the climate change agenda left the offices of scientists and environmentalists, and moved onto the agendas of heads of governments, world leaders on a large scale recognised the need to contain the human-induced global warming to a maximum of 2 °C above pre-industrial levels. On the basis of that recognition, world leaders agreed to take action to meet this challenge.

The path to this recognition was not without obstacles; it was a steep climb, but a climb inspired and fuelled by the increasing force of the scientific findings mounting and developing. In 2007, the Intergovernmental Panel on Climate Change (IPCC) published its Fourth Assessment Report, which gave a thorough and comprehensive review of the science of climate change. This report played an immensely important role in creating global awareness of the urgency of a global response to climate change. However, scientists produce new results and publish new findings every day.

It was thus very timely that the International Alliance of Research Universities (IARU) in March 2009, only nine months before COP15, organised the congress 'Climate Change: Global Risks, Challenges & Decisions'. A uniquely wide scope of scientific disciplines focusing on climate change was represented at this congress. The discussions emphasised the vast knowledge base available regarding climate change, and provided a forum in which to present and discuss the newest scientific results. The scope of the global challenge clearly requires the combined efforts of scientific disciplines; natural climate science integrated with the social, political and economic sciences in order to be addressed.

In many areas, new results presented at the congress and in this book have continued to document trends of climate change, as well as its current and anticipated impacts. The global community must deal effectively with climate change, both through mitigation and adaptation. Fortunately, we already have a large variety of tools at hand to do so. This book provides updated information on the existing tools as well as potential pathways to reach our climate goals, including that of limiting the human-induced increase in global temperature to a maximum of 2 °C.

Addressing the climate change challenge is not only an issue for natural scientists, engineers and economists. It is a task that cannot be detached from the geopolitical context of energy and climate security. Contributions on this issue from relevant areas of the social sciences and humanities are a great asset of this book.

It is important that we now make our utmost effort to retain climate change issues at the very top of the political agenda. With every year of delayed progress, there is the danger that societies will continue to invest in outdated technologies.

This book is more than a testimony of just another congress or climate event. It comprises an essential resource in explaining the current scientific understanding of climate change. The book is underpinned by an unprecedented breadth of scientific disciplines and expertise and, as such, constitutes a solid source of incentives for politicians and others who wish to develop a thorough understanding of climate change. It conveys the call for humankind to take action.

Lars Løkke Rasmussen
Prime Minister of Denmark

Preface

Human activities impact many of the Earth's natural functions and cycles. Local and regional impacts of human activity on the planet are easily seen, while global impacts are not so immediately obvious. Nevertheless, studies in Earth System science carried out in recent decades have unequivocally demonstrated impacts of human activity that reverberate at the global level.

This recognition has led to the suggestion that we may have moved out of the geological period referred to as the Holocene – an epoch that covers the past approximately 12 000 years and in which human societies have flourished – and have now entered a new era, where human impacts are changing Earth System functioning. The knowledge that human activities can and do influence planetary functioning implies an obligation to actively monitor and manage the relationship between humans and the planet.

This paradigm shift in the relationship between humans and the planet actually started with the Montreal Protocol (ratified in 1989), which limits the global emission of chlorofluorocarbon (CFC) gases that lead to a reduction in the ozone layer that surrounds the Earth and shields it from dangerous ultraviolet radiation. Dealing with human-induced climate change can be viewed as the next step in this redefinition of the human–Earth relationship.

Managing the human activities that lead to climate change is more difficult than regulating the emission of CFC gases, as it will require radical changes in the very fabric of most societies: a change in attitude with respect to energy use as well as changes in global society's primary energy source, use of natural resources, methods of food production and modes of transport. How we as a global society deal with the knowledge that human activities influence the Earth's climate system can be viewed as a harbinger for our species' future relationship with the planet.

The Intergovernmental Panel on Climate Change (IPCC) has, in four reports, systematically assessed and presented the evolving scientific understanding of human-induced climate change. Especially the latest of the reports (from 2007) has been instrumental in increasing public and political awareness concerning climate change. Thanks to extensive research activities across the globe, the scientific understanding of climate change has continued to advance since the last IPCC assessment.

This book summarises the highlights of this new research and provides an up-to-date overview of the current state of scientific understanding of climate change, its known and projected impacts, and the options we have available for responding to the challenges it presents. While not being a complete report of the proceedings, this book is developed from presentations and discussions that took place at the open scientific congress, *CLIMATE CHANGE: Global Risks, Challenges and Decisions*, which was held in Copenhagen, 10–12 March 2009, and organised by the International Alliance of Research Universities.[1] Although the presentations made at the congress were not peer reviewed, they are sometimes used as examples in the book to illustrate more generic points being made. In addition to drawing upon contributions to the congress, the book draws upon peer-reviewed papers that have appeared in the scientific literature in recent years and subsequent to the congress. The book has been written by a team of authors led by Katherine Richardson, Will Steffen, and Diana Liverman. Each author has contributed to the sections of the book where he/she has expertise.

---

[1] Australian National University, ETH Zürich, National University of Singapore, Peking University, University of California – Berkeley, University of Cambridge, University of Copenhagen, University of Oxford, The University of Tokyo, Yale University. For further information, please visit http://www.IARUni.org.

Acronyms and abbreviations

ACF	autocorrelation function
AIDS	acquired immunodeficiency syndrome
AIM	Action Impact Matrix
AR4	IPCC Fourth Assessment Report
AR5	IPCC Fifth Assessment Report
ASE	Amundsen Sea Embayment
AWG	Ad hoc Working Group
C	carbon
CAFE	Corporate Average Fuel Economy
CCS	carbon capture and storage
CDM	Clean Development Mechanism
CDR	carbon dioxide removal
CER	certified emission reduction
CFC	chlorofluorocarbon
CGIAR	Consultative Group on International Agricultural Research
CH_4	methane
CHP	combined heat and power
CO_2	carbon dioxide
CO_2-e	carbon dioxide equivalent
COP15	15th Conference of the Parties
CRU	Climate Research Unit
CWC	cumulative warming commitment
DALY	disability-adjusted life-years
DFA	detrended fluctuation analysis
DNA	deoxyribonucleic acid
EAIS	East Antarctic Ice Sheet
EBAMM	Energy Resources Group Biofuel Meta-Analysis Model
EE	energy efficiency
EEP	eastern equatorial Pacific
EETS	European Emissions Trading System

EMIC	Earth System Model of Intermediate Complexity
ENGO	environmental non-governmental organisations
ENSO	El Niño Southern Oscillation
EPA	Environmental Protection Agency
ESSP	Earth System Science Partnership
EU	European Union
FAO	United Nations Food and Agriculture Organization
FIT	feed-in tariff
GATT	General Agreement on Tariffs and Trade
GCM	General Circulation Model; in non-technical discussions, often replaced by Global Climate Model
GDP	gross domestic product
GECAFS	Global Environmental Change and Food Systems
GFC	global financial crisis
GHG	greenhouse gas
GIS	Greenland Ice Sheet
Gl	gigalitres
Gt	gigatonne
GWI	global warming intensity
HIV	human immunodeficiency virus
HKH	Hindu Kush–Himalaya
HKHT	Hindu Kush–Himalaya–Tibetan
HVAC	heating, ventilation and cooling
IARU	International Alliance of Research Universities
ICU	initial condition uncertainty
IMF	International Monetary Fund
IOD	Indian Ocean Dipole
IPCC	Intergovernmental Panel on Climate Change
IR	infrared
ISA	Integrated Sustainability Assessment
ISM	Indian Summer Monsoon
K	kelvin
kW_e	kilowatts of electric power
LED	light-emitting diode
LPJ	Lund-Potsdam-Jena (dynamic global vegetation model)
MEF	Major Economies Forum
MT	megatonnes
MW	megawatt
MW_e	megawatts of electric power
NADW	North Atlantic deep water
NAMA	nationally appropriate mitigation action
NAO	North Atlantic Oscillation

NAPA	National Adaptation Programmes of Action
NATO	North Atlantic Treaty Organization
NBER	National Bureau of Economic Research
NGO	non-governmental organisation
NO_x	nitrogen oxide
NRDC	Natural Resources Defense Council
OECD	Organisation for Economic Co-operation and Development
OEED	Office of Economic Employment and Development
OPEC	Organization of the Petroleum Exporting Countries
PACE	Property Assessed Clean Energy
PACJA	Pan African Climate Justice Alliance
P–E	precipitation–evaporation
PEAC	Pacific ENSO Applications Center
PETM	Paleocene–Eocene Thermal Maximum
p.p.m.	parts per million
PTC	production tax credit
R&D	research and development
RE	renewable energy
REDD	Reducing Emissions From Deforestation and Forest Degradation
RPS	Renewable Energy Portfolio Standards
SAM	Southern Annular Mode
SO_x	sulphur oxide
SRM	solar radiation management
SST	sea surface temperature
SWNA	Southwest North America
THC	thermohaline circulation
UCDP	Uppsala Conflict Dataset Program
UEA	University of East Anglia
UN	United Nations
UNDP	United Nations Development Programme
UNEP	United Nations Environment Programme
UNFCCC	United Nations Framework Convention on Climate Change
UK	United Kingdom
USA	United States of America
VA	vulnerability areas
VMT	vehicle miles travelled
WAIS	West Antarctic Ice Sheet
WAM	West African Monsoon
WTO	World Trade Organization
WWF	World Wildlife Fund
XBT	eXpendable BathyThermographs

Part I

Climatic trends

For many key parameters, the climate system is moving beyond the patterns of natural variability within which civilisations have developed and thrived...

1

Identifying, monitoring and predicting change in the climate system

'Without the willow, how to know the beauty of the wind' [1]

Weather directly impacts our lives on a minute-to-minute basis. Radio and television channels are devoted to keeping us up to date on current and future weather conditions. These include temperature, barometric pressure, precipitation, severe storms, humidity, wind and more. When we refer to 'climate', we mean average patterns in weather. Thus, climate change is a deviation from the weather patterns that have prevailed over a given period. Taken together, the weather we experience and the weather patterns across the globe are the product of processes occurring in the Earth's 'climate system', which is composed of interactions between the atmosphere, the hydrosphere (including the oceans), the cryosphere (ice and snow), the land surface and the biosphere. Ultimately, this system is controlled by the amount of energy stored as heat at the Earth's surface and the redistribution of this heat energy. Because we humans live in the atmosphere at the surface of the Earth, we (wrongly) assume that changes in air temperature are the only and best indicator of climate change. In fact, relatively little (<5%) of the change in the amount of heat energy stored at the Earth's surface that has taken place in recent decades has occurred in the atmosphere (IPCC, 2007a).

To understand changes in the climate system, the changes in the heat energy content of compartments other than the atmosphere also need to be considered. Regardless of where in the climate system we are looking for evidence of possible change, however, there is one common rule: identifying changes in the climate system requires data series that are collected over several decades – three at a minimum but five or more are better. Newspaper headlines have, in recent years, been eager to declare global warming as a thing of the past on the basis of one or a few years that have been colder than the immediately preceding. However, the presence or absence of global warming cannot be identified on the basis of one or a few years' data.

[1] Attributed to Lao She, Chinese writer (1899–1966).

Climate Change: Global Risks, Challenges and Decisions, Katherine Richardson, Will Steffen and Diana Liverman *et al*. Published by Cambridge University Press. © Katherine Richardson, Will Steffen and Diana Liverman 2011.

1.1 The interaction of many different processes make up the climate system

The factors that influence the global climate system are called 'climate forcings'. The most important of these is, of course, the sun and the energy that it transmits to Earth. Human activities do not directly influence the amount of energy produced by the sun. However, there are natural variations in the sun's activity and these variations obviously influence the global climate system.

Also the shape of the Earth's orbit around the sun influences the amount of solar energy reaching the Earth. Much of the climate variability recorded during the Earth's almost 5 billion-year history can be explained by changes in several characteristics of the Earth's orbit around the sun. For example, during the past approximately 12 000 years, the Earth's orbit has been more circular than at any other time during the last half-million years. This has resulted in a particularly stable and – in comparison to a period of many thousands of preceding years prior to this – warm climate on Earth, and it has been suggested that this comparatively warm and stable climate may have been a contributing factor to the rapid development of human societies (van der Leeuw, 2008). Scientists predict that the Earth's orbit around the sun will continue to have a similar orbit for thousands of years into the future (Berger and Loutre, 2002). We can, then, expect that in the absence of other changes in the climate system, the climate will remain relatively warm and stable in the foreseeable future.

In addition to the sun itself, there are a number of climate forcings that influence the amount of solar energy that reaches the Earth's surface, the amount that is retained on the Earth itself, and the amount that is radiated back into the atmosphere and retained there as heat (Figure 1.1). Many of these climate forcings are found in the atmosphere either as greenhouse gases (GHGs) or aerosols (fine particles or liquid droplets).

A part of the sun's energy reaching the Earth is radiated back into the atmosphere in the form of infrared (IR) radiation (heat). Greenhouse gases absorb this radiation and the process results in heat retention in the atmosphere. This 'greenhouse effect' has been well understood since the nineteenth century and it is not unique to the Earth. Every planet with an atmosphere containing greenhouse gases experiences a greenhouse effect; the extreme surface temperatures (440 °C) on Venus, for example, can only be explained by the high concentration of CO_2 in the atmosphere there. Without the greenhouse effect, the average temperature on Earth would be about −19 °C, i.e. approximately 34 °C colder than today.

The most important greenhouse gas in Earth's atmosphere is actually water vapour, which accounts for about 60% of the natural greenhouse effect for clear skies (Kiehl and Trenberth, 1997). Human activities have not directly resulted in a significant change in the absolute amount of water vapour in the atmosphere (Gordon *et al.*, 2005). Fast feedbacks within the climate system can, however, change the amount and distribution of water vapour. Because there is no significant *direct* human influence on water vapour concentrations, the importance of water vapour as a greenhouse gas is seldom a topic of discussion in the public debate concerning climate change. Instead, most of the discussion focuses on the greenhouse gases whose concentrations have been directly influenced by human

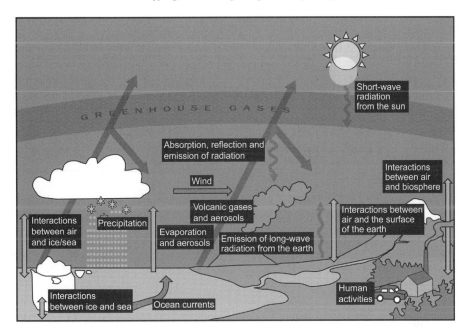

Figure 1.1 Components of the climate system.

activities. Of these, CO_2 is the single most important gas.[2] Human-induced changes in other greenhouse gases are, however, also quantitatively important in terms of their impact on the climate system and it is as important to focus on their reduction of emissions as it is to focus on CO_2 reductions (Chapter 8).

The accumulation of greenhouse gases is not the only change in the atmosphere as a result of human activities. Different types of aerosols are also introduced to the atmosphere via our activities. These different aerosol types have different influences on the climate system – some result in net warming of the planet (because they absorb energy and retain it as heat) and others in net cooling (because they scatter or reflect incoming radiation). Figure 1.2 illustrates the various climate forcings influenced directly by human activities and how they are estimated to have changed in the period 1750–2000. People are often surprised to realise that human-induced changes in aerosol concentrations have had a net cooling effect since the industrial revolution. The flip side of that coin is, however, that any reduction in emission of the aerosol types which have a net cooling effect in the climate

---

[2] It has become a convention in much of the political discussion concerning greenhouse gas emission to refer to the impact of all greenhouse gases emitted through anthropogenic activities under the heading of 'CO_2'. To do so, the impact of the non-CO_2 greenhouse gases on warming is converted to the amount of CO_2 that would be required to give the same warming effect. Thus, the concentrations of the non-CO_2 greenhouse gases are converted to 'CO_2 equivalents' (CO_2-e). While this convention is convenient, and appears in some chapters of this book, it is important to remember that CO_2 is not the only greenhouse gas impacted by human activities. Other GHGs influenced by human activities include methane, nitrous oxide and ozone.

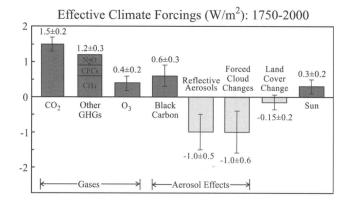

Figure 1.2 Greenhouse gas and aerosol climate forcings influenced by human activities and how they are estimated to have changed in the period 1750–2000. Black carbon is material (e.g. soot) produced by the incomplete combustion of fossil fuels, biomass and biofuels. Source: Hansen *et al.* (2005). Reproduced/modified by permission of American Geophysical Union.

system will result in an additional warming. This means that efforts to reduce many types of air pollution will have as a consequence an increase in global warming. Precisely how much global temperature will increase as a result of a reduction in aerosols stemming from human activities is not yet well known, and is thus currently a topic of enormous interest and active research (Box 1.1).

Box 1.1

Potentially strong sensitivity of late twenty-first century climate to projected changes in short-lived air pollutants

HIRAM LEVY II, M. DANIEL SCHWARZKOPF, DREW SHINDELL, LARRY HOROWITZ, V. RAMASWAMY AND KIRSTEN FINDELL

Previous projections of future climate change have focused primarily on long-lived greenhouse gases such as carbon dioxide, with much less attention paid to the projected emissions of short-lived pollutants, i.e. aerosols and their precursors. When interpolated out to 2050 with a middle-of-the-road IPCC emission scenario (A1B), two of the three climate models in a recent study found that changes in short-lived pollutants contributed 20% of the predicted global warming (CCSP, 2008).

Interpolating out to 2100 with this emission scenario (A1B), one climate model (GFDL – CM2.1) found that projected changes in the emissions of short-lived pollutants (primarily a sharp decrease in sulphur dioxide and a doubling of black carbon) were responsible for

30–40% of the summertime (June–July–August) warming predicted over central North America and southern Europe by the end of the twenty-first century. This leads to a significant decrease in precipitation and increase in soil drying in the summertime central USA. Moreover, the primary increase in radiative forcing from the changing levels of these short-lived pollutants is over Asia (Figure 1.3) while the primary summertime climate response in degrees centigrade (Figure 1.4) is over the central USA and southern Europe (Levy *et al.*, 2008).

   While this general disconnect between the regional locations of the pollutants and their radiative forcing and the regional climate response is quite robust across a range of climate models (Shindell *et al.*, 2008 and references therein), and the particular climate sensitivities of the summertime central USA and southern Europe are also robust across the climate models in the IPCC (2007a), there are two important caveats:

- the magnitude of the climate response to changing emissions of short-lived pollutants depends critically on their projected trajectories, and these projections are highly uncertain (Shindell *et al.*, 2008),

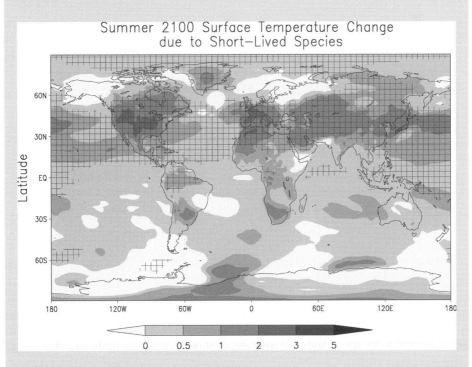

Figure 1.3 Summer (JJA) temperature changes (K) from the 2000s (years 2001–2010) to the 2090s (years 2091–2100) for the forcing only by the short-lived species (A1B(2090s) – A1B*(2090s)) – (A1B(2000s) – A1B*(2000s)). Source: modified from Levy *et al.* (2008).

Box 1.1 (*cont.*)

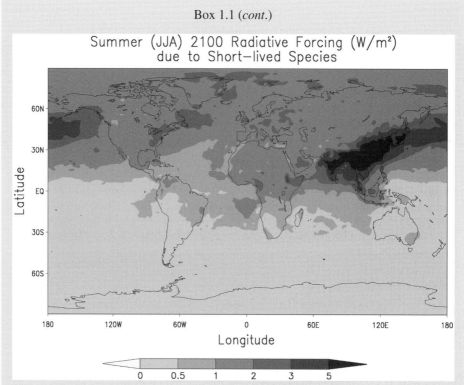

Figure 1.4 Changes in summer-mean adjusted radiative forcing (W m⁻²) between years 2100 and 2000 for the contribution from short-lived species only. Source: modified from Levy *et al*. (2008).

- this calculation only considered the direct radiative effect of aerosols (i.e. indirect effects such as cloud formation were not considered), and the aerosols were treated as external mixtures (i.e. each particle is assumed to be composed of one pure compound).

Changes on the Earth's surface itself can also influence the climate system. Alterations in land or sea cover can, for example, alter the Earth's ability to reflect sunlight (i.e. cause a change in 'albedo'). The mechanisms causing albedo change are described in Box 1.2. With respect to human-induced albedo changes on the Earth, one of the most worrying of those occurring is the melting of sea-ice (Box 7.2), as ice reflects most of the light that impinges upon it and water absorbs the majority of this energy. In addition, boreal forests are moving northwards into tundra ecosystems as the climate warms. This means that during winter, less snow is exposed directly to the sky because more of it lies underneath the forest canopy than before. These two phenomena lead to a decrease in the reflectance of incoming solar radiation, and thus act to accelerate regional warming.

Box 1.2
Albedo

KEVIN J. NOONE

The overall energy balance of the Earth is very simple: at steady state, the amount of incoming energy from the sun that is absorbed by the Earth is matched by the amount of energy that the Earth radiates back to space. Mathematically, this simple energy balance is written as:

$$(1-A)S_0 = 4\sigma T_e^4 \tag{1.1}$$

where S_0 is the solar constant (the amount of energy per square metre from the sun that reaches the Earth's orbit; 1360 W m^{-2}), σ is the Stephan–Boltzmann constant ($5.67*10^{-8}$ W m^{-2} K^{-4}), T_e is the Earth's *effective temperature* (related to, but not the same as, the surface temperature) and A is the planetary albedo – the fraction of incoming solar radiation that is reflected back to space (making $1-A$ the amount that the Earth absorbs).

If the albedo increases, less energy is absorbed, and the effective temperature will decrease; if the albedo decreases, more energy is absorbed and the effective temperature will increase.

The Earth's albedo is determined by how reflective different land and ocean surfaces are, as well as by clouds and particles in the atmosphere. Feedbacks and interactions between incoming solar radiation and snow and ice cover on both land and oceans are thought to be main drivers of the natural glacial/interglacial oscillations that we have observed over the past 800 000 years. However, human activities have also altered the Earth's albedo and affected the energy balance.

Human-induced land-use changes and desertification have changed the reflectivity of the land surface. Most of our land-use activities tend to increase the reflectivity of the surface. Replacing trees with grasslands or crops through deforestation, and replacing grasslands with bare soil or sand due to overgrazing and desertification, are examples of land-use changes that increase the surface albedo. Betts *et al.* (2007) estimated that the present global mean radiative forcing due to surface albedo changes relative to 'natural' (i.e. pre-1750) conditions is −0.2 W m^{-2}. This value is comparable with the global mean radiative forcing due to N_2O of 0.16 W m^{-2} (IPCC, 2007b). While this global mean cooling may be relatively small compared to the warming of all the greenhouse gases, on a regional basis the forcing may be much larger (Betts *et al.*, 2007).

As the planet warms due to human activities, the surface area covered by snow and ice is decreasing. Arctic sea-ice has already been observed to be decreasing in area much faster than expected (Box 7.2). Ice is very reflective (A is between about 0.7 and 0.9), while ocean water absorbs most of the incoming solar radiation (A is less than about 0.1). Replacing a very reflective surface with a very absorptive one results in more energy being absorbed and retained by the surface, further accelerating surface warming. This becomes an amplifying feedback, as a warmer surface will result in even less snow and ice cover. Humans can also change the albedo of snow and ice surfaces through the deposition of absorbing particles. Black carbon deposited on snow and ice surfaces has been estimated to have a global mean warming effect of about +0.1 W m^{-2} (IPCC, 2007b); however, this value remains uncertain.

In addition to changes in surface albedo, human activities have also changed the overall reflectivity of the atmosphere by modifying the properties of aerosol particles and clouds

Climatic trends

Box 1.2 (*cont.*)

Figure 1.5 Image of 'ship tracks' in clouds off the coast of California, USA. Source: adapted from Noone *et al.* (2000).

(Box 1.1). 'Ship tracks' (Figure 1.5) are a striking example of how human particulate emissions (from individual ships) can increase the albedo of marine clouds (Noone *et al.*, 2000). The global result of this 'indirect cloud albedo forcing' is estimated to be −0.7 W m^{-2} (range of −0.3 to −1.8; IPCC, 2007b). Particles can directly alter the planetary albedo by absorbing or scattering incoming solar radiation. The overall global direct radiative forcing due to aerosols is estimated to be negative (−0.5 W m^{-2}, range of −0.1 to −0.9; IPCC, 2007b). However, as particles can both scatter and absorb solar radiation depending on their chemical composition, and as composition varies widely across the globe, the regional radiative effects of aerosols can be very different (Ramanathan and Feng, 2008). These regional differences can lead to different outcomes if particle abatement strategies are implemented. As one example, particulate pollution over the eastern seaboard of North America has a net cooling effect on both the atmosphere and the surface by scattering incoming solar radiation back to space. 'Atmospheric brown clouds' over parts of Asia containing high concentrations of black carbon aerosols can cool the surface by absorbing incoming radiation, but heat the atmosphere (Ramanathan and Carmichael, 2008). Reducing the amount of particulate emissions in these regions by the same amount could lead to very different results in terms of the energy balance in the regions.

A complete description of the climate system would require a book in itself. Therefore, the focus in this brief discussion has been on the climate forcings that are influenced by human activities. These are many and, as we have seen, influence the climate system in opposing ways. In the following sections, we examine how scientists identify changes in the functioning of the climate system and their probable causes.

1.2 Identifying changes in the climate system

That large-scale burning of fossil fuels (coal) would lead to global warming was already pre-dicted by the Swedish chemist, Svante Arrhenius, in 1896 (Arrhenius, 1896).[3] Thus, the theo-retical understanding of the potential for humans to influence the climate system has been in place for over a century. However, because of natural variability in the system, demonstrating that a change has occurred requires the accumulation of data time series that span decades.

Perhaps the most convincing evidence that there is a strong relationship between atmos-pheric CO_2 concentration and temperature was published in 1999 when a Russian and French research team published an ice core record from Antarctica that provided a time-series of data describing the atmospheric concentration of CO_2 and methane, as well as a proxy for temperature, from 420 000 years ago until the near-present (Petit *et al.*, 1999). A more recent, longer ice core pushes the data record back to almost 800 000 years (EPICA community members, 2004). When put together with data collected in the recent past, the picture clearly emerges of a major, rapid change in the atmospheric concentrations of CO_2 and other important greenhouse gases (Chapter 4) since the advent of the industrial revolu-tion. Several lines of evidence confirm that the additional CO_2 that has accumulated in the atmosphere since the beginning of the industrial revolution is almost entirely caused by human activities (Prentice *et al.*, 2001).

However, because there are so many climate forcings and they can work in opposite directions, identifying a change in one forcing – in this case, greenhouse gas forcing – does not necessarily imply a significant change in the climate system as a whole. Identifying such a change requires demonstration of a change in the total amount of heat stored at the surface of the Earth. As we live in the surface atmosphere and are directly affected by its temperature, we have long been routinely measuring surface air temperature. These data are now extremely valuable in identifying significant changes in the climate system. Figure 1.6a shows proxy-based estimates of the near-surface air temperature in the northern hemisphere since about 200 AD. There is considerable uncertainty in temperature records in the period before the thermometer was invented. Nevertheless, it is clear that there has been rapid warming in recent decades.

Despite an upward trend in temperatures in recent decades, year-to-year temperature fluctuations are obvious when we look closely at the annual global average for near-surface air temperature from 1970 to the present (inset in Figure 1.6b). The causes of these year-to-year temperature differences are, in many cases, well understood. It is well known, for example, that strong El Niño and La Niña events can influence global air temperature in a

---

[3] A timeline for the discovery of human-induced climate change is found in Chapter 17.

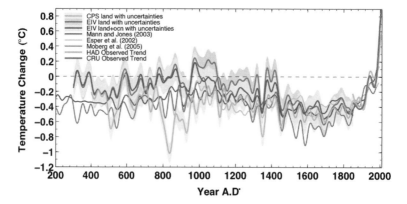

Figure 1.6a Northern hemisphere temperature change since 200 AD (reconstructed for the period up to the initiation of actual measurement). Source: http://www.copenhagendiagnosis.com. (In colour as Plate 1.)

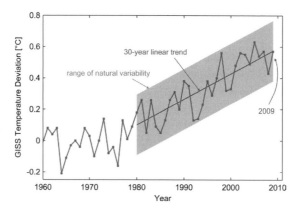

Figure 1.6b Global average annual temperature for the period 1960–2009. The line shows the 30-year linear trend 1980–2009. The shaded area shows ± two standard deviations of the detrended data of the past 30 years, i.e. ± 0.19 °C. Data are GISS data (NASA Goddard Institute for Space Studies, USA); http://data.giss.nasa.gov/gistemp/). Analysis performed by S. Rahmstorf.

positive and negative direction, respectively. Likewise, volcanic eruptions – because they lead to an increase in atmospheric aerosols – can lead to cooler global temperatures for a period of a year or two. Because of these and other causes of year-to-year variations in global temperature, a single cooler year, such as 2008, cannot be taken as evidence of a reversal in the upward trend of air temperatures recorded over recent decades. To be able to identify a statistically significant change in the climate system using air temperature data, a time-series of several decades is necessary. Thus, it is only recently that time-series of sufficient length have been available to identify climate change, and it was as recently as 2007 that the UN Intergovernmental Panel on Climate Change (IPCC) stated that there is unequivocal evidence of a warming in the climate system and estimated that there is a better than 90% probability that human activities are the primary cause (IPCC, 2007a).

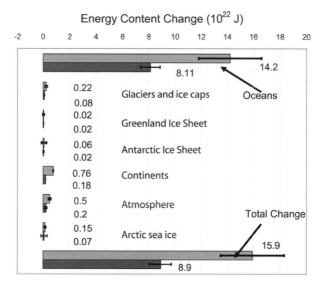

Figure 1.7 Energy content changes in different components of the Earth System for two periods (1961–2003 and 1993–2003). Light grey bars are for 1961 to 2003, grey bars for 1993 to 2003. Positive energy content change means an increase in stored energy (i.e. heat content in oceans, latent heat from reduced ice or sea-ice volumes, heat content in the continents excluding latent heat from permafrost changes, and latent and sensible heat. All error estimates are 90% confidence intervals. Source: IPCC (2007b).

1.3 Other indicators of change in the climate system

From our anthropocentric viewpoint, we tend to believe that changes in surface air temperature are the only indicator of changes in the climate system. However, surface air temperature is not the only – or even necessarily the most robust – indicator of a change in the climate system.

Global climate conditions are dictated by the amount of energy stored as heat at the Earth's surface and the redistribution of this energy. Heat is stored in many other compartments in the Earth System. By far the greatest amount of the additional heat energy that has accumulated on Earth in recent decades is stored in the oceans. Figure 1.7 shows the results of the IPCC's analysis of the magnitude of change in heat content in various compartments of the surface of Earth in the period 1961–2003. That over half of the change in all compartments is estimated to have taken place during the last 10 years of the study period indicates an increasing rate of warming in recent decades.

To date, the global surface air temperature has risen by about 0.7 °C over the pre-industrial temperature (IPCC, 2007a). However, the heat absorbed on the Earth does not immediately influence all compartments equally. Thus, for example, heat that recently has become stored in the ocean has not yet influenced surface air temperature and the climate conditions we experience on land. Because of this time lag, it is predicted (IPCC, 2007a) that the Earth is committed to a warming of approximately 1.3 °C above the pre-industrial period – even if all anthropogenic emission of greenhouse gases would cease as of today.

Given the huge amount of heat stored in the ocean we can, in essence, say that air temperature change nearly as large as the one the Earth has experienced since the industrial revolution is already stored in the ocean.

The ocean, then, is the major heat reservoir of relevance for the development of Earth's climate system and it is here that we may expect to find the best evidence for change in the global climate system. Collecting data on the global ocean temperature is not, however, a simple undertaking and it is only recently that ocean temperature datasets have been of sufficient length and quality to make robust conclusions concerning trends in ocean temperature. There have been significant new analyses of the state of the ocean since the 2007 IPCC report across a broad range of parameters (Box 1.3). These have improved our understanding and confidence in the changing state of the oceans and indicate, among other things, that the ocean temperature has been rising faster (possibly up to 50% faster) in recent decades than was realised at the time of the previous IPCC report.

Box 1.3
Ocean warming and sea-level rise

JOHN A. CHURCH, NEIL J. WHITE, CATIA M. DOMINGUES, PAUL M. BARKER, SUSAN E. WIJFFELS AND JEFF R. DUNN

Increasing atmospheric greenhouse gas concentrations result in the trapping of more solar energy in the global Earth System and its warming. For 1961–2003 and 1993–2003, about 90% of this additional energy was stored in the ocean (Bindoff *et al.*, 2007), with the largest amount in the upper 750 m.

Ocean heat content estimates are based largely on high-quality observations made from research ships that have been combined with observations from merchant, fishing and navy vessels using eXpendable BathyThermographs (XBTs) and, since 2000, by autonomous (Argo) profiling floats (Gould and the Argo Science Team, 2004). All these datasets indicate a warming ocean. However, two major issues have complicated the determination of an accurate time history of global averaged heat content. First, the historical record is incomplete, particularly in the southern hemisphere (before the widespread deployment of Argo floats starting about 2005) and in the deep ocean. As a result, there is a need for statistical approaches to interpolate between the sparse (in space and time) observations. Second, biases are present in some observational datasets (Gouretski and Koltermann, 2007). For XBTs, the largest biases are the result of small errors in the estimated rate at which these instruments fall through the water column (Wijffels *et al.*, 2008; Ishii and Kimoto, 2009). For the Argo data, initial problems have been corrected (Willis *et al.*, 2009) but more subtle biases have been identified in the pressure sensors on the floats (Barker *et al.*, in revision) and are being corrected.

Domingues *et al.* (2008) addressed both instrumental and sampling biases by applying the Wijffels *et al.* (2008) XBT corrections to the global ocean dataset of Ingleby and Huddleston (2007) and used a reduced-space, optimal-interpolation technique (Kaplan *et al.*, 2000) to

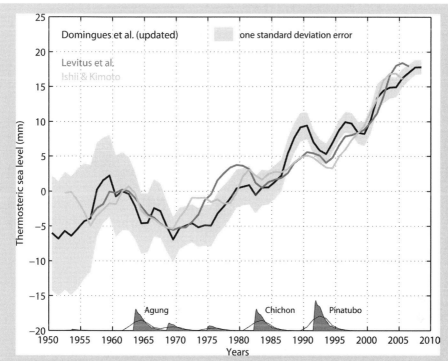

Figure 1.8 Updated estimates of ocean thermal expansion. The updated Domingues *et al.*
(2008) time series is shown in black, and one standard deviation uncertainty estimates are
indicated by the grey shading. The estimates for Ishii and Kimoto (2009) and Levitus *et al.*
(2009) are shown as light grey and mid-grey lines respectively. The estimated stratospheric
aerosol loading (arbitrary scale) from the major volcanic eruptions is shown at the bottom.

statistically interpolate across data voids. For the period 1961 to 2003, Domingues *et al.*'s
(2008) estimate of the heat content increase of the upper 700 m of the ocean was $16 \pm 3 \times 10^{22}$ J,
equivalent to a thermal expansion of 0.52 ± 0.08 mm yr^{-1} (i.e. a linear trend about 50%
larger than earlier results). Climate models that include all factors which lead to a climate
change report a slightly smaller trend, and indicate that the variability in the observations is
at least partly a result of volcanic eruptions (Domingues *et al.*, 2008). The results reveal little
thermosteric rise prior to the mid 1970s, then a steady increase in ocean heat content and
thermal expansion from the mid 1970s.

   Updated (from Domingues *et al.*, 2008) estimates of ocean heat-content and thermal-
expansion changes using carefully checked and corrected Argo data of Barker *et al.*
(submitted) indicate multi-decadal warming and expansion continues to the end of the
record in December 2008 (Figure 1.8). The equivalent results from Ishii and Kimoto (2009)
and Levitus *et al.* (2009), using different corrections and interpolation schemes, generally
agree but there are some small differences during the 1970s, 1990s and in the early 2000s.
Further analysis and careful quality control of data are required to refine the multi-decadal
estimates.

Box 1.3 (*cont.*)

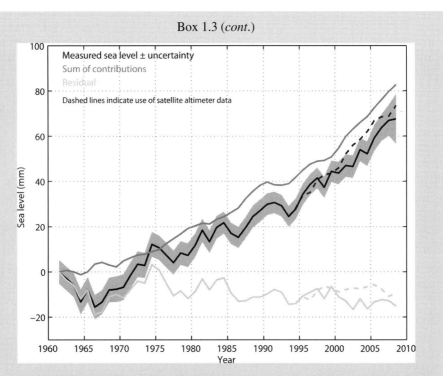

Figure 1.9 Global average sea level. The updated estimates from coastal tide gauges (following Church and White, 2006) are shown by the solid black line (with one standard deviation error estimates shaded), and the satellite altimeter estimates (from 1993) are shown by the dotted black line. The sum of contributions (not including any changes in terrestrial storage) is shown as the upper, mid-grey line. The uncertainty of these estimates (not shown) would include contributions from deep- and upper-ocean thermal expansion, glaciers and ice sheets. The residuals (the differences between the observed and the sum of contributions) are shown as the lower, light grey lines.

Recent analyses have revealed widespread warming of the ocean below 3 000 m (Fukasawa *et al.*, 2004; Johnson and Doney, 2006; Johnson *et al.*, 2007, 2008). However, this deep warming is even less well constrained by observations than the upper ocean warming.

Global averaged sea level has been independently estimated from coastal and island sea-level measurements, as well as by satellite altimeters since 1993 (Church and White, 2006). Updates of these estimates (Figure 1.9) show sea level has continued to rise through to the end of the record in December 2008.

Domingues *et al.* (2008) combined their estimate of upper-ocean thermal expansion with estimates for the deep-ocean warming, and contribution from glaciers and ice sheets, to provide a quantitative explanation of the observed sea-level rise from 1961 to 2003. These updated estimates of ocean thermal expansion are combined with updated estimates of contributions from glaciers (Cogley, 2009), the Greenland Ice Sheet (Rignot *et al.*, 2008) and a constant 0.2 mm yr^{-1} from Antarctica. The sum of contributions shows a similar rate of rise to the observations (Figure 1.9). The largest discrepancy is in the early 1960s, when the sum of contributions does not explain the small fall in sea level inferred from the coastal and island measurements. Note that this sum of contributions does not include any allowance for changes in the terrestrial storage of water (e.g. either from the building of dams (Chao *et al.*, 2008) or the mining of ground water). Both the observed sea levels and the sum of contributions show a substantially faster rate of rise since the mid 1990s.

1.4 Linking cause to effect

There is now good empirical evidence that the amount of heat stored at the Earth's surface has been increasing in recent decades. There is also indisputable evidence that human activities (burning of fossil fuels) have increased the climate forcing by certain greenhouse gases to well above the level that the Earth's climate system has experienced during the last nearly 1 million years. However, the climate system is complex and influenced by many factors. Noting that one climate forcing (greenhouse gases) has been influenced by human activities in a manner that is known to cause warming does not necessarily mean that this change in the climate system is responsible for all of the warming observed at the Earth's surface. To understand how the many different climate forcings interact with one another, scientists develop models.

Using these models, it is possible for scientists to 'test' how well they understand the climate system and the relative importance of the different components that comprise it. There are different types of models that are based on different assumptions and methodologies, and there is no model that is 'correct' in the sense that it totally describes what occurs in the natural world. There are various levels of uncertainty associated with almost every parameter included in models, and quantifying the uncertainty associated with model-generated results is an important field of research (Box 1.4). Despite these uncertainties, models are important tools to help us understand how various forcings within the climate system interact with one another. In order to test how well a given model recreates the interaction between different climate forcings and the state of the climate system, one can compare the temperature predicted by the model to the actual temperature under the conditions assumed in the model.

Box 1.4
Quantifying and dealing with uncertainty in climate-related models

DAVID A. STAINFORTH

Computer models are key tools in research efforts to understand the climate system, and how it could respond to both anthropogenic greenhouse gas emissions and geoengineering actions. The science of climate prediction is still in its infancy, however, and faces many significant challenges. Foremost is the lack of any means to verify, or more correctly confirm (Oreskes

Box 1.4 (*cont.*)

et al., 1994), a climate forecast many decades ahead (Stainforth *et al.*, 2007a). This is in stark contrast to other fields where computer modelling is also a critical tool; fields such as weather forecasting or computer-aided drug or building design where model predictions can be paired with new real-world observations that cannot have been used in the model development process. The lack of such 'out-of-sample' data with which to judge model-based predictions significantly increases the requirement for uncertainty exploration, the evaluation of relevant information content, and research into the basis for confidence in climate forecasts across a range of variables and scales.

Climate itself is described by the IPCC as 'the state, including a statistical description, of the climate system' (Solomon *et al.*, 2007). This system is highly nonlinear. Indeed many common examples of chaotic behaviour are derived from the study of components of the climate system (Lorenz, 1963, 1984). When addressing the climate prediction problem this nonlinear perspective provides an essential context. Consequences include an expected sensitivity to initial conditions, an expectation that the details of any prediction could depend on the details of the forcing pathway, and an acknowledgement of spatial interactions such that we cannot expect those models that simulate a chosen region or variable best in the past to necessarily be most informative about the changes in such variables or regions under climate change in the future.

It is helpful to discuss uncertainty in the climate of the twenty-first century in terms of five different sources (Stainforth *et al.*, 2007a). These sources demand different approaches to their quantification, characterisation and communication. They are uncertainty in (i) anthropogenic actions that affect climate (principally greenhouse gas emission but also land cover changes and geoengineering activities), (ii) current climatic conditions on rapidly varying scales (as this includes most small-scale phenomena it has been termed 'microscopic initial condition uncertainty (ICU)'), (iii) current conditions on slowly varying scales ('macroscopic ICU'), (iv) model inadequacy, and (v) model uncertainty.

Uncertainty in anthropogenic actions that affect climate is largely a consequence of uncertainty in political and policy decisions, and their socio-economic consequences. The latter are studied with integrated assessment models (see Wahba and Hope, 2006). The former, although open to academic study and influence, are also open to informed judgement and debate by those seeking to use climate predictions. Climate models can portray scenarios of the possible consequences of such decisions. A series of *emission* scenarios is used for this task (Nakicenovic and Swart, 2000). For pragmatic reasons these are often applied in global climate models (GCMs) as scenarios for atmospheric GHG *concentrations*. A consequence of this is that model uncertainties in the carbon cycle (Friedlingstein *et al.*, 2006) may be obscured, which is one reason to expect current model predictions to underestimate the uncertainty in the real world response.

Given its nonlinear nature, we expect high sensitivity to initial conditions in the climate system; and this is the case on timescales from hours to centuries. On climate timescales, ICU is critical for two reasons. First, the internal variability it highlights is key to many adaptation decisions; presenting only the mean response to changes in forcing risks misleading many such decisions. Second, ICU is known to significantly affect long-term seasonal means in certain regions, even in models with no ocean variability (Stainforth *et al.*, 2007a). Such variations could have regional land surface feedbacks (e.g. on glaciers, snowfields or soil quality), which could affect the timing and possibly the character of twenty-first century climate change in the region. Such impacts will, of course, feed back on the larger-scale response, and it is plausible that such feedbacks could be significant. These effects have not yet been well investigated, as initial condition ensembles of simulations of the twenty-first century are small – typically four or five members. Microscopic ICU differs from macroscopic ICU in terms of policy and research responses. Microscopic ICU is irreducible and 'simply' requires sufficient ensemble sizes to quantify the effect on the scales of interest. Macroscopic ICU similarly requires sufficient ensemble sizes to investigate the effect, but also provides the possibility of reducing uncertainty through directed observations.

A context for all model-based predictions is model inadequacy. All climate models are, of necessity, limited in scope; we cannot model everything. The key question is whether processes important on our timeframe of interest are missing. With current models these might include ice sheets, a stratosphere, the effects of methane clathrates, etc. For processes that are included there is uncertainty over whether they are correctly simulated, particularly in terms of their response to forcing changes outside the range for which we have good observations. This is termed model uncertainty and to some extent is amenable to quantification through multi-model (Meehl *et al.*, 2007) and perturbed physics ensembles (Allen and Stainforth, 2002; Murphy *et al.*, 2004; Stainforth *et al.*, 2005). Suitable methods for analysis and interpretation of such ensembles is still the subject of debate (Murphy *et al.*, 2007; Stainforth *et al.*, 2007a), and further work is needed on how to link them to impact models for direct decision support (Stainforth *et al.*, 2007b). The impact of model inadequacy and the extent to which model uncertainty has been quantified are both amenable to subjective judgement; such second-order uncertainty is a critical part of the communication of model-based results.

The quantification of uncertainty in model-based climate predictions is still a very young research field. At scales relevant for most adaptation decisions we should expect uncertainty to increase before we have sufficient data and understanding to reduce it. The characteristics of this uncertainty strengthen, rather than undermine, the rationale for very significant mitigation measures.

The IPCC (2007a) compared the temperature predicted by a number of models describing climate system interactions with the actual temperature recorded for the period 1900–2000. The result (Figure 1.10a) demonstrates clearly that the models have not captured all details in the climate system. Nevertheless, the trend in the actual temperature development over this period is well described by the suite of models the IPCC examined. When so many different models arrive at such similar conclusions, we can be reasonably certain that we have a good overall understanding of the fundamental physics of the climate system and the interactions of the processes occurring within it. The IPCC then ran the same models, with the same fundamental physics, but removed the climate forcing created by anthropogenic activities (i.e. release of greenhouse gases, felling of natural vegetation, changes in Earth albedo, etc.) (Figure 1.10b). In this case, none of the models was able to recreate

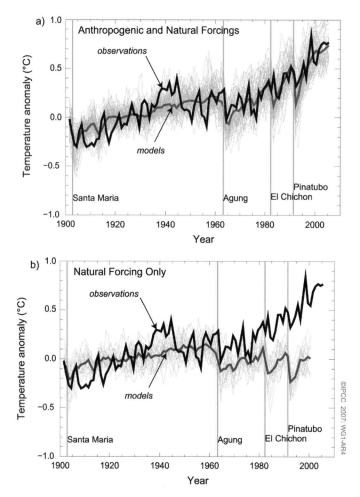

Figure 1.10 (a) Comparison of the annual global temperature predicted by a series of climate models for the period 1900–2000 and the actual observed temperatures; and (b) as for (a) except that the models were modified such that the climate forcing created by anthropogenic activities (i.e. release of greenhouse gases, felling of natural vegetation, changes in Earth albedo, etc.) have been removed. Source: IPCC (2007b).

the warming observed in surface air temperatures in recent decades. Such studies provide strong support for the argument that human activities are the primary source of the global warming currently being experienced.

1.5 Predicting future climate conditions

Current models have demonstrated that they are able to recreate the pattern of air temperature development during the twentieth century. We do not know precisely what temperatures will be in the twenty-first century. However, on the basis of their performance for data in the twentieth century, we can have some confidence in what these same models predict for air temperature development in the future. On that basis, it is predicted that unless the human-induced increase in greenhouse gas climate forcing is drastically curtailed, the average global temperature at the end of this century may be 2.3–4.5 °C above that experienced just prior to the onset of the industrial revolution (IPCC, 2007a).

Most popular media deal with future climate change as a prediction problem – will it happen or not? In reality, however, climate change is a problem of risk: how great a risk is society willing to take that human activities will lead to a significantly warmer climate than that which human societies have previously experienced? Usually, when a particular activity or phenomenon would give rise to large disruption of society, great effort (and expense) is directed towards reducing the risk of the phenomenon occurring. The engineer designing a nuclear power plant, for example, is not asked to predict whether or not an accident will occur, but to reduce and contain the potential risk as far as possible. Similarly, the probability of a terrorist attack at any individual airport is not large. Nevertheless, the consequences of such an attack are considered unacceptable for society and considerable expense is incurred in reducing this presumably already small risk.

Oddly, climate change is not viewed in this perspective by the majority of people, despite the fact that a large majority of the relevant scientific community (Chapter 17) share an understanding of the climate system that suggests a significant probability of a global temperature increase of up to 4.5 °C or more by the end of this century unless action is taken to reduce the anthropogenic emissions of greenhouse gases. Such a temperature rise would have drastic consequences for human society and life on Earth. While the public debate most often focuses on whether one 'believes' in climate change or not, the more pertinent question really is whether society is willing to accept the risks associated with a significant probability of such a large temperature increase – probabilities that are much larger than those associated with a terrorist attack or a nuclear power accident.

One of the major weaknesses in today's models is that they – for good reason – cannot take into account how various climate forcings respond to global warming. For example, the ocean has, until now, taken up about one third of the CO_2 released to the atmosphere through the combustion of fossil fuels (Canadell *et al.*, 2007; Le Quéré *et al.*, 2007, 2009). The future ocean – as a consequence of both temperature increases and chemical changes – will not be able to remove as much CO_2 from the atmosphere as has been the case until now (Chapter 4). Current development of global models is focusing on incorporating the 'feedbacks' that climate change induces on the various climate forcings.

Most climate modelling until now has been developed with the purpose of trying to discern whether or not there is a human-induced component to the global warming currently observed and how great this component might be. Now that it has, with a high degree of probability, been established that there is a human component – and that it is likely to be the primary cause of the global warming currently observed – the focus of climate modelling will move from the global level to the regional. This is because better understanding of the factors interacting to create regional climatic conditions is becoming increasingly important as it becomes necessary to develop regional strategies for societal adaptation to the impacts of climate change.

1.6 Learning from prehistoric climate change

The Earth and its climate system have existed for nearly 5 billion years. In contrast, scientists have been able to make direct measurements on the climate system's components for just over 100 years. By using proxies for various signals in the climate system, it is possible to reconstruct the state of the climate system for much longer periods of time into the Earth's past. These palaeoclimate studies help us to develop far longer time series of climate data than we have from direct measurements – time-series that extend to long before humans inhabited the planet. Understanding what conditions preceded or precipitated major climate changes in the past (Box 1.5) can help us to understand how the climate may respond to changes in climate forcings in the future.

Box 1.5.
Estimating the Earth System sensitivity using an integrated model-data approach

DANIEL J. LUNT, ALAN M. HAYWOOD, ULRICH SALZMANN, PAUL J. VALDES AND HARRY DOWSETT

Climate sensitivity is defined as the equilibrium global mean surface air temperature response to a doubling of atmospheric CO_2 concentration. It is a useful benchmark for comparing different climate models in idealised circumstances, and has been one of the central concepts used by the IPCC in their assessments of future climate change. However, the models that have been used to estimate climate sensitivity typically neglect many feedbacks associated with changes to ice sheets, vegetation, non-greenhouse gases, various aerosols such as desert dust, and other components of the Earth System (Box 8.1).The recent development of 'Earth System models', which aim to simulate as many processes as possible, is a step towards estimating the true long-term response of the system to elevated CO_2 – the 'Earth System sensitivity'. However, many of the missing feedbacks are not well understood, and therefore a challenge to model. Furthermore, many of the missing processes act on very long timescales, making simulations that allow equilibrium to be reached computationally unfeasible with state-of-the-art models.

Records of past climate change can be used as an indicator of states of the Earth System in equilibrium with various levels of CO_2. Hansen *et al.* (2008), making use of records of CO_2 and global mean temperature derived from Antarctic ice core records, estimated the true Earth System sensitivity to be twice as large as the traditional climate sensitivity (with its restricted range of feedbacks). However, this approach also has its weaknesses. During the period of the ice-core record, atmospheric CO_2 concentrations were never significantly higher than the pre-industrial period (Siegenthaler *et al.*, 2005), making their extrapolation to climates warmer than modern questionable. Furthermore, simple estimates of global temperature derived from a single ice core are problematic (see Sime *et al.*, 2009).

Combining modelling and data approaches is also possible. Provided that the CO_2 forcing and palaeoenvironmental boundary conditions (e.g. ice-sheet extent) from some period in the past are sufficiently known, they can be *prescribed* in a climate model, allowing relatively short integration times to reach equilibrium. The model can therefore be used to provide an estimate of long-term equilibrium surface temperature for the given CO_2 forcing. It has the advantage over the pure modelling approach in that the palaeoenvironmental boundary conditions associated with the feedbacks are prescribed rather than calculated, and has the advantage over the pure observational approach in that the model allows a truly global estimate of temperature change.

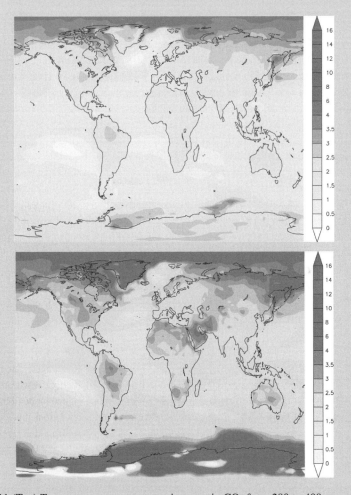

Figure 1.11 (Top) Temperature response to an increase in CO_2 from 280 to 400 p.p.m., as traditionally calculated. (Bottom) Earth System sensitivity to the same CO_2 forcing. Source: Lunt *et al.* (2010). Reprinted by permission from Macmillan Publishers Ltd: *Nature Geoscience*, copyright 2010. (In colour as Plate 3.)

Box 1.5 (*cont.*)

The mid-Pliocene warm period (*c.* 3.3 to 3 million years ago) had relatively high atmospheric CO_2 concentrations (~ 400 p.p.m.; see Raymo *et al.*, 1996), temperatures were significantly higher than the pre-industrial period (see Dowsett *et al.*, 2009) and published datasets exist (see Dowsett *et al.*, 1999) that allow at least two of the important longer-term feedbacks – vegetation and ice-sheet extent – to be addressed. Changes in continental configuration compared to modern were negligible, and the main external forcings were orographic changes (e.g. tectonic uplift in the Rocky Mountains since the Pliocene) and elevated CO_2. The climatic response induced by these forcings included vegetation and ice sheet changes, and because both the forcings and vegetation and ice sheet changes are somewhat constrained by the geological record, they can be imposed independently as boundary conditions in a model, allowing a calculation of the Earth System sensitivity.

Lunt *et al.* (2010) took this approach to simulating the mid Pliocene, and used the model to correct for the effects of the orographic (physical geography of mountains) forcing, to leave the response to the CO_2 forcing. This allows the regional patterns of the Earth System sensitivity to be calculated and compared with the traditional climate sensitivity (Figure 1.11). The impact on temperature of the reduced Antarctic and Greenland ice sheets is clearly visible (Figure 1.11b), with local warming of more than 16 °C. The vegetation changes also have a considerable effect on the global scale. Overall, we calculated the Earth System sensitivity to be about 40% greater than the traditional climate sensitivity. The uncertainty in the estimate (30%–50%) comes from a consideration of the model's ability to simulate the mid Pliocene sea surface temperatures compared to observations (Dowsett *et al.*, 2009), and uncertainties in the vegetation and ice-sheet reconstructions.

The IPCC has focused on the traditional climate sensitivity (Box 8.1), and groups have used this sensitivity to determine the degree of emissions likely to lead to 'dangerous' climate change (e.g. Chapter 8; Meinshausen *et al.*, 2009). Our work suggests that the equilibrium climate change associated with an increase of CO_2 is likely to be significantly larger than has traditionally been estimated. Given the uncertainties in the timescale for vegetation and ice sheet responses, the most prudent and responsible course of action is to base estimates of the impacts of long-term greenhouse gas stabilisation scenarios on estimates of Earth System sensitivity rather than traditional climate sensitivity.

The prehistoric record also informs us about how our ancestors dealt with climate changes. As Figure 1.12 indicates, there appears to be a close relationship between major human migrations and dramatic climate change events. An obvious response to climate change and the resulting change in living conditions is to move on to other areas where the conditions are more conducive to survival. The same response in human societies to current and future climate change also seems likely. However, there are many more humans on Earth today than at the time when our ancestors confronted dramatic climate change, so the amount of unoccupied territory to which climate refugees can move is very limited (Chapter 5).

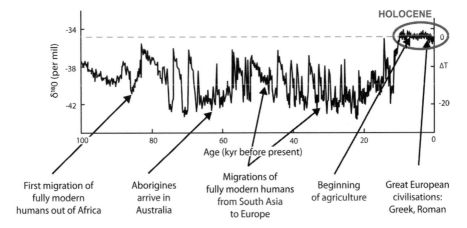

Figure 1.12 Temperature variation for the last glacial cycle, as estimated from the ^{18}O record from the Greenland ice core (GRIP – Johnsen *et al.*, 1992), and selected events in human history. Source: draws on information from Oppenheimer (2004).

1.7 Summary and conclusions

The climate system is complex and comprised of many different interacting components. Nevertheless, model 'experiments' suggest that the current scientific understanding of the factors influencing the climate system and the way in which these factors interact is generally robust. This understanding identifies human activities – particularly the combustion of fossil fuels – as the primary cause of the global warming observed over recent decades. Models that accurately describe temperature development in the twentieth century indicate that we can expect a global temperature increase of up to around 6 °C in the twenty-first century unless the human emission of greenhouse gases, in particular CO_2, is curtailed.

Some studies put a 50:50 chance on the planet reaching a 4 °C average warming as early as 2065 if greenhouse gas emissions are not curtailed and there is a concomitant small shift in the ability of the natural carbon sinks (Chapter 4) on Earth to take up carbon (New *et al.*, 2009). Even if society does take steps to reduce emissions, the average global temperature increase may exceed a 2 °C guardrail this century while societies face the challenges of transforming infrastructure and economies (Chapters 11 and 12) to lower carbon systems (Lowe *et al.*, 2009).

On the face of it, a temperature rise of over 2 °C may not sound too worrying. However, it is important to remember that when global warming is discussed, the changes referred to are *average* temperature changes. No organism lives at average temperature. Behind seemingly small changes in average temperature are often very large changes in temperature extremes, and thus in living conditions. This can be illustrated with the analogy of one having their head in the oven and feet in the freezer. Such a situation might result in quite a comfortable average temperature. However, the conditions experienced by the extremities would present a challenge to the survival of the organism as a whole.

Tremendous research effort is currently being expended in obtaining a better understanding of future temperature and climate conditions in various regions of the planet. However, predicting future climate conditions – both in terms of average temperature changes and extremes – is hampered by an incomplete understanding of how the various processes influencing the climate system are changing in response to climate change itself. The following chapters consider potential feedbacks in the climate system generated by changing climate conditions and the risks that climate change poses for life on Earth.

References

Allen, M. R. and Stainforth, D. A. (2002). Towards objective probabalistic climate forecasting. *Nature*, **419**, 228.

Arrhenius, S. (1896). On the influence of carbonic acid in the air upon the temperature of the Earth ground, London. *Edinburgh, and Dublin Philosophical Magazine and Journal of Science (fifth series)*, **41**, 237–75.

Barker, P. M., Dunn, J. R., Wijffels, S. E. and Domingues, C. M. (in revision). Correcting pressure sensor drifts in Argo's APEX floats and evaluating their impact on estimates of ocean temperature, salinity, and thermosteric sea level. *Journal of Atmospheric and Oceanic Technology*.

Berger, A. and Loutre, M. F. (2002). An exceptionally long interglacial ahead? *Science*, **297**, 1287–88.

Betts, R. A., Falloon, P. D., Goldewijk, K. K. and Ramankutty, N. (2007). Biogeophysical effects of land use on climate: Model simulations of radiative forcing and large-scale temperature change. *Agricultural and Forest Meteorology*, **142**, 216–23.

Bindoff, N., Willebrand, J., Artale, V. *et al.* (2007). Observations: Oceanic climate change and sea level. In *Climate Change 2007: The Physical Science Basis. Contribution of Working Group I to the Fourth Assessment Report of the Intergovernmental Panel on Climate Change*, eds. S. Solomon, D. Qin, M. Manning, M. Marquis, K. Averyt, M. M. B. Tignor, H. L. Miller and Z. Chen. Cambridge, UK and New York, NY: Cambridge University Press, pp. 385–432.

Canadell, J. G., Le Quéré, C., Raupach, M. R. *et al.* (2007). Contributions to accelerating atmospheric CO_2 growth from economic activity, carbon intensity, and efficiency of natural sinks. *Proceedings of the National Academy of Sciences (USA)*, **104**, 18866–70.

CCSP (2008). *Climate Projections Based on Emissions Scenarios for Long-Lived and Short-Lived Radiatively Active Gases and Aerosols. Report by the U.S. Climate Change Science Program and the Subcommittee on Global Change Research*, eds. H. Levy II, D. Shindell, A. Gilliland, L. W. Horowitz and M. D. Schwarzkopf. Department of Commerce, NOAA's National Climatic Data Center, Washington, D.C.

Chao, B. F., Wu, Y. H. and Li, Y. S. (2008). Impact of artificial reservoir water impoundment on global sea level. *Science*, **320**, 212–14.

Church, J. A. and White, N. J. (2006). A 20th century acceleration in global sea-level rise. *Geophysical Research Letters*, **33**, L01602.

Cogley, J. G. (2009). Geodetic and direct mass-balance measurements: Comparison and joint analysis. *Annals of Glaciology*, **50**, 96–100.

Domingues, C. M., Church, J. A., White, N. J. *et al.* (2008). Improved estimates of upper-ocean warming and multi-decadal sea-level rise. *Nature*, **453**, 1090–93.

Dowsett, H. J., Barron, J. A., Poore, R. Z. *et al.* (1999). *Middle Pliocene Paleoenvironmental Reconstruction: PRISM2*. U.S. Geological Survey Open File Report, 99–535, http://pubs.usgs.gov/of/1999/of99-535/.

Dowsett, H. J., Robinson, M. M. and Foley, K. M. (2009). Pliocene three-dimensional global ocean temperature reconstruction. *Climate of the Past Discussions*, **5**, 769–83.

EPICA community members (2004). Eight glacial cycles from an Antarctic ice core. *Nature*, **429**, 623–28.

Friedlingstein, P., Cox, P., Betts, R. *et al.* (2006). Climate–carbon cycle feedback analysis: Results from the C4MIP model intercomparison. *Journal of Climate*, **19**, 3337–53.

Fukasawa, M., Freeland, H., Perkin, R. *et al.* (2004). Bottom water warming in the North Pacific Ocean. *Nature*, **427**, 825–27.

Gordon, L. J., Steffen, W., Jönsson, B. F. *et al.* (2005). Human modification of global water vapor flows from the land surface. *Proceedings of the National Academy of Sciences (USA)*, **102**, 7612–17.

Gould, J. and The Argo Science Team (2004). Argo profiling floats bring new era of in situ ocean observations. *Eos, Transactions of the American Geophysical Union*, **85**.

Gouretski, V. and Koltermann, K. P. (2007). How much is the ocean really warming? *Geophysical Research Letters*, **34**, L01610.

Hansen, J., Sato, M., Ruedy, R. *et al.* (2005). Efficacy of climate forcings. *Journal of Geophysical Research*, **110**, D18104.

Hansen, J., Sato, M., Kharecha, P. *et al.* (2008). Target atmospheric CO_2: Where should humanity aim? *The Open Atmospheric Science Journal*, **2**, 217–31.

Ingleby, B. and Huddleston, M. (2007). Quality control of ocean temperature and salinity profiles – Historical and real-time data. *Journal of Marine Systems*, **65**, 158–75.

IPCC (2007a). *Climate Change 2007: Synthesis Report. Contribution of Working Groups I, II and III to the Fourth Assessment Report of the Intergovernmental Panel on Climate Change*, eds. Core Writing Team, R. K. Pachauri and A. Reisinger. Geneva, Switzerland: IPCC.

IPCC (2007b). Summary for policymakers. In *Climate Change 2007: The Physical Science Basis. Contribution of Working Group I to the Fourth Assessment Report of the Intergovernmental Panel on Climate Change*, eds. S. Solomon, D. Qin, M. Manning, Z. Chen, M. Marquis, K. B. Averyt, M. Tignor and H. L. Miller. Cambridge, UK and New York, NY: Cambridge University Press.

Ishii, M. and Kimoto, M. (2009). Reevaluation of historical ocean heat content variations with time-varying XBT and MBT depth bias corrections. *Journal of Oceanography*, **65**, 287–99.

Johnsen, S. J., Clausen, H. B., Dansgaard, W. *et al.* (1992). Irregular glacial interstadials recorded in a new Greenland ice core. *Nature*, **359**, 311–13.

Johnson, G. C. and Doney, S. C. (2006). Recent western South Atlantic bottom water warming. *Geophysical Research Letters*, **33**, L14614.

Johnson, G. C., Mecking, S., Sloyan, B. M. and Wijffels, S. E. (2007). Recent bottom water warming in the Pacific Ocean. *Journal of Climate*, **20**, 5365–75.

Johnson, G. C., Purkey, S. G. and Bullister, J. L. (2008). Warming and freshening in the abyssal southeastern Indian Ocean. *Journal of Climate*, **21**, 5351–63.

Kaplan, A., Kushnir, Y. and Cane, M. A. (2000). Reduced space optimal interpolation of historical marine sea level pressure: 1854–1992. *Journal of Climate*, **13**, 2987–3002.

Kiehl, J. T. and Trenberth, K. E. (1997). Earth's annual global mean energy budget. *Bulletin of the American Meteorological Society*, **78**, 197–208.

Le Quéré, C., Rodenbeck, C., Buitenhuis, E.T. *et al.* (2007) Saturation of the Southern Ocean CO$_2$ sink due to recent climate change. *Science*, **316**, 1735–38.

Le Quéré, C., Raupach, M. R., Canadell, J. G. *et al.* (2009). Trends in the sources and sinks of carbon dioxide. *Nature Geoscience*, **2**, 831–36.

Levitus, S., Antonov, J. I., Boyer, T. P. *et al.* (2009). Global ocean heat content 1955–2008 in light of recently revealed instrumentation problems. *Geophysical Research Letters*, **36**, L07608.

Levy II, H., Schwarzkopf, M. D., Horowitz, L., Ramaswamy, V. and Findell, K. L. (2008). Strong sensitivity of late 21st century climate to projected changes in short-lived air pollutants. *Journal of Geophysical Research*, **113**, D06102.

Lorenz, E. N. (1963). Deterministic nonperiodic flow. *Journal of Atmospheric Sciences*, **20**, 130–41.

Lorenz, E. N. (1984). Irregularity: A fundamental property of the atmosphere. *Tellus Series A-Dynamic Meteorology and Oceanography*, **36**, 98–110.

Lowe, J. A., Huntingford, C., Raper, S. C. B. *et al.* (2009). How difficult is it to recover from dangerous levels of global warming? *Environmental Research Letters*, **4**, 014012.

Lunt, D. J., Haywood, A. M, Schmidt, G. A. *et al.* (2010). Earth system sensitivity inferred from Pliocene modelling and data. *Nature Geoscience*, **3**, 60–64.

Meehl, G. A., Covey, C., Delworth, T. *et al.* (2007). The WCRP CMIP3 multimodel dataset – A new era in climate change research. *Bulletin of the American Meteorological Society*, **88**, 1383–94.

Meinshausen, M., Meinshausen, N., Hare, W. *et al.* (2009). Greenhouse-gas emission targets for limiting global warming to 2 °C. *Nature* **458**, 1158–62.

Murphy, J. M., Booth, B. B. B., Collins, M. *et al.* (2007). A methodology for probabilistic predictions of regional climate change from perturbed physics ensembles. *Philosophical Transactions of the Royal Society A – Mathematical, Physical & Engineering Sciences*, **365**, 1993–2028.

Murphy, J. M., Sexton, D. M. H., Barnett, D. N. *et al.* (2004). Quantification of modelling uncertainties in a large ensemble of climate change simulations. *Nature*, **430**, 768–72.

Nakicenovic, N. and Swart, R. (eds.) (2000). *IPCC Special Report on Emissions Scenarios*. IPCC.

New, M., Liverman, D. and Anderson, K. (2009). Mind the gap. *Nature Reports Climate Change*, 143–44.

Noone, K. J., Ostrom, E., Ferek, R. J. *et al.* (2000). A case study of ships forming and not forming tracks in moderately polluted clouds. *Journal of the Atmospheric Sciences*, **57**, 2729–47.

Oppenheimer, S. (2004). *Out of Eden: Peopling the World*. London: Constable and Robinson.

Oreskes, N., Shraderfrechette, K. and Belitz, K. (1994).Verification, validation, and confirmation of numerical models in the earth sciences. *Science*, **263**, 641–46.

Petit, J. R., Jouzel, J., Raynaud, D. *et al.* (1999). Climate and atmospheric history of the past 420,000 years from the Vostok ice core, Antarctica. *Nature*, **399**, 429–36.

Prentice, I. C., Farquhar, G. D., Fasham, M. J. R. *et al.* (2001). The carbon cycle and atmospheric carbon dioxide content. In *Contribution of Working Group I to the Third Assessment Report of the Intergovernmental Panel on Climate Change*, eds.

J. T. Houghton, Y. Ding, D. J. Griggs, M. Noguer, P. J. van der Linden and D. Xiaosu. Cambridge, UK and New York, NY: Cambridge University Press, pp. 184–238.

Ramanathan, V. and Carmichael, G. (2008). Global and regional climate changes due to black carbon. *Nature Geoscience*, **1**, 221–27.

Ramanathan, V. and Feng, Y. (2008). On avoiding dangerous anthropogenic interference with the climate system: Formidable challenges ahead. *Proceedings of the National Academy of Sciences (USA)*, **105**, 14245–50.

Raymo, M. E., Grant, B., Horowitz, M. and Rau, G. H. (1996). Mid-Pliocene warmth: Stronger greenhouse and stronger conveyor. *Marine Micropaleontology*, **27**, 313–26.

Rignot, E., Box, J. E., Burgess, E. and Hanna, E. (2008). Mass balance of the Greenland ice sheet from 1958 to 2007. *Geophysical Research Letters*, **35**, L20502.

Shindell, D., Levy II, H., Schwarzkopf, M. D. *et al.* (2008). Multimodel projections of climate change from short-lived emissions due to human activities. *Journal of Geophysical Research*, **113**, D11109.

Siegenthaler, U., Stocker, T. F., Monnin, E. *et al.* (2005). Stable carbon cycle--climate relationship during the late Pleistocene. *Science*, **310**, 1313–17.

Sime, L. C., Wolff, E. W., Oliver, K. I. C. and Tindall, J. C. (2009). Evidence for warmer interglacials in East Antarctic ice cores. *Nature*, **462**, 342–45.

Solomon, S., Qin, D., Manning, M. *et al.* (eds.) (2007). *Climate Change 2007: The Physical Science Basis. Contribution of Working Group I to the Fourth Assessment Report of the Intergovernmental Panel on Climate Change*. Cambridge, UK and New York, NY: Cambridge University Press.

Stainforth, D. A., Aina, T., Christensen, C. *et al.* (2005). Uncertainty in predictions of the climate response to rising levels of greenhouse gases. *Nature*, **433**, 403–06.

Stainforth, D. A., Allen, M. R., Tredger, E. R. and Smith, L. A. (2007a). Confidence, uncertainty and decision-support relevance in climate predictions. *Philosophical Transactions of the Royal Society A – Mathematical, Physical & Engineering Sciences*, **365**, 2145–61.

Stainforth, D. A., Downing, T. E., Washington, R., Lopez, A. and New, M. (2007b). Issues in the interpretation of climate model ensembles to inform decisions. *Philosophical Transactions of the Royal Society A – Mathematical, Physical & Engineering Sciences*, **365**, 2163–77.

van der Leeuw, S. E. (2008). Climate and society: lessons from the past 10 000 years. *Ambio*, **1437**, 476–82.

Wahba, M. and Hope, C. (2006). The marginal impact of carbon dioxide under two scenarios of future emissions. *Energy Policy*, **34**, 3305–16.

Wijffels, S. E., Willis, J., Domingues, C. M. *et al.* (2008). Changing expendable bathythermograph fall rates and their impact on estimates of thermosteric sea level rise. *Journal of Climate*, **21**, 5657–72.

Willis, J. K., Lyman, J. M., Johnson, G. C. and Gilson, J. (2009). In situ data biases and recent ocean heat content variability. *Journal of Atmospheric and Oceanic Technology*, **26**, 846–52.

2

The oceans and the climate system

'Water water every where nor any drop to drink...'[1]

As described in Chapter 1, the oceans represent, by far, the largest reserve of heat energy found at the surface of the Earth. They also are instrumental in the transport and redistribution of heat on the planet. Thanks, for example, to the North Atlantic Current – which transports heat from tropical regions northwards – residents in the western reaches of Northern Europe experience much warmer temperatures than inhabitants living at the same latitudes in Siberia or northern Canada. Climate change can potentially disrupt some of the current systems that transport heat, thus drastically influencing regional climate patterns (Chapter 7).

In addition to storing and transporting heat, the oceans contain more than 97% of the water on our planet. This means they comprise the dominant component of the Earth's water cycle ('hydrological cycle') which, itself, is critical for the functioning of the climate system. As noted in Chapter 1, water vapour is the most important greenhouse gas, and rainfall patterns are an obvious and direct way in which the functioning of the climate system impacts on human societies. The sheer volume of water in the oceans (approximately 1370 million km^3) and the response of this mass of water to global warming makes a significant contribution to sea-level rise (Box 1.3 and Chapter 3), as warm water simply occupies a larger volume than cold water. This thermal expansion varies as a function of the water's temperature and salinity; the volume change of sea water heated from 20 °C to 21 °C will increase by more than four times as much as if it is heated from 0 °C to 1 °C. So, for sea-level rise, it really matters whether the temperature changes most in tropical seas or the polar oceans. Understanding of climate change requires an appreciation of the oceans' importance in forming the climate conditions we experience and that different regions of the ocean respond and contribute to climate change differently.

---

[1] From Rime of the Ancient Mariner, Samuel Taylor Coleridge

Climate Change: Global Risks, Challenges and Decisions, Katherine Richardson, Will Steffen and Diana Liverman *et al*. Published by Cambridge University Press. © Katherine Richardson, Will Steffen and Diana Liverman 2011.

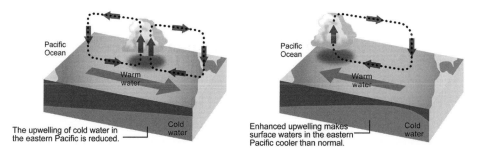

Figure 2.1 The El Niño–Southern Oscillation (ENSO) phenomenon originates in the equatorial Pacific Ocean, where an air–sea interaction involving trade winds and surface ocean circulation drive cyclical changes in sea surface temperature across the region. In the El Niño state (left panel), warm waters move eastward across the Pacific towards the South American coast, with a rise of up to 2–4 °C in SST in the eastern Pacific. In the La Niña state (right panel), the warm surface waters contract westward towards Australia, with cooler surface waters returning to the eastern Pacific. An El Niño brings torrential rains to the coast of Ecuador and Peru and wet conditions to parts of North America, while severe drought comes to northeast Brazil, Australia, Indonesia and southern Africa. Source: redrawn from USA TODAY, http://www.usatoday.com/weather/resources/graphics/2008–09–25-el-nino-la-nina-affect-us-weather_N.htm.

2.1 Modes of variability

It is well recognised that climate conditions vary from year to year. Although we experience inter-annual changes at the local level, much of the variability recorded is due to changes that occur further afield; that is, changes in global or regional patterns in air pressure distributions, storm tracks and trade winds. Very often, these changes are due to changes in ocean–atmosphere interactions that are driven by subtle changes in the surface ocean currents and/or temperature. We recognise a number of specific ocean–atmosphere systems that exhibit changes and influence climate conditions. The best known of these is the El Niño Southern Oscillation (ENSO), where El Niño represents the warm phase of the oscillation and La Niña the cold phase (Figure 2.1). While ENSO describes a warming or cooling in the eastern tropical Pacific, its implications are felt worldwide in the form of changes in rainfall patterns and temperatures. That global air temperature in 1998, for example, was unusually warm (Figure 1.4) is believed in large part to be related to a strong El Niño. Likewise, the comparatively cold temperatures recorded in 2008 are, in large part, explained by a strong La Niña event. Palaeoclimate studies indicate that the behaviour of the ENSO system has varied during the history of the Earth. From such studies, we can get an indication of how the system might react in a warmer climate (Box 2.1).

Box 2.1

Lessons of the mid Pliocene: planet Earth's last interval of greater global warmth

ALAN M. HAYWOOD, SARAH G. BONHAM, DANIEL J. HILL, DANIEL J. LUNT AND ULRICH SALZMANN

The mid Pliocene warm period (~3.3 to 3 million years ago) was the last time in Earth history when global mean temperatures were significantly higher than present-day temperatures and

Box 2.1 (*cont.*)

were caused, at least in part, by higher than pre-industrial levels of carbon dioxide in the atmosphere (estimated to range from 360 to 425 p.p.m.; see Kürschner *et al.*, 1996; Raymo *et al.*, 1996). The period provides an unparalleled opportunity to examine the long-term response of the Earth System to elevated greenhouse gas concentrations (Box 1.5) and has been recognised by the IPCC as 'an accessible example of a world that is similar in many respects to what models estimate could be the Earth of the late 21st century' (Jansen *et al.*, 2007). Through a combination of geological data acquisition; geological data synthesis; and numerical climate, vegetation and ice sheet modelling the nature of climate and environmental change during this unique period is being revealed.

Evidence derived from palaeo-shorelines (Dowsett and Cronin, 1990), shallow marine sedimentary records (Naish and Wilson, 2009), geochemical analyses of foraminifera and ostracods (Lisiecki and Raymo, 2005; Dwyer and Chandler, 2009), coupled with climate–ice sheet model simulations (Lunt *et al.*, 2008), suggest that global ice volume during that period was significantly reduced, with sea levels approximately 25 m higher than modern. Modelling

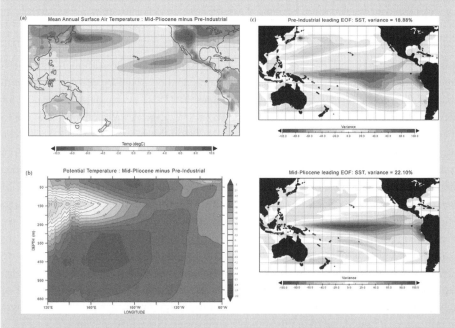

Figure 2.2 HadCM3-coupled simulations for the mid-Pliocene (MP) and the pre-industrial (PI) period: (a) mean annual surface air temperature (°C) for the MP minus PI over the Pacific, with MP surface temperatures 2–3°C warmer in the EEP; (b) mean annual potential temperature (°C) to 650 m depth along 120°E to 80°W, averaged over 5°N to 5°S, for the MP minus PI, highlighting the warmer subsurface ocean temperatures and upwelling water in the EEP; and (c) leading empirical orthogonal function (EOF) for December, January, February and March Pacific SSTs for the PI (top) and the MP (bottom), showing a ~15% increase in ENSO variability in the mid-Pliocene simulation from the pre-industrial period. (In colour as Plate 2.)

indicates that Greenland was largely deglaciated at this time, with ice restricted to high-altitude regions of East Greenland (Lunt *et al.*, 2008). Margins of the East Antarctic Ice Sheet may also have retreated, especially in the Wilkes sub-glacial and Aurora basin sectors (Hill *et al.*, 2007). Recent climate and ice-sheet sensitivity studies have indicated that the most important forcing in driving a largely deglaciated Greenland at this time was the higher than pre-industrial levels of CO_2 (Lunt *et al.*, 2008).

Pollen records and other plant remains from around the world indicate large changes in biomes compared to today. For example, the extent of arid deserts appears to have been greatly reduced. Tropical savannah and woodlands were extended and coniferous forests replaced tundra in the northern hemisphere (Salzmann *et al.*, 2008).

Geological proxies for sea surface temperature (SST) indicate that the mean state of the eastern equatorial Pacific (EEP) may have been 2–3 °C warmer than modern, substantially reducing the SST gradient between the west and the east Pacific to a situation similar to that which occurs during a modern El Niño event (Wara *et al.*, 2005). Owing to the temporal resolution of the data sampling, it is not possible to obtain information on the interannual variability of the SST, hence the period has been characterised as displaying a permanent El Niño-like mean state, or El Padre to differentiate it from a modern El Niño (Ravelo, 2008). Coupled ocean–atmosphere climate models are able to reproduce a reduced SST gradient across the Pacific and warmer EEP SSTs (Haywood *et al.*, 2007; Figure 2.2), but still display SST variability over ENSO timescales. Predictions from the Hadley Centre coupled climate model version 3 (HadCM3), show that the frequency and magnitude of El Niño events differed from today (i.e. became more frequent and stronger), suggesting that the warmer EEP SSTs may not have been caused by a perennial El Niño condition (Bonham *et al.*, 2009; Wünsch, 2009; Figure 2.2). The reason why ENSO frequency changed is currently being explored.

While ENSO is the best known and strongest of the ocean–atmosphere interactions leading to year-to-year differences in climate conditions, several other such systems are also recognised. These include the North Atlantic Oscillation (NAO), Polar Vortex, Indian Ocean Dipole (IOD) and Southern Annular Mode (SAM). Recent research indicates that climate change may interact with these ocean–atmosphere modes of climate variability. These systems influence patterns of temperature and rainfall on inter-annual timescales. Also seasonal changes in rainfall and temperature (i.e. monsoon systems) can, however, be driven by ocean–atmosphere interactions. Common for many of these ocean–atmosphere interactions is that the phenomena may be affected by climate change in ways that drive a larger amplitude in climate fluctuations in coming decades (Box 2.2).

Box 2.2
Changing modes of ocean–atmosphere variability

MATTHEW ENGLAND

While ocean heat uptake significantly buffers the magnitude of anthropogenic warming of the atmosphere (Levitus *et al.* 2001), there are major risks associated with this ocean warming. Apart

Box 2.2 (*cont.*)

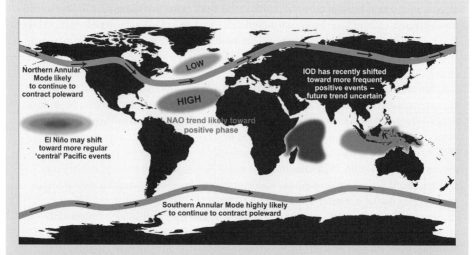

Figure 2.3 Schematic diagram showing the major modes of climate variability and how they are likely to change in the future. The high-latitude modes have already undergone significant change over the past century. Trends in the tropical modes (ENSO, IOD) have been detected in the more recent climatological record. (Full version of this figure in colour as Plate 4.)

from raising sea levels (Box 1.3 and Chapter 3), changing the ocean's thermohaline circulation (Chapter 7) and reducing carbon uptake (Chapter 4), ocean surface warming will also alter coupled ocean–atmosphere modes such as El Niño. Other ocean–atmosphere modes of variability are also likely to be affected, including the Indian Ocean Dipole. This will occur as human-induced warming directly changes the temperature of the surface mixed layer and also alters the ocean's thermocline depth; both are key properties that determine the magnitude, nature and frequency of coupled ocean–atmosphere modes of variability. The higher latitude modes of variability will also be affected, such as the North Atlantic Oscillation and the Southern Annular Mode, although these modes will primarily be altered via an internal atmospheric response to increased greenhouse gases and depleted polar stratospheric ozone. Changes in these major hemisphere-scale patterns of variability will significantly impact regional climate, such as mid-latitude rainfall rates. Figure 2.3 shows an overview of the major climate modes and their projected trends.

Model projections (Chapter 7) suggest that anthropogenic climate change will force El Niño events to drift toward a different preferred state in coming decades (Latif and Keenlyside, 2008; Yeh *et al.*, 2009). There is even some evidence that change is already underway – with several studies noting a greater number of 'central', as opposed to 'eastern', Pacific El Niño events occurring in recent years (see Ashok *et al.*, 2007; Taschetto and England, 2009). One such central Pacific event occurred in 2002–03 when ocean warming developed most markedly in the central equatorial Pacific instead of in the east, which is atypical for ENSO episodes. The anomaly pattern has been tentatively classified as an independent mode, termed the El Niño 'Modoki' (or pseudo-El Niño). This mode is characterised by warmer than usual central Pacific waters and associated cooler SST anomalies on the eastern and western portions of the basin (Ashok *et al.*, 2007). While the peak Pacific warming in 2002–03 was only modest, the rainfall response in some regions was significant. For example, eastern Australia experienced some of its most severe reductions in rainfall during this Modoki event. This unusual El Niño event has raised questions as to exactly how regional rainfall is controlled by the location and magnitude of tropical Pacific SST variability, and importantly how this will be transformed in the future.

A major coupled mode of Indian Ocean variability also appears to be changing, with a reported increased incidence of positive phase Indian Ocean Dipole events over the last few decades (Abram *et al.*, 2008; Cai *et al.*, 2009). This coincides with recent non-uniform warming trends in the Indian Ocean (Alory *et al.*, 2007; Ihara *et al.*, 2008), which have likely affected the average background climate of the region. Like shifts in El Niño, this could have significant impacts on regional climate if the trend continues, affecting rainfall over East Africa (Latif *et al.*, 1999; Black *et al.*, 2003), India, Indonesia (D'Arrigo and Wilson, 2008), and Australia (England *et al.*, 2006; Ummenhofer *et al.*, 2008).

While trends are becoming evident in the coupled climate modes in the observational records, it remains unclear how these trends will evolve over the coming century. It is, for example, unclear how El Niño and the Indian Ocean Dipole will be influenced by the emergent changes in the large-scale meridional and zonal circulation of the atmosphere (e.g. the Hadley and Walker circulations). Regardless, against the background of ongoing ocean warming and continuing trends in the large-scale atmospheric circulation, it is highly likely that the major coupled ocean–atmosphere modes of variability will change significantly in the future, with unknown consequences for regional rainfall and climate.

The most prominent feature of variability in the extratropical regions is the so-called annular modes in both hemispheres, which regulate the latitude of the polar front jets and subpolar westerly winds. In the south is the Southern Annular Mode, a pressure oscillation between Antarctica and the subtropical high pressure belt, which controls variations in the strength and latitude of the subpolar westerly winds. In the north is the Northern Annular Mode, which is linked closely to the North Atlantic Oscillation (Visbeck *et al.*, 2001); the latter is effectively a regional manifestation of the circumpolar annular mode. These extratropical oscillations are largely internal (uncoupled) modes of atmospheric variability, controlled by interactions between the mean and transient atmospheric circulation; in particular the mid latitude cyclones. The annular modes exhibit significant variability on synoptic, seasonal and interannual timescales. Yet as climate change is impacting both the mean state of the atmosphere and the transient eddies at mid latitudes, the annular modes are also undergoing change, gradually contracting toward a more poleward mean position, particularly in the southern hemisphere.

This is having a significant impact on regional weather systems and mean climate. For example, over the southern hemisphere, the frequency of extratropical low pressure systems in the latitude band 40 °S to 60 °S has decreased over the past few decades. This has been linked to both increasing greenhouse gases (Fyfe, 2003) and depleted stratospheric ozone (Gillett and Thompson, 2003); these two anthropogenic causes have combined to force a poleward shift in the southern hemisphere polar front jet. This significant adjustment in the extratropical regions is already having a profound impact on regional climate; for example, affecting drought over parts of Australia (Cai and Cowan, 2006). Furthermore, this trend is projected to continue as greenhouse gas concentrations rise this century (Sen Gupta *et al.*, 2009), with a similar shift likely for the Northern Annular Mode. Such a change in the atmospheric circulation in the extratropical regions would have a profound impact on regional rainfall and climate.

2.2 Oceans as a sink/source for greenhouse gases

Another important role the oceans play in the climate system is as a carbon reservoir or 'sink'. The oceans cover 71% of the Earth's surface. With the exception of where they are ice-covered, the oceans are in direct contact with the atmosphere and gases exchange between these two Earth System compartments. Since the industrial revolution, there has been a net flux of CO_2 from the atmosphere into the ocean. The oceans have taken up around 27%–34% of the CO_2 produced by humankind since the industrial revolution (Canadell *et al.*, 2007; Le Quéré *et al.*, 2007, 2009). Both physical/chemical and biological processes are responsible for this transfer of CO_2 from the atmosphere to the ocean. Had the oceans not removed this CO_2 from the atmosphere, then the atmospheric concentration of CO_2 would be somewhere around 420 p.p.m. rather than the 387 p.p.m. that is recorded today, and the magnitude of climate change would be correspondingly greater than observed today.

Thus, one of the great but undervalued services that the ocean provides human societies is that it slows the progression of human-induced global warming by removing anthropogenic carbon from the atmosphere. Unfortunately, however, most of the processes influencing the oceans' ability to semi-permanently store ('sequester') the carbon they remove from the atmosphere are being influenced by climate change in such a manner that the oceans' ability to take up and store carbon from the atmosphere is being reduced (Chapter 4). In addition to being a sink for CO_2, the oceans serve as a source (i.e. releases to the atmosphere) of some other greenhouse gases such as methane and nitrous oxide. Climate and other global changes – for example, human interference with the global nitrogen cycle, which leads to coastal eutrophication and the production of nitrous oxides – may increase the importance of the oceans as a source for GHGs to the atmosphere.

2.3 Ocean acidification

As the ocean and atmosphere are in direct contact over most of the Earth's surface, an increase in the amount of CO_2 in the atmosphere will automatically lead to an increase in the concentration of CO_2 in the surface waters of the ocean. We cannot say that this CO_2 is sequestered in the ocean as there is free exchange between the surface ocean and the atmosphere. Thus, if the concentration of CO_2 in the atmosphere were to be reduced suddenly, the CO_2 would leave the surface ocean and return to the atmosphere such that a balance was maintained between the concentration in the air and in the water. When CO_2 is dissolved in water, carbonic acid is formed and the solution acidifies.

This means that ocean acidification is a direct response to increasing CO_2 in the atmosphere. Thus, while the ocean uptake of CO_2 helps reduce the human impact on the climate system, this service comes at the price of a dramatic change to ocean chemistry. In particular, and of great concern, are the observed changes in ocean pH, and carbonate and bicarbonate ion concentrations. Evidence is accumulating that suggests ocean acidification is a serious threat to many organisms and may have implications for food webs and ecosystems, as well as the multi-billion dollar services they provide (Box 2.3).

Box 2.3

Ocean acidification: examples of potential impacts

CAROL TURLEY, JEAN-PIERRE GATTUSO AND OVE HOEGH-GULDBERG

Ocean acidification is caused by ocean uptake of anthropogenic CO_2; it is a global issue and is happening now, it is measurable, and it will continue as more CO_2 is emitted. Already ocean acidity has increased by 30% since the industrial revolution and by 2100, if we continue emitting CO_2 at the same rate, ocean acidity would have increased by as much as 150%. Such a substantial alteration in basic ocean chemistry is likely to have wide implications for ocean life, especially for many organisms that require calcium carbonate to build shells or skeletons (Turley and Findlay, 2009).

Tropical coral reef ecosystems (Box 6.4) depend heavily on the calcifying activities of reef-building corals and other organisms such as calcareous red algae. Together, these organisms deposit large amounts of the calcium carbonate that builds up over time to create the three-dimensional structure typical of healthy coral reefs, providing the habitat within which an estimated one million species live. The calcifying abilities of reef-building corals are sensitive to the carbonate ion concentration (Kleypas *et al.*, 1999; Kleypas and Langdon, 2006), which has decreased by around 16% as oceans have become more acidic. Research indicates a major decrease in the calcification of key coral species, suggesting that the combined impact of thermal stress and ocean acidification is already having a major impact on the ability of corals to build and maintain coral reefs (De'ath *et al.*, 2009; Tanzil *et al.*, 2009). Relatively small changes (15%–30%) can tip the carbonate balance in favour of physical and biological erosion, potentially leading to the net disintegration of coral reefs over time (Figure 2.4). If this happens, the biological diversity and ecological services (e.g. food, income, coastal protection) of coral reefs will decrease, undermining the livelihood and well-being of over 500 million people throughout tropical coastal areas of the world (Hoegh-Guldberg *et al.*, 2007).

Among the areas most threatened by ocean acidification are the polar and sub-polar regions as well as the deep sea. Very few perturbation experiments have been conducted on organisms and communities living in these areas. The first results were collected on pteropods (pelagic marine snails) and deepwater corals, both playing essential roles in their respective ecosystems (Orr *et al.*, 2005; Turley *et al.*, 2007). The pteropod *Limacina helicina* has an important role in the food chain and functioning of the Arctic and subarctic marine ecosystems. Its calcium carbonate shell provides vital protection. However, a recent study has shown that the shell of this mollusc is precipitated at a rate that is 30% slower when it is kept in sea water with the characteristics anticipated in 2100 (Figure 2.5; Comeau *et al.*, 2009). An even more marked reduction (50%) has been reported in the deepwater coral, *Lophelia pertusa* (Maier *et al.*, 2009). While tropical coral reefs are built by a large number of species, coral communities in cold waters are constructed by one or two species but provide shelter for many others.

Fish (including shellfish) represent 15% of animal protein for three billion people worldwide, and a further one billion rely upon fisheries for their primary protein (Millennium Ecosystem Assessment, 2003; see also Box 5.2). Research is producing a growing list of organisms, many of which provide food, which may be affected by ocean

Box 2.3 (*cont.*)

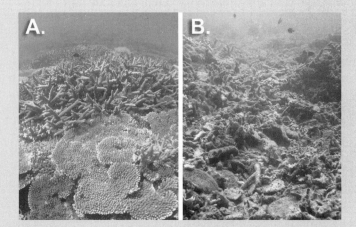

Figure 2.4 Coral reefs face an uncertain future if their ability to form carbonate skeletons is diminished under high atmospheric carbon dioxide. A. Reef-building coral reefs like this one at Heron Island (on the southern Great Barrier Reef) have relatively healthy populations of reef-building corals and red coralline algae that are able to maintain the three-dimensional framework of the reef. This framework provides habitat for an estimated million species and ultimately food, livelihoods and coastal protection to hundreds of millions of people worldwide. B. Increasing concentrations of atmospheric carbon dioxide, however, will lead to a sharp decline in the rate of calcification, causing a decrease in the ability of reefs to calcify and maintain reefs relative to the rate of physical and biological erosion. Under these circumstances, reefs may crumble and disappear. This example shows a section of a reef near Karimunjawa in the Java Sea, Indonesia, where the reef framework has been weakened by the loss of reef calcifiers. While a different set of drivers (i.e. poor water quality and destructive fishing) lie behind the loss of coral in this particular example, the outcome may be similar to that anticipated under the impact of warming and acidifying seas on corals, coralline algae and reef calcification if current rates of CO_2 emissions continue. (In colour as Plate 5.)

acidification in a number of ways at some stage of their life history (Turley and Findlay, 2009). The worry for future food security is how ocean acidification will directly affect those organisms consumed by humans, and further, and indirectly, how food webs and ecosystems that support such organisms might be adversely affected to the detriment of both ecology and food supply.

Acknowledgements

Contribution of the European Project on Ocean Acidification (EPOCA) with funding from the European Community (grant agreement 211384) and the French Polar Institute (IPEV).

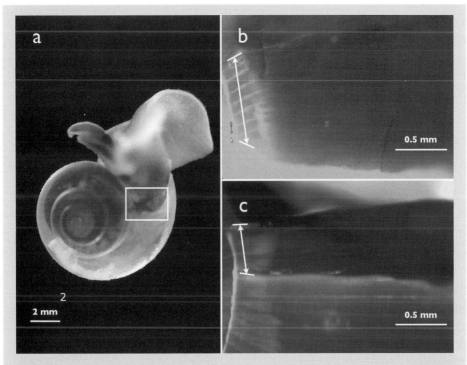

Figure 2.5 Representative example of live Arctic *Limacina helicina* (a) stained with calcein, and subsequently maintained at pH_T 8.09 (b) and 7.8 (c). Most calcification occurs near the shell opening (white rectangle). The arrow indicates the five days linear extent of the shell. Quantitative estimates obtained with a radiotracer indicate a significant 30% reduction of the calcification rate. Source: Comeau *et al.* (2009) *Biogeosciences* (open access). (In colour as Plate 6.)

Surface ocean pH has not been lower in the past 23 million years (Figure 2.6). The predicted rate of change in ocean surface pH if atmospheric CO_2 concentrations continue to increase in coming decades is dramatic, and would be faster than that rate recorded in over the past 65 million years (Ridgwell and Schmidt, 2010). While it is unclear exactly how marine ecosystems will respond to this acidification, previous ocean acidification events have been identified as potentially causative agents in mass extinctions of marine species (Veron, 2008). Ocean acidification will continue to track future CO_2 emissions to the atmosphere so emission reductions are the only way of reducing the increase and impact of ocean acidification.

2.4 The oceans and the hydrological cycle

Evaporation from the oceans is important for the formation of rain and clouds, i.e. in controlling the amount of water vapour in the atmosphere. As water vapour is the most important GHG, increased evaporation from the ocean due to warmer temperatures is an important

Climatic trends

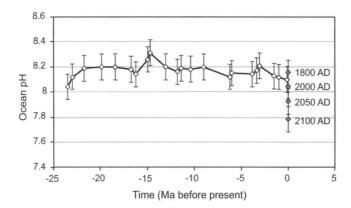

Figure 2.6 Ocean acidity (pH) over the past 25 million years and projected to 2100. The lower the pH, the more acidic the ocean becomes. Prehistoric surface-layer pH values reconstructed using boron-isotope ratios of ancient planktonic foraminifer shells. Future pH values derived from models based on IPCC mean scenarios. Source: Turley *et al.* (2007).

'positive feedback' in the climate system, i.e. a process whereby global warming leads to an even greater increase in global warming. Thus, while human activities have not directly changed the amount of the GHG water vapour in the atmosphere, human activities that drive global warming have indirectly increased the amount of water in the atmosphere.

Measuring salinity in the ocean tells us something about the patterns and mechanisms whereby freshwater (rainfall and runoff from land) and salt water in the ocean mix. Thus, changes in salinity can be used to inform us about changes in the global hydrological (water) cycle. Comprehensive analyses of ocean salinity show the freshening of high latitudes due to either increased rainfall, or increased water runoff from land due to ice melt, or both. Regions exhibiting excess evaporation over precipitation have become saltier. These salinity changes are due to a strengthening of the hydrological cycle. The patterns of salinity change noted are also consistent with regional circulation and inter-basin exchanges, and there is increased evidence that the patterns of rainfall change over the global ocean as reflected in salinity can be attributed to climate change (Box 2.4).

Box 2.4
Detecting changes in the hydrological cycle through changes in ocean salinity

NATHANIEL L. BINDOFF AND KIERAN P. HELM

Changes in precipitation have been detected on land (Zhang *et al.*, 2007) and in the short satellite records (Wentz *et al.*, 2007). Covering approximately 70% of the globe, the oceans also act as an historical rain gauge, with changes in precipitation being reflected in the *in situ* observations of salinity (Wong *et al.*, 1999; Boyer *et al.*, 2005; Bindoff *et al.*, 2007). Traditionally, ocean salinity has been used as a tracer, or dye, for understanding the large-scale global ocean circulation, and ocean salinity is a key variable that shows the large-scale 'conveyor belt' or, equivalently, the global overturning circulation. However, the pattern of global surface salinity can also be used to understand the Earth's hydrologic cycle.

The pattern of surface salinity can be attributed largely to the distribution of precipitation and evaporation over the globe (Figure 2.7) and also, to a lesser extent, the continental runoff from rivers and melt of ice sheets in Greenland and Antarctica. The pattern of salinity in the interior of the ocean beneath the surface is created by the circulation that spreads the surface waters into the ocean interior. Much of the ocean temperature and salinity signal at the surface is largely preserved as the waters spread into the ocean interior far from their source region. For example, in the near-surface waters of the subtropics (10° to 35° northern or southern hemisphere) evaporation dominates precipitation, resulting in high-salinity surface waters in the upper 300 m of the ocean (Figure 2.7). These high-salinity near-surface and surface waters are found in both the northern hemisphere and southern hemisphere oceans. By contrast, precipitation dominates evaporation in the mid to high latitudes, resulting in fresher surface waters. A fresh, low-salinity layer of water sinking at high latitudes and spreading towards the equator can be seen in both hemispheres (at depths of 500–1000 m). In the hydrographic section in the central Pacific Ocean, the fresh layer at these depths is called Antarctic Intermediate Water and is characterised by a lower salinity than the waters above or below. In the North Pacific, the fresh layer originates near the Okhotsk Sea and circulates around the large-scale gyres in the North Pacific Ocean. These remarkable freshwater layers are found in every ocean (except the North Atlantic) and all originate in the polar regions in either hemisphere.

The changes in ocean salinity of the shallow, high-salinity water and Antarctic Intermediate Water are shown in Figure 2.8. The near-surface high-salinity waters found in the subtropics have become more saline as shown in the upper panel. The pattern of increase in salinity in these near-surface saline waters has been occurring almost everywhere in the subtropics

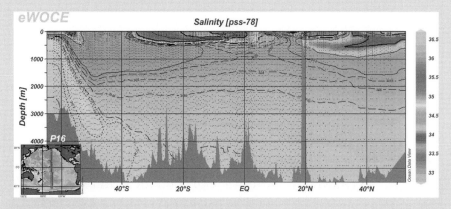

Figure 2.7 An oceanographic section of salinity from top to bottom from the Pacific Ocean from the Southern Ocean to Alaska. See the insert panel for the precise location of the section. Sources: http://www.ewoce.org/ and Schlitzer (2000). 'PSS' stands for 'practical salinity scale', using the UNESCO 1978 equation between salinity and conductivity. Salinity is dimensionless and reported in parts per thousand. (In colour as Plate 7.)

Box 2.4 (*cont.*)

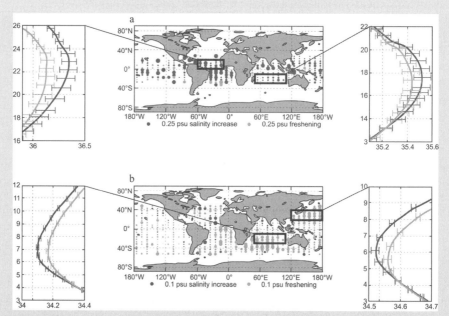

Figure 2.8 Salinity changes within the ocean: (a) the salinity maximum layer; (b) the salinity minimum layer (AAIW or NPIW). Each dot represents the average change inside the 5*10 degree grid cell. The top left and right panels show a temperature–salinity plot for the corresponding boxed region and are centered on the salinity maximum (approximately 100 m depth). The bottom left and right panels are for the salinity minimum (approximately 600 to 1000 m depth). The black line is temperature–salinity for the recent data (~2005) and the grey line is for the period centred on 1970. Reproduced/modified by permission of American Geophysical Union (Helm *et al.*, 2010).

since the 1970s (Helm, 2008). This pattern of salinity increase suggests that in the subtropics evaporation has increased relative to the contribution of rain. The low-salinity waters (Antarctic Intermediate Water and North Pacific Intermediate Water) have freshened between 1970 and 2005 (Figure 2.8b). This freshening with modified properties reaches as far as the equatorial region via low-latitude western boundary currents, particularly in the south-west Pacific where large changes are observed that are consistent with these known circulation pathways (Figure 2.8b) (Bindoff *et al.*, 2007). The North Pacific Intermediate Water also shows a broad pattern of freshening between 20 °N and 35 °N. This freshening of the low-salinity layers suggests there has been an increase in rain in the surface waters in the polar regions.

The hydrological changes from these observed salinity changes can be estimated from the salt content change (Wong *et al.*, 1999; Helm 2008; Helm *et al.*, 2010). The estimated surface freshwater flux from this approach reveals a pattern of the strong addition of freshwater south of 45° S of up to 100 mm yr^{-1} near Antarctica, a smaller addition of freshwater north of 45° N and with statistically significant but smaller decreases in the mid-latitudes in both hemispheres. Based on climatologies of precipitation minus evaporation, this observed freshwater flux change represents about a 16±6% increase in the Southern Ocean, a 7±4% increase in the northern hemisphere and a decrease of about 3±2% in the subtropical gyres of both hemispheres. These estimates do not include the surface ocean and are, therefore, likely to change with new surface (and deep) measurements.

Coupled global climate models show that the hydrological cycle has intensified over the past 50 years and is projected to intensify further by 2100 (Meehl *et al.*, 2007), resulting in more saline surface waters in the low latitudes and fresher surface waters at higher latitudes. The salinity as used here to estimate precipitation–evaporation (P–E) from ocean observations can provide an independent comparison with simulations of climate change for the same period as the observations. The correspondence of the estimated surface freshwater fluxes for the 1970 to 2005 period with the simulated patterns of P–E changes in the atmosphere over the same period lends strong support for conclusion that the global hydrological cycle has accelerated (Bindoff *et al.*, 2007).

2.5 Tropical storms

Cyclones[2] have tremendous impact on human societies (Chapter 5). Therefore, it is important to understand what causes them and the role of ocean temperature in their formation. Tropical cyclones form following a 'fault' – for example, a thunderstorm – over the ocean. Warm, moist air rises from the ocean surface and cools as it moves away from the ocean surface. This causes water vapour to condense. The heat released during the condensation process causes a continued rapid upward motion of the air, as this internal heating process maintains a warmer air temperature in the column of rising air than the air around it. Note, though, that the ultimate source of energy for this process is the heat in the ocean below. The rising air leaves a 'hole' near the surface of the ocean, and air flows from all directions into the 'hole' as a replacement for the rising air. The Earth's rotation causes this air to be deflected and the system starts to rotate.

The resulting circulatory system (Figure 2.9) can be sustained for several days until it runs out of 'fuel' (i.e. heat energy from the surface ocean). This happens when the tropical cyclone reaches colder waters or land. Alternatively, a cyclone can be destroyed under certain wind conditions. Fortunately, relatively few of these storms make landfall. Tropical cyclones do not occur near the equator (because of the special wind conditions there). However, away from the equator, they can occur anywhere where ocean surface temperatures are over 26.5 ° C. Warming of the surface waters of the ocean thus potentially expands the area over which these storms can form and provides more energy to fuel such storms. Just as for every other component or expression of the climate system, there is considerable natural variability with respect to the number of tropical cyclones recorded in a given year.

---

[2] When wind speeds exceed a certain level, tropical cyclones are referred to as hurricanes in the Atlantic and the northeast Pacific. In the northwest Pacific, they are referred to as typhoons. Elsewhere, they are simply called cyclones.

Figure 2.9 GOES image of Hurricane Luis, 1995. Courtesy of NASA Goddard Space Flight Center.

Therefore, relatively long datasets (Box 5.5) are required in order to identify changes in the frequency or amplitude of cyclones.

In the case of cyclones, the accuracy of datasets and, therefore, data analysis is also hampered by the fact that only a relatively small number of these storms reach land. Until the advent of continuous remote sensing of weather conditions on Earth (in practice, this means satellite-based sensors), many of the cyclones at sea went unrecorded. Thus, reliable datasets describing the frequency of cyclone formation are relatively short. Examination of these datasets indicates that there has been a decrease in the number of cyclones and cyclone days in the North Pacific, Indian and south-west Pacific oceans and an increase in the number in the North Atlantic in recent decades. At the same time, there has been an increase in the intensity of these storms (as evidenced by the number of storms reaching category 4 or 5). The largest increases in the numbers of storms reaching category 4 or 5 occurred in the ocean basins where a decrease in total number of storms was recorded (Webster *et al.*, 2005). Much research activity is currently focused on developing a better understanding of the possible interactions between climate change, and cyclone frequency and intensity.

2.6 Summary and conclusions

Humans can only survive in habitats found at the surface of the land-covered regions on Earth. Thus, although the oceans cover the vast majority of the planet, we have little first-hand experience of how they function, or appreciation for the global services they provide. Nevertheless, the oceans are a dominating driver in the global climate system. Not only are they critical in storing and transporting heat on the planet, but they are also critical in determining rainfall distributions on land. Thus, the oceans are, in many ways, the architect behind the habitats we are dependent upon for our very existence. Given that such a large percentage of the heat energy stored on Earth is found in the oceans, changes in ocean heat content are a robust indicator of global climate change. During the past few years, it has been documented that ocean temperatures have been increasing about 50% faster in recent decades than was previously believed (Box 1.3 and references therein).

In addition to temperature changes, climate change is causing changes in the physical chemistry of the oceans. These include changes in salinity distributions that signal changes in the global hydrological cycle as well as acidification of surface waters. This acidification potentially threatens many different types of marine organisms and influences the functioning of the global carbon cycle (Chapter 4).

The changes in the oceans generated by elevated CO_2 concentrations and global warming not only impact living conditions for marine organisms, they will ultimately also alter the living conditions humans experience on land. This is because many of these changes will reduce the capacity of the oceans to sequester CO_2 from the atmosphere, thereby accelerating the rate of global warming – unless anthropogenic CO_2 emissions are drastically reduced. Prediction of future climate and its impacts requires a better scientific understanding of the changes occurring in the oceans. Public appreciation of the potential impacts of climate change might be enhanced if there were a greater general recognition of the role that the oceans play in the climate system, the effect that climate change is having on the oceans and how the climate change-induced changes in the oceans can influence the climate conditions experienced by organisms living on land.

References

Abram, N. J., Gagan, M. K., Cole, J. E., Hantoro, W. S. and Mudelsee, M. (2008). Recent intensification of tropical climate variability in the Indian Ocean. *Nature Geoscience*, **1**, 849–53.

Alory, G., Wijffels, S. and Meyers, G. (2007). Observed temperature trends in the Indian Ocean over 1960–1999 and associated mechanisms. *Geophysical Research Letters*, **34**, L02606.

Ashok, K., Behera, S. K., Rao, S. A., Weng, H. and Yamagata, T. (2007). El Niño Modoki and its possible teleconnection. *Journal of Geophysical Research*, **112**, C11007.

Bindoff, N. L., Willebfrand, J., Artale, V. *et al.* (2007). Observations: Oceanic climate change and sea level. In *Climate Change 2007: The Physical Science Basis. Contribution of Working Group I to the Fourth Assessment Report of the Intergovernmental Panel on Climate Change*, eds. S. Solomon, D. Qin, M. Manning, Z. Chen, M. Marquis, K. B. Averyt, M. Tignor and H. L. Miller. Cambridge, UK and New York, NY: Cambridge University Press, pp. 385–432.

Black, E., Slingo, J. and Sperber, K. R. (2003). An observational study of the relationship between excessively strong short rains in coastal East Africa and Indian Ocean SST. *Monthly Weather Review*, **131**, 74–94.

Bonham, S. G., Haywood, A. M., Lunt, D. J., Collins, M. and Salzmann, U. (2009). El Niño–Southern Oscillation, Pliocene climate and equifinality. *Philosophical Transactions of the Royal Society A – Mathematical, Physical & Engineering Sciences*, **367**, 127–56.

Boyer, T. P., Levitus, S., Antonov, J. I., Locarnini, R. A. and Garcia, H. E. (2005). Linear trends in salinity for the World Ocean. *Geophysical Research Letters*, **32**, L01604.

Cai, W., Cowan, T. and Sullivan, A. (2009). Recent unprecedented skewness towards positive Indian Ocean Dipole occurrences and its impact on Australian rainfall. *Geophysical Research Letters*, **36**, L11705.

Cai, W. and Cowan, T. (2006). SAM and regional rainfall in IPCC AR4 models: Can anthropogenic forcing account for southwest Western Australian winter rainfall reduction? *Geophysical Research Letters*, **33**, L24708.

Canadell, J. G., Le Quéré, C., Raupach, M. R. *et al.* (2007). Contributions to accelerating atmospheric CO_2 growth from economic activity, carbon intensity, and efficiency of natural sinks. *Proceedings of the National Academy of Sciences (USA)*, **104**, 18866–70.

Comeau, S., Gorsky, G., Jeffree, R., Teyssié, J.-L. and Gattuso, J.-P. (2009). Impact of ocean acidification on a key Arctic pelagic mollusc (*Limacina helicina*). *Biogeosciences*, **6**, 1877–82.

D'Arrigo, R. and Wilson, R. (2008). El Niño and Indian Ocean influences on Indonesian drought: Implications for forecasting rainfall and crop productivity. *International Journal of Climatology*, **28**, 611–16.

De'ath, G., Lough, J. M. and Fabricius, K. E. (2009). Declining coral calcification on the Great Barrier Reef. *Science*, **323**, 116–19.

Dowsett, H. J. and Cronin, T. M. (1990). High eustatic sea level during the middle Pliocene: Evidence from the southeastern U.S. Atlantic Coastal Plain. *Geology*, **18**, 435–38.

Dwyer, G. S. and Chandler, M. A. (2009). Mid-Pliocene sea level and continental ice volume based on coupled benthic Mg/Ca palaeotemperatures and oxygen isotopes. *Philosophical Transactions of the Royal Society A – Mathematical, Physical & Engineering Sciences*, **367**, 157–68.

England, M. H., Ummenhofer, C. C. and Santoso, A. (2006). Interannual rainfall extremes over southwest Western Australia linked to Indian Ocean climate variability. *Journal of Climate*, **19**, 1948–69.

Fyfe, J. C. (2003). Extratropical southern hemisphere cyclones: Harbingers of climate change? *Journal of Climate*, **16**, 2802–05.

Gillett, N. P. and Thompson, D. W. J. (2003). Simulation of recent southern hemisphere climate change. *Science*, **302**, 273–75.

Haywood, A. M., Valdes, P. J. and Peck, V. L. (2007). A permanent El Niño-like state during the Pliocene. *Paleoceanography*, **22**, 1–A1213.

Helm, K. (2008). Decadal Ocean Water Mass Changes: Global Observations and Interpretation, PhD Thesis, University of Tasmania, Australia.

Helm, K. P., Bindoff, N. L. and Church, J. A. (2010). Changes in the global hydrological-cycle inferred from ocean salinity. *Geophysical Research Letters*, **37**, L18701, doi: 10.1029/2010GL044222.

Hill, D. J., Haywood, A. M., Hindmarsh, R. C. A. and Valdes, P. J. (2007). Characterizing ice sheets during the mid Pliocene: Evidence from data and models. In *Deep-time Perspectives on Climate Change: Marrying the Signal from Computer Models and Biological Proxies*, eds. M. Williams, A. M. Haywood, F. J. Gregory and D. N. Schmidt. London: The Geological Society of London, pp. 517–38.

Hoegh-Guldberg, O., Mumby, P. J., Hooten, A. J. *et al.* (2007). Coral reefs under rapid climate change and ocean acidification. *Science*, **318**, 1737–42.

Ihara, C., Kushnir, Y. and Cane, M. A. (2008). Warming trend of the Indian Ocean SST and Indian Ocean Dipole from 1880 to 2004. *Journal of Climate*, **21**, 2035–46.

Jansen, E., Overpeck, J., Briffa, K. R. *et al.* (2007). Palaeoclimate. In *Climate Change 2007: The Physical Science Basis. Contribution of Working Group I to the Fourth Assessment Report of the Intergovernmental Panel on Climate Change*, eds. S. Solomon, D. Qin, M. Manning, Z. Chen, M. Marquis, K. B. Averyt, M. Tignor and H. L. Miller. Cambridge, UK and New York, NY: Cambridge University Press, pp. 432–97.

Kleypas, J. A. and Langdon, C. (2006). Coral reefs and changing seawater carbonate chemistry. In *Coral Reefs and Climate Change: Science and Management*, eds. J. Phinney, O. Hoegh-Guldberg, J. Kleypas, W. Skirving and A. E. Strong, AGU Monograph Series, Coastal and Estuarine Studies, **61**, Washington, D.C.: American Geophysical Union, pp. 73–110.

Kleypas, J. A., Buddemeier, R. W., Archer, D. *et al.* (1999). Geochemical consequences of increased atmospheric CO_2 on coral reefs. *Science*, **284**, 118–20.

Kürschner, W. M., van der Burgh, J., Visscher, H. and Dilcher, D. L. (1996). Oak leaves as biosensors of late Neogene and early Pleistocene paleoatmospheric CO_2 concentrations. *Marine Micropaleontology*, **27**, 299–312.

Latif, M. and Keenlyside, N. S. (2008). El Niño/Southern Oscillation response to global warming. *Proceedings of the National Academy of Sciences (USA)*, **106**, 20578–83.

Latif, M., Dommenget, D., Dima, M. and Grötzner, A. (1999). The role of Indian Ocean sea surface temperature in forcing East African rainfall anomalies during December–January 1997/98. *Journal of Climate*, **12**, 3497–504.

Le Quéré, C., Rodenbeck, C., Buitenhuis, E. T. *et al.* (2007). Saturation of the Southern Ocean CO_2 sink due to recent climate change. *Science*, **316**, 1735–38.

Le Quéré, C., Raupach, M. R., Canadell, J. G. *et al.* (2009). Trends in the sources and sinks of carbon dioxide. *Nature Geoscience*, **2**, 831–36.

Levitus, S., Antonov, J. I., Wang, J. *et al.* (2001). Anthropogenic warming of Earth's climate system. *Science*, **292**, 267–70.

Lisiecki, L. E. and Raymo, M. E. (2005). A Pliocene-Pleistocene stack of 57 globally distributed benthic $\delta 18O$ records. *Paleoceanography*, **20**, PA1003.

Lunt, D. J., Foster, G. L., Haywood, A. M. and Stone, E. J. (2008). Late Pliocene Greenland glaciation controlled by a decline in atmospheric CO_2 levels. *Nature*, **454**, 1102–05.

Maier, C., Hegeman, J., Weinbauer, M. G. and Gattuso, J.-P. (2009). Calcification of the cold-water coral *Lophelia pertusa* under ambient and reduced pH. *Biogeosciences*, **6**, 1671–80.

Meehl, G. A., Stocker, T. F., Collins, W. D. *et al.* (2007). Global climate projections. In *Climate Change 2007: The Physical Science Basis. Contribution of Working Group I to the Fourth Assessment Report of the Intergovernmental Panel on Climate Change*, eds. S. Solomon, D. Qin, M. Manning, Z. Chen, M. Marquis, K. B. Averyt, M. Tignor and H. L. Miller. Cambridge, UK and New York, NY: Cambridge University Press, pp. 747–846.

Millennium Ecosystem Assessment (MA) (2003). *Ecosystems and Human Well-being: A Framework for Assessment*. Chapter 2: Ecosystems and their services, http://www.millenniumassessment.org.

Naish, T. R. and Wilson, G. S. (2009). Constraints on the amplitude of Mid-Pliocene (3.6–2.4Ma) eustatic sea-level fluctuations from the New Zealand shallow-marine sediment record. *Philosophical Transactions of the Royal Society A – Mathematical, Physical & Engineering Sciences*, **367**, 169–87.

Orr, J. C., Fabry, V. J., Aumont, O. *et al.* (2005). Anthropogenic ocean acidification over the twenty-first century and its impact on calcifying organisms. *Nature*, **437**, 681–86.

Ravelo, A. C. (2008). Lessons from the Pliocene warm period and the onset of northern hemisphere glaciation. *American Geophysical Union*, Fall Meeting 2008, abstract #PP23E-01.

Raymo, M. E., Grant, B., Horowitz, M. and Rau, G. H. (1996). Mid-Pliocene warmth: Stronger greenhouse and stronger conveyor. *Marine Micropaleontology*, **27**, 313–26.

Ridgwell, A. and Schmidt, D. N. (2010). Past constraints on vulnerability of marine calcifiers to massive carbon dioxide release. *Nature Geoscience*, doi: 10.1038/NGE0755.

Salzmann, U., Haywood, A. M., Lunt, D. J., Valdes, P. J. and Hill, D. J. (2008). A new global biome reconstruction and data-model comparison for the Middle Pliocene. *Global Ecology and Biogeography*, **17**, 432–47.

Schlitzer, R. (2000). Electronic atlas of WOCE hydrographic and tracer data now available, *Eos, Transactions, American Geophysical Union*, **81**, 45.

Sen Gupta, A., Santoso, A., Taschetto, A. S. *et al.* (2009). Projected changes to the southern hemisphere ocean and sea-ice in the IPCC AR4 climate models. *Journal of Climate*, **22**, 3047–78.

Tanzil, J. T. I., Brown, B. E., Tudhope, A. W. and Dunne, R. P. (2009). Decline in skeletal growth of the coral *Porites lutea* from the Andaman Sea, South Thailand between 1984 and 2005. *Coral Reefs*, **28**, 519–28.

Taschetto, A.S. and England, M. H. (2009). El Niño Modoki impacts on Australian rainfall. *Journal of Climate*, **22**, 3167–74.

Turley, C. M. and Findlay, H. S. (2009). Ocean acidification as an indicator of global change. In *Climate Change: Observed Impacts on Planet Earth*, ed. T. Letcher. Elsevier, pp. 367–90.

Turley, C. M., Roberts, J. M. and Guinotte, J. M. (2007). Corals in deep-water: Will the unseen hand of ocean acidification destroy cold-water ecosystems? *Coral Reefs*, **26**, 445–8.

Ummenhofer, C. C., Sen Gupta, A., Pook, M. J. and England, M. H. (2008). Anomalous rainfall over southwest Western Australia forced by Indian Ocean sea surface temperatures. *Journal of Climate*, **21**, 5113–34.

Veron, J. E. N. (2008). Mass extinctions and ocean acidification: Biological constraints on geological dilemmas. *Coral Reefs*, **27**, 459–72.

Visbeck, M. H., Hurrell, J. W., Polvani, L. and Cullen, H. M. (2001). The North Atlantic Oscillation: Past, present, and future. *Proceedings of the National Academy of Sciences (USA)*, **98**, 12876–77.

Wara, M. W., Ravelo, A. C. and Delaney, M. L. (2005). Permanent El Niño-like conditions during the Pliocene warm period. *Science*, **309**, 758–61.

Webster, P. J., Holland, G. J., Curry, J. A. and Chang, H.-R. (2005). Changes in tropical cyclone number, duration, and intensity in a warming environment. *Science*, **309**, 1844–46.

Wentz, F. J., Ricciardulli, L., Hilburn, K. and Mears, C. (2007). How much more rain will global warming bring? *Science*, **317**, 233–35.

Wong, A. P. S., Bindoff, N. L. and Church, J. A. (1999). Large-scale freshening of intermediate waters in the Pacific and Indian oceans. *Nature*, **400**, 440–43.

Wunsch, C. (2009). A perpetually running ENSO in the Pliocene? *Journal of Climate*, **22**, 3506–10.

Yeh, S.-W., Kug, J.-S., Dewitte, B. *et al.* (2009). El Niño in a changing climate. *Nature*, **461**, 511–14.

Zhang, X., Zwiers, F. W., Hegerl, G. C. *et al.* (2007). Detection of human influence on twentieth-century precipitation trends. *Nature*, **448**, 461–65.

3

Sea-level rise and ice-sheet dynamics

'You melt, we drown'[1]

Sea-level rise has emerged as one of the most intensely studied and discussed aspects of climate change in recent years. The period of relative stability of sea level over the past 6000–7000 years (Harvey and Goodwin, 2004) has now ended, and sea level is undoubtedly rising in the post-industrial period (IPCC, 2007a). Given the massive heat capacity of the ocean, the Earth is already committed to many more centuries of sea-level rise due to thermal expansion alone. The dynamics of the large polar ice sheets and the rapid retreat of glaciers and ice caps will significantly add to the magnitude of sea-level rise. The critical questions are: how much and how fast? The implications for long-lived coastal infrastructure, coastal ecosystems, and low-lying urban areas and settlements are significant. This chapter explores our current understanding of sea-level rise, including observations of sea-level rise in the more recent past as well as insights from deeper in Earth's history, projections of sea-level rise out to the end of this century, the dynamics of the large polar ice sheets in Greenland and Antarctica, and the consequences of sea-level rise for contemporary society.

3.1 Observations of sea-level rise

Observations of sea level from 1870 to 2001 (Figure 3.1) show an increase of about 20 cm over the period (Church and White, 2006). An extension of the record with more recent data shows that the rate of sea-level rise has increased within the past two decades, from 1.6 mm yr^{-1} in the 1961–2003 period to 3.1 mm yr^{-1} in the 1993–2003 period (Church and White, 2006; Domingues *et al.*, 2008). As shown in Figure 3.2, from 1990 to 2008, the period for which model-based projections of sea-level rise are available, the observed increase in sea level has tracked at or near the upper limit of the envelope of IPCC projections.

---

[1] Mr Ronny Jumeau, Minister for Environment, Seychelles Islands in response to a media question regarding his participation in the 7th Royal Colloquium "Arctic Under Stress: A Thawing Tundra," convened by HM King Carl XVI Gustaf, Sweden, May 2005.

Climate Change: Global Risks, Challenges and Decisions, Katherine Richardson, Will Steffen and Diana Liverman *et al*. Published by Cambridge University Press. © Katherine Richardson, Will Steffen and Diana Liverman 2011.

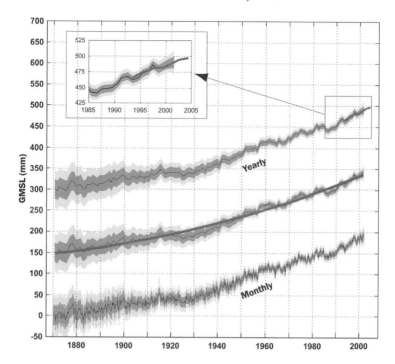

Figure 3.1 Global mean sea level rise for January 1870 to December 2001. The monthly global average, the yearly average with a quadratic fit to the yearly values and the yearly averages from the satellite altimeter data superimposed and offset by 150 mm. The one (dark shading) and two (light shading) standard deviation error estimates are shown. Inset: Observational records of recent global mean sea level rise, showing the acceleration since the 1990s to the present. Source: Church and White (2006). Reproduced/modified by permission of American Geophysical Union.

Over most of the record from 1870 to the present, the thermal expansion of sea water due to the increased heat content of the ocean (Chapter 1) has been the most important component of sea-level rise. This is reflected in an analysis of the sea-level rise budget for the 1961–2003 period (Domingues *et al.*, 2008), the period for which accurate estimates of change in ocean heat content are available. This analysis includes independent estimates of four components of sea-level rise, with their proportional importance: thermal expansion of the ocean (*c*. 40%), glaciers and ice caps (*c*. 35%), the large polar ice sheets on Greenland and Antarctica (*c*. 25%), and terrestrial storage in dams and mining of groundwater (negligible). The sum of these four is 1.5 ± 0.4 mm yr^{-1}, very close to the independent observations from satellite altimeter and tide gauges of sea-level rise for the same period of 1.6 ± 0.2 mm yr^{-1} (Domingues *et al.*, 2008).

For the 1961–2003 period, the contribution to sea-level rise of the large polar ice sheets has been relatively small. However, over the past decade the magnitude of their contribution has risen significantly (Section 3.5), and they are likely to become the dominant factor in sea-level rise later this century and beyond.

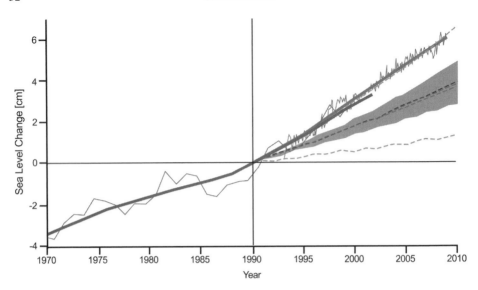

Figure 3.2 Sea-level change from 1970 to 2008. The envelope of IPCC projections is shown for comparison. Source: after Rahmstorf *et al.* (2007), based on data from Cazenave and Narem (2004); Cazenave (2006) and A. Cazenave for 2006–08 data. Reproduced with permission of AAAS.

3.2 Projections of sea-level rise from the IPCC Fourth Assessment Report (2007)

Projections of sea-level rise from 1990 to 2100 have generated considerable controversy, largely due to misinterpretation of the presentation of projections in the IPCC Fourth Assessment Report (2007a). Unlike the sea-level projections to 2100 reported in the IPCC Third Assessment Report (2001), which included an estimate of the contribution from the large polar glaciers within their graphical representation to give a range of 0.11–0.88 m, the IPCC (2007a) excluded changes within the polar ice sheets in their model-based projections, as the dynamics of these ice sheets are not understood well enough to model them quantitatively. The resulting 2007 projections showed an apparently narrower range, from 0.18 m to 0.59 m, compared to the 2001 projections. However, the IPCC (2007a) estimated the possible contribution from the large polar ice sheets as – 0.01–0.17 m (see the last row of Table 10.7 of IPCC, 2007a), giving a projected range of 0.18–0.76 m. This is not significantly different from the IPCC (2001) projection of 0.11 to 0.88 m. Importantly, the IPCC (2007a) report clearly stated that higher sea-level rises could not be ruled out.

This rather complicated approach to communicating the sea-level rise projections is simplified graphically in Figure 3.3. For comparison, the inset shows the observed sea-level rise from 1990 to the present. When all of the factors treated in the IPCC (2007a) projections are included in the graph, the projections are indeed in agreement with the IPCC (2001) and with the observations. Thus, observations and the IPCC projections, both 2001

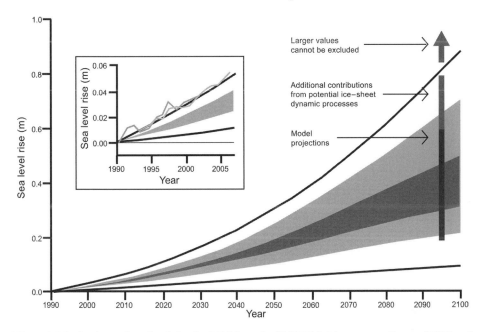

Figure 3.3 Projections of sea-level rise for 2100 from the IPCC Third Assessment Report (TAR) and the Fourth Assessment Report (AR4). The TAR projections are indicated by the shaded regions and the curved lines are the upper and lower limits. The AR4 projections are the bars plotted at 2095. The inset shows sea level observed with satellite altimeters from 1993 to 2006 (orange) and observed with coastal sea-level measurements from 1990 to 2001 (blue). Source: ACE CRC (2008).

and 2007, consistently show that sea level is currently tracking towards a rise of 0.8–1.0 m in 2100 compared to the 1990 value.

3.3 Semi-empirical models to predict sea-level rise

Given the difficulties in quantitatively modelling the dynamical processes in the large polar ice sheets, an alternative is to use semi-empirical models that relate the observed sea-level rise to the observed temperature rise over the past 120 years (Rahmstorf, 2007; Box 3.1). These models do not explicitly simulate any of the individual processes that contribute to sea-level rise, which is both a strength and a weakness with this approach. It is a strength in that it gets around the lack of process understanding by simply incorporating the effects of complex processes on sea level in the observed relationship between sea-level rise and temperature increase. The weakness is that semi-empirical models cannot, by definition, include changes in the temperature–sea-level rise relationship that have not yet been observed. These might include, for example, a shift in the dynamical processes in large polar ice sheets when a temperature threshold is crossed. To be fair, it is currently extremely difficult for *any* type of model to capture threshold-abrupt change behaviour or a strongly nonlinear shift in dynamics as new processes are activated.

Box 3.1
Statistical sea level models

STEFAN RAHMSTORF

Sea-level changes cannot yet be predicted with confidence using models based on physical processes, as the dynamics of ice sheets and glaciers – and to a lesser extent also that of oceanic heat uptake – are not sufficiently understood. The previous IPCC assessment report did not include rapid ice-flow changes in its projected sea-level ranges, arguing that these could not yet be modelled. Partly for this reason, observed sea-level rise exceeded that predicted by models by about 50% for the periods 1990–2006 (Rahmstorf *et al.*, 2007) and likewise for 1961–2003 (IPCC, 2007a).

An alternative is to analyse past behaviour of sea level to establish how it is related to global temperature changes. Typically such models assume that the rate of sea-level rise is related in a linear way to changes in global mean temperature. In the simplest form (Rahmstorf, 2007), the relation is:

$$dH / dt = a\left(T - T_0\right), \tag{3.1}$$

where H is the global mean sea level, T is the global mean temperature and T_0 is a temperature in equilibrium with sea level (i.e. for $T = T_0$ sea level is stable). The above equation is a first-order approximation for the initial response of sea level to a rapid change in temperature away from a previous equilibrium. It assumes the timescale of interest is short so that a 'saturation' of the sea level response does not matter.

This approach has been applied to GCM projections to project future sea-level rise (Horton *et al.*, 2008) or only its thermal expansion component (Vellinga *et al.*, 2009). Several refinements have been proposed, i.e. including a finite timescale for the sea-level response (Grinsted *et al.*, 2009), using radiative forcing instead of global temperature as the driver (Jevrejeva *et al.*, 2009) or adding a rapid-response term (Vermeer and Rahmstorf, 2009).

The key calibration of the method is the twentieth-century global sea level and temperature data (Figure 3.4), but older data have also been used (Grinsted *et al.*, 2009). Similar results are obtained using either the GISS or HadCRUT temperature data, combined with either the sea-level data of Church and White (2006) or Jevrejeva *et al.* (2006).

Projections for the twenty-first century using this method are generally higher than those of IPCC (2007a) (Table 3.1). This is because inherent in Eq. (3.1) is an acceleration of sea-level rise as it gets warmer, as seen in the twentieth century data (Figure 3.4). A constant rise at the rate found for the past 16 years in the satellite altimeter data (3.4 mm yr^{-1}; see Section 3.1), without any acceleration, would lead to a twenty-first century rise by 34 cm, close to the central estimate of the IPCC.

An inherent limitation of all statistical models is that one cannot be certain that statistical links between sea level and temperature found for the past continue to hold in future, especially as global warming progresses well outside the range for which the models have been tested. One concern is that the contribution from glacier melt, about 35% of sea-level rise over 1961–2003 (Domingues *et al.*, 2008), will decline as mountain glaciers are progressively lost. Total glacier mass has been estimated as ~60 cm in sea-level equivalent (Radić and Hock, 2010),

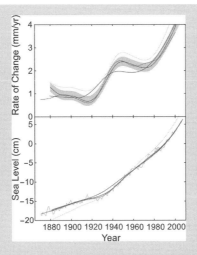

Figure 3.4 (Top) Observations-based rate of sea-level rise (with tectonic and reservoir effects removed, red) compared to that predicted by Eq. 1 (grey) and a refined version (blue with uncertainty estimate) using observed global mean temperature data. Also shown is the refined estimate using only the first half of the data (green) or the second half of the data (light blue). (Bottom) The integral of the curves in the top panel, i.e. sea level proper. In addition to the smoothed sea level used in the calculations, the annual sea level values (light red) are also shown. The dark blue prediction almost obscures the observed sea level due to the close match. Source: Vermeer and Rahmstorf (2009). (In colour as Plate 8.)

Table 3.1. *Estimates for twenty-first-century sea-level rise using statistical models. For exact definitions of the time periods and emissions scenarios considered, see the original references*

Study	Sea-level range (cm)	Notes
IPCC AR4 (2007)	18 – 59	Excludes rapid dynamical changes in ice flow
Rahmstorf (2007)	50 – 140	
Horton *et al.* (2008)	47 – 100	Warmest emission scenario considered is A2
Grinsted *et al.* (2009)	30 – 215 (72 – 160)	Numbers in brackets are valid for the best-fitting data constraint out of several investigated options
Vermeer and Rahmstorf (2009)	75 – 190	Twentieth-century sea level is adjusted for reservoir storage (11)
Jevrejeva *et al.* (2010)	59 – 180	Uses radiative forcing rather than global temperature as input

Box 3.1 (cont.)

so if their share remained constant at 35% they would have run out of ice at 170 cm of total sea-level rise.

Another concern is the potential nonlinear response of the large ice sheets in Greenland and Antarctica, which contributed 25% to sea-level rise during 1961–2003 (Domingues *et al.*, 2008). The data indicate a recent acceleration of ice loss well beyond that suggested if Eq. (3.1) were applied to ice sheets separately, and their contribution to total sea-level rise appears to have increased recently to 40% over 2003–08 (Cazenave *et al.*, 2008a, b).

Note, however, that the statistical models do not distinguish between glaciers and ice sheets; they only assume that the grand total of sea-level rise is linearly related to global temperature. This could still be a good assumption if the ice-melt contribution progressively shifts from glaciers to ice sheets.

Nevertheless, semi-empirical models provide useful information on likely sea-level rise over the coming decades and perhaps out to the end of the twenty-first century. As these models are developed further (Box 3.1), they are consistent with other lines of evidence – primarily the observed rate of sea-level rise over the past few decades and the upper range of the IPCC projections – which show that a sea-level rise towards 1 m by 2100 is a distinct possibility and that somewhat higher values cannot be excluded, depending on the level of temperature increase that is actually realised this century.

3.4 Palaeo-observations of sea-level rise

Current and future climatic trends suggest that the Earth is moving into a 'super-interglacial' (Overpeck *et al.*, 2005), so examination of the last interglacial (the Eemian, or OIS 5e) 130 000 – 116 000 years ago, as well as other periods in the past (Box 3.2), can give insights into what might lie ahead. Global mean temperature in the Eemian was about 1.5–2 °C higher than present (Clarke and Huybers, 2009) but probably at least double that in the high latitudes (Turney and Jones, submitted). Earlier estimates that maximum global mean sea-level during the Eemian was 4–6 m higher than present (Rostami *et al.*, 2000; Muhs *et al.*, 2002) have been revised upwards. Applying a novel statistical approach that assesses sea-level proxies from 47 individual localities suggests that Eemian sea level peaked at 6–9 m above the present level (Kopp *et al.*, 2009). It is likely that both the Greenland and West Antarctic ice sheets contributed significantly to the sea-level rise (Overpeck *et al.*, 2006), whether it was 4–6 m or 6–9 m above the present level. During the Eemian, when mean sea level was within 10 m of its present value – so arguably a good analogue for sea-level rise over the next few centuries – the 1000-year average rate of sea-level rise reached about 1 m per century (Fairbanks, 1989; Lambeck *et al.*, 2002). This accords well with recent estimates from statistical models of projected sea-level rise to 2100 (Rahmstorf, 2007; Box 3.1). However, even higher rates due to meltwater events were observed for some periods during the transition from the Last Glacial Maximum to the Holocene (Box 3.2; Hanebuth *et al.*, 2000).

Box 3.2
Going back to the future: sea-level rise in the past

CHRIS TURNEY

Sea level is increasingly driven upwards by rising temperatures (Jevrejeva *et al.*, 2009) and is currently tracking around the IPCC's 'business-as-usual' scenario projections of more than 3 mm per year (Rahmstorf *et al.*, 2007; Allison *et al.*, 2009), implying higher sea levels in the future. Unfortunately, the magnitude of this change remains an area of considerable uncertainty, with estimates ranging from 0.18 m (Meehl *et al.*, 2007) to 1.9 m (Vermeer and Rahmstorf, 2009) higher than today. A significant part of this large range is the associated uncertainty in the dynamical behaviour of polar ice sheets to future warming, particularly the Greenland and Antarctic ice sheets (Section 3.5; Boxes 3.3 and 3.4). The question is: How might other evidence be brought to bear to narrow projections of their contribution to future sea-level rise?

Critically, past changes in sea level can provide valuable insights into the future (PALSEA, 2010). While past periods may not be complete analogues for anthropogenically driven climate change, the mechanisms that operated at different times can provide an improved understanding of future climate processes. Although there has been relatively little net global sea-level rise over the past two millennia (Church *et al.*, 2008), a wealth of geomorphological, chemical and biological records clearly indicate that large-scale, abrupt and irreversible (centennial to millennial in duration) shifts in sea level took place over the last glacial–interglacial cycle (Lambeck and Chappell, 2001; Siddall *et al.*, 2003). The timing, magnitude and rate of change in sea level over this period provides valuable insights into ice sheet sensitivity to regional and global climate change. For instance, at the peak of the Last Glacial Maximum around 21 000 years ago (Mix *et al.*, 2001), Greenland and Antarctic ice volume was significantly larger than today, with the development of other extensive ice sheets over North America and Eurasia, resulting in a global sea level some 130 m lower than today (Clark *et al.*, 2009). At this time, global mean temperatures were estimated to have been some 6.4 °C cooler than today (Hostetler *et al.*, 2006). Between 20 000 and 19 000 years ago, however, increasing insolation over high latitudes drove a decline in global ice volume (Clark *et al.*, 2009), with the effective long-term extinction of major ice sheets outside Greenland and Antarctica. The rate of change in global sea level was highly variable, however, with periodic growth in ice and rapid melting in different regions over the past 20 000 years (Gehrels, 2010). The most prominent melting was MeltWater Pulse-1A (MWP-1A) around 14 600 years ago, with a global sea-level rise of 16 m over three centuries (Hanebuth *et al.*, 2000) (equivalent to approximately 50 mm per year). The origin of MWP-1A is not precisely known but seems likely to be Antarctica (Clark *et al.*, 2002, 2009; Carlson, 2009), with potentially the greatest contribution coming from the Antarctic Peninsula (Bassett *et al.*, 2007).

The cause of polar ice sheet retreat is an area of considerable debate. Of particular concern is the West Antarctic Ice Sheet (WAIS), which is fringed by floating ice shelves that buttress fast-flowing inland ice streams (Section 3.5.2; Box 3.4). It was originally thought that the size of the WAIS was controlled by sea level, driven principally by the extent of ice in the northern hemisphere (Penck, 1928). Although this may be true of changes in sea level of the order of 100 m or so, recent work has recognised the significance of subglacially derived sediments

Box 3.2 (*cont.*)

for helping stabilise the WAIS against past and future sea-level rise (Alley *et al.*, 2007). Other controls, however, can play a significant role on the extent of ice, most importantly sea temperatures around Antarctic grounding lines (Walker *et al.*, 2008; the 'grounding line' is the point at which a land-based ice stream begins to float on seawater and becomes an ice shelf). In contrast to the East Antarctic Ice Sheet (EAIS), the WAIS grounding lines are several hundred metres below sea level with deepening beds upstream, raising the potential for future runaway retreat (Box 3.4) and significant global sea-level rise (Box 3.4; Mercer, 1978). Modelling work by Pollard and DeConto (2009) demonstrates the past vulnerability of the WAIS to ocean warming and suggests a collapse when ocean temperatures warm by some 5 °C.

The potential importance of polar ice sheets to future sea-level rise was recently demonstrated by Kopp *et al.* (2009) who demonstrated that during the last interglacial (130 000 – 116 000 years ago) global sea level was most likely between 6.6 m and 9.4 m higher than today, with Greenland and Antarctica (most probably the WAIS) contributing at least 5 m. An analysis of published datasets suggests global temperatures during the last interglacial were some 1.3 °C warmer (relative to 1961–1990) with warming >5 °C at high latitudes (Turney and Jones, 2010). The global pattern of warming at this time was comparable to projections for the end of this century (Overpeck *et al.*, 2006), with sea levels rising some 6–9 mm per year (Kopp *et al.*, 2009), at least double the current global average.

There is some evidence that these estimates are informing government planning for the future, e.g. the Australian Government Department of Climate Change (2009) reported the likely displacement of 250,000 Australians from their homes as a result of a sea-level rise exceeding 1 m by the end of this century, an estimate informed by the past.

The past can truly help narrow projections and help plan for the future.

Further back in time, the Pliocene (about 3 million years ago) probably experienced atmospheric CO_2 concentrations of about 400 p.p.m. (at the mid Pliocene), close to today's concentration of 387 p.p.m. (Box 2.1; Haywood *et al.*, 2009). The temperature at the mid-Pliocene was estimated to be about 3 °C higher than present, with greater increases in temperature at high latitudes. Peak sea levels during the Pliocene were estimated to be about 25 m higher than present (Dowsett and Cronin, 1990; Shackleton *et al.*, 1995), which agrees well with the long-term relationship between global average temperature and sea level (Figure 3.5). The mid Pliocene is not a direct analogue for the near-term future, however, as temperatures and sea levels then had reached equilibrium, with slow feedbacks (e.g. changes in ice sheet extent and vegetation distribution) as well as greenhouse gases contributing to the climatic changes observed. We may experience similar conditions when the anthropogenically driven excursion in climate finally reaches equilibrium, at least a millennium or two from now (Solomon *et al.*, 2009).

3.5 The dynamics of the large polar ice sheets

The palaeo-evidence clearly shows the importance of the large polar ice sheets as a major factor in sea-level rise at temperatures similar to those almost surely to be realised in this

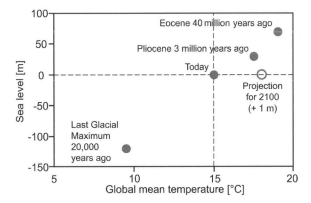

Figure 3.5 Actual vs. projected sea level relative to global mean temperature over geological time. Text: German Advisory Council on Global Change (WBGU) 2006, after Archer (2006).

century. Contemporary observational evidence confirms that the polar ice sheets may already be playing a larger role in sea-level rise since the beginning of this century. A recent analysis of the components of sea-level rise for the 2003–08 period shows that the large polar ice sheets are now contributing 40% to the overall budget, with glaciers and ice caps contributing another 40% and thermal expansion 20% (Cazenave *et al.*, 2008a, b). At present, then, glaciers and small ice caps contribute at about the same proportion as the large polar ice sheets, and these temperate and tropical glaciers and ice caps will likely continue to make a major contribution to sea-level rise through much of this century at least. However, in the longer term, the massive volume of ice stored in the polar ice sheets means that they will eventually play a dominant role in the rate and amount of sea-level rise.

The rapid increase in the relative importance of the polar ice sheets suggests that some significant changes are occurring in the processes by which they lose ice to the ocean. Both of the large polar ice sheets discussed here – the Greenland and West Antarctic ice sheets – are subject to thermal thresholds, beyond which loss of most or all of the ice will be unstoppable and possibly rapid. This is discussed further in Chapter 7.

New measurement techniques such as satellite altimetry and very sensitive airborne measurements of gravity fields (the GRACE satellites) have revolutionised our ability to detect changes in the mass balance of the large polar ice sheets, and confirm their increasing role in sea-level rise (Cazenave, 2006). Figure 3.7 shows a summary of these measurements for the Greenland Ice Sheet. The observations show that the mass of the Greenland Ice Sheet was approximately in balance through the 1990s but since 2000 has begun to decline rapidly (Box 3.3). Although the average rate of ice loss was 160 Gt yr^{-1} for the 2003–08 period, towards the end of the record, the ice sheet was losing mass at the rate of about 200 Gt yr^{-1}. The observations for Antarctica are more complex. Through the 1990s, the continent's ice cover was about in balance, with small losses from West Antarctica balanced by accretion of ice in East Antarctica. Since 2000, the rate of accretion of ice in East Antarctica has increased slightly, but is more than compensated by an increasing rate of loss in West Antarctica, yielding a small net loss of mass for the continent as a whole. The most recent observations, however, show that the contribution from the Antarctic ice sheets to sea-level rise is now the same order of magnitude as that from Greenland (Velicogna, 2009).

Box 3.3
Dynamics of the Greenland Ice Sheet

DORTHE DAHL-JENSEN AND KONRAD STEFFEN

The Greenland Ice Sheet is an awakening giant in relation to loss of mass and sea-level rise. During the 1970–80 period, the Ice Sheet was in mass balance within the accuracy of our measurements – so the mass received as snow fall was in balance with the mass lost by melt and runoff, and by ice discharge (Rignot *et al*., 2007; Hanna, 2008; AMAP, 2009; van den Broecke *et al*., 2009). Since 1995 we have observed an increasing loss of mass from the Ice Sheet with an average value of 160 Gt yr^{-1} for the 2003–08 period and even higher values for the most recent years (Rignot *et al*., 2007; Wouters *et al*., 2007; AMAP, 2009; van den Broecke *et al*., 2009; Velicogna, 2009). The high latitudes of the northern hemisphere are the regions that have experienced the strongest changes during the past 30 years. While the global annual temperature increase has been 0.7 °C during the past century (1901–2005), the increase over the Arctic has been twice the global value (IPCC, 2007a).

The interior of the Ice Sheet (the region above 2000 m) is thickening due to increased snow accumulation, but this is by far counteracted by the increased loss of mass along the margin (Mote, 2007; Wouters *et al*., 2007; Hanna, 2008; Thomas *et al*., 2008; Chen *et al*., 2009; Ettema *et al*., 2009; van den Broecke *et al*., 2009). By gravitational force, the ice from the interior slowly flows down into the Ice Sheet, and then (dynamically) quasi-parallel to the bedrock towards the margin. In the lower part of the Ice Sheet close to its margin, called the ablation region, mass is lost by melt and runoff, and by ice discharge. Loss of mass by melt and runoff accounts for 45% of the total loss, and has increased by 30% between 2000 and 2008 (Hanna, 2008; AMAP, 2009; van den Broecke *et al*., 2009). In the summer months melt-water forms rivers, which feed lakes on the ice surface – some draining through moulins to the bedrock under the ice (Zwally *et al*., 2002). This melt-water under the ice lubricates the base, which allows the ice to flow more rapidly towards the ice margin (Zwally *et al*., 2002; Alley *et al*., 2008; Joughin *et al*., 2008). MODIS satellite images of the surface velocities show that ice has been flowing faster along the whole west coast of Greenland for the past decade (Fahnestock *et al*., 1993; Rignot and Kanagaratnam, 2006; Joughin and Fahnestock in AMAP, 2009). We know from earlier observations that the region with fast-flowing ice has moved further north since 2000 (Rignot and Kanagaratnam, 2006).

Along the margin, very high ice velocities are seen to coincide with fast-flowing glaciers that terminate in deep fjords (Figure 3.6). The glaciers discharge icebergs that flow into the ocean to melt, and the ocean water melts the underside of the floating ice tongues (Reeh *et al*., 1994; Holland *et al*., 2008; Rignot *et al*., 2007, 2010). Loss of mass by ice discharge and by

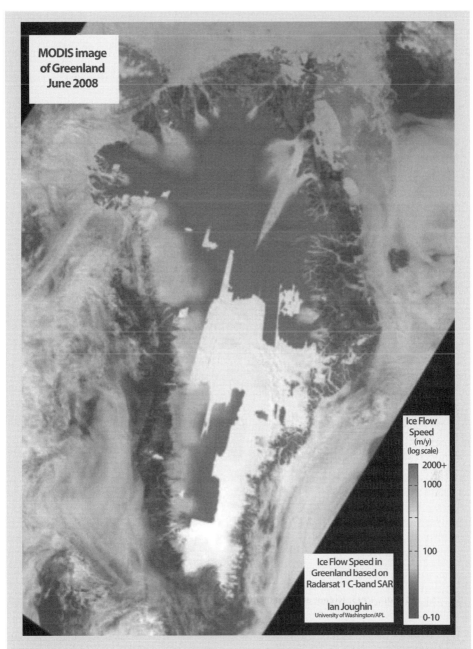

MODIS image
of Greenland
June 2008

Ice Flow
Speed
(m/y)
(log scale)

2000+

1000

100

Ice Flow Speed in
Greenland based on
Radarsat 1 C-band SAR

Ian Joughin
University of Washington/APL

0-10

Figure 3.6 Velocity (m yr^{-1}) mosaic of the Greenland Ice Sheet derived from radar satellite data. The mosaic shows how the slow-flowing inland ice transitions into the fast-moving outlet glaciers, and the large variability of flow along the margin. Source: I. Joughin and M. Fahnestock, unpublished, presented in AMAP (2009). (In colour as Plate 10.)

Box 3.3 (*cont.*)

ocean melt accounts for 55% of the loss of mass from the Greenland Ice Sheet (AMAP, 2009). Since 2002, many of the fast-flowing glaciers have increased their velocity, thereby increasing ice discharge (Howat *et al.*, 2007; Rignot *et al.*, 2007; Moon and Joughin, 2008; Pfeffer *et al.*, 2008). The Jakobshavn Isbræ near Ilulissat has increased its velocity from 7 km yr^{-1} to 15 km yr^{-1} (Dietrich *et al.*, 2007; Amundson *et al.*, 2008). The increased velocities have increased the loss of mass by ice discharge by 30% between 2000 and 2008 (AMAP, 2009). Most glaciers that have increased their velocity so dramatically terminate in deep fjords and, as noted above, the cause is likely the warmer water in the fjords rather than the surface melt-water lubricating the base of the ice sheet (Howat *et al.*, 2007; Holland *et al.*, 2008; Straneo *et al.*, 2010).

The Greenland Ice Sheet has the potential to produce significant sea-level rise in the future due to its large volume (Pfeffer *et al.*, 2008; AMAP, 2009). Predictions for future ice loss are hampered by the inability to include processes such as enhanced ice discharge and basal lubrication in ice sheet models (IPCC, 2007a; AMAP, 2009). If we take the present rate of loss of mass (Rignot *et al.*, 2007; Wouters *et al.*, 2007; Thomas *et al.*, 2008; AMAP, 2009; van den Broecke *et al.*, 2009; Velicogna, 2009) as a fixed rate, the Greenland Ice Sheet would contribute 5 cm to sea-level rise in 2100. This can be seen as a minimum value. Maximum

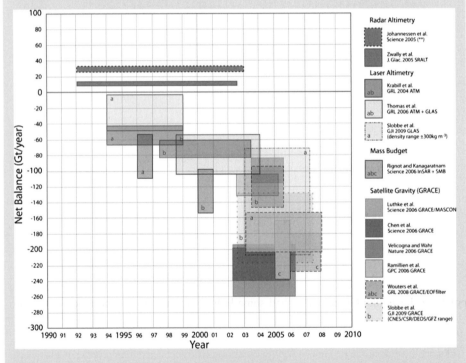

Figure 3.7 Results from the recent large area total balance measurements, placed into common units and displayed, versus the time intervals of the observations. The heights of the boxes cover the published error bars or ranges in mass change rate over those intervals. Source: AMAP (2009). (In colour as Plate 9.)

values of 50 cm sea-level rise in 2100 would be obtained by accelerating the Greenland surface mass balance ice loss at present-day rates of change; dynamic discharge has been calculated by accelerating outlet glacier velocities by an order of magnitude in the first decade (Pfeffer *et al*., 2008). A value of 20 cm contribution to sea-level rise in 2100 is as good an estimate as can be made at present. Greenland Ice Sheet and climate modellers are engaged in producing improved model predictions based on new insight and findings about ice dynamics (Figure 3.7), which will aim to reduce the uncertainty in sea-level projections for the IPCC AR5 report.

3.5.1 The Greenland Ice Sheet

Accumulation, melting and ice discharge are all sensitive to the strong increase in regional temperature and are important for the overall dynamics of the Greenland Ice Sheet (Box 3.3). The area of surface melting, defined by the area that reaches melting point temperature at least one day during the summer, has increased 50% from 1979 to 2008 in a roughly linear manner but with much interannual variation (K. Steffen and R. Huff, personal communication, 2009). The apparent inconsistency with the mass balance data, which show no change in Greenland's mass balance through the 1990s, is explained by the increase in snowfall in the interior of the island, which has approximately balanced the loss by melting through the 1979–2000 period. As temperatures rise further, this balance may slowly shift in favour of melting. Model projections show that if global mean temperature rises to within the 1.9–4.6 °C range, well within the suite of IPCC projections, and is maintained at those elevated levels or higher for several millennia, the Greenland Ice Sheet would disappear through surface melting alone (Gregory and Huybrechts, 2006). A recent expert elicitation suggests that if the global average temperature rises above 4 °C, there is a high probability of crossing a threshold for the loss of the Greenland Ice Sheet (Chapter 7; Kriegler *et al*., 2009).

The rapid change in mass balance after 2000, however, is not due to a sudden change in the relative importance of melting and snow accumulation, but to the activation of processes that have increased the discharge of ice from the large outlet glaciers along the edges of the ice sheet (Das *et al*., 2008; van den Broeke *et al*., 2009). Discharges of large blocks of land-based ice cause an instantaneous rise in sea level through displacement of sea water when they move from the land to the sea. Box 3.3 presents a more detailed analysis of the dynamics of the Greenland Ice Sheet.

A theoretical analysis of the kinematic constraints on rapid discharge of ice from large polar ice sheets like Greenland's shed some light on the possible importance of this process into the future (Pfeffer *et al*., 2008). The study has concluded that a plausible estimate of total sea-level rise to 2100, including the most likely contribution from dynamical ice discharge from both Greenland and Antarctica, is about 0.8 m. The maximum possible sea-level rise according to this analysis is about 2.0 m, but this would occur only with the most extreme levels of climate forcing.

Given the potential importance of dynamical ice discharge for sea-level rise, yet the considerable lack of understanding of processes, the topic is the subject of a much enhanced research effort in Greenland. New techniques being used in this research include use of interferometric synthetic aperture radar to quantify variations in ice sheet flow; exploration of subsurface hydrologic channels and cavities using video and digital cameras, and a rotating laser on a tethered autonomous system; an autonomous explorer probe that logs temperature, pressure and water flow rate; and observation and modelling of moulin distribution. Enhanced understanding derived from such research will contribute to the development of more reliable models of dynamic ice sheet discharge (Shepherd and Wingham, 2007; Doherty *et al.*, 2009; Mernild *et al.*, 2009).

3.5.2 *The West Antarctic Ice Sheet*

Although much of the research on the polar ice sheets has focused more strongly on Greenland, there is some concern that the West Antarctic Ice Sheet might be more vulnerable than Greenland to the degree of climatic change expected this century (Box 3.4). The total volume of ice contained in the West Antarctic Ice Sheet (equivalent to about 6 m of sea-level rise) is slightly less than that contained in the Greenland Ice Sheet (about 7 m), but it is the rapidity with which the West Antarctic Ice Sheet can change that is the grounds for concern. A partial collapse could lead relatively quickly to an average global sea-level rise of several metres.

Box 3.4
West Antarctic Ice Sheet dynamics

JONATHAN BAMBER

The WAIS is unique. Some two-thirds of it rests on bedrock substantially below sea level, unlike most of East Antarctica and Greenland (Bamber *et al.*, 2007). Along with the fact that the bedrock deepens inland (Figure 3.8), this has resulted in the hypothesis that the WAIS is in an inherently unstable configuration (Weertman, 1974).

  The original hypothesis has been borne out by more recent theoretical treatment of the problem (Schoof, 2007) and recent observations of grounding line retreat in the Amundsen Sea Embayment (ASE) of the WAIS (Rignot, 1998). The ice streams and outlet glaciers that drain the WAIS (Figure 3.9) are characterised by relatively low surface slopes and, therefore, a low gravitational driving force. The majority of the motion is due to sliding at the bed rather than internal deformation (or creep) of the ice. In this situation, small changes in the balance between the driving force (gravity), and the opposing forces of basal and lateral drag and ice shelf buttressing, can have a dramatic impact on ice dynamics (Joughin *et al.*, 2002).

  As is evident from Figure 3.9, these low-profile, fast-flowing ice streams drain into the two largest ice shelves in Antarctica – the Filchner-Ronne and Ross – as well as smaller floating tongues fringing the continent. This floating ice is particularly susceptible and sensitive to changes in ocean circulation and temperature (Rignot and Jacobs, 2002). The recent rapid speed-up of glaciers in the ASE is believed to be due to localised ocean warming (Payne *et al.*,

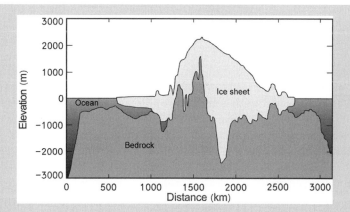

Figure 3.8 A cross-section of surface and bedrock topography for the WAIS from the Filchner-Ronne to Ross ice shelves. At its deepest, the bed reaches to 2.5 km below sea level.

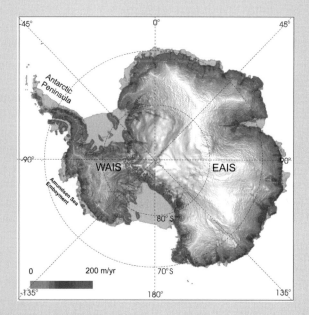

Figure 3.9 Surface topography (shaded grey) and ice velocities (colour) for Antarctica. Purple to red colours indicate ice streams where flow is generally dominated by sliding at the bed. Source: adapted from Bamber *et al.* (2007). (In colour as Plate 11.)

2004). Thus, although surface air temperatures may be insufficient to have any impact on ice-melt in West Antarctica over the next century or more, oceanic warming will likely play a key role in determining the rate of ice loss. The Filchner-Ronne and Ross ice shelves are large and thick (>1000 m thick near their grounding lines) and it has been suggested

Box 3.4 (*cont.*)

that global temperature increases of 5–6 °C would be required to threaten their existence (Pollard and DeConto, 2009). However, the smaller shelves that, in particular, buttress the ASE glaciers, appear to be already responding to recent changes in oceanic forcing and this sector of the WAIS contains enough ice alone to raise global sea level by around 1.5 m (Vaughan, 2008). The numerical models that have been used, to date, to predict the future response of Antarctica and, in particular the WAIS, lacked the physics necessary to capture (the now observed) grounding line migration and acceleration of ice streams due to changes in force balance. One-dimensional models have been developed that convincingly reproduce the observations along a flowline (Payne *et al.*, 2004; Nick *et al.*, 2009) and progress has been made in incorporating more realistic grounding line migration in an ice-sheet model (Pollard and DeConto, 2009), but we still lack the tools to adequately explore the sensitivity of the WAIS to external forcing. As a consequence, there remains uncertainty over the future response of the WAIS to climate change.

The larger vulnerability of the West Antarctic Ice Sheet is based on the different dynamical processes that can potentially cause such rapid and large rises in sea level (Box 3.4). The buttressing ice shelves that protect the West Antarctic Ice Sheet are threatened by rises in both air and sea surface temperatures. Rapid disintegration of the Larsen A and B ice shelves has already occurred along the Antarctic Peninsula, an area that has experienced much more rapid warming than the bulk of the continent. However, warming rates of about 0.1 °C per decade have been observed in the West Antarctic region over the past half-century, probably related to increases in sea surface temperature (Steig *et al.*, 2009). Palaeo-evidence reinforces concerns about the vulnerability of the West Antarctic Ice Sheet. Climate oscillations in the past have been accompanied by relatively rapid periods of retreat and advance of the ice sheet (Pollard and DeConto, 2009). During the Pliocene when global mean temperature was about 3 °C higher than today, the entire West Antarctic Ice Sheet appears to have collapsed (Naish *et al.*, 2009), as noted above.

3.6 Impacts of sea-level rise

The impacts of rising sea levels present profound challenges for humans and ecosystems (see IPCC, 2007b for a more detailed analysis of the impacts of sea-level rise). Direct impacts on our societies include inundation of human settlements, threats to key pieces of infrastructure situated along coastlines, and erosion of sandy beaches and soft coastlines. Natural ecosystems are also under threat as sea levels rise faster than some of them are likely capable of adapting, with consequent flow-on effects to humans in terms of the provision of ecosystem services. Small island states are especially at risk; a sea-level rise of about one metre will eliminate the most vulnerable of them. An estimated 10% of the global population – over 650 million people – will be directly impacted by a sea-level rise of between 0.5 m and 1.0 m, which now may represent a best-case scenario.

Developing countries are especially at risk because of their lower adaptive capacity. For example, several of the large Asian mega-cities are located on very low-lying river deltas, along with much of the population of Bangladesh. The options to cope with rising sea level may be much less for these countries and cities – given their limited adaptive capacity – leading to the potential migration of millions of people to other areas of their own countries or to other parts of the world (Chapter 5).

Australia, with a high proportion of its population and infrastructure situated along the coastline, is one of the most vulnerable of developed countries to sea-level rise and, consequently, has already invested much time and many resources in analysing and mapping vulnerability and planning for adaptation. Much of this research, often carried out in close collaboration with stakeholders, now goes beyond simple inundation analyses and is broadly applicable to many parts of the world.

Figure 3.10 is based on one analysis of vulnerability of Australia's coastline, in terms of the multiplying effect of sea-level rise on the incidence of extreme sea-level events (ACE CRC, 2008). These are defined as inundation events associated with high tides and storm surges, exacerbated by the slowly rising level of the sea. A modest sea-level rise of 0.5 m,

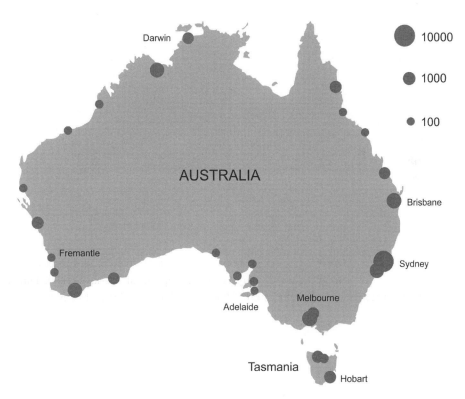

Figure 3.10 Estimated multiplying factor for the increase in frequency of occurrence of high sea-level events along the Australian coast for a sea-level rise of 0.5 m. Source: ACE CRC (2008).

at the lower end of the projections for 2100, leads to surprisingly high multiplying factors of 100 or even 1000 for extreme sea-level events, affecting some of Australia's largest cities. A multiplying factor between 100 and 1000 means that an extreme sea-level event that would occur, on average, once in every hundred years at the sea level of 1990 would occur several times a year with only a 0.5 m increase in that sea level. With the observed increase in sea level of about 0.2 m since the late 1880s, two-fold and three-fold increases in extreme sea-level events have already been observed in some areas along the Australian coast.

A more detailed analysis of the vulnerability of Australia's coast has been carried out by the Australian Government (DCC, 2009) in support of adaptation measures. The study has examined in fine detail several low-lying areas of the coast as examples of what might be expected later this century. A worst-case scenario of a 1.1 m sea-level rise compared to 1990 was applied. Figure 3.11 shows two examples of what might be expected under this scenario, one example focusing on a major piece of infrastructure (Brisbane airport) and the other on an area of private dwellings in an urban settlement. In both cases, the implications of sea-level rise will force difficult and costly decisions

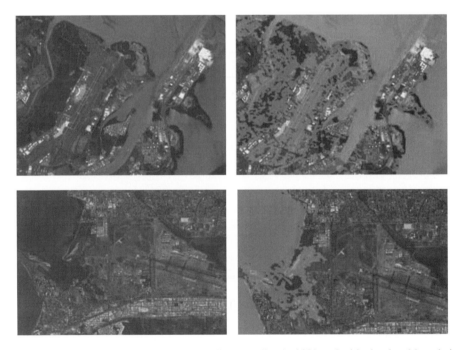

Figure 3.11 Images at places along the Australian coastline in 2009 and with simulated inundation from a sea-level rise of 1.1 m using medium-resolution elevation data: (top pair) Brisbane airport; (bottom pair) Darwin, Northern Territory. Copyright CNES 2009/Imagery supplied courtesy of SPOT Imaging Services and Geospatial Intelligence Pty Ltd. Source: Australian Government Department of Climate Change (2009).

on property owners, town planners, local councils, and the owners and operators of major infrastructure.

Most developed countries already have protective structures, such as levees and sea walls, in place to avoid the worst of these inundation events. With adequate resources and foresight, some of these structures could be raised by 0.5–1.0 m through the twenty-first century to account for the increased likelihood of high sea-level events. In the longer-term, as urban and coastal infrastructure is renewed, it can be located in less vulnerable locations higher above sea level. These actions, however, require the political acceptance of the reality of climate change-driven sea-level rise, the allocation of significant resources for adaptation, and the societal support for major changes in the location of critical infrastructure and, indeed, in the form of existing coastal urban centres.

3.7 Summary and conclusions

Several lines of evidence point towards a sea-level rise by 2100 of perhaps a metre higher than 1990, which is at the upper limit of the IPCC (2007a) range of projections. First, the mid-range of projections of semi-empirical models is centred around 1 m (Rahmstorf, 2007; Grinsted *et al.*, 2009). Second, observed sea-level rise is currently tracking at or near the upper limit of the IPCC projections (Rahmstorf *et al.*, 2007; Domingues *et al.*, 2008). Third, recent observations show an increasing rate of mass loss from both the Greenland and West Antarctic ice sheets (Cazenave, 2006; Rignot and Kanagaratnam, 2006). Finally, an analysis of the kinematic constraints on dynamical ice loss suggests a plausible increase in sea level of 0.8 m by 2100 (Pfeffer *et al.*, 2008).

Nevertheless, considerable uncertainty surrounds many of the processes that contribute to sea-level rise, especially concerning the dynamics of the large polar ice sheets. All of these uncertainties point in the same direction – towards more rapid and severe sea-level rise. Thus, rises higher than 1 m by 2100 cannot be ruled out.

References

ACE CRC (Antarctic Climate & Ecosystems Cooperative Research Centre) (2008). *Briefing: A Post-IPCC AR4 Update on Sea-level Rise.* Hobart, Tasmania: ACE CRC.

Alley, R. B., Fahnestock, M. and Joughin, I. R. (2008). Understanding glacier flow in changing times. *Science*, **322**, 1061–62.

Alley, R. B., Anandakrishnan, S., Dupont, T. K., Parizek, B. R. and Pollard, D. (2007). Effect of sedimentation on ice-sheet grounding-line stability. *Science*, **315**, 1838–41.

Allison, I., Bindoff, N. L., Bindschadler, R. A. *et al.* (2009). *The Copenhagen Diagnosis, 2009: Updating the World on the Latest Climate Science.* Sydney: University of New South Wales Climate Change Research Centre.

AMAP (2009). *The Greenland Ice Sheet in a Changing Climate: Snow, Water, Ice and Permafrost in the Arctic (SWIPA).* Oslo: Arctic Monitoring and Assessment Programme (AMAP).

Amundson, J. M., Truffer, M., Lüthi, M. P. *et al.* (2008). Glacier, fjord, and seismic response to recent large calving events, Jakobshavn Isbræ, Greenland. *Geophysical Research Letters*, **35**, L22501.

Archer, D. (2006). *Global Warming: Understanding the Forecast*. Oxford: Blackwell Publishers.

Bamber, J. L., Alley, R. B. and Joughin, I. (2007). Rapid response of modern day ice sheets to external forcing. *Earth and Planetary Science Letters*, **257**, 1–13.

Bassett, S. E., Milne, G. A., Bentley, M. J. and Huybrechts, P. (2007). Modelling Antarctic sea-level data to explore the possibility of a dominant Antarctic contribution to melt-water pulse IA. *Quaternary Science Reviews*, **26**, 2113–27.

Carlson, A. E. (2009). Geochemical constraints on the Laurentide Ice Sheet contribution to Meltwater Pulse 1A. *Quaternary Science Reviews*, **28**, 1625–30.

Cazenave, A. (2006). How fast are the ice sheets melting? *Science*, **314**, 1250–52.

Cazenave, A. and Narem, R. S. (2004). Present-day sea level change: Observations and causes. *Reviews of Geophysics*, **42**, doi: 10.1029/2003RG000139.

Cazenave, A., Dominh, K., Guinehut, S. *et al.* (2008a). Sea level budget over 2003–2008: A reevaluation from GRACE space gravimetry, satellite altimetry and Argo. *Global and Planetary Change*, **65**, 83–88.

Cazenave, A., Lombard, A. and Lovel, W. (2008b). Present-day sea level rise: A synthesis. *Comptes Rendus Geoscience*, **340**, 761–70.

Chen, L.-L., Johannessen, O. M., Khvorostovsky, K. and Wang, H.-J. (2009). Greenland Ice Sheet elevation change in winter and influence of atmospheric teleconnections in the northern hemisphere. *Atmospheric and Oceanic Science Letters*, **2**, 376–80.

Church, J. A. and White, N. J. (2006). A 20th century acceleration in global sea-level rise. *Geophysical Research Letters*, **33**, L01602.

Church, J. A., White, N. J., Aarup, T. *et al.* (2008). Understanding global sea levels: Past, present and future. *Sustainability Science*, **3**, 9–22.

Clarke, P. U. and Huybers, P. (2009). Global change: Interglacial and future sea level. *Nature*, **462**, 856–57.

Clark, P. U., Mitrovica, J. X., Milne, G. A. and Tamisiea, M. E. (2002). Sea-level finger-printing as a direct test for the source of Global Meltwater Pulse IA. *Science*, **295**, 2438–41.

Clark, P. U., Dyke, A. S., Shakun, J. D., *et al.* (2009). The Last Glacial Maximum. *Science*, **325**, 710–14.

Das, S. B., Joughin, I., Behn, M. D. *et al.* (2008). Fracture propagation to the base of the Greenland Ice Sheet during supraglacial lake drainage. *Science*, **320**, 778–81.

Department of Climate Change (DCC) (2009). *Climate Change Risks to Australia's Coast: A First Pass National Assessment*. Canberra: Australian Government.

Dietrich, R., Maas, H.-G., Baessler, M. *et al.* (2007). Jakobshavn Isbræ, West Greenland: Flow velocities and tidal interaction of the front area from 2004 field observations. *Journal of Geophysical Research*, **112**, F03S21.

Doherty, S. J., Bojinksi, S., Henderson-Sellers, A. *et al.* (2009). Lessons learned from IPCC AR4: Future scientific developments needed to understand, predict, and respond to climate change. *Bulletin of the American Meteorological Society*, **90**, 497–513.

Domingues, C. M., Church, J. A., White, N. J. *et al.* (2008). Improved estimates of upper-ocean warming and multi-decadal sea-level rise. *Nature*, **453**, 1090–93.

Dowsett, H. J. and Cronin, T. M. (1990). High eustatic sea level during the middle Pliocene: Evidence from the southeastern U.S. Atlantic Coastal Plain. *Geology*, **18**, 435–38.

Ettema, J., van den Broeke, M. R. *et al.* (2009). Higher surface mass balance of the Greenland ice sheet revealed by high-resolution climate modeling. *Geophysical Research Letters*, **36**, L12501.

Fahnestock, M. A., Bindschadler, R., Kwok, R. and Jezek, K. (1993). Greenland Ice Sheet properties and ice dynamics from ERS-1 SAR imagery. *Science*, **262**, 1530–34.

Fairbanks, R. G. (1989). A 17,000-year glacio-eustatic sea level record: Influence of glacial melting rates on the Younger Dryas event and deep-ocean circulation. *Nature*, **342**, 637–42.

Gehrels, R. (2010). Sea-level changes since the Last Glacial Maximum: an appraisal of the IPCC Fourth Assessment Report. *Journal of Quaternary Science*, **25**, 26–38.

German Advisory Council on Global Change (WBGU) (2006). *The Future Oceans: Warming Up, Rising High, Turning Sour*. Berlin: WBGU.

Gregory, J. M. and Huybrechts, P. (2006). Ice-sheet contributions to future sea-level change. *Philosophical Transactions of the Royal Society A – Mathematical, Physical & Engineering Sciences*, **364**, 1709–31.

Grinsted, A., Moore, J. C. and Jevrejeva, S. (2009). Reconstructing sea level from paleo and projected temperatures 200 to 2100 AD. *Climate Dynamics*, doi 10.1007/s00382–008–0507–2.

Hanebuth, T., Stattegger, K. and Grootes, P. M. (2000). Rapid flooding of the Sunda Shelf: A late-glacial sea-level record. *Science*, **288**, 1033–35.

Hanna, E., Huybrechts, P., Steffen, K. *et al.* (2008). Increased runoff from melt from the Greenland Ice Sheet: A response to global warming. *Journal of Climate*, **21**, 331–41.

Harvey, N. and Goodwin, I. (2004). Impacts of sea level rise. In *Global Change and the Earth System: A Planet Under Pressure*, eds. W. Steffen, R. A. Sanderson, P.D. Tyson, J. Jäger, P.A. Matson, B. Moore III, F. Oldfield, K. Richardson, H.-J. Schellnhuber, B.L. Turner and R.J. Wasson, 1st edn. Berlin: Springer-Verlag, pp. 154–55.

Haywood, A. M., Dowsett, H. J., Valdes, P. J. *et al.* (2009). Introduction. Pliocene climate, processes and problems. *Philosophical Transactions of the Royal Society A – Mathematical, Physical & Engineering Sciences*, **367**, 3–17.

Holland, D. M., Thomas, R. H., de Young, B., Ribergaard, M. H. and Lyberth, B. (2008). Acceleration of Jakobshavn Isbræ triggered by warm subsurface ocean waters. *Nature Geoscience*, **1**, 659–64.

Horton, R., Herweijer, C., Rosenzweig, C. *et al.* (2008). Sea level rise projections for current generation CGCMs based on the semi-empirical method. *Geophysical Research Letters*, **35**, L02715.1–L02715.5.

Hostetler, S., Pisias, N. and Mix, A. (2006). Sensitivity of Last Glacial Maximum climate to uncertainties in tropical and subtropical ocean temperatures. *Quaternary Science Reviews*, **25**, 1168–85.

Howat, I. M., Joughin, I. and Scambos, T. A. (2007). Rapid changes in ice discharge from Greenland outlet glaciers. *Science*, **315**, 1559–61.

Intergovernmental Panel on Climate Change (IPCC) (2001). *Climate Change 2001: The Scientific Basis. Contribution of Working Group I to the Third Assessment Report of the Intergovernmental Panel on Climate Change*, eds. J. T. Houghton, Y. Ding, D.J. Griggs, M. Noguer, P. J. van der Linden and D. Xiaosu. Cambridge, UK and New York, NY: Cambridge University Press.

Intergovernmental Panel on Climate Change (IPCC) (2007a). *Climate Change 2007: The Physical Science Basis. Contribution of Working Group I to the Fourth Assessment Report of the Intergovernmental Panel on Climate Change*, eds. S. Solomon, D. Qin,

M. Manning, Z. Chen, M. Marquis, K. Averyt, M. M. B. Tignor, H. L. Miller Jr and
Z. Chen. Cambridge, UK and New York, NY: Cambridge University Press.

Intergovernmental Panel on Climate Change (IPCC) (2007b). *Climate Change 2007.
Impacts, Adaptation and Vulnerability. Contribution of Working Group II to the Fourth
Assessment Report of the Intergovernmental Panel on Climate Change*, eds. M. Parry,
O. Canziani, J. Palutikov, P. van der Linden and C. Hanson. Cambridge, UK and New
York, NY: Cambridge University Press.

Jevrejeva, S., Grinsted, A. and Moore, J. C. (2009). Anthropogenic forcing dominates sea
level rise since 1850. *Geophysical Research Letters*, **36**, L20706.

Jevrejeva, S., Grinsted, A., Moore, J. C. and Holgate, S. (2006). Nonlinear trends and mul-
tiyear cycles in sea level records. *Journal of Geophysical Research*, **111**, C09012.

Jevrejeva, S., Moore, J. C., and Grinsted, A. (2010). How will sea level respond to changes
in natural and anthropogenic forcings by 2100? *Geophysical Research Letters*, **37**,
L07703, 5 pp., doi: 10.1029/2010GL042947.

Joughin, I. R., Das, S. B., King, M. A. *et al.* (2008). Seasonal speedup along the western
flank of the Greenland Ice Sheet. *Science*, **720**, 781–83.

Joughin, I., Tulaczyk, S., Bindschadler, R. and Price, S. F. (2002). Changes in West Antarctic
ice stream velocities: Observation and analysis. *Journal of Geophysical Research –
Solid Earth*, **107**, 2289.

Kopp, R. E., Simons, F. J., Mitrovica, J. X., Maloof, A. C. and Oppenheimer, M. (2009). Probabilistic assessment of sea level during the last interglacial stage. *Nature*, **462**, 863–67.

Kriegler, E., Hall, J. W., Held, H., Dawson, R. and Schellnhuber, H.-J. (2009). Imprecise
probability assessment of tipping points in the climate system. *Proceedings of the
National Academy of Sciences (USA)*, **106**, 5041–46.

Lambeck, K. and Chappell, J. (2001). Sea level change through the last glacial cycle.
Science, **292**, 679–86.

Lambeck, K., Yokoyama, Y. and Purcell, T. (2002). Into and out of the Last Glacial
Maximum: Sea-level change during oxygen isotope stages 3 and 2. *Quaternary
Science Reviews*, **21**, 343–60.

Meehl, G. A., Stocker, T. F., Collins, W. D. *et al.* (2007). Global climate projections. In
*Climate Change 2007: The Physical Science Basis. Contribution of Working Group I
to the Fourth Assessment Report of the Intergovernmental Panel on Climate Change*,
eds. S. Solomon, D. Qin, M. Manning, Z. Chen, M. Marquis, K.B. Averyt, M. Tignor
and H.L. Miller. Cambridge, UK and New York, NY: Cambridge University Press.

Mercer, J. H. (1978). West Antarctic ice sheet and CO_2 greenhouse effect: A threat of disaster. *Nature*, **271**, 321–25.

Mernild, S. H., Liston, G. E., Hiemstra, C. A. *et al.* (2009). Greenland Ice Sheet surface
mass-balance modeling and freshwater flux for 2007, and in a 1995–2007 perspective.
Hydrological Processes, **23**, 2470–84.

Mix, A.C., Bard, E. and Schneider, R. (2001). Environmental processes of the ice age: Land,
oceans, glaciers (EPILOG). *Quaternary Science Reviews*, **20**, 627–57.

Moon, T. and Joughin, I. (2008). Changes in ice front position on Greenland's outlet glaciers from 1992 to 2007. *Journal of Geophysical Research*, **113**, F02022.

Mote, T. L. (2007). Greenland surface melt trends 1973–2007: Evidence of a large increase
in 2007. *Geophysical Research Letters*, **34**, L22507.

Muhs, D. R., Simmons, K. R. and Steinke, B. (2002). Timing and warmth of the last interglacial period: New U-series evidence from Hawaii and Bermuda and a new fossil
compilation for North America. *Quaternary Science Review*, **21**, 1355–83.

Naish, T., Powell, R., Levy, R. *et al.* (2009). Obliquity-paced Pliocene West Antarctic ice
Sheet oscillations. *Nature*, **458**, 322–28.

Nick, F. M., Vieli, A., Howat, I. M. and Joughin, I. (2009). Large-scale changes in Greenland outlet glacier dynamics triggered at the terminus. *Nature Geoscience*, **2**, 110–14.

Overpeck, J. T., Otto-Bliesner, B. L., Miller, G. H. *et al.* (2006). Paleoclimate evidence for future ice-sheet instability and rapid sea-level rise. *Science*, **311**, 1747–50.

Overpeck, J. T., Sturm, M., Francis, J. A. *et al.* (2005). Arctic system on trajectory to new, seasonally ice-free state. *Eos*, **86**, 309–13.

PALSEA (2010). The sea-level conundrum: Case studies from palaeo-archives. *Journal of Quaternary Science*, **25**, 19–25.

Payne, A. J., Vieli, A., Shepherd, A. P., Wingham, D. J. and Rignot, E. (2004). Recent dramatic thinning of largest West Antarctic ice stream triggered by oceans. *Geophysical Research Letters*, **31**, L23401.

Penck, A. (1928) Die Ursachen der Eiszeit. *Sitzungsberichte der Preussischen Akademie der Wissenschaften, Phys.-Math. Klasse*, **6**, 76–85.

Pfeffer, W. T., Harper, J. T. and O'Neel, S. (2008). Kinematic constraints on glacier contributions to 21st-century sea-level rise. *Science*, **321**, 1340–43.

Pollard, D. and DeConto, R. M. (2009). Modelling West Antarctic ice sheet growth and collapse through the past five million years. *Nature*, **458**, 329–32.

Radić, V. and Hock, R. (2010). Regional and global volumes of glaciers derived from statistical upscaling of glacier inventory data. *Journal of Geophysical Research*, **115**, F010101. Doi: 10.1029/2009JF001373.

Rahmstorf, S. (2007). A semi-empirical approach to projecting future sea-level rise. *Science*, **315**, 368–70.

Rahmstorf, S., Cazenave, A., Church, J. A. *et al.* (2007). Recent climate observations compared to projections. *Science*, **316**, 709.

Reeh, N., Bøggild, C. E. and Oerter, H. (1994). Surge of Storstrømmen, a large outlet glacier from the inland ice of north-east Greenland. *Rapport Grønlands Geologiske Undersøgelse*, **162**, 201–09.

Rignot, E. J. (1998). Fast recession of a West Antarctic glacier. *Science*, **281**, 549–51.

Rignot, E. and Jacobs, S. S. (2002). Rapid bottom melting widespread near Antarctic ice sheet grounding lines. *Science*, **296**, 2020–23.

Rignot, E. and Kanagaratnam, P. (2006). Changes in the velocity structure of the Greenland Ice Sheet. *Science*, **311**, 986–90.

Rignot, E., Box, J. E., Burgess, E. and Hanna, E. (2007). Mass balance of the Greenland ice sheet from 1958 to 2007. *Geophysical Research Letters*, **35**, L20502.

Rignot, E., Koppes, M. and Velicogna, I. (2010). Rapid submarine melting of the calving faces of West Greenland glaciers. *Nature Geoscience*, doi: 10.1038/NGEO765.

Rostami, K., Peltier, W. R. and Mangini, A. (2000). Quaternary marine terraces, sea-level changes and uplift history of Patagonia, Argentina: Comparisons with predictions of ICE-4G (VM2) model of the global process of glacial isostatic adjustment. *Quaternary Science Reviews*, **19**, 1495–1525.

Schoof, C. (2007). Ice sheet grounding line dynamics: Steady states, stability, and hysteresis. *Journal of Geophysical Research*, **112**, F03S28.

Shackleton, N. J., Hall, J. C. and Pate, D. (1995). Pliocene stable isotope stratigraphy of Site 846. *Proceedings of the Ocean Drilling Program, Scientific Results*, **138**, 337–56.

Shepherd, A. and Wingham, D. (2007). Recent sea-level contributions of the Antarctic and Greenland ice sheets. *Science*, **315**, 1529–32.

Siddall, M., Rohling, E. J. *et al.* (2003). Sea-level fluctuations during the last glacial cycle. *Nature*, **423**, 853–58.

Solomon, S., Plattner, G.-K., Knutti, R. and Friedlingstein, P. (2009). Irreversible climate change due to carbon dioxide emissions. *Proceedings of the National Academy of Sciences (USA)*, **106**, 1704–09.

Steig, E. J., Schneider, D. P., Rutherford, S. D. *et al.* (2009). Warming of the Antarctic ice-sheet surface since the 1957 International Geophysical Year. *Nature*, **457**, 459–62.

Straneo, F., Hamilton, G. S., Sutherland, D. A. *et al.* (2010). Rapid circulation of warm subtropical waters in a major glacial fjord in East Greenland. *Nature Geoscience*, doi: 10.1038/NGEO764.

Thomas, R., Davis, C., Frederick, E. *et al.* (2008). A comparison of Greenland ice-sheet volume changes derived from altimetry measurements. *Journal of Glaciology*, **54**, 203–12.

Turney, C. S. M. and Jones, R. T. (2010). Agulhas current amplifies warming during super-interglacials. *Journal of Quaternary Science*, **25**, 839–843.

van den Broeke, M., Bamber, J., Ettema, J. *et al.* (2009). Partitioning recent Greenland mass loss. *Science*, **329**, 984–86.

Vaughan, D. (2008). West Antarctic Ice Sheet collapse – the fall and rise of a paradigm. *Climatic Change*, **91**, 65–79.

Velicogna, I. (2009). Increasing rates of ice mass loss from the Greenland and Antarctic ice sheets revealed by GRACE. *Geophysical Research Letters*, **36**, L19503.

Vellinga, P., Katsman, C., Sterl, A. *et al.* (2009). *Exploring High-end Climate Change Scenarios for Flood Protection of the Netherlands*. International Scientific Assessment carried out at the request of the Delta Committee. KNMI publication WR-2009–05. De Bilt: KNMI.

Vermeer, M. and Rahmstorf, S. (2009). Global sea level linked to global temperature. *Proceedings of the National Academy of Sciences (USA)*, **106**, 21527–32.

Walker, R. T., Dupont, T. K., Parizek, B. R. and Alley, R. B. (2008). Effects of basal melting distribution on the retreat of ice shelf grounding lines. *Geophysical Research Letters*, **35**, L17503.

Weertman, J. (1974). Stability of the junction between an ice sheet and an ice shelf. *Journal of Glaciology*, **13**, 3–11.

Wouters, B., Chambers, D. and Schrama, E. J. O. (2007). GRACE observes small-scale mass loss in Greenland. *Geophysical Research Letters*, **35**, L20501.

Zwally, H. J., Abdalati, W., Herring, T. *et al.* (2002). Surface melt-induced acceleration of Greenland ice-sheet flow. *Science*, **297**, 218–22.

4

Carbon cycle trends and vulnerabilities

'We are only a tool in the cycle of things … (we) go out into the world and help keep the balance of nature. It's a big cycle of living with the land, and eventually going back to it …'[1]

The Earth's element cycles – nitrogen, carbon, phosphorus, sulphur, silicon and others – are central to the functioning of the climate system, and to life itself. In the context of climate change, the carbon cycle has assumed centre stage, primarily through the rapid rise in human-induced emissions of the important greenhouse gases carbon dioxide (CO_2) and methane (CH_4). The political debate on responses to the climate change challenge has focused primarily on one aspect of the carbon cycle – reducing the emissions of CO_2 to the atmosphere. However, the carbon cycle is very complex and human activities affect other parts of the cycle – for example, the ability of natural processes on the land and in the ocean (carbon 'sinks') to take up a significant fraction of the CO_2 emitted to the atmosphere. The human imprint also operates indirectly on the carbon cycle via climate change itself, as several important feedback processes are predicted to be activated as the planet warms. For example, pools of carbon in the natural world, such as the CH_4 stored in frozen soils in the northern high latitudes, that have hitherto been stable could become an important new source of a powerful greenhouse gas as the planet warms. Understanding how the effects of human perturbations on the carbon cycle cascade through the Earth System is crucial for effective management of our interaction with the carbon cycle.

4.1 Trends in human emissions of carbon dioxide

The increase in atmospheric concentration of CO_2 since the industrial revolution has been large in both magnitude and rate compared to the natural variation in the late Quaternary period. For the past 400 000 years, which encompasses the entire period that fully modern humans have been in existence, atmospheric CO_2 concentration has varied naturally

---

[1] Vilma Webb, Noongar People. Quote from *Elders: Wisdom from Australia's Indigenous Leaders*, edited by Peter McConchie (2003).

Climate Change: Global Risks, Challenges and Decisions, Katherine Richardson, Will Steffen and Diana Liverman *et al.* Published by Cambridge University Press. © Katherine Richardson, Will Steffen and Diana Liverman 2011.

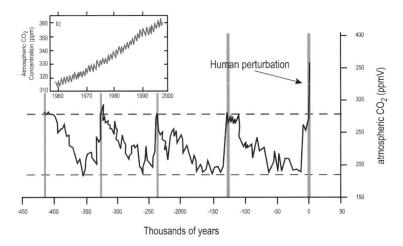

Figure 4.1 The human perturbation to atmospheric carbon dioxide (red line) compared to the 420 000-year record of carbon dioxide variation from an Antarctic ice core. Sources: Petit *et al.* (1999); Keeling and Whorf (2000); http://cdiac.ornl.gov/trends/trands.htm.

between ice ages and warm periods, with a minimum of 180 p.p.m. during ice ages, and a maximum that varies between 280 and 300 p.p.m. during the shorter and warmer inter-glacial periods (Petit *et al.*, 1999). Current CO_2 concentration is approaching 390 p.p.m., which has effectively doubled the normal operating range of CO_2 in only 200 years. This rate of increase is at least 10 times greater than any rate of change observed in the pre-industrial period over the past 600 000 years, and possibly 100 times greater (Figure 4.1; Falkowski *et al.*, 2000).

The change in the overall carbon budget since the industrial revolution, which includes both the human-induced emissions of CO_2 and the uptake of a significant fraction of these emissions by natural processes in the ocean and on land, is shown in Figure 4.2. The rate of human emissions (shown as Gt of C per year in the figure) has been highly nonlinear, with a sharp increase in the rate around 1950, coincident with the rapid economic development of OECD countries after the Second World War. This nonlinearity is reflected in the atmos-pheric CO_2 concentration, which rose from 270–275 p.p.m. in the pre-industrial period to about 310 p.p.m. in 1950, but then to 369 p.p.m. in 2000 and to 387 p.p.m. in 2008. About half of the rise since the pre-industrial period has occurred in the past 30 years (Etheridge *et al.*, 1996; Keeling and Whorf, 2000; http://cdiac.ornl.gov/trends/trends.htm).

Although the multi-decadal trend in human-induced CO_2 emissions has been inexorably up, some long-term changes in the sources of emissions can be seen. For example, the pri-mary source of emissions from deforestation slowly shifted from temperate forests in the 1800s to become dominated by tropical forests in the past 40–50 years. In addition, some variations due to socio-economic factors can be seen on shorter timescales, for example, the slowdown in emissions after 1990 with the collapse of the Soviet Union. The post-2000 observations of emissions as well as of atmospheric CO_2 concentration, described in more

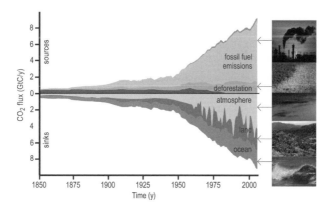

Figure 4.2 Changes in the global carbon budget from 1850 to 2000. CO_2 emissions to the atmosphere (sources) as the sum of fossil fuel combustion, land-use change, and other emissions (upper); the fate of the emitted CO_2, including the increase in atmospheric CO_2 plus the sinks of CO_2 on land and in the oceans (lower). Flux is in gigatonnes: Gt yr^{-1} carbon. For the deforestation source the light brown colour refers to tropical forests and the dark brown temperate forests. Sources: adapted from Canadell *et al.* (2007), with additional data from C. Le Quéré and the Global Carbon Project. Copyright 2007 National Academy of Sciences, USA. (In colour as Plate 12.)

detail in Box 4.1, both show a further increase in rate over the 1950–2000 period due to several factors. Another socio-economic event – the global financial crisis of 2008–09 – may also be observable in the most recent emission and CO_2 concentration observations (Le Quéré *et al.*, 2009).

Box 4.1
Changes in anthropogenic CO_2 emissions

JOSEP G. CANADELL AND MICHAEL R. RAUPACH

Human activities have released half a trillion tonnes of carbon to the atmosphere in the form of carbon dioxide (CO_2) since the beginning of the industrial revolution. These emissions have led to an increase of 38% in atmospheric CO_2, from 280 p.p.m. in 1750 to 387 p.p.m. in 2009, and contributed 63% of all anthropogenic radiative forcing to date. Currently, CO_2 emissions are responsible for 80% of the growth of anthropogenic radiative forcing (Butler, 2009).

The total CO_2 emission flux from human activities was almost 10 Gt C yr^{-1} in 2008 (Le Quéré *et al.*, 2009), composed of 8.7 Gt C yr^{-1} (1 Gt = 1 gigatonne = 10^9 tonnes = 1 Pg = petagram = 10^{15} grams) from fossil fuel combustion and industrial activities, and the remainder from net land-use change. Since the start of the industrial revolution around 1750, the cumulative anthropogenic CO_2 emission from both sources is over 500 Gt C (a point passed in about 2006).

The dominant CO_2 emissions contribution from fossil fuels has accelerated in recent years, increasing at an average annual growth rate of 3.4% per year during the period 2000–08,

Box 4.1 (*cont.*)

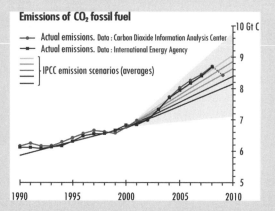

Figure 4.3 Fossil fuel emissions of CO_2 in Gt C yr^{-1}, plotted together with average emissions for IPCC scenario families. Dotted line is a projection based on the expected GDP decline in 2009. Grey band indicates range of individual IPCC scenarios. Source: redrawn from Le Quéré *et al.* (2009).

compared to 1.1% per year during the 1990s and an average of 2% per year since 1970. This growth rate was above most IPCC scenarios published in 2000, emphasising the unprecedented and largely unexpected emissions growth of the past decade (Figure 4.3; Raupach *et al.*, 2007; Le Quéré *et al.*, 2009).

An important aspect of this acceleration has been an increase in the global share of emissions from the combustion of coal for electricity generation in emerging economies, particularly China and India. Coal was the single largest source of fossil fuel emissions in 2007 and 2008, after more than 40 years of oil supremacy.

In 2008, developing countries emitted more fossil-fuel CO_2 (55% of the total) than the developed world (45%); however, per capita emissions remained much higher in developed countries. Apart from small oil states such as Qatar, Kuwait and the United Arab Emirates with about 9 kg C per person per year, the US and Australia top the list with about 5 kg C per person per year. The latter value is double the average for European countries, four times that of China, and thirteen times that of India and the least developed countries in Africa (Raupach *et al.*, 2007; Canadell *et al.*, 2009).

The global financial crisis that began in 2008 is expected to have a discernable impact on global fossil-fuel CO_2 emissions, but a scarcely detectable effect on the growth of the concentration of atmospheric CO_2 because of the large interannual variability in natural CO_2 sinks. The emission growth in 2008 was 2.0% per year, less than the average of 3.6% per year for 2000–07, suggesting an incipient slowdown from the global financial crisis. The projection for 2009, based on a contraction of the gross world product by 1.8%, is a decline of 2.8% per year in fossil-fuel emissions. This is likely to be followed by a rapid return to emissions growth as the gross world product begins to increase again in 2010 (Figure 4.3; IMF, 2009; Le Quéré *et al.*, 2009).

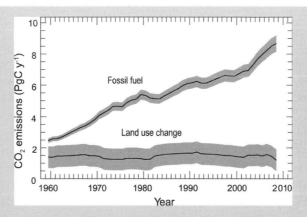

Figure 4.4 CO_2 emissions from the combustion of fossil fuel (including a small fraction of emissions from industrial processes, mainly cement production) and from land-use change. Source: Le Quéré *et al.* (2009).

Emissions from land-use change, the second most important source of anthropogenic CO_2 after the combustion of fossil fuels, are the net balance of multiple fluxes in and out of terrestrial ecosystems including those from deforestation, degradation and reforestation. This net flux was 1.5 ± 0.7 Gt C yr^{-1} during the period 1990–2005. There is an indication that land-use change emissions decreased in 2007–08, but the trend is not yet clear because of large uncertainty in the estimation of land-use change emissions. Averaged since the start of the industrial revolution, land-use change emissions account for about one third of total cumulative CO_2 emissions from human activities, but this fraction has decreased progressively, to 12% in 2008 (Figure 4.4; Le Quéré *et al.*, 2009). As emissions from fossil fuel continue to grow, the relative importance of land-use change emissions will decline further (Canadell *et al.*, 2007).

4.2 Trends in natural carbon sinks

The observation that, on average, less than half of the human emissions of CO_2 remain in the atmosphere highlights the importance of the natural carbon sinks in the ocean and on land (Figure 4.2). In fact, the so-called airborne fraction of CO_2 emissions has been remarkably stable from 1959 to the present at a value tightly constrained between 0.4 and 0.5 (Canadell *et al.*, 2007; Raupach *et al.*, 2007). The primary reason for this stability is that the land and ocean carbon sinks together have responded nearly linearly to the increase in CO_2 above the pre-industrial level.

This remarkable stability in the airborne fraction over a half century of increasing CO_2 emissions gives some confidence in the capability of these natural buffering processes to continue to operate for some time into the future. It would take a very significant change in sink strength to affect atmospheric CO_2 growth rate, which is currently dominated by three socio-economic factors – population, per capita income and the total carbon intensity of the global economy (Canadell *et al.*, 2007; Raupach *et al.*, 2007). Thus, many analyses of the emission reductions required to limit climate change to a given level of temperature increase tacitly assume that the strength of these sinks will not change significantly, and will thus keep pace with the changes in emissions so that the fraction absorbed will remain the same. However, some small changes in sink strength are possibly being observed now (Le Quéré *et al.*, 2007; Knorr, 2009) and there is concern about the behaviour of the sinks as the Earth continues to warm.

The natural carbon sinks are approximately equally distributed between land and ocean. The ocean sink is due to two processes: (i) the increasing dissolution of CO_2 directly in sea water as partial pressure of CO_2 in the atmosphere increases (the 'physical pump'), and (ii) the uptake of CO_2 by phytoplankton and the subsequent sinking of faecal matter or dead organisms deeper into the ocean (the 'biological pump'). The former is directly affected by water temperature (CO_2 is less soluble in warmer waters) and indirectly by ocean acidity (Goodwin *et al.*, 2009; Hofmann and Schellnhuber, 2009). Regarding the latter, marine productivity is affected by the concentration of dissolved CO_2, water temperature, the provision of macronutrients such as nitrogen and phosphorus, and the availability of micronutrients such as iron.

The land sinks reside in living biomass (both above- and below-ground), dead biomass and soil carbon. These land sinks are ultimately driven by increased photosynthesis of terrestrial vegetation, the rate of which is affected by many factors. These include the CO_2 fertilisation effect due to increasing atmospheric CO_2 concentration, availability of nutrients such as nitrogen and phosphorus, effects of temperature and precipitation changes on growth and respiration, the ratio of plants with different photosynthetic pathways (C3 and C4 plants), and the impact of disturbances such as fire and insect outbreaks. In addition, the legacy of past land-use changes also influences the strength of the land sink.

There is considerable interannual variability in the strength of these sinks, particularly the land sink, which dominates the pattern of change from year to year in the atmospheric CO_2 concentration. Figure 4.5 shows the ~50-year trend in the behaviour of the two sinks, and in the change in atmospheric CO_2 concentration; all are expressed as a fraction of the total anthropogenic emissions of CO_2 for any particular year (Canadell *et al.*, 2007). The land sink, despite its high interannual variability, shows no long-term trend. But the ocean sink has weakened proportionately, from absorbing about 32% of anthropogenic emissions in 1960 to about 26% now. This means that the fraction of emissions that remains in the atmosphere has increased slowly over the past half-century, at about 0.2% per year, to a value of about 45% now (but has still remained within the 0.4 to 0.5 range).

Changes in the ocean sink are described in more detail in Box 4.2, and include the observation that the sink is weakening in some regions, such as the Southern Ocean (Le Quéré *et al.*, 2007) but may be strengthening elsewhere. The future of the ocean sink as the climate changes is subject to some uncertainty, particularly because the response of the biological pump to increasing temperature is largely unknown. Recent measurements from

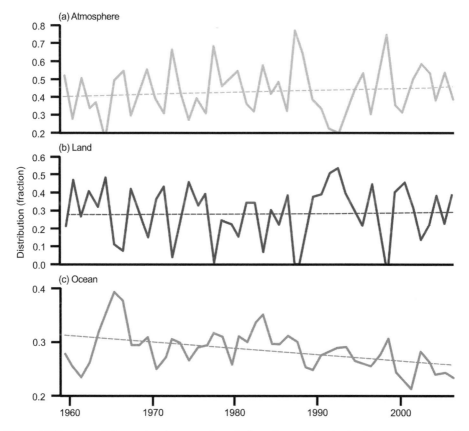

Figure 4.5 Fraction of the total anthropogenic emissions that remains in the (a) atmosphere, (b) the land biosphere, and (c) the ocean. Source: Canadell *et al.* (2007). Copyright 2007 National Academy of Sciences, USA.

around the world's oceans, however, are beginning to provide some observational evidence for a temperature effect on the potential strength of the biological pump (Box 4.3).

Box 4.2
Changes in the oceanic carbon sink: observations, processes and future prospects

NICOLAS GRUBER

Over the course of the Anthropocene (Crutzen and Stoermer, 2000; Crutzen, 2002; Steffen *et al.*, 2007), i.e. roughly over the past 250 years, the ocean has taken up about one third of the carbon that has been emitted into the atmosphere as a result of human activities. To this date, this amounts to about 140±25 Gt C (Sabine *et al.*, 2004; Khatiwala *et al.*, 2009). Without this

Box 4.2 (*cont.*)

service, atmospheric CO_2 today would already have crossed 450 p.p.m., which is deemed to be approximately the level above which Earth's surface temperature is no longer likely to stay within the 2 °C maximum warming target. It is therefore of utmost importance to know whether the oceanic sink will continue, or whether global climate change will affect it in such a way that a substantially smaller fraction of the emitted CO_2 will be removed from the atmosphere by the ocean.

The primary process by which the ocean has been taking up CO_2 from the atmosphere over the past 250 years is the gas transfer across the air–sea interface, driven by the chemical disequilibrium across this interface induced by the addition of anthropogenic CO_2 to the atmosphere. This chemical–physical process would soon reach its capacity to remove anthropogenic CO_2 from the atmosphere if this CO_2 were not transported away from the near-surface waters into the ocean's interior. In fact, this latter process is the primary rate-limiting step for the oceanic uptake of anthropogenic CO_2 (Sarmiento *et al.*, 1992) and explains why the ocean is currently only realising about one third of its sink potential. Thus, when considering how climate change may impact the oceanic sink for CO_2, the first process to consider is a change in the rate by which anthropogenic CO_2 is transported from the surface to depth. Climate change simulations rather consistently suggest that this transport will decrease in the future, primarily by an increase in upper-ocean stratification due to warming and/or an excess of evaporation over precipitation leading to salinisation. The magnitude of this effect may be as large as 20% (Sarmiento *et al.*, 1998).

When considering a changing ocean, the fate of the huge pool of natural CO_2 present in the ocean needs to be carefully investigated as well. With the ocean roughly containing 60 times more inorganic carbon than the atmosphere, a small change in this inventory would have a large impact on atmospheric CO_2. A first driving force is ocean warming which, by reducing the solubility of CO_2 in seawater, will cause CO_2 to be expelled from the ocean. Assuming the amount of warming expected from a business-as-usual emission scenario, this effect may cause more than 100 Gt C to be lost from the ocean. A second driving element is the aforementioned reduction in the surface-to-deep exchange. Absent of any change in the rate of biological production and biological export of carbon from the near-surface ocean, stratification would actually lead to an increase in the uptake of carbon from the atmosphere, primarily because there would be an excess of the downward transport of organic carbon from biological sources over the upward transport of inorganic carbon by transport and mixing. This causes a deficit of carbon in the upper ocean, which is then compensated by uptake of carbon from the atmosphere. The final, and least well constrained, change is the activity of ocean biology in a changing ocean. Will it benefit from ocean warming (e.g. by an increase in light availability) or will biological productivity and the export of organic carbon decrease (e.g. by the reduced upward supply of nutrients)? Model simulations suggest that the overall balance of changes in circulation and biology on the air–sea CO_2 balance is a slight gain of carbon by the ocean, although with very high uncertainty.

Summing up, while the last process might cause an increase in the uptake of carbon from the atmosphere, all others will cause a reduction in the uptake. An assessment of the possible magnitude of these effects (Gruber *et al.*, 2004), as well as model simulations, all suggest that the reduction in uptake will overwhelm the increase, so that the global effect will likely be reduced CO_2 uptake by the ocean. Thus, the ocean is likely to become a positive feedback in the climate system.

Are we seeing such feedbacks already today? Results from the investigation of long-term trends in the ocean carbon cycle are sparse, primarily because the observational database for oceanic carbon parameters is limited. The first global survey of the ocean interior carbon content with high-quality measurements was conducted only in the late 1980s and early 1990s (Wallace, 2001), and efforts to repeat this survey are underway. The initial results from these repeat surveys reveal substantial changes, but most of the changes can be attributed to the uptake of anthropogenic CO_2 or internal redistributions of the natural carbon pool (Sabine and Tanhua, 2010).

Measurements of the surface ocean partial pressure of CO_2 (Figure 4.6) extend further back in time, permitting a more extensive assessment of whether the ocean carbon sink has already been altered by feedbacks. While there is mounting evidence that certain regions have indeed experienced a reduction in the net uptake of CO_2 from the atmosphere in recent decades, e.g. the Southern Ocean (Le Quéré *et al.*, 2007; Lovenduski *et al.*, 2007) and parts of the North Atlantic (Schuster and Watson, 2007), there are other regions where the sink has increased (northwestern North Pacific) (Takahashi *et al.*, 2006). An attempt to assess such regional trends globally in a homogenous manner reveals that no global trend can be identified yet, in part due to the sparse sampling, but also because the spatiotemporal heterogeneity is substantial, masking potentially underlying global trends (Oberpriller and Gruber, personal communication).

With data from recent decades largely being insufficient to confirm the ocean's nature as a positive feedback in the climate system, a view back in time strongly suggests such behaviour.

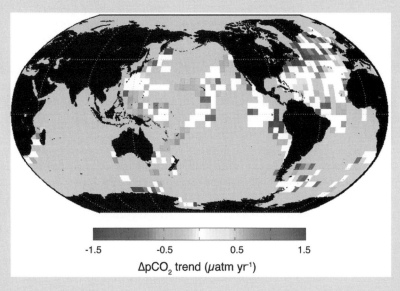

Figure 4.6 Map of long-term trends in the difference of the partial pressure of CO_2 between the atmosphere and ocean surface. A positive value (warm colours) indicates that the oceanic pCO_2 is increasing at a faster rate than that in the atmosphere, i.e. indicating a reduction of the oceanic sink strength. Negative values (cold colours) indicate an increase in the oceanic sink strength. Source: figure from Oberpriller and Gruber, personal communication. (In colour as Plate 13.)

Box 4.2 (*cont.*)

In the millennial timescale transitions from cold to warm climates during the past million years, the ocean likely lost more than 700 Gt C to the atmosphere and land biosphere (Sigman and Boyle, 2000). Arguably, this occurred on timescales much longer than those associated with current global warming, but several of the processes that operated then – such as ocean warming – are also acting today.

In summary, the ocean will continue to take up CO_2 from the atmosphere with very high likelihood. It is also likely that the sink strength is vulnerable to global climate change, leading to an increase in the fraction of the anthropogenic CO_2 emissions remaining in the atmosphere. Thus, the ocean cannot provide the 'magic bullet' required to stabilise atmospheric CO_2.

Box 4.3

Increased bacterial activity in a warmer ocean will reduce the amount of organic carbon transported from the surface to the ocean interior

JØRGEN BENDTSEN AND KATHERINE RICHARDSON

The magnitude of photosynthesis occurring annually on land and in the ocean is approximately equal (Field *et al.*, 1998). Thus, plant activity in the ocean is also important for removal of CO_2 from the atmosphere. The vast majority of the plant activity in the ocean is carried out by microscopic plants (phytoplankton) in surface waters. When the organic material they produce is broken down (remineralised) in surface waters, the CO_2 incorporated by photosynthesis is released and is free to exchange with the atmosphere. However, some percentage of the organic material produced by phytoplankton sinks out of the surface layer. If it is transported all the way to the interior of the deep ocean, the CO_2 released upon its remineralisation is retained ('sequestered') in the bottom waters and can no longer exchange freely with the atmosphere. The organic material and CO_2 in the interior of the ocean can remain out of contact with the atmosphere for hundreds to around 1000 years (Stuiver *et al.*, 1983) and, thus, is not of immediate concern with respect to climate change.

A warmer ocean will be expected to influence the magnitude of the transfer of organic material from ocean surface to interior (i.e. the biological pump). Of particular interest is the relative role of temperature on production and respiration of organic matter, and analysis of the metabolic balance in the euphotic zone (i.e. where light is sufficient for photosynthesis to occur) shows that changes in community respiration tend to exceed primary production in a warmer surface layer of the ocean, thereby resulting in a reduced ocean CO_2 demand (López-Urrutia *et al.*, 2006). These findings have been supported by mesocosm experiments on plankton dynamics (Wohlers *et al.*, 2009). Cycling of organic carbon below the productive surface layer is determined primarily by heterotrophic activity from zooplankton and the microbial biomass. Despite this branch of the biological pump regulating the cycling of about 10 Gt C yr^{-1} (Dunne *et al.*, 2005), few studies have so far been undertaken on the sensitivity of these processes to climate-change induced temperature changes.

To explore the influence from changing temperatures on the microbial respiration in the ocean upper mesopelagic zone (100–400 m), a comprehensive field study covering all major ocean basins was carried out on the circumnavigating Galathea 3 expedition (2006–07) where incubation experiments analysed the temperature sensitivity of microbial respiration (Bendtsen *et al.*, submitted). Incubation results from the Namibia upwelling area show that oxygen utilisation increases significantly when respiration takes place at higher temperatures. The averaged temperature sensitivity implies that microbial respiration increases by a factor of 3 for a 10 °C temperature increase (Figure 4.7). Analysis of ocean measurements reveals

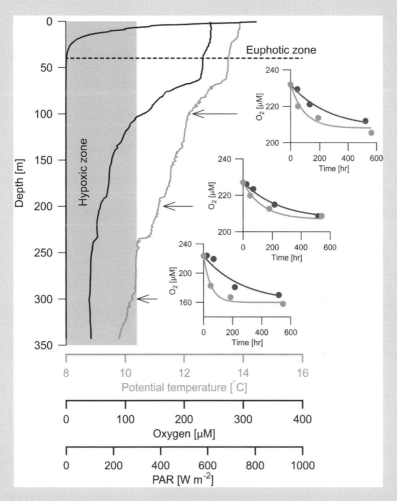

Figure 4.7 Measurements from the Galathea 3 expedition of (graph at left): temperature (light grey), oxygen (black) and photosynthetic available radiation (PAR) (dark grey) in the Benguela Current upwelling area in the South Atlantic (24° 9.05′ S; 13° 17.96′ E; 12–10–2006 08:41 UTC). Primary production is limited to the upper 40 m due to the rapid attenuation of PAR (dashed line). Below the euphotic zone, heterotrophic respiration regulates the cycling of organic carbon. Concentrated material from 11–30 litres of water sampled at 100, 200 and 300 m depths (graphs on right) was incubated at 5 °C below (dark grey) and above (light grey) *in situ* temperature for about 25 days. Increased respiration was found in all incubations (made in triplicate) when stored at higher temperatures. Source: Bendtsen *et al.* (submitted); data from Galathea 3 expedition.

Box 4.3 (*cont.*)

a general increase in upper ocean temperature during the past 50 years (Box 1.3; Levitus *et al.*, 2005), and coupled data and model-based basin-scale analysis indicate that temperature has increased by 0.1–0.2 °C in the upper 500 m of the South Atlantic since 1960 (Barnett *et al.*, 2005). This temperature change only increases remineralisation rates by 1%–2% (Q_{10} = 3) and therefore only minor changes associated with changed respiration patterns in the ocean interior have occurred. However, when considering results from climate change simulations suggesting a relatively large temperature change of about 3 °C in the upper ocean by the end of this century (IPCC, 2007), the influence from changed respiration rates in the mesopelagic becomes significant. A 3 °C temperature increase corresponds to an increased remineralisation rate of about 40%. Such significant increases in the remineralisation rate would impact upon the cycling of carbon in the upper mesopelagic zone through (i) sinking particulate organic carbon, which would be respired faster and thereby resulting in less material being exported to the deep ocean, (ii) increased cycling of carbon and nutrients which, in turn, would increase oxygen demand in the upper mesopelagic zone, and (iii) increased acidification (Box 6.3) due to elevated upper-ocean dissolved inorganic carbon concentrations. Model sensitivity studies of the influence from remineralisation depth on the air–sea CO_2 exchange show that decreased organic matter transport to the deep ocean decreases the ocean CO_2 uptake significantly (Schneider *et al.*, 2008). Corresponding model calculations of the impact from relatively small changes in the global remineralisation depth of about 24 m resulted in changes in atmospheric pCO_2 levels of 10–27 p.p.m. (Kwon *et al.*, 2009).

Oxygen-minimum zones are characterised by O_2 levels less than 120 µM, where higher life-forms become stressed or die. Therefore, O_2 levels directly influence ecosystem structure in the large oxygen-minimum zone areas of the world ocean, e.g. high productive equatorial or coastal upwelling areas. Increased respiration due to higher ocean temperatures would reduce oxygen levels and thereby contribute to the current expansion of oxygen-minimum zones in the world ocean (Stramma *et al.*, 2008).

The future of the land sink is subject to at least as much uncertainty as the ocean sink. Model simulations using dynamic global vegetation models driven by climate scenarios produce a very wide range of results, ranging from an additional 20 p.p.m. of CO_2 in the atmosphere by 2100 to about 200 p.p.m., with most of the models simulating an addition of between 50 p.p.m. and 100 p.p.m. CO_2 (Friedlingstein *et al.*, 2006). The major factors in these model simulations affecting the behaviour of the land sink are the change in heterotrophic respiration (oxidation of soil carbon) as temperature and precipitation change, and the dieback in tropical forests – particularly in the Amazon Basin – as the climate warms and dries in some of those regions. Increased disturbance such as fires also plays a role in the dieback of tropical forests. Business-as-usual climate scenarios suggest that large-scale dieback of the Amazon rainforest due to drought stress is possible (Cook and Vizy, 2008). Other model results suggest that nitrogen limitation may constrain terrestrial primary production and thus limit the size of the land sink (Wang *et al.*, 2007).

Observations of the behaviour of regional carbon sinks in response to present-day climate, including extreme events, provide some evidence with which to test the model projections. For example, the carbon sink in the Canadian boreal forest has weakened significantly in recent decades as the climate has warmed, primarily due to an increase in disturbances such as insect outbreaks and wildfires (Kurz and Apps, 1999; Kurz *et al.*, 2008a, b). The extreme heat wave in central Europe in 2003 affected the carbon sink there through tree mortality and, as a legacy, has left a weakened sink for subsequent years, as the forests take considerable time to recover (Ciais *et al.*, 2005). The dynamics of tropical forests have changed with increasing temperature, showing increasing productivity but also higher mortality (Hilbert *et al.*, 2009). In addition, the severe 2005 drought in Amazonia changed the area into a significant carbon source instead of its usual sink behaviour (Phillips *et al.*, 2009). These observations suggest, as the models simulate, an increasing vulnerability of tropical sinks to drought. Increases in the rate of deforestation of any of the major tropical forest areas in the Amazon, central Africa and Indonesia would also lead to a relative weakening of the land sink (Chapin *et al.*, 2008).

4.3 Possible new sources of carbon emissions to the atmosphere

In addition to changes in the proportional strength of the ocean and land carbon sinks, the activation of hitherto inactive pools of carbon may also affect the global carbon budget (Field *et al.*, 2007). Even less is known about these possible new sources of carbon emissions than about the behaviour of existing sources and sinks, so any potential new sources are not yet included in analyses of emission reduction targets and timetables (Chapter 8) or in coupled carbon cycle–climate modelling.

Most of these possible new emissions of carbon involve CH_4, either as CH_4 clathrates stored in marine sediments on continental shelves or potential CH_4 emissions from carbon stored in anaerobic environments in terrestrial ecosystems (either in tropical peatlands or in frozen soils (permafrost) in the northern high latitudes).

An abrupt, catastrophic emission of CH_4 clathrates from marine sediments is one of the possible explanations for the spike in global temperature at the Paleocene–Eocene Thermal Maximum (PETM) about 65 million years ago (Lenton *et al.*, 2008). However, the base temperature and atmospheric CO_2 concentration were much higher around the time of the PETM than now, or predicted for the next century at least, so the marine clathrates are thought to be stable under warming scenarios for this century (Chapter 7; Krey *et al.*, 2009). Thus, most of the concern about the vulnerability of new carbon pools focuses on the terrestrial environments.

Tropical peatlands are undoubtedly vulnerable to change this century, as they are most affected by human processes such as forest clearing and drainage rather than to a changing climate. These environments contain up to 70 Gt C (Page *et al.*, 2002, 2004); by comparison, anthropogenic emissions are currently about 10 Gt C yr^{-1}.

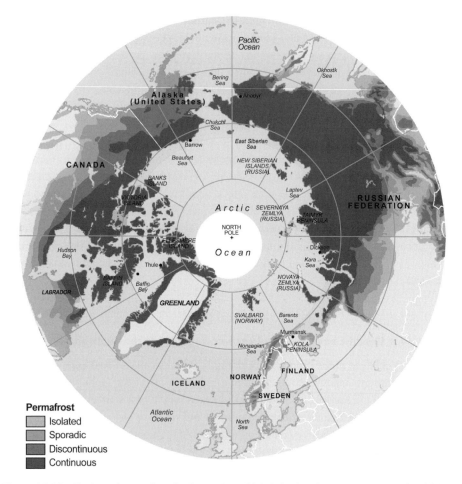

Figure 4.8 Distribution of permafrost in the northern high latitudes. Source: Brown *et al.* (1998).

Tropical peatlands, however, pale into insignificance compared to the amount of carbon stored in permafrost (Figure 4.8). The most recent estimates (Tarnocai *et al.*, 2009) report that about 1650 Gt C is stored in permafrost, about double the previous estimate. This amounts to about half the estimated total below-ground organic carbon pool on the planet, and is about twice the amount of carbon currently in the atmosphere. The processes by which the carbon stored in permafrost can be transferred to the atmosphere as the permafrost thaws are not straightforward (Schuur *et al.*, 2008). Depending on whether the carbon is decomposed under aerobic or anaerobic conditions, it can be emitted as either CO_2 or CH_4, respectively. In addition, the thawing of permafrost can occur slowly over a long period of time, or permafrost can be lost abruptly, leading to rapid releases of greenhouse gases. In fact, there is some concern that the higher end of IPCC warming scenarios could lead to a runaway collapse of the soil carbon pool in the northern high latitudes, thus acting as a planetary tipping element (Chapter 7; Khvorostyanov *et al.*, 2008a, b).

As the climate warms, there are also some ecological processes in the northern high latitudes that lead to a net uptake of carbon from the atmosphere; these include higher productivity of existing vegetation and the migration of boreal forests northward to replace tundra ecosystems. However, these processes appear to be too slow to significantly counteract the potential loss of large amounts of CH_4 or CO_2 from melting permafrost over short timeframes. Thus, the most likely scenario is a net increase in carbon-based greenhouse gases from the northern high latitudes as the climate continues to warm.

Thawing of permafrost has begun in some parts of the far north already (IPCC, 2007), coincident with the recently observed increase in atmosphere CH_4 concentration (Figure 4.9). This latest increase cannot be attributed clearly to any other known source of CH_4, such as a significant increase in wildfires, and so could signal the onset of increased loss of CH_4 from thawing permafrost. Some recent observations suggest an increase in CH_4 emissions over 2003–07 from temperate and high latitude wetlands (Bloom *et al.*, 2010).

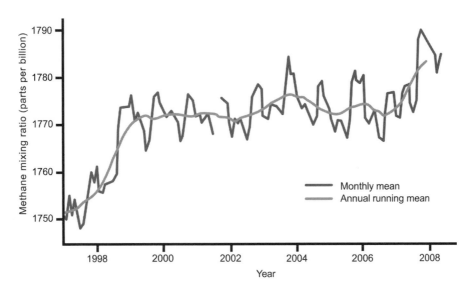

Figure 4.9 The global-average surface mixing ratio of methane from 1997 to 2008, showing a sharp increase in 2007. Source: M. Rigby, Massachusetts Institute of Technology, based on data from the Advanced Global Atmospheric Gases Experiment (AGAGE) and Australian Commonwealth Scientific and Industrial Research Organisation (CSIRO) monitoring networks. Original figure from Rigby *et al.* (2008).

A summary of the future of the terrestrial carbon sink, including the possibility of new sources of carbon such as from the melting of permafrost, is presented in more detail in Box 4.4.

Box 4.4
Future of the terrestrial carbon sink

JOSEP G. CANADELL AND MICHAEL R. RAUPACH

The terrestrial CO_2 sink is responsible for removing the equivalent of 30% of all CO_2 emissions from human activities, with a further 25% being removed by the ocean sink (Canadell *et al.*, 2007; Le Quéré *et al.*, 2009). These fractions have changed only slowly over the period 1959–2008. The magnitude of the terrestrial sink has increased together with anthropogenic CO_2 emissions over this period, reaching an average of 3.0±0.2 Gt C yr^{-1} over 2000–08 (Le Quéré *et al.*, 2009). Most models project continued growth during the first part of this century (Friedlingstein *et al.*, 2006; Sitch *et al.*, 2008). A failure in achieving this growth would lead to faster CO_2 accumulation in the atmosphere and therefore larger mitigation requirements to achieve a given CO_2 stabilisation target.

Current Earth System models attribute most of the observed and future growth of the terrestrial sink to the CO_2 fertilisation effect on plant photosynthesis. Although this is based on robust knowledge of plant physiology, a number of field experiments have failed to show higher plant biomass at elevated CO_2 (Mooney *et al.*, 1999; Körner, 2006; Norby *et al.*, 2009), demonstrating the non-universality of the response. The IPCC has assigned the highest uncertainty to the CO_2 fertilisation effect when assessing the magnitude of carbon–climate feedback (Denman *et al.*, 2007). The uncertainty also includes the possibility of a bigger CO_2 sink through the synergistic effect of elevated CO_2, and higher nitrogen availability through increased mineralisation of soil organic nitrogen under warmer conditions.

In addition to sink processes, there are a number of carbon pools and processes vulnerable to future climate change and land use that would lead to higher carbon fluxes to the atmosphere and therefore a decreased terrestrial sink. The decomposition of currently frozen organic matter in northern regions is considered one of the most important vulnerabilities. A recent study estimated that 1650 Gt C is buried in frozen organic soil carbon in the northern circumpolar permafrost region (Tarnocai *et al.*, 2009). This is about double the carbon content in the atmosphere and a much larger estimate than previously published because of the inclusion of carbon in the vast yedoma sediments of Eastern Siberia and northern river deltas, such as the Yukon (Alaska), Mackenzie (Canada) and Lena (Russia). Although a large amount of frozen carbon is not a sufficient condition for potential vulnerability, climate models show that up to 90% of the near-surface permafrost can thaw with the A2-IPCC emission scenario (Lawrence *et al.*, 2005). Estimates based on models and field measurements suggest that between 50 and 100 Gt C could be released from thawing permafrost during this century (Zhuang *et al.*, 2006; Khvorostyanov *et al.*, 2008; Schuur *et al.*, 2009). The finding that up to 80% of all carbon respired from thawed soils in Alaska comes from old organic matter (thousands of years old) shows the high sensitivity of this process to rising temperatures (Schuur *et al.*, 2009).

Of particular importance is the possibility that permafrost thawing, and associated carbon emissions, could be an irreversible process. Experimental and modelling work both show the possibility that thawing could become a self-sustaining process driven by the heat generated in microbial decomposition. This would lead to irreversible carbon emissions over hundreds of years, independent of the human-induced trajectory of warming (Zimov *et al.*, 2006; Khvorostyanov *et al.*, 2008).

There are also large deposits of carbon stored in wetlands under low-oxygen conditions that prevent decomposition. It is estimated that cold regions store about 400 Gt C and tropical regions about 70 Gt C (Sabine *et al.*, 2004; Jaenicke *et al.*, 2008). Human activities including drainage, deforestation and burning are the main drivers of significant fluxes from these carbon stores to the atmsophere. In South-East Asia alone, it is estimated that over the next 60 years tropical peatlands will emit about 30 Gt C from peat decomposition (10 Gt C) and fires (20 Gt C) (Hooijer *et al.*, 2009).

There are other potential significant vulnerabilities but large uncertainties exist. These include the partial savanisation of the Amazon forest due to declined precipitation brought about by climate change (Cox *et al.*, 2000; Nobre and De Simone, 2009) or the decline in sinks in temperate regions due to increasing temperatures as shown during the 2003 heatwave in Europe (Ciais *et al.*, 2005). Finally, little is known about the vast deposits of methane hydrates in continental shelves and permafrost regions (estimates range between 1000 and 10 000 Gt C). Although warming may threaten a relatively small fraction of this store in permafrost regions, resource extraction for additional gas supply could become the biggest threat unless geosequestration is part of the technology (Krey *et al.*, 2009).

4.4 Implications of changing sources and sinks of carbon for projections of climate change

The long-term nature of committed climate change based on anthropogenic emissions of CO_2 over this century and beyond is well established (IPCC, 2007; Solomon *et al.*, 2009). Substantial commitments for both elevated temperatures (centuries) and sea-level rise (a millennium or more) exist after a stabilisation of atmospheric greenhouse gas concentrations. Thus, warming is virtually irreversible for at least 1000 years after emissions cease. Further modelling work using Earth System Models of Intermediate Complexity (EMICs) have explored the long-term dynamics of the climate system if coupled carbon cycle–climate feedbacks are taken into account (Box 1.5; Plattner, 2009). The results show that emission reductions have to be substantially larger than currently envisaged to limit climate change to a given target level because of the 'slow feedbacks' to climate, such as changing vegetation distribution and the loss of ice from the large polar ice sheets. Uncertainties related to the carbon cycle are substantial, in addition to the well-known uncertainties that surround climate sensitivity.

Unlike these EMIC studies, in many model studies that couple the carbon cycle and climate over shorter timeframes, the most important relationships that link the carbon cycle to climate in terms of terrestrial ecosystems are the effects of changing temperature and moisture on productivity and respiration. In particular, positive feedbacks usually result from a decrease in productivity accompanied by an increase in carbon release from higher respiration rates. As noted above, however, substantial uncertainties relate to the carbon cycle model, especially if extended to longer timescales. A synthesis of experimental evidence has explored some of these uncertainties, showing that the observed responses of ecosystems to warming are often much more complicated than a direct productivity–respiration balance (Luo, 2007). The productivity and respiration responses are often overridden by other processes, such as changes in species composition, phenology, length of growing season, nutrient dynamics and ecohydrology. This has implications for the level of sophistication of the dynamic vegetation model component used in coupled carbon cycle–climate model experiments.

Another EMIC-based modelling study (Eby *et al.*, 2009) suggests that extremely long timescales may be associated with the oceanic carbon sink. The results of this study show that the total amount of future carbon emissions strongly influences the time required for the oceans to completely absorb the anthropogenic CO_2. That is, for total emissions close to the currently known level of fossil fuel reserves (nearly 4000 Gt of carbon; IPCC, 2007), the time to absorb just half of the emitted CO_2 is more than 2000 years. Given the logarithmic relationship between atmospheric temperature and CO_2 concentration, the maximum temperature perturbation will persist much longer than the lifetime of anthropogenic CO_2. In fact, two thirds of the maximum temperature perturbation may persist for 10 000 years or longer.

4.5 Summary and conclusions

Humanity's most important lever in influencing the climate system is via the carbon cycle, so understanding the behaviour of the cycle – from the changing nature of the human emissions of carbon compounds to the natural processes that move carbon between the land, atmosphere and ocean – is crucial to be able to intervene in the carbon cycle in responsible and effective ways. Research on the carbon cycle has intensified over the past few decades, yielding important insights into current trends and future possibilities.

- The rate of anthropogenic emissions of CO_2 to the atmosphere increased sharply around 1950 and again after 2000. The latter increase is reflected in a jump in the atmospheric CO_2 growth rate from 1.5 p.p.m. per year in the 1980–2000 period to 2.0 p.p.m. per year in the 2000–07 period.
- Natural carbon sinks in the ocean and on land continue to absorb slightly over half of anthropogenic CO_2 emissions, although the ocean sink has shown a slight proportional weakening over the past several decades.
- Model-based estimates of the magnitude of carbon cycle feedbacks to a changing climate – both positive (reinforcing change) and negative (slowing change) – show a net positive feedback, accelerating climate change at 2100 by increasing atmospheric CO_2 concentration by 20%–30% over what it would otherwise have been (range: 10%–50%).

- There are several potential new natural sources of carbon emissions to the atmosphere that could be activated by a warming climate; the most significant of these that could be activated this century is probably the carbon stored in permafrost in the northern high latitudes.
- Simulation of carbon cycle–climate links over very long timeframes has confirmed that the human perturbation of the climate system will be appreciable for millennia into the future, and thus irreversible in any societally relevant timeframe.

References

Barnett, T. P., Pierce, D. W., AchutaRao, K. M. *et al.* (2005). Penetration of human-induced warming into the world's oceans. *Science*, **309**, 284–87.

Bendtsen, J., Hilligsøe, K. M., Hansen, J. L. S. and Richardson, K. (submitted). Temperature sensitive remineralisation rates of organic matter in the mesopelagic zone.

Bloom, A. A., Palmer, P. I., Fraser, A., Reay, D. S. and Frankenberg, C. (2010). Large-scale controls of methanogenesis inferred from methane and gravity spaceborne data. *Science*, **327**, 322–25.

Brown, J., Ferrians, O. J. Jr,, Heginbottom, J. A. and Melnikov, E. S. (1998). *Circum-Arctic Map of Permafrost and Ground-ice Conditions*. Revised February 2001. Boulder, CO: National Snow and Ice Data Center/World Data Center for Glaciology.

Butler, J. (2009). *The NOAA Annual Greenhouse Gas Index (AGGI)*, http://www.esrl.noaa.gov/gmd/aggi.

Canadell, J. G., Le Quéré, C., Raupach, M. R. *et al.* (2007). Contributions to accelerating atmospheric CO_2 growth from economic activity, carbon intensity, and efficiency of natural sinks. *Proceedings of the National Academy of Sciences (USA)*, **104**, 18866–70.

Canadell, J. G., Raupach, M. R. and Houghton, R. A. (2009). Anthropogenic CO_2 emissions in Africa. *Biogeosciences*, **6**, 463–68.

Chapin III, F. S., Randerson, J. T., McGuire, A. D., Foley, J. A. and Field, C. B. (2008). Changing feedbacks in the climate-biosphere system. *Frontiers in Ecology and the Environment*, **6**, 313–20.

Ciais, P., Reichstein, M., Viovy, N. *et al.* (2005). Europe-wide reduction in primary productivity caused by the heat and drought in 2003. *Nature*, **437**, 529–33.

Cook, K. H. and Vizy, E. K. (2008). Effects of 21st century climate change on the Amazon rainforest. *Journal of Climate*, **21**, 542–60.

Cox, P. M., Betts, R. A., Jones, C. D., Spall, S. A. and Totterdell, I. J. (2000). Acceleration of global warming due to carbon-cycle feedbacks in a coupled climate model. *Nature*, **408**, 184–87.

Crutzen, P. J. (2002). Geology of mankind: The Anthropocene. *Nature*, **415**, 23.

Crutzen, P. J. and Stoermer, E. F. (2000). The 'Anthropocene'. *Global Change Newsletter*, **41**.

Denman, K. L., Chidthaisong, G. B. A., Ciais, P. *et al.* (2007). Couplings between changes in the climate system and biogeochemistry. In *Climate Change 2007: The Physical Science Basis. Contribution of Working Group I to the Fourth Assessment Report of the Intergovernmental Panel on Climate Change*, eds. S. Solomon, D. Qin, M. Manning, Z. Chen, M. Marquis, K. B. Averyt, M. Tignor and H. L. Miller. Cambridge, UK and New York, NY: Cambridge University Press, pp. 499–587.

Dunne, J. P., Armstrong, R. A., Gnanadesikan, A. and Sarmiento, J. L. (2005). Empirical and mechanistic models for the particle export ratio. *Global Biogeochemical Cycles*, **19**, GB4026.

Eby, M., Zickfeld, K., Montenegro, A. *et al.* (2009). Lifetime of anthropogenic climate change: Millennial time scales of potential CO_2 and surface temperature perturbations. *Journal of Climate*, **22**, 2501–11.

Etheridge, D. M., Steele, L. P., Langenfelds, R. L. *et al.* (1996). Natural and anthropogenic changes in atmospheric CO_2 over the last 1000 years from air in Antarctic ice and firn. *Journal of Geophysical Research*, **101**, 4115–28.

Falkowski, P., Scholes, R. J., Boyle, E. *et al.* (2000). The global carbon cycle: A test of our knowledge of Earth as a system. *Science*, **290**, 291–96.

Field, C. B., Behrenfeld, M. J., Randerson, J. T. and Falkowski, P. (1998). Primary production of the biosphere: Integrating terrestrial and oceanic components. *Science*, **281**, 237–40.

Field, C. B., Lobell, D. B., Peters, H. A. and Chiariello, N. R. (2007). Feedbacks of terrestrial ecosystems to climate change. *Annual Review of Environment and Resources*, **32**, 1–29.

Friedlingstein, P., Cox, P., Betts, R. *et al.* (2006). Climate-carbon cycle feedback analysis: Results from the C4MIP model intercomparison. *Journal of Climate*, **19**, 3337–53.

Goodwin, P., Williams, R. G., Ridgwell, A. and Follows, M. J. (2009). Climate sensitivity to the carbon cycle modulated by past and future changes in ocean chemistry. *Nature Geoscience*, **2**, 145–50.

Gruber, N., Friedlingstein, P., Field, C. B. *et al.* (2004). The vulnerability of the carbon cycle in the 21st century: an assessment of carbon–climate–human interactions. In *The Global Carbon Cycle: Integrating Humans, Climate, and the Natural World*, eds. C. B. Field and M. R. Raupach. Washington, D.C.: Island Press, pp. 45–76.

Hilbert, D. W., Canadell, J. G., Metcalfe, D. and Bradford, M. (2009). New observations suggest vulnerability of the carbon sink in tropical rainforests. *IOP Conference Series: Earth and Environmental Science*, **6**, 042003.

Hofmann, M. and Schellnhuber, H.-J. (2009). Ocean acidification affects marine carbon pump and triggers extended marine oxygen holes. *Proceedings of the National Academy of Sciences (USA)*, **106**, 3017–22.

Hooijer, A., Page, S., Canadell, J. G. *et al.* (2009). Current and future CO_2 emissions from drained peatlands in Southeast Asia. *Biogeosciences Discussions*, **6**, 7207–30.

IMF (2009). *World Economic Outlook: Sustaining the Recovery*. Washington, D.C.: International Monetary Fund.

Intergovernmental Panel on Climate Change (IPCC) (2007). *Climate Change 2007: The Physical Science Basis. Contribution of Working Group I to the Fourth Assessment Report of the Intergovernmental Panel on Climate Change*, eds. S. Solomon, D. Qin, M. Manning, M. Marquis, K. Averyt, M. M. B. Tignor, H. L. Miller Jr and Chen, Z., Cambridge, UK and New York, NY: Cambridge University Press.

Jaenicke, J., Rieley, J. O., Mott, C., Kimman, P. and Siegert, F. (2008). Determination of the amount of carbon stored in Indonesian peatlands. *Geoderma*, **147**, 151–58.

Keeling, C. D. and Whorf, T. P. (2000). Atmospheric CO_2 records from sites in the SIO air sampling network. In *Trends: A Compendium of Data on Global Change*. Oak Ridge, TN: Carbon Dioxide Information Analysis Center, Oak Ridge National Laboratory, US Department of Energy.

Khatiwala, S., Primeau, F. and Hall, T. (2009). Reconstruction of the history of anthropogenic CO_2 concentrations in the ocean. *Nature*, **462**, 346–49.

Khvorostyanov, D. V., Krinner, G., Ciais, P., Heimann, M. and Zimov, S. A. (2008a). Vulnerability of permafrost carbon to global warming. Part I Model description and role of heat generated by organic matter decomposition. *Tellus*, **60**, 250–64.

Khvorostyanov, D. V., Ciais, P., Krinner, G. *et al.* (2008b). Vulnerability of permafrost carbon to global warming. Part II: sensitivity of permafrost carbon stock to global warming. *Tellus*, **60B**, 265–75.

Knorr, W. (2009). Is the airborne fraction of anthropogenic CO_2 emissions increasing? *Geophysical Research Letters*, **36**, L21710.

Körner, Ch. (2006). Plant CO_2 responses: An issue of definition, time and resource supply. *New Phytologist*, **172**, 393–411.

Krey, V., Canadell, J. G., Nakicenovic, N. *et al.* (2009). Gas hydrates: entrance to a methane age or climate threat? *Environmental Research Letters*, **4**, 034007.

Kurz, W. A. and Apps, M. J. (1999). A 70-year retrospective analysis of carbon fluxes in the Canadian forest sector. *Ecological Applications*, **9**, 526–47.

Kurz, W. A., Dymond, C. C., Stinson, G. *et al.* (2008a). Mountain pine beetle and forest carbon feedback to climate change. *Nature*, **452**, 987–90.

Kurz, W. A., Stinson, G., Rampley, G. J., Dymond, C. C. and Neilson, E. T. (2008b). Risk of natural disturbances makes future contribution of Canada's forests to the global carbon cycle highly uncertain. *Proceedings of the National Academy of Sciences (USA)*, **105**, 1551–55.

Kwon, E. Y., Primeau, F. and Sarmiento, J. L. (2009). The impact of remineralization depth on the air–sea carbon balance. *Nature Geoscience*, **2**, 630.

Lawrence, D. M. and Slater, A. G. (2005). A projection of severe near-surface permafrost degradation during the 21st century. *Geophysical Research Letters*, **32**, L24401.

Le Quéré, C., Rodenbeck, C., Buitenhuis, E. T. *et al.* (2007). Saturation of the Southern Ocean CO_2 sink due to recent climate change. *Science*, **316**, 1735–38.

Le Quéré, C., Raupach, M. R., Canadell, J. G. *et al.* (2009). Trends in the sources and sinks of carbon dioxide. *Nature Geoscience*, **2**, 831–36.

Lenton, T. M., Held, H., Kriegler, E. *et al.* (2008). Tipping elements in the Earth's climate system. *Proceedings of the National Academy of Sciences (USA)*, **105**, 1786–93 (supporting information).

Levitus, S., Antonov, J. and Boyer, T. (2005) Warming of the world ocean, 1955–2003. *Geophysical Research Letters*, **32**, L02604.

López-Urrutia, A., San Martin, E., Harris, R. P. and Irigoien, X. (2006). Scaling the metabolic balance of the oceans. *Proceedings of the National Academy of Sciences (USA)*, **103**, 8739–44.

Lovenduski, N. S., Gruber, N., Doney, S. C. and Lima, I. D. (2007). Enhanced CO_2 outgassing in the Southern Ocean from a positive phase of the Southern Annular Mode. *Global Biogeochemical Cycles*, **21**, GB2026.

Luo, Y. (2007). Terrestrial carbon-cycle feedback to climate warming. *Annual Review of Ecology, Evolution, and Systematics*, **38**, 683–712.

McConchie, P. (ed.) (2003). *Elders: Wisdom from Australia's Indigenous Leaders*. Cambridge University Press.

Mooney, H., Canadell, J., Chapin, F. S. *et al.* (1999). Ecosystem physiology responses to global change. In *The Terrestrial Biosphere and Global Change. Implications for Natural and Managed Ecosystems*, eds. B. H. Walker, W. L. Steffen, J. Canadell and J. S. I. Ingram. Cambridge, UK: Cambridge University Press, pp. 141–89.

Nobre, C. A. and De Simone, L. (2009). Tipping points for the Amazon forest. *Current Opinion in Environmental Sustainability*, **1**, 28–36.

Norby, R. J., Warren, J. M., Iversen, C. M. *et al.* (2009). CO_2 enhancement of forest productivity constrained by limited nitrogen availability. *Nature Proceedings*, http://hdl.handle.net/10101/npre.2009.3747.1.

Page, S. E., Siegert, F., Rieley, J. O. *et al.* (2002). The amount of carbon released from peat and forest fires in Indonesia during 1997. *Nature*, **420**, 61–65.

Page, S. E., Wust, R. A. J., Weiss, D. *et al.* (2004). A record of late Pleistocene and Holocene carbon accumulation and climate change from an equatorial peat bog (Kalimantan, Indonesia): implications for past, present and future carbon dynamics. *Journal of Quaternary Science*, **19**, 625–35.

Petit, J. R., Jouzel, J., Raynaud, D. *et al.* (1999). Climate and atmospheric history of the past 420,000 years from the Vostok ice core, Antarctica. *Nature*, **399**, 429–36.

Phillips, O.L., Aragão, L. E. O. C., Lewis, S. L. *et al.* (2009). Drought sensitivity of the Amazon rainforest. *Science*, **323**, 1344–47.

Plattner, G.-K. (2009). Long-term commitment of CO_2 emissions on the global carbon cycle and climate. *IOP Conference Series: Earth and Environmental Science*, **6**, 042008.

Raupach, M. R., Marland, G., Ciais, P. *et al.* (2007). Global and regional drivers of accelerating CO_2 emissions. *Proceedings of the National Academy of Sciences (USA)*, **104**, 10288–93.

Rigby, M., Prinn, R. G., Fraser, P. J. *et al.* (2008). Renewed growth of atmospheric methane. *Geophysical Research Letters*, **35**, L22805.

Sabine, C. L. and Tanhua, T. (2010). Estimation of anthropogenic CO_2 inventories in the ocean. *Annual Review of Marine Science*, **2**, 175–98.

Sabine, C. L., Feely, R. A., Gruber, N. *et al.* (2004). The oceanic sink for anthropogenic CO_2. *Science*, **305**, 367–71.

Sarmiento, J. L., Orr, J. C. and Siegenthaler, U. (1992). A perturbation simulation of CO_2 uptake in an ocean general circulation model. *Journal of Geophysical Research*, **97**, 3621–45.

Sarmiento, J. L., Hughes, T. M. C., Stouffer, R. J. and Manabe, S. (1998). Simulated response of the ocean carbon cycle to anthropogenic climate warming. *Nature*, **393**, 245–49.

Schneider, B., Bopp, L., Gehlen, M. *et al.* (2008). Climate-induced interannual variability of marine primary and export production in three global coupled climate carbon cycle models. *Biogeosciences*, **5**, 597–614.

Schuster, U. and Watson, A. J. (2007). A variable and decreasing sink for atmospheric CO_2 in the North Atlantic. *Journal of Geophysical Research*, **112**, C11006.

Schuur, E. A. G., Bockheim, J., Canadell, J. G. *et al.* (2008). Vulnerability of permafrost carbon to climate change: Implications for the global carbon cycle. *Bioscience*, **58**, 701–14.

Schuur, E. A. G., Vogel, J. G., Crummer, K. G. *et al.* (2009). The effect of permafrost thaw on old carbon release and net carbon exchange from tundra. *Nature*, **459**, 556–59.

Sigman, D. M. and Boyle, E. A. (2000). Glacial/interglacial variations in carbon dioxide: Searching for a cause. *Nature*, **407**, 859–69.

Sitch, S., Huntingford, C., Gedney, N. *et al.* (2008). Evaluation of the terrestrial carbon cycle, future plant geography and climate–carbon cycle feedbacks using five Dynamic Global Vegetation Models (DGVMs). *Global Change Biology*, **14**, 2015–39.

Solomon, S., Plattner, G.-K., Knutti, R. and Friedlingstein, P. (2009). Irreversible climate change due to carbon dioxide emissions. *Proceedings of the National Academy of Sciences (USA)*, **106**, 1704–09.

Stramma, L., Johnson, G. C., Sprintall, J. and Mohrholz, V. (2008). Expanding oxygen-minimum zones in the tropical oceans. *Science*, **320**, 655–58.

Steffen, W., Crutzen, P. J. and McNeill, J. R. (2007). The Anthropocene: Are humans now overwhelming the great forces of nature? *Ambio*, **36**, 614–21.

Stuiver, M., Quay, P. D. and Ostlund, H. G. (1983). Abyssal water carbon-14 distribution and the age of the world oceans. *Science*, **219**, 849–51.

Takahashi, T., Sutherland, S. C., Feely, R. A. and Wanninkhof, R. (2006). Decadal change of the surface water pCO_2 in the North Pacific: a synthesis of 35 years of observations. *Journal of Geophysical Research*, **111**, C07S05.

Tarnocai, C., Canadell, J. G., Schuur, E. A. G. *et al.* (2009). Soil organic carbon pools in the northern circumpolar permafrost region. *Global Biogeochemical Cycles*, **23**, GB2023.

Wallace, D. W. R. (2001). Ocean measurements and models of carbon sources and sinks. *Global Biogeochemical Cycles*, **15**, 3–11.

Wang, Y.-P., Houlton, B. Z. and Field, C. B. (2007). A model of biogeochemical cycles of carbon, nitrogen, and phosphorus including symbiotic nitrogen fixation and phosphatase production. *Global Biogeochemical Cycles*, **21**, GB1018.

Wohlers, J., Engel, A., Zöllner, E. *et al.* (2009). Changes in biogenic carbon flow in response to sea surface warming. *Proceedings of the National Academy of Sciences (USA)*, **106**, 7067–72.

Zhuang, Q., Melillo, J. M., Sarofim, M. C. *et al.* (2006). CO_2 and CH_4 exchanges between land ecosystems and the atmosphere in northern high latitudes over the 21st century. *Geophysical Research Letters*, **33**, L17403.

Zimov, S. A., Schuur, E. A. G. and Chapin III, F. S. (2006). Permafrost and the global carbon budget. *Science*, **312**, 1612–13.

Part II

Defining 'dangerous climate change'

Defining 'dangerous climate change' is an issue for societies to discuss and determine but the climate change research community is providing much more information to support this process...

5

The impact of climate change on human societies

> *'At many church services in Denmark during the world climate conference three special things will be brought into church and placed on the altar: a stone, a corn cob, and a fragment from a coral reef. The stone is from Greenland, the corn cob is from Malawi, and the coral from the Pacific Ocean ... All three things remind us that climate change caused by human activity affects living conditions around the world ...'*[1]

Scientists have now documented numerous anthropogenic influences on several components of the climate system and identified an overwhelming probability that the changing climatic conditions recorded on Earth over the past decades are primarily a response to this human interference of the climate system (Zhang *et al.*, 2007). All of this would, however, be of purely academic interest but for the fact that these changes in climate conditions influence the conditions for all life on Earth. Of particular and immediate interest to all concerned citizens of the world is, of course, *how does or will climate change affect me, my family and my community?*

Understanding the potential impacts of climate change on our societies is directly relevant to the question of what constitutes 'dangerous climate change'. As the magnitude and rate of climate change increases through this century, the consequences for us and our societies also escalate. At some point, the impacts of climate change will become too large and costly – economically, socially, environmentally – for societies to adapt. Determining what is dangerous climate change is a societal value judgement, but must ultimately be based on a risk assessment informed by knowledge of potential impacts, vulnerability and adaptive capacity.

This chapter examines the potential impacts of climate change on several key sectors of direct importance for human well-being, and concludes with insights on what proved to be dangerous, or not, for past societies that collapsed or transformed in the face of climatic shifts during their time.

---

[1] Rev. Anders Gadegaard, Dean, Danish Lutheran Evangelical Cathedral, Denmark, 13 December 2009 (with reference to the COP15 in Copenhagen).

Climate Change: Global Risks, Challenges and Decisions, Katherine Richardson, Will Steffen and Diana Liverman *et al.* Published by Cambridge University Press. © Katherine Richardson, Will Steffen and Diana Liverman 2011.

5.1 Agriculture and food security

About 10 000 years ago, when our ancestors discovered how to domesticate wild species to improve food supplies, the relationship between humans and the Earth changed dramatically. That humans now had the ability to, at least in part, control and increase the availability of their own food released a previous limitation on the numbers of our species that the planet could support. Clearly, there are many more humans on the planet today than would be possible if we still were hunter-gatherers. The flip side of this story, however, is that human societies are, today, highly dependent upon agricultural systems for their survival. Agricultural systems are influenced directly by climatic conditions and, thus, potentially vulnerable to climate change. Agriculture, in turn, is part of complex food systems that are also vulnerable – so that food security is threatened not only by climate impacts on crop yields, livestock and fisheries, but also by climate impacts on the processing, transport and consumption of food.

In some ways, agricultural systems are analogous to the climate system in that they are comprised of several different interacting components. Climate change can positively or negatively affect some or all of these components in ways that lead to both net increases or decreases in the overall productivity of the system. Among the components of agricultural systems that are directly or indirectly impacted by climate change are growth temperature, access to water, the ambient CO_2 concentration, plant–pathogen interactions and nutrient availability. As in the climate system, the interactions between these various components of agricultural systems are complex and they are influenced by many other factors than just climate. This makes it difficult to predict how agricultural systems will be influenced by climate change.

Individual countries and regions are currently conducting studies aimed at predicting the potential climate change effects on future food availability for their inhabitants. Not surprisingly, it appears that large countries/regions are less vulnerable to overall production losses than small ones. Thus, although a geographic redistribution of regions with the highest rice productivity is anticipated in China, the overall rice production there is expected to increase in response to climate change (Huang *et al.*, 2009). In contrast, along the Mekong delta, a substantial decrease in rice production is projected (Agarwal and Babel, 2009), which would negatively impact food security in the region. IPCC (2007a) suggested that some temperate regions would see increased production from initial warming, which might allow agriculture to extend growing seasons or expand poleward. As some regions will benefit and others lose in terms of agricultural productivity as a result of climate change, and because farmers can be very adaptable, there are uncertainties in how the overall global productivity of agricultural systems will be influenced by climate change at least over the next couple of decades. If temperatures rise above 2 °C, however, agriculture may meet limits to adaptation, and IPCC (2007a) suggests crop production would decline in both temperate and tropical regions. However, in terms of water availability, a general rule of thumb identified by the IPCC (2007b) concerning climate change and precipitation is that the areas of the Earth that currently are 'wet' will get wetter while those that are 'dry' will get drier, with the northern high latitudes becoming wetter everywhere.

This means that countries and regions where access to water is already a limiting condition for agriculture will be particularly vulnerable to climate change. Figure 5.1 illustrates among other things the regions where water shortages due to desiccation and drought are predicted to result from changes in the climate system. In many cases, these regions coincide with regions where food security is already an issue. Lobell *et al.* (2008) examined the projected climate change effect on staple crops from 12 different regions where food production is already a serious limitation for the local societies. Their study suggested that in a region such as central Africa the productivity of all major staples will decline. Owing to other factors (e.g. population growth), it has been suggested (Alcamo, 2009) that meeting Africa's food demands in 2050 will require a 2.5 to 3 times increase in agricultural productivity. Climate change will exacerbate insecurity in already tenuous food supply systems in regions like central Africa, and thus pose a serious threat to food security in this and other developing regions of the world (Box 5.1).

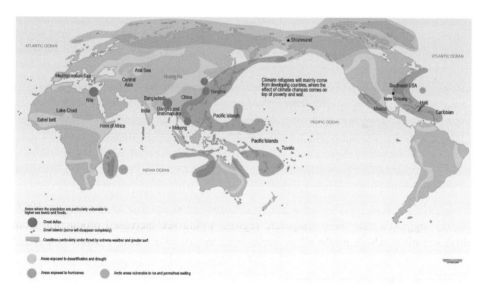

Figure 5.1 Regions predicted to be particularly vulnerable to different aspects of climate change, such as higher sea levels/floods, extreme weather, desertification and drought, ice/permafrost melting and hurricanes. Source: http://maps.grida.no/go/graphic/fifty-million-climate-refugees-by-2010. Reproduced with permission of UNEP/GRID-Arendal, cartography by Emmanuel Bournay. (In colour as Plate 14.)

Box 5.1
Climate change and food systems: the GECAFS approach

DIANA LIVERMAN, POLLY ERICKSEN AND JOHN INGRAM

Most research linking climate change and food security focuses solely on agriculture – either the impacts of climate change on agricultural production, or emissions and sequestration of greenhouse gases by croplands and livestock systems. But food security depends on much more than production, and is affected by other steps in the food system such as food processing, packaging, transporting and consumption – as well as by factors that influence access to good-quality food, including income and entitlements. Thus, emissions from transport and disruption of distribution by extreme events are linked to climate change. Food security is defined by the UN Food and Agriculture Organization (FAO) as 'when all people, at all times, have physical and economic access to sufficient, safe and nutritious food to meet their dietary needs and food preferences for an active and healthy life' (FAO, 1996). According to the FAO, almost a billion people are food insecure, with large populations of hungry people in Africa, India and China (FAO, 2009). The Global Environmental Change and Food Systems (GECAFS) program is one of the joint projects of the Earth System Science Partnership (ESSP; http://www.essp.org) and has used a food system approach to understand the links between global environmental changes – including climate – and food systems. Summarised in Figure 5.2, the approach examines the links between food system activities (production, processing, distribution, retailing and consumption) and outcomes that include food security and environmental conditions. Food security includes factors such as the nutritional and social value of food, affordability and allocation.

Climate change has the potential to transform the geography of food systems, especially the patterns and productivity of crop, livestock and fishery systems. Climate change poses risks to many aspects of food systems beyond production, including transport (e.g. through extreme weather) and food quality (e.g. spoilage at higher temperatures) with consequences for food security. If climate change alters crop mix across complete landscapes, there are also risks to the livelihoods of producers – as well as to cultural traditions and preferences, which are tied to regional varieties and diets. Food security also faces risks from changes in vulnerability that intersect with climate change, such as poverty associated with the loss of employment in climate-sensitive sectors or loss of livelihoods in natural disasters. All poor households that are net buyers of food (i.e. they consume more than they produce) are vulnerable to increases in food prices. The impacts of climate change on producers are also more complicated than some studies suggest – because prices often emerge from a global trading system where a decline in yields can lead to higher prices for those who can still produce. And it is important to remember that food security is not only about the amount of food, but also depends on the nutritional quality, safety and cultural appropriateness of foods (Ericksen, 2008b).

The complex relationships between global environmental change and food systems became clear in a series of regional studies and scenario exercises that showed the significance of shifts in global trade and economic policy, and the ways in which climate change in producing regions affected consumption in other parts of the world. For example, in the Caribbean, local

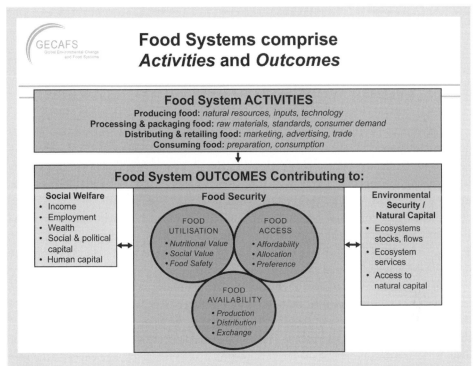

Figure 5.2 The GECAFS 'Food System' concept integrates a set of Food System Activities and the Outcomes of these activities. Source: GECAFS (2009), adapted from Ericksen (2008a).

stakeholders outlined their vulnerability to trade policy in Europe and to global production variations, in addition to local climate variability (Trotman *et al.*, 2009).

The sudden price rises in the cost of basic food commodities in 2007–08 increased interest in food systems and the possible role of climate, especially when these price increases undermined food security for poor populations who spend the bulk of their income on food. Some analysts blamed the price rises on climate change – especially drought in grain-producing areas such as Australia – and on policies to expand biofuels as an alternative to carbon-intensive fossil fuels, which were competing with food production. Although these were important, food system analyses showed a number of other key factors including increasing consumption of meat and dairy as incomes rose in Asia, a decline in food reserves traditionally used to stabilise prices, and speculation in food commodities during financial crises (FAO, 2008).

There are a number of challenges for research and policies that link climate change and food security. These include (i) the need to learn from past experience in using plant breeding to adapt to climate change, (ii) the importance of trade-offs (such as those between food and biofuels), (iii) the risk to cultural values of food as climate changes, and in designing responses, (iv) the links between employment in food systems and incomes needed to purchase food, (v) the potential of regional trading systems to reduce vulnerability to climate change,

---

Box 5.1 (*cont.*)

and (vi) the roles of non-state actors, including business and non-governmental organisations in managing food security in relation to climate change (Ingram *et al.*, 2010). Some of these challenges are being taken up by the new Climate Change Agriculture and Food Security joint program of the ESSP and CGIAR (http://www.ccafs.cgiar.org).

In later chapters, examples are given as to how farmers in some regions are already adapting to climate change – adjusting the crops they plant, altering their use of agricultural inputs or finding alternative employment. In addition, the international community is organising to consider how to deal with these threats to food security. For example, efforts are underway to develop climate change insurance programs for small-scale farmers in developing countries, which would provide index-based drought insurance where indemnity is paid out on the basis of the local rainfall index (Hochrainer *et al.*, 2009; IIASA, 2009). In this manner, poor farmers can obtain capital to invest in the seeds of more drought-resistant species and strains.

While most human food security is based on agricultural systems, there are some societies that are heavily dependent on hunting or fishing (Box 5.2) for the majority of their protein supply. The success of both hunters and fishers depends upon the distributions of their prey in the natural environment, and climate is an important factor in determining the abundance and distribution of species (Chapter 6). Thus, climate change can alter the very fundamentals of food security for societies dependent on hunting or fishing. Furthermore, changes in weather patterns and storm intensity (Chapter 2) can threaten the physical security of – especially – fishers. It is reported, for example, that small-scale coastal fishers in Bangladesh operating in the Bay of Bengal perceive that fishing has become more risky and costly due to increasingly turbulent weather. The monsoon season is the peak period for coastal fishing and also the period with the greatest storm activity. In addition to threatening human life, cyclones and sea storms threaten the infrastructure of small-scale fishers, i.e. boats, houses and nets. Fishing on species associated with coral reefs supports local communities in both the south Pacific and the Indian oceans. Thus, climate change-induced degradation of coral reefs (Box 6.4) will directly impact these communities.

Box 5.2

Climate change, fisheries and food security

STEPHEN J. HALL

Current climate projections strongly suggest that effects on coasts, lakes and rivers, and on the fisheries they support, will bring new challenges for these systems and to the people who depend upon them. Rising sea temperatures, changing sea levels, increasing ocean acidification, altered rainfall patterns and river flows, and increased incidence of extreme weather events will all take their toll (WorldFish, 2007). The inescapable conclusion is that we can expect often profound effects on the productivity, distribution and seasonality of fisheries, and the quality and availability of the habitats that support them.

As for many other sectors, climate-related impacts on fisheries, brought on primarily by the rich in the developed world, will be especially problematic for the poor in developing countries. Three statistics illustrate the importance of fisheries as sources of food, nutrition and income for these people. First, fish and other aquatic products provide at least 20% of protein intake for a third of the world's population. This dependence is generally highest in developing countries. Second, small-scale fisheries are by far the most important for food security. They supply more than half of the protein and minerals for over 400 million people in the poorest countries in Africa and South Asia. Third, fisheries and aquaculture directly employ over 36 million people worldwide, 98% of them in developing countries. They also indirectly support nearly half a billion more people as dependents or in ancillary occupations.

Figure 5.3 shows where the potential for climate-induced changes in fishery systems pose the greatest threat to food security. Out of the 33 countries identified as having economies that are highly vulnerable (Allison *et al.*, 2009), two have been assessed as alarmingly food insecure, 11 as extremely insecure and 11 as seriously insecure (von Grebmer *et al.*, 2009).

Figure 5.3 also highlights the Pacific Island states. For the most part, these are not currently food insecure and were not included in Allison *et al.*'s (2009) analysis. However, coral reef and tuna fisheries play a vital role in the food security and economies of these countries.

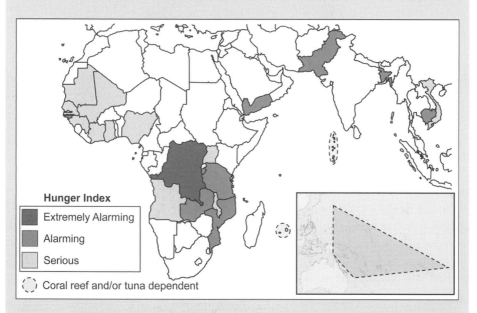

Figure 5.3 . Main figure: countries with national economies that are highly vulnerable to climate-induced changes in fisheries (Allison *et al.*, 2009) and that are ranked in the 'Serious Concern' category or above on the Global Hunger Index (von Grember *et al.*, 2009). Main figure and inset: shaded areas show other countries that are highly dependent on coral reef and tuna fisheries and likely to suffer food insecurity through climate change-induced effects on fisheries.

Box 5.2 (*cont.*)

To maintain 2009 fish consumption in 2030, even well-managed coastal fisheries operating at current production levels will only be able to meet food demands in 6 of 22 Pacific Island countries and territories (Bell *et al.*, 2008). Add to this the high likelihood of wholesale effects on coral reefs through bleaching and ocean acidification, and prospects look even bleaker.

The presence of several inland African countries in Figure 5.3 also shows that, although often forgotten when we think about fisheries, freshwater fish are essential to the food security of millions of the world's poorest people. As river flows become more unpredictable and water demands increase, diversion of water into irrigation and hydropower schemes is likely to profoundly affect both fish and fishers.

Given these prospects, vulnerable governments need to ask how scientifically guided adaptation of fishery and aquaculture management can sustain these vital sources of livelihoods and protein, and build resilience within the communities that depend on them. Policy for adaptation involves supporting measures to reduce exposure of fishing people to climate-related risks, reducing dependence of people's livelihoods on climate-sensitive resources, and increasing people's capacity to anticipate and cope with climate-related changes. Measures will need to include disaster response planning, mangrove rehabilitation, early warning systems, diversifying livelihood portfolios, improving fisheries management and governance systems, and many others.

Deciding which measures are appropriate for a given location can only be done by local fishers, fish farmers and others who understand the prevailing circumstances, but there are many lessons to share. The United Nations Framework Convention on Climate Change (UNFCCC), for example, provides access to a database of local adaptation measures that have been adopted.[2] We also need to increase our knowledge of the complex links between climate change, livelihoods and food security so that we can improve policies and management strategies for fisheries and aquaculture.

The impacts of natural disasters illustrate the ways in which climate change can affect food systems throughout the supply chains that link production to consumption. When climate variability brings drought or floods to vulnerable regions it can destroy not only crops and animals, but also infrastructure such as roads that is essential to the distribution of food. A loss of refrigeration or contamination of water can degrade the quality of food processed commercially or in the home, and increases in energy costs can make it difficult for households to cook foods. When crop failure results in loss of agricultural employment and incomes, increases in poverty mean that people cannot afford to buy food or decide to migrate, thus creating further risks to food security (Ericksen, 2008b).

5.2 Access to water resources

Climate change impacts on water systems are already observable, and many are likely to accelerate and intensify in the following decades irrespective of any political commitments

---

[2] http://maindb.unfccc.int/public/adaptation/

to reduce greenhouse gas emissions. One of the most notable climate change impacts on water systems is expressed via change in rainfall patterns (Boxes 2.2 and 2.4). Rainfall can be linked directly to economic status and growth of many societies in developing countries as both reductions (droughts) as well as excesses in rainfall (floods) have a significant impact on development. In a study focusing on the impacts of climate variability and water availability on the development of countries in Sub-Saharan Africa, annual data on rainfall and total and agricultural gross domestic product (GDP) growth rates from 1979 until 2001 were examined (Ludwig *et al.*, 2009). Climate variability here had a clear impact on GDP growth rates.

Most African countries outside the central tropical zone exhibited the highest total and agricultural GDP growth rates during years with average or slightly above average rainfall (Figure 5.4). The impact of climate variability on GDP growth was more pronounced during dry than during wet years. In below-average rainfall years, growth was severely reduced and, generally, the drier the year, the lower the GDP growth rate. In years with above-average rainfall, growth rates tended to be similar. If this same relationship between rainfall and GDP growth continues into the future, then for all African regions, except Central Africa, a climate with increased rainfall variability would reduce GDP growth.

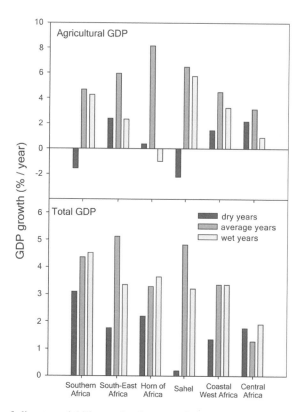

Figure 5.4 Impact of climate variability on development. Source: Ludwig *et al.* (2009).

While rainfall is the primary water source for developing countries in most of the world, populations in some mountainous regions in Asia and Andean South America depend on melt-water from glaciers for a major part of their water supply. Global warming presents an obvious and direct threat to these water supplies. The situation is particularly acute in South America, where Andean glaciers are small and receding rapidly now, but in terms of the size of the vulnerable population, is far more serious in Asia. Most of the great rivers in South and East Asia, on which billions of humans are dependent for much of their water resources, arise from the glaciers and ice sheets of the great Hindu Kush–Himalayan ranges. These hitherto reliable water resources are now of concern as the rate of loss of the great Asian glaciers increases (Box 5.3).

Box 5.3
A need for mountain perspectives: Impacts of climate change on ecosystem services in the greater Hindu Kush–Himalayan region

ROBERT J. ZOMER, EKLABYA SHARMA AND NAKUL CHETTRI

The Hindu Kush–Himalaya (HKH) region of Asia (approx. 4.3 million km²), with the highest and most remote mountains in the world, is comprised of various ranges spanning an arc from Afghanistan to southern China, and includes the whole of the Tibetan Plateau (Figure 5.5). This vast region is endowed with rich natural resources, and high levels of natural and agricultural biodiversity, providing a broad array of ecosystem services directly to the 200 million people living within the HKH, and indirectly to the 1.3 billion people living downstream (Eriksson *et al.*, 2009). The HKH is also one of the world's poorest regions, facing enormous challenges ranging from global change to rapid population growth, poor infrastructure, industrialisation, pollution, illiteracy and political instability. Despite the importance of the HKH region, climate change across this vast area is poorly understood (Xu *et al.*, 2007), sparsely monitored and generally under-researched (IPCC, 2007b; Eriksson *et al.*, 2009), so that there is great uncertainty regarding impacts, drivers and trends; their magnitude; and even the direction of change.

This region is sometimes referred to as the 'Third Pole', due to the great quantity of water stored in these glaciers (Qiu, 2008). There is over 116 000 km² of glacial ice covering these mountain regions (Owen *et al.*, 2002; Li *et al.*, 2008), the largest area of ice outside of the polar region. Many of these glaciers, though not all, are losing volume at rapid rates (Yao *et al.*, 2004; Barnett *et al.*, 2005; Nogues-Bravo *et al.*, 2007; Xu *et al.*, 2007). Studies in Nepal and China have shown that temperatures are rising at higher rates in high altitude areas than in other areas (Shrestha *et al.*, 1999; Liu and Chen, 2000). The Tibetan Plateau, in particular, has been shown to be warming at a rate three times the global average (Liu and Chen, 2000; Xu *et al.*, 2009). The observed warming in Nepal was 0.6 °C per decade between 1977 and 2000 (Shrestha *et al.*, 1999). In many higher altitude areas, a greater proportion of total annual precipitation appears to be falling as rain, rather than snow (Sharma *et al.*, 2009). As a result, snow-melt begins earlier and winters are shorter. This affects river regimes and impacts on water supply, agro-ecological adaptations and livelihoods, as well as causing natural disasters.

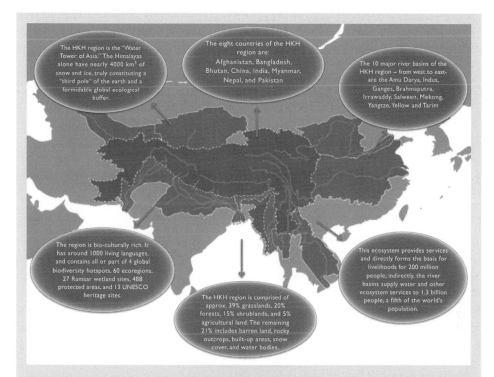

Figure 5.5 The HKH region and its river basins provide a broad array of ecosystem services for over 20% of the world's population. Source: Schild (2008). This figure was first published in *Mountain Research and Development* (MRD), vol. 28, no. 3/4, pp. 328–331. The rights of reproduction remain with the co-copyright holders: the International Mountain Society (IMS) and the United Nations University (UNU), c/o MRD Editorial Office, Bern, Switzerland (http://www.mrd-journal.org).

In addition, there are significant implications for biodiversity and conservation efforts (Myers *et al.*, 2000), as species' ranges may shift outside historical limits or existing protected areas set up to conserve those species. This is equally true for the finely adapted and highly diversified agricultural systems, and the high agricultural biodiversity (including livestock adapted to climatic conditions and nomadic herding patterns) upon which mountain subsistence communities depend for their food security (Zomer, 2008).

The livelihoods and sustenance of large numbers of people in the HKH region are dependent upon glacial melt-water for dry-season flows, pre-monsoonal plantings, and to avert catastrophic drought when the monsoon is weak, delayed or fails to materialise (Meehl, 1997). This is especially true for the drier regions of the HKH, notably the upper Indus basin. Changes in seasonality, temperature, rainfall, onset of monsoon rains and glacial melt-water will all have a profound effect on mountain farming systems, as well as on the forests, wetlands and other natural ecosystems of this region. In addition, the impact of increased temperatures and a reduction in water supplies from the HKH on downstream food production is of great concern

Box 5.3 (*cont.*)

(Battisti and Naylor, 2009), whether resulting from melting of snow and ice, or changes in the Asian Monsoon.

A better understanding of the impact of climate change in the HKH on food security, livelihoods and national economies of the region is urgently required (Schild, 2008). Systematic monitoring, analysis and modelling of ongoing change of the region's cryosphere and hydrology is weak, preliminary and fragmented by political boundaries. An improved understanding of the role of snow and ice in regional hydrological budgets, and the potential downstream hydrological impacts of projected change, is urgently required to inform regional planning and adaptation strategies, both within the HKH region and in the downstream basins.

Water scarcity brought on by climate change and increased demand is not a problem that is confined only to developing countries. Large regions in the developed world, such as the Mediterranean basin and the southwest USA, are also suffering from reduced water supplies. None has been impacted more dramatically than Australia, however. The most populous and agriculturally productive regions of the continent have suffered the most severe impacts. Urban water supplies are under pressure in the drying climate. Perth experienced an abrupt drop in inflow to its dams in the mid-1970s and possibly another drop at the beginning of the century (data from Western Australian Water Corporation, personal communication, 2009). The water supply for Melbourne, Australia's second largest city, has dropped precipitously from near full capacity in 1996 to well below 50% capacity in the 2006–08 period (data from Melbourne Water, personal communication, 2009). Both Perth and Melbourne are building desalination plants in response to the acute shortages.

Australia's largest river basin– the Murray–Darling, which supports the country's largest agricultural region – is also experiencing a severe drying trend. The average inflow into the river system dropped from a post-1950 average of 12 300 Gl yr^{-1} to 4150 Gl yr^{-1} in the 2000–07 period, and reached a record low of only 770 Gl yr^{-1} in the April 2006–March 2007 period (Cai and Cowan, 2008). The actual amount of water stored in the basin is now vastly below capacity (Figure 5.6). The abrupt drying trends that have reduced the Perth and Melbourne water supplies can be confidently attributed in large part to anthropogenic climate change, and the evidence is strengthening that a significant climate change signal can now be discerned in the drying of the Murray–Darling Basin (Steffen, 2009).

5.3 Impacts on human health

Human health is an integrator of all the stressors interacting on the individual and is, therefore, intricately related to food and water security. Undernourished or dehydrated individuals are weakened and less able to withstand other stressors, including those induced by climate change. Thus, many of the impacts of climate change on human health are indirect (Figure 5.7).

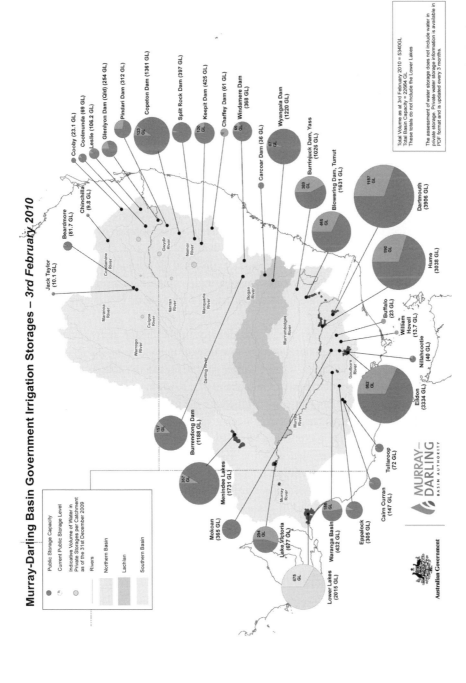

Figure 5.6 Actual storage of water (light grey wedges) compared to storage capacity (dark grey circles) in government dams in the Murray–Darling Basin, Australia, as at 3 February 2010. Source: data from Murray–Darling Basin Authority, Australian Government. © MDBA.

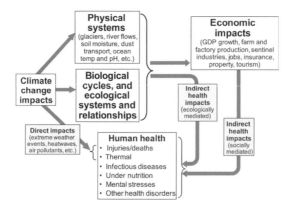

Figure 5.7 Diagram of impacts on human health related to natural disasters and climate change. Source: A. J. McMichael.

Even before temperatures reach extremes where mortality increases, high temperatures can have an effect on human well-being. For example, increased temperatures in the workplace have been linked to decreases in work productivity (Ebi *et al.*, 2005a,b), that is, workers are no longer able to perform their tasks or duties at the normal rate due to physiological stress caused by heat. It is becoming increasingly common for workplaces without air-conditioning to close when the temperature reaches a particular extreme value, often around 35 °C (Kjellstrom, 2000).

Another way in which climate change can directly impact health is by changing the distribution of organisms that either directly cause disease in humans or are vectors in the transmission of disease to humans. As an example of a change in the abundance of organisms directly threatening human health, Onozuka *et al.* (2010) have presented an analysis suggesting that warmer ambient temperatures increase the incidence of bacterially induced gastroenteritis. More attention has, however, been given to climate change-induced impact on the distributions of major vector-borne diseases such as malaria and schistosomiasis, (otherwise known as 'bilharzia' and which is a widespread parasitic disease transferred to humans via flukes hosted by freshwater snails). These diseases are discussed further in Box 5.4. While malaria and schistosomiasis are now confined primarily to the developing world, the distributions of vector-borne diseases in other parts of the world are also influenced by climate change. The spread of tick-borne encephalitis in Europe, for example, appears to be related to changing climate conditions (Rizzoli *et al.*, 2009).

Box 5.4
Risks to human health, present and future

ANTHONY J. MCMICHAEL AND ROBERTO BERTOLLINI

Risks to human health from climate change are many and varied; they will increase in future. Climate change affects human health both directly (e.g. by amplifying temperature extremes and weather disasters) and – of greater long-term concern – indirectly by changes to ecological and geophysical systems and associated social disruption. Particular sets of health risks affect different regions and populations. Some impacts emerge before others, and some are easier to detect than others.

Evidence of health impacts, to date, includes: (i) an upward trend over recent decades in deaths, injuries and other adverse health impacts from cyclones, storms, wild-fires, flooding; (ii) an increase in annual deaths from heatwaves in several regions; (iii) shifts in the range and seasonality of some climate-sensitive infectious diseases; (iv) adverse mental health consequences in farming communities affected by drying; and (vi) impairment of food yields and hence risk of malnutrition-related child development in some already food-insecure populations (Costello *et al.*, 2009).

Specific examples include the following.

- The two- to three-fold increase in average annual excess, heatwave-related, deaths over the past two decades in Hungary, accompanying a three-fold increase in the average annual frequency of heatwaves (Paldy, 2009).
- Biomathematical modelling of vector-borne disease transmission, which shows that a small temperature rise can significantly affect transmission (Pascual *et al.*, 2006). Hence, where malaria has increased alongside gains in climate-determined suitability for transmission – as in parts of eastern Africa recently – it is reasonable to assume that the climate trend partly explains the observed increases in disease.
- Northwards extension of the temperature-limited schistosomiasis transmission zone in eastern China, in association with warming since the 1960s, putting 21 million more persons at risk (Zhou *et al.*, 2008).
- The likelihood that up to two-thirds of the approximately 50 000 excess deaths during Europe's August 2003 heatwave were attributable to underlying regional warming. That proportion reflects the estimated warming-related doubling in probability of occurrence of such a severe heatwave (Stott *et al.*, 2004) – plus allowance for, first, the health impacts of coexistent elevated air pollution (itself due partly to warming) and second, the curvilinear temperature–mortality relationship.

Quantifying climate-attributable health impacts in humans is inherently difficult compared to attributing ice-sheet melting or earlier bird-nesting to associated changes in climate. The 'background noise' of cultural attributes and behaviours that influence most human health outcomes can readily obscure early climate-related health impacts. Counting road deaths due to speeding is much easier than quantifying under-nutrition (and consequent child stunting) due to climate change, or estimating mental health problems in rural communities facing warming, drying and declining incomes.

The first formal estimation of the burden of deaths and chronic disabling disease attributable to climate change (for the year 2000) focused on four selected climate-sensitive health outcomes: under-nutrition, malaria, diarrhoeal disease and flooding (McMichael *et al.*, 2004). Nearly all of the estimated 155 000 deaths per year from those four causes attributed to climate change (as it stood in the year 2000) were in low-income countries, especially sub-Saharan Africa and South Asia. The approximate annual numbers, mostly in children, were: malnutrition (~80 000 deaths), diarrhoea (~50 000), malaria (~20 000) and flooding (~3000). The Global Humanitarian Forum (Geneva) recently estimated that at least 300 000

Box 5.4 (*cont.*)

climate-attributable deaths occur annually in developing countries from those four causes (Human Impact Report, 2009).

Estimates of future health risks, for internationally agreed scenarios of greenhouse gas emissions, have been conducted at national and global levels. Examples include:

- substantial increases in the population at risk of malaria in Africa by 2015 (Hay *et al.*, 2005b) – as also illustrated by the modelled extension of the high-probability transmission zone for malaria (currently limited to the lowlands) to most of Zimbabwe by 2050 (Ebi *et al.*, 2005b);
- increased potential for transmission of dengue in much of sub-Saharan Africa, South Asia and South America during this century (Hales *et al.*, 2002);
- marked future increases in the population at risk of injury, death or other health disorders from flooding in the UK (Kovats, 2008).

Overall, the current and projected impacts of climate change on human health provide a clear and very important signal – a signal that underscores the stakes that humankind is *really* playing for. Further improvements and completeness in that 'signal', spanning the attributable 'body count' and the projection of likely future health impacts, will make even clearer the non-sustainable nature of the path down which we are currently, perilously, travelling.

Mitigation (true primary prevention by climate change abatement) and adaptation (a spectrum of health-risk management strategies) can reduce current and future climate-related disease burdens. Many mitigation (emissions-reduction) strategies by governments will also provide a relatively prompt, sometimes large, 'bonus' health benefit to their population via localised environmental enhancements: cleaner urban air; substantial elimination of traditional heavy indoor air pollution (from inefficient cook stoves – many of them in confined, unventilated spaces) in poor developing country populations; more walking and cycling in cities; and reduced production and consumption of animal-source foods (*The Lancet*, 2009).

Overall, the extent to which adverse health impacts actually occur as a result of climate change will depend on the timing and effectiveness of these two linked strategies.

Recently, a new human health concern related to climate change has been raised: The melting of Arctic ice in response to climate change may release the deposits of various toxic pollutants, such as mercury, to the surrounding marine environment. Some of these can, potentially, be concentrated in the marine food web prior to human consumption of marine organisms (Grandjean *et al.*, 2008). Such a possible climate change health effect would likely only have a local or regional impact on human health. Nevertheless, the example illustrates the intricate relationships in the Earth System that can be affected by climate change, and thus the possibility for far-reaching and, as yet, unrecognised impacts on human health.

5.4 Physical security (climate-related natural disasters)

Some studies suggest that climate change will result in an increasing probability of more frequent and/or intense extreme climate-related events, i.e. severe storms and droughts (Meehl *et al.*, 2007; Sun *et al.*, 2007). In addition, increases in sea level are inevitable in a warmer climate (Chapter 3). Intense storms alone, or in combination with increased sea level, threaten not only the physical security of individuals, but also infrastructure (buildings, power supply system, access to clean water, etc.) (Section 3.6). The prediction that an increase in climate-related natural disasters should occur in response to change in the climate system is now well established (Rosenzweig *et al.*, 2007). Indeed, there are now enough observations around the world of increases in extreme events to lend credence to this prediction (Box 5.5).

Box 5.5

How can we relate changes in extreme events to climate change?

MARTIN STENDEL

While much of the discussion concerning climate change is based on consideration of mean values (for example of temperature), the extremes are of considerable scientific and societal importance. In recent decades, increases in the numbers of extreme events, e.g. heat waves, heavy precipitation and flooding, storms (both tropical and extra-tropical), etc., have been observed (Trenberth *et al.*, 2007). The area affected by such events has also increased. A factor that has to be taken into account in relation to the observation of increases in extreme events is that technological improvements enable people in most parts of the worlds within a short period to take notice of such events. Thus, observation of extreme events from remote regions has become more accessible. There is also a dramatic increase in the damage related to these events, partly because the concentration of population and infrastructure in the affected regions like coasts or along rivers.

Obviously, one of the problems when dealing with extremes is that they are so extreme, that means that the statistical sample upon which one can draw is very small. One way to circumvent this problem is to consider the tails of the probability density function, that means that are exceeded for, for example, 1%, 5%, or 10% of the time.

Quite often extreme values of a quantity behave differently from mean values under external forcing. Furthermore, the tails of the distribution do not generally change similarly. As an example, daily minimum temperatures have increased about three times faster than daily maxima for several regions of the Earth (Beniston *et al.*, 1997; Easterling *et al.*, 1997; Jones *et al.*, 1999), and the number of cold days – no matter how exactly they are defined – has decreased faster than the number of hot days has increased. For several regions of the Earth, e.g. southern Europe, most climate models agree on a decrease of mean precipitation, but project at the same time an increase of heavy precipitation events, in particular for the most extreme percentiles (Christensen and Christensen, 2003).

Box 5.5 (*cont.*)

The fact that the statistical properties of extreme events behave differently from the mean value can, in part, be explained by the quantity under consideration.

- In some cases, behaviour of the quantity may change if a certain threshold is crossed. This is the case, for example, for SST. An increase in the area with SSTs above a certain level favours the development of strong hurricanes and, therefore, relates hurricane activity and anthropogenic climate change. While such a relationship for maximum wind speeds exists (Emanuel, 2005; Mann and Emanuel, 2006), the connection of the total number of Atlantic hurricanes or their overall intensity to increasing SSTs is disputed (Knutson *et al.*, 2008; Zhao *et al.*, 2009).
- Over Greenland, a huge increase in variability of temperature and precipitation and more extreme precipitation events are projected where sea-ice disappears and is replaced by open water (Stendel *et al.*, 2008).
- For mid-latitude and tropical rainfall, following Pall *et al.* (2007), the maximum precipitable amount of water is given by the Clausius–Clapeyron equation, which is much more directly related to temperature than is mean precipitation. In these cases, an increase in frequency and/or intensity of extreme events can directly be related to anthropogenic climate change.
- For other variables, like the northern hemisphere winter storm track activity, which also has experienced an increase during recent decades, the situation is less clear because these are related to interannual to intradecadal variability of the North Atlantic Oscillation.

The question is, then, whether individual extreme events can be explained by anthropogenic warming. This is normally not the case, as – per definition – extreme events also occur under stable climate conditions. Furthermore, extreme events such as floods or heat waves generally result from more than one factor. For example, following IPCC (2007b), several factors contributed to the record-breaking hot European summer of 2003 when a quasi-persistent high pressure system led to soil drying, which left more solar energy available to heat the land.

To address the question of whether anthropogenic activities have led to a change in frequency of extreme events, one usually follows a 'likelihood' approach, for example by running ensembles of simulations with a climate model including changes in both natural forcings only (volcanic activity, solar irradiation changes) and including changes in both natural and anthropogenic forcing. Many studies show that the climate evolution during the twentieth century cannot be reproduced with natural forcings alone, but only when anthropogenic forcings are included (Broccoli *et al.*, 2003; Meehl *et al.*, 2004; Stott *et al.*, 2006, see discussion in IPCC, 2007b) (see Figure 1.1). Based on these experiments, it has been estimated that human influences markedly increased the risk for a European summer as hot as 2003 and that without human influences, such an event would have had an estimated return period of thousands or tens of thousands of years (Schär *et al.*, 2004). The value of such a probability-based approach is that it can be used to estimate the influence of external factors (e.g. increase in greenhouse gas concentration) on the frequency of specific events (e.g. heat waves) via changes in the probability density function.

The climate has not only warmed, but is also becoming more variable in many regions. Fischer and Schär (2009) have decomposed changes in the uppermost percentiles in summer temperature over Europe for ten regional climate models into changes in interannual variability, changes in intraseasonal variability and changes in the seasonal cycle. The underlying processes differ for these components and are of different relative importance in individual models, but, generally, relate to stronger soil drying and, as a result, a reduction in cloudiness and an increased variability of surface radiation. Such variability increases are also predicted by climate models for future climate (Schär *et al.*, 2004). According to these studies, there is an increased risk of extreme summer temperatures, both due to warming on average and enhanced variability on all time scales. However, a careful statistical analysis is necessary, since such conclusions are derived from model-based estimates of climate variability, a quantity that is underestimated in several climate models.

The frequency and extent of large, intense wildfires have also increased over the past half-century, especially in the vast boreal forests of Canada and Russia and also in the western USA, in the region around the Mediterranean Sea and in eastern Australia (Westerling *et al.*, 2006; Le Page *et al.*, 2008). Such fires have become more probable with a warming climate, as there is a higher number of extreme fire weather days, and the forests and woodlands are often more prone to burning because they are drier due to higher evaporation rates (Bowman *et al.*, 2009). The incidence of extreme rainfall events and consequent flooding and landslides has also been increasing over the past several decades, consistent with the projections of climate models (IPCC, 2007a).

Economic losses from weather-related natural hazards are rising, averaging roughly US$100 billion per annum in the past decade (MunichRe, 2007). Losses to productive capacity in developing countries due to extreme weather events may be even larger (Linnerooth-Bayer *et al.*, 2009). Over the past 25 years, 95% of deaths from natural disasters occurred in developing countries. While these data include non-climate related disasters such as earthquakes, it is estimated by the United Nations International Strategy for Disaster Reduction (UNISDR, 2007) that more than three-quarters of recent economic losses can be attributed to wind storms, floods, droughts and other climate-related incidents. Although a large proportion of the economic losses may result from higher land and property values, and from people moving into hazardous zones, these events may well be exacerbated by climate change (Hoeppe and Gurenko, 2006).

5.5 Geopolitical relationships

Climate change, then, alters the living conditions for populations and affects food, water and physical security. These changes may lead to local conflicts due to competition for

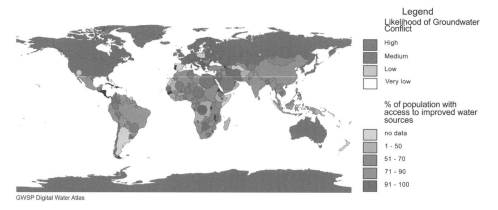

GWSP Digital Water Atlas

Figure 5.8 Figure showing the percentage of a population in a given country with access to improved water resources and identifying regions where there is a possible likelihood of groundwater conflicts. Source: modified from GWSP Digital Water Atlas (2008). Map 23: Likelihood of Groundwater Conflict 1 (V1.0). Available online at http://atlas.gwsp.org. (In colour as Plate 15.)

resources and to international conflicts, mainly through migration or power shifts. As noted above, for much of Africa, the rate of growth of GDP can be related to rainfall – where reductions in rainfall lead to significant reductions in the rate of GDP growth. The underlying cause for these correlations is a reduction in access to resources brought on by desiccation. A reduction in the availability of resources may be expected to increase local competition for these resources and, indeed, studies do suggest that the amount of rainfall, in addition to being related to GDP, also can be related to local conflicts (Figure 5.8 and Box 5.6).

Box 5.6
Does climate change lead to more deadly conflicts?

METTE KILDEGAARD GRAVERSEN

A recent study empirically analyses how climate change affects conflict severity, based on hypotheses: (1) that short-term changes in rainfall levels affect income levels, and (2) that income levels affect conflict severity. From these relationships, we can evaluate the long-term effects of climate change.

Ciccone (2008) studied the relationship between rainfall and income with a focus on conflict risk. Because income is endogenous, he used rainfall levels as a driver for changes in income. Ciccone found that lagged income changes (rainfall-driven) increase the risk of conflict. This is only one of several possible pathways for climate change to increase conflict risk with migration usually placed as a main intermediary variable (Gleditsch *et al.*, 2007). The plausibility of that link is well established, i.e. climate change can lead to migration, and migration can result in conflict, although the complete chain is not proven.

The study is based on the assumption that income levels and conflict severity are linked in poor societies. Rebel recruitment is easier because income is low (Collier and Hoeffler, 2004). Hence, rebel groups which are able to provide a relatively high reward to their recruits in poor societies are more likely to attract individuals who seek short-term economic gain and are less committed to the long-term objective of the movement. In some cases, the recruits are from outside the region and are less likely to share ethnicity or ideological values with the local population. This can also mean that the rebel leadership will have less influence on recruits behaviour towards civilians, making the situation more violent (Weinstein, 2007).

Where rainfall levels (above and below normal) are a key aspect of climate change, more extreme rainfall patterns may have dramatic consequences for livelihood, although only low levels of rainfall relative to the average level have an effect on income (Ciccone, 2008). According to Miguel *et al.* (2004), Sub-Sahara Africa is the only area where there is a relationship between rainfall and income due to the importance of rain-fed agricultural production.

Conflicts over food scarcity will often be between local non-state groups, which are not part of the government. Suliman (1999) argued that current African conflicts over renewable resources are mostly local.

The Uppsala Conflict Dataset Program (UCDP, 2008) on non-state conflicts was used to analyse links between climate and local conflict.[3] The dataset covers the period 2002–05 and records conflicts with a threshold of 25 battle deaths per year. As there is only one observation for each region, it is necessary to use a cross-sectional dataset ($N = 70$). To account for regional differences, the UCDP dataset was combined with geographical data (Theisen and Brandsegg, 2007) and conflicts that spread over a range of 600 km were excluded (similar to Raleigh and Urdal, 2007). Data on rainfall and population were taken from Thiesen and Brandsegg (2007), while data on income were from G-Econ (2008).

To determine conflict severity, the number of fatalities was normalised to the region's population size in 2000. As a measure of regional rainfall levels, the annual level of rainfall in 1999 was normalised to the average level of rainfall in previous years (1997–98). As an estimate of income levels, the GDP per capita, adjusted for differences in purchasing power parity in 2000 was used. An adjustment of GDP at the regional level to GDP at the national level was used in order to measure the impact of the income level relative to the national country-specific income level.

An OLS regression, with robust standard errors with clusters at the country level, was used in order to reduce the potential impact of within-state dependence between units (Table 5.1).

Models 1, 1a and 1b in Table 5.1 relate rainfall levels and GDP per capita. Rainfall levels, in general, were found to be insignificant. But making a distinction similar to that of Ciccone

---

[3] Countries included in the sample are Burundi, Côte d'Ivoire, Democratic Republic Congo, Ethiopia, Ghana, Kenya, Madagascar, Nigeria, Somalia, Sudan and Uganda.

Box 5.6 (*cont.*)

Table 5.1. *Ordinary least squares regressions of conflict severity in non-state conflicts, 2002–2005*

	Per capita GDP, 2000			Conflict severity, 2002–2005	
	(1)	(1a)	(1b)	(2)	(3)
Rainfall, $t-1$	−0.108				
	(0.27)				
Increasing levels in rainfall, $t-1$		−0.107			
		(0.09)			
Decreasing levels in rainfall, $t-1$			0.173*		−0.798
			(0.09)		(0.55)
Per capita GDP, 2000				−2.056**	−2.097***
				(0.65)	(0.62)
R^2	0.007	0.03	0.04	0.11	0.13

Note. The table reports coefficients with standard errors in parenthesis. $*p < 0.1$; $**p < 0.05$; $***p < 0.001$.

(2008), between above- and below-average levels of rainfall, the analysis shows a significant positive relationship between decreasing rainfall levels and regional income levels. Model 2 relates regional GDP per capita and conflict severity, and found a significant negative relationship. Model 3 shows that rainfall only affected severity through income.

Analysing regional non-state conflicts in Sub-Saharan Africa, the analysis suggests that conflicts are more severe in low-income regions and that below-average levels of rainfall (drought) reduce regional income. These results suggest that permanent reductions in rainfall levels may increase conflict severity due to its negative impact on income levels. These results are similar to the findings by Ciccone (2008).

In parallel to this work, Burke *et al.* (2009) have done studies on African conflict trying to isolate the temperature factor. They found that a 1 °C increase implied a 4.5% increase in civil war in the same year and 0.9% increase in the next year – a 49% relative increase in civil war risk. Interestingly, they controlled for the rainfall factor, and thus we can see that climate change increases conflict in Africa both through the route of rainfall and income levels and through temperature increases (presumably working mainly through agriculture).

Not only is conflict at the local level impacted by climate change. Throughout history, the major human response to climatic changes beyond local adaptation capacity has been migration (Figure 1.12 and Chapter 13). Earlier in history, the world was not carved up into tightly regulated territorial states, and there was new unoccupied and unclaimed land to which people could move. Today, the world is crowded – there are now over 6 billion people and the global population is expected to top 9 billion by 2050. Thus, there are very few new areas to which to move, and migration is more likely to lead to conflict than was the case when our early ancestors fled from changing climate conditions.

Increasing numbers of people migrating in response to changing climate conditions are already being reported. The UN estimated that there were approximately 20 million climate refugees in 2008 and that number is predicted to increase to 50 million in 2010 (UNEP). Large-scale migration is usually resisted by states and becomes a conflict issue between them (Spring, 2009; Stern, 2009).

It has been argued that a correlation between climate change and conflict is not well supported by quantitative data (Linnerooth-Bayer *et al.*, 2009). This is partly because of the rather recent concern with the issue and the challenges of data collection (Gleditsch and Nordås, 2009; Scheffran, 2009). Research is currently aimed at producing data better focused on measuring these relationships, thereby also preparing international society for managing the resulting conflicts (Buhaug *et al.*, 2009).

Meanwhile, (often unpublished) intelligence service and military reports place climate change ever more centrally in preparations for future conflicts (Brauch, 2009; Wright, 2009a). If major powers become involved in conflicts, political cooperation on climate policy will become much more difficult (Chapter 13). If international climate policy comes to be seen as manifestly failing, unilateral attempts to deal with the emergency situation could lead to conflicts, for example, over geoengineering (deliberate human manipulation of the climate system; see Chapters 7 and 17 for further discussion). Generally, when issues are cast in security terms, leaders get increased latitude to take dramatic measures. It is crucial that this 'security-driven empowerment' in the case of climate change gets 'channelled' into strengthening international institutions, and not into unilateral emergency acts (Gleditsch and Nordås, 2009; Scheffran, 2009; Wright, 2009b).

Factoring security into the climate change equation runs the risk of escalating vicious circles. In the parts of the world where health and well-being are most negatively impacted by climate change, the likelihood of conflict will increase most, and these conflicts will further reduce living standards. More privileged parts of the world are likely to first feel the spill-over effects from these conflicts, such as refugees and diseases, and at higher temperature increases see their own security agenda reorganised around climate change.

5.6 Defining dangerous climate change: Lessons from the past

Over the 200 000 or so years that modern humans have been on Earth, they have experienced one complete glacial–interglacial cycle as well as many modes of natural climate variability, some of them severe and abrupt. Uncovering the ways in which our ancestors dealt with the climatic changes that they experienced can inform the discussion surrounding the definition of dangerous climate change as well as provide clues as to how these early societies adapted to a changing climate (Chapter 14).

Until very recently, humans basically had two options when their local climate made life more difficult for them: they could migrate to another area with a more favourable climate; or they could remain where they were and change their lifestyles or technologies to cope with the changing climate. As described in the previous section, the option of migration, commonly used in the past, will not be possible in most cases in the crowded world of today – at least not without a high probability of conflict. What, then, can we learn about dangerous climate change from those of our ancestors who remained in place and attempted to adapt to the changing climate around them?

In some cases, earlier societies were able to adapt successfully to significant changes in their climate, often through the three major determinants of social evolution – catastrophe (associated with a changing climate), communication and cooperation. For some societies, climate change was a trigger for revolutionary change and transformation. For example, at the climatic transition into the Holocene about 11 700 years ago, people in south-western France adapted by exploiting a new resource that was previously unavailable – warren-based hunting of the wild European rabbit (Jones, 2006), thus anticipating the domestication of animals that was to come 3000 years later. In the American Southwest, the Navajo peoples have survived in the same region for centuries, despite a highly variable climate in terms of precipitation, by adopting a very flexible subsistence strategy (Towner and Dean, 1996).

In other cases, however, earlier societies were unable to adapt to a changing climate and eventually collapsed. The Norse colonies in Greenland are a classic example; they disappeared around 1400 AD after an extended cooling trend in the regional climate (Arneborg, 2008). Long, dry periods – often in combination with other stresses – have been implicated in the demise of several past societies. Examples include the classic Mayan civilisation of meso-America, which disappeared around 850 AD (Hodell *et al.*, 1995, 2001), and the Akkadian Empire of present-day Syria, which collapsed around 4000 years b.p. when the regional climate suffered an abrupt, severe shift to an arid climate (Cullen *et al.*, 2000).

Examples of both transformation and collapse point towards the characteristics of societies that confer resilience or rigidity as the critical elements in determining the outcome when they are faced with a changing climate. A necessary characteristic for all societies that survived significant environmental stress is the capability for deep change; that is, the ability to overcome the great reluctance to change behaviours that have worked well in the past (Hetherington and Reid, 2010). Societal change was probably driven by a small number of people or groups who were more open to diversity and novelty, and could embrace difference and change. They could innovate, think outside the box, grab new ideas and drive cooperation rather than conflict. Furthermore, they could provide the leadership that encouraged others to follow. What was dangerous climate change for some societies in the past was transformative for others.

5.7 Defining dangerous climate change: synthesis of impacts analyses

Taken together, the analysis of impacts described in the preceding sections provides an information source on which to base a discussion of what constitutes dangerous climate change. A synthesis of such information was carried out by the Third Assessment Report of the IPCC and presented as the Reasons for Concern diagram (sometimes also called the 'burning embers diagram') (Figure 5.9; Smith *et al.*, 2009). The information is presented as a relationship between five areas of potential impacts and the increase in global average temperature above pre-industrial levels.

Using the same methodology, a group of researchers updated the Reasons for Concern diagram using the rapidly growing body of research on climate change impacts that has been created since the IPCC Third Assessment Report was published (Smith *et al.*, 2009). The results (right-hand panel) show a clear pattern; more severe potential impacts are now expected at lower levels of temperature increase. For example, for the widely accepted 2 °C guardrail, there will already be severe risks to unique and threatened species, and the risks of extreme weather events will be high in many parts of the world. Even the risk of

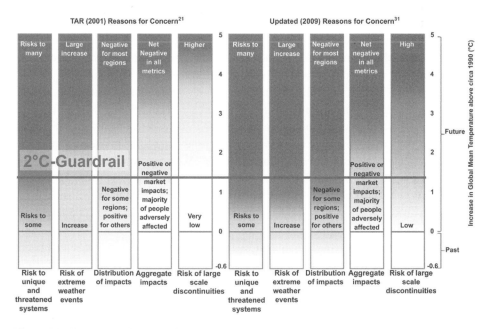

Figure 5.9 The potential impacts of climate change shown as a function of the rise in global average temperature. Zero on the temperature scale corresponds approximately to 1990 average temperature (i.e. 0.6 °C above pre-industrial levels), and the bottom of the temperature scale to pre-industrial average temperature. The level of risk or severity of potential impacts increases with the intensity of shading. The panel on the left is from Smith *et al.* (2001) as published in the IPCC Third Assessment Report. The panel on the right is an updated version from Smith *et al.* (2009), using the same methodology as the Third Assessment Report, based on expert judgement. The 2 °C guardrail is shown for reference. (In colour as Plate 16.)

triggering abrupt change in large-scale tipping elements in the climate system (Chapter 7) is no longer assessed to be insignificant. The bottom line is that the 2 °C guardrail carries higher risks of serious impacts than was thought likely at the beginning of the century. It is clear that temperature rises above 2 °C would be considered dangerous by most societies around the world, but it has become equally clear that increases in the 1–2 °C range carry significant risks for many peoples and regions.

5.8 Summary and conclusions

The UN Framework Convention on Climate Change (UNFCCC) states that 'dangerous' climate change is to be avoided. Exactly what 'dangerous' means is, however, not defined. At the level of (some) individuals, climate change is already undermining living conditions by influencing access to food and water, human health, and physical security.

Relationships between people and societies are also potentially impacted by climate change, as it increases competition for natural resources or territory. All of these types of climate change impacts are already occurring and are well documented. At present, a relatively small proportion of the global population is profoundly impacted – and most of the affected are living in developing regions of the world (Chapter 9). However, as human-induced climate change increases, ever increasing numbers of people will be impacted and the changes in living conditions will become more extreme.

In Chapter 1, it was argued that the popular debate concerning climate change often (mistakenly) considers the challenge as a prediction issue (i.e. are we 100% certain it is happening?) and that it would be more appropriate to consider human-induced climate change as a risk issue (how large a risk is society willing to take that human activities will lead to drastically altered living conditions?). It is a job for society as a whole, and not scientists alone, to define what constitutes 'dangerous' climate change. Human-induced climate change interacts with many other stressors that affect human societies. Thus, in the same way that it will never be possible to attribute an individual storm event directly and completely to human-induced climate change (Chapter 2), there will, at least in the short term, be few – if any – responses in human societies that can be directly attributed to climate change alone.

This is a challenge in defining 'dangerous' climate change. Nevertheless, predicting the probable impacts on human societies from climate change is relatively straightforward. Many of these impacts relate to obtaining sufficient food and water and, thus, challenge individuals at the level of their very survival. Even though it is difficult or impossible to quantify the exact effects of climate change on human societies, climate change will likely increase the number of individuals on Earth whose very existence is threatened. The challenge in defining 'dangerous' climate change is deciding at what level these impacts on individuals and societies are not acceptable. On top of that, it must be recognised that it is all living organisms – and not only human individuals and societies – that are affected by climate change.

References

Agarwal, A. and Babel, M. S. (2009). Forecasting rice yield under climate change scenarios and evaluation of agro-adaptation measures for Mekong basin region: a simulation study. *IOP Conference Series: Earth and Environmental Sciences*, **6**, 472006.

Alcamo, J. (2009). Simulating the interactions between climate change, food, and water in future Africa. *IOP Conference Series: Earth and Environmental Sciences*, **6**, 512002.

Allison, E. H., Perry, A. L., Badjeck, M.-C. *et al.* (2009). Vulnerability of national economies to the impacts of climate change on fisheries. *Fish and Fisheries*, **10**, 173–96.

Arneborg, J. (2008). The Norse settlements in Greenland. In *The Viking World*, eds. S. Brink and N. Price. London: Routledge, pp. 558–97.

Barnett, T. P., Adam, J. C. and Lettenmaier, D. P. (2005). Potential impacts of a warming climate on water availability in snow-dominated regions. *Nature*, **438**, 303–309.

Battisti, D. S. and Naylor, R. L. (2009). Historical warnings of future food insecurity with unprecedented seasonal heat. *Science*, **323**, 240–44.

Bell, J. D., Kronen, M., Vunisea, A. *et al.* (2008). Planning the use of fish for food security in the Pacific. *Marine Policy*, **33**, 64–76.

Beniston, M., Diaz, H. F. and Bradley, R. S. (1997). Climatic change at high elevation sites: An overview. *Climatic Change*, **36**, 233–51.

Bowman, D. M. J. S., Balch, J. K., Artaxo, P. *et al.* (2009). Fire in the earth system. *Science*, **324**, 481.

Brauch, H. G. (2009). Climate change impacts on migration: Conflict and cooperation in the Mediterranean. *IOP Conference Series: Earth and Environmental Sciences*, **6**, 562004.

Broccoli, A. J., Dixon, K. W., Delworth, T. L. *et al.* (2003). Twentieth-century temperature and precipitation trends in ensemble climate simulations including natural and anthropogenic forcing. *Journal of Geophysical Research*, **108**, 4798.

Buhaug, H., Gleditsch, N. P. and Theisen, O. M. (2009). Implications of climate change for armed conflict. In *The Social Dimensions of Climate Change: Equity and Vulnerability in a Warming World*, eds. R. Mearns and A. Norton. Washington, D.C.: World Bank, pp. 75–102.

Burke, M. B., Miguel, E., Satyanath, S., Dykema, J. A. and Lobell, D. B. (2009). Warming increases the risk of civil war in Africa. *Proceedings of the National Academy of Sciences (USA)*, **106**, 20670–74.

Cai, W. and Cowan, T. (2008) Evidence of impacts from rising temperature on inflows to the Murray-Darling Basin. *Geophysical Research Letters*, **35**, L07701.

Christensen, J. H. and Christensen, O. B. (2003). Severe summertime flooding in Europe. *Nature*, **421**, 805–06.

Ciccone, A. (2008). *Transitory Economic Shocks and Civil Conflict*. Economics Working Paper 1127. Barcelona: Universitat Pompeu Fabra.

Collier, P. and Hoeffler, A. (2004). Greed and grievance in civil war. *Oxford Economic Papers*, **56**, 563–95.

Costello, A., Abbas, M., Allen, A. *et al.* (2009). Managing the health effects of climate change. *The Lancet*, **373**, 1693–1733.

Cullen, H. M., DeMenocal, P. B., Hemming, S. *et al.* (2000). Climate change and the collapse of the Akkadian empire: Evidence from the deep sea. *Geology*, **28**, 379–82.

Easterling, D. R., Horton, B., Jones, P. D. *et al.* (1997). Maximum and minimum temperature trends for the globe. *Science*, **277**, 364–67.

Ebi, K. L., Smith, J. B. and Burton, I. (eds.) (2005). *Integration of Public Health with Adaptation to Climate Change*. New York, NY: Taylor and Francis.

Ebi, K. L., Hartman, J., Chan, N. *et al.* (2005). Climate suitability for stable malaria transmission in Zimbabwe under different climate change scenarios. *Climatic Change*, **73**, 375–93.

Emanuel, K. (2005). Increasing destructiveness of tropical cyclones over the past 30 years. *Nature*, **436**, 686–88.

Ericksen, P. J. (2008a). Conceptualizing food systems for global environmental change research. *Global Environmental Change*, **18**, 234–45.

Ericksen, P. (2008b). What is the vulnerability of a food system to global environmental change? *Ecology and Society*, **13**, 14.

Ericksen, P. J., Ingram, J. S. I. and Liverman, D. M. (2009). Food security and global environmental change: emerging challenges. *Environmental Science & Policy*, **12**, 373–77.

Eriksson, M., Jianchu, X., Shrestha, A. B. *et al.* (2009). *The Changing Himalayas: Impact of Climate Change on Water Resources and Livelihoods in the Greater Himalayas*. Kathmandu: ICIMOD.

FAO (1996). World Food Summit Plan of Action. Rome: Food and Agriculture Organization of the United Nations.

FAO (2008). *The State of Food Insecurity in the World: High Food Prices and Food Security – Threats and Opportunities*. Rome: FAO.

FAO (2009). *The State of Food Insecurity in the World: Economic Crises – Impacts and Lessons Learned*. Rome: FAO.

Fischer, E. M. and Schär, C. (2009). Future changes in daily summer temperature variability: driving processes and role for temperature extremes. *Climate Dynamics*, **33**, 917–35.

GECAFS (2009). *A Food Systems Approach to Food Security and Global Environmental Change Research*. Oxford, UK: Global Environmental Change and Food Systems International Project Office.

G-Econ (2008). *G-Econ Project*. New Haven, CT: Yale University.

Gleditsch, N. P. and Nordås, R. (2009). IPCC and the climate-conflict nexus. *IOP Conference Series: Earth and Environmental Sciences*, **6**, 562007.

Gleditsch, N. P., Nordås, R. and Salehyan, I. (2007). *Climate Change and Conflict: The Migration Link: Coping with Crisis*. International Peace Academy.

Grandjean, P., Bellinger, D., Bergman, A. *et al.* (2008). The Faroes statement: Human health effects of developmental exposure to chemicals in our environment. *Basic & Clinical Pharmacology & Toxicology*, **102**, 73–75.

Hales, S., de Wet, N., Maindonald, J. and Woodward, A. (2002). Potential effect of population and climate changes on global distribution of dengue fever: An empirical model. *The Lancet*, **360**, 830–34.

Hay, S. I., Tatem, A. J., Guerra, C. A. and Snow, R. W. (2005). *Foresight on Population at Malaria Risk in Africa: 2005, 2015 and 2030. Scenario Review Paper Prepared for the Detection and Identification of Infectious Diseases Project (DIID)*. London: Foresight Project, Office of Science and Innovation.

Hetherington, R. and Reid, R. G. B. (2010). Climate change and modern human evolution. *IOP Conference Series: Earth and Environmental Sciences*, **6**, 072027.

Hochrainer, S., Mechler, R. and Pflug, G. (2009). Climate change and climate insurance. The case of Malawi. *IOP Conference Series: Earth and Environmental Sciences*, **6**, 422003.

Hodell, D. A., Curtis, J. H. and Brenner, M. (1995). Possible role of climate in the collapse of Classic Maya civilization. *Nature*, **375**, 391–94.

Hodell, D. A., Brenner, M., Curtis, J. H. and Guilderson, T. (2001). Solar forcing of drought frequency in the Mayan lowlands. *Science*, **292**, 1367–70.

Hoeppe, P. and Gurenko, E. N. (2006). Scientific and economic rationales for innovative climate insurance solutions. *Climate Policy*, **6**, 607–20.

Huang, Y., Zhang, W., Yu, Y. *et al.* (2009). A primary assessment of climate change impact on rice production in China. *IOP Conference Series: Earth and Environmental Sciences*, **6**, 472003.

Human Impact Report (2009). *Climate Change. The Anatomy of a Silent Crisis*. Geneva: Global Humanitarian Forum.

IIASA (2009). *Climate Change and Extreme Events: What Role for Insurance?* Policy Brief #04. Laxenburg, Austria: International Institute for Applied Systems Analysis, http://www.iiasa.ac.at/Admin/PUB/policy-briefs/pb04.html

Ingram, J. Ericksen, P. and Liverman, D. eds. (2010). *Food Security and Global Environmental Change*. Earthscan London, 352 pp.

Intergovernmental Panel on Climate Change (IPCC) (2007a). *Climate Change 2007: Impacts, Adaptation and Vulnerability. Contribution of Working Group II to the Fourth Assessment Report of the Intergovernmental Panel on Climate Change*, eds. M. Parry, O. Canziani, J. Palutikovf, P. van der Linden and C. Hanson. Cambridge, UK: Cambridge University Press.

Intergovernmental Panel on Climate Change (IPCC) (2007b). *Climate Change 2007: The Physical Science Basis. Contribution of Working Group I to the Fourth Assessment Report of the Intergovernmental Panel on Climate Change*, eds. S. Solomon, D. Qin, M. Manning, Z. Chen, M. Marquis, K. B. Averyt, M. Tignor and H. L. Miller. Cambridge, UK and New York, NY: Cambridge University Press.

Jones, E. L. (2006). Prey choice, mass collecting, and the wild European rabbit (*Oryctolagus cuniculus*). *Journal of Anthropological Archaeology*, **25**, 275–89.

Jones, P. D., New, M., Parker, D. E., Martin, S. and Rigor, I. G. (1999). Surface air temperature and its changes over the past 150 years. *Reviews of Geophysics*, **37**, 173–99.

Kjellstrom, T. (2000). Climate change, heat exposure and labour productivity. *Epidemiology*, **11**, S144.

Knutson, T. R., Sirutis, J. J., Garner, S. T., Vecchi, G. A. and Held, I. M. (2008). Simulated reduction in Atlantic hurricane frequency under twenty-first century warming conditions. *Nature Geoscience*, **1**, 359–64.

Kovats, S. (ed.) (2008). *Health Effects of Climate Change in the UK Health Protection Agency. An Update of the Department of Health Report 2001/2002*. London: Department of Health.

Le Page, Y., Pereira, J., Trigo, R. *et al.*(2008). Global fire activity patterns (1996–2006) and climatic influence: An analysis using the World Fire Atlas. *Atmospheric Chemistry and Physics*, **8**, 1911–24.

Li, X., Cheng, G., Jin, H. *et al.* (2008). Cryospheric change in China. *Global and Planetary Change*, **62**, 210–18.

Linnerooth-Bayer, J., Warner, K., Bals, C. *et al.* (2009). Insurance, developing countries and climate change. *The Geneva Papers on Risk and Insurance Issues and Practice*, **34**, 381–400.

Liu, X. and Chen, B. (2000). Climatic warming in the Tibetan Plateau during recent decades. *International Journal of Climatology*, **20**, 1729–42.

Lobell, D. B., Burke, M. B., Tebaldi, C. *et al.* (2008). Prioritizing climate change adaptation needs for food security in 2030. *Science*, **319**, 607–10.

Ludwig, F., Kabat, P., Hagemann, S. and Dorlandt, M. (2009). Impacts of climate variability and change on development and water security in Sub-Saharan Africa. *IOP Conference Series: Earth and Environmental Science*, **6**, 292002.

Mann, M. E. and Emanuel, K. A. (2006). Atlantic hurricane trends linked to climate change. *EOS, Transactions, American Geophysical Union*, **87**, 233–44.

McMichael, A. J., Campbell-Lendrum, D., Kovats, R. S. *et al.* (2004). Climate change. In *Comparative Quantification of Health Risks: Global and Regional Burden of Disease Due to Selected Major Risk Factors*, Vol. 2, eds. M. Ezzati, A. D. Lopez, A. Rodgers and C. Mathers. Geneva: World Health Organization, pp. 1543–1650.

Meehl, G. A. (1997). The South Asian Monsoon and the Tropospheric Biennial Oscillation. *Journal of Climate*, **10**, 1921–43.

Meehl, G. A., Stocker, T. F., Collins, W. D. *et al.* (2007). Global climate projections. In *Climate Change 2007: The Physical Science Basis. Contribution of Working Group I to the Fourth Assessment Report of the Intergovernmental Panel on Climate Change*, eds. S. Solomon, D. Qin, M. Manning, Z. Chen, M. Marquis, K. B. Averyt, M. Tignor and H. L. Miller. Cambridge, UK and New York, NY: Cambridge University Press, pp. 747–846.

Meehl, G. A., Washington, W. M., Ammann, C. M. *et al.* (2004). Combinations of natural and anthropogenic forcings in 20th century climate. *Journal of Climate*, **17**, 3721–27.

Miguel, E., Satyanath, S. and Sergenti, E. (2004). Economic shocks and civil conflict: An instrumental variable approach. *Journal of Political Economy*, **112**, 725–53.

MunichRe (2007). *Topics: Natural Disasters. Annual Review of Natural Disasters 2006.* Munich: Munich Reinsurance Group.

Myers, N., Mittermeier, R. A., Mittermeier, C. G., da Fonseca, G. A. B. and Kent, J. (2000). Biodiversity hotspots for conservation priorities. *Nature*, **403**, 853–58.

Nogués-Bravo, D., Araújo, M. B., Errea, M. P. and Martínez-Rica, J. P. (2007). Exposure of global mountain systems to climate warming during the 21st century. *Global Environmental Change*, **17**, 420–28.

Onozuka, D., Hashizume, M. and Hagihara, A. (2010). Effects of weather variability on infectious gastroenteritis. *Epidemiology and Infection*, **138**, 236–43.

Owen, L. A., Finkel, R. C. and Caffee, M. W. (2002). A note on the extent of glaciation throughout the Himalaya during the global Last Glacial Maximum. *Quaternary Science Reviews*, **21**, 147–57.

Paldy, A. (2009). Climate change and health – challenges for Hungary. In *Proceedings of the Seventh Asia–Europe Meeting (ASEM)*. 4–5 November, Hanoi, Vietnam, p. 78.

Pall, P., Allen, M. R. and Stone, D. A. (2007). Testing the Clausius–Clapeyron constraint on changes in extreme precipitation under CO_2 warming. *Climate Dynamics*, **28**, 351–63.

Pascual, M., Ahumada, J. A., Chaves, L. F., Rodó, X. and Bouma, M. (2006). Malaria resurgence in the East African highlands: temperature trends revisited. *Proceedings of the National Academy of Sciences (USA)*, **103**, 5829–34.

Qiu, J. (2008). China: The third pole. *Nature*, **454**, 393–96.

Raleigh, C. and Urdal, H. (2007). Climate change, environmental degradation and armed conflict. *Political Geography*, **26**, 674–94.

Rizzoli, A., Hauffe, H. C., Tagliapietra, V., Neteler, M. and Rosà, R. (2009). Forest structure and roe deer abundance predict tick-borne encephalitis risk in Italy. *PLOS One*, **4**, e4336.

Rosenzweig, C., Casassa, G., Karoly, D. J. *et al.* (2007). Assessment of observed changes and responses in natural and managed systems. *Climate Change 2007: Impacts, Adaptation and Vulnerability. Contribution of Working Group II to the Fourth Assessment Report of the Intergovernmental Panel on Climate Change*, eds. M. L. Parry, O. F. Canziani, J. P. Palutikof, P. J. van der Linden and C. E. Hanson. Cambridge, UK and New York, NY: Cambridge University Press, pp. 79–131.

Schär, C., Vidale, P. L., Lüthi, D. *et al.* (2004). The role of increasing temperature variability for European summer heatwaves. *Nature*, **427**, 332–36.

Scheffran, J. (2009). Climate-induced instabilities and conflicts. *IOP Conference Series: Earth and Environmental Sciences*, **6**, 562010.

Schild, A. (2008). ICIMOD's position on climate change and mountain systems: The case of the Hindu Kush–Himalayas. *Mountain Research and Development*, **28**, 328–31.

Sharma, E., Chettri, N., Tsering, K. *et al.* (2009). *Climate Change Impacts and Vulnerability in the Eastern Himalayas*. Kathmandu, Nepal: ICIMOD.

Shrestha, A. B., Wake, C. P., Mayewski, P. A. and Dibb, J. E. (1999). Maximum temperature trends in the Himalaya and its vicinity: An analysis based on temperature records from Nepal for the period 1971–94. *International Journal of Climatology*, **12**, 2775–86.

Smith, J.B., Schellnhuber, H.-J., Mirza, M. M. Q. *et al.* (2001). Vulnerability to climate change and reasons for concern: A synthesis. In *Climate Change 2001: Impacts, Adaptation, and Vulnerability*, eds. J. McCarthy, O. Canziana, N. Leary, D. Dokken and K. White. New York, NY: Cambridge University Press, pp. 913–67.

Smith, J. B., Schneider, S. H., Oppenheimer, M. *et al.* (2009). Assessing dangerous climate change through an update of the Intergovernmental Panel on Climate Change (IPCC) 'reasons for concern'. *Proceedings of the National Academy of Sciences (USA)*, **106**, 4133.

Spring, U. O. (2009). Social vulnerability and geopolitical conflicts due to socio-environmental migration in Mexico. *IOP Conference Series: Earth and Environmental Sciences*, **6**, 562005.

Steffen, W. (2009). *Climate Change 2009: Faster Change and More Serious Risks*. Canberra: Department of Climate Change, Australian Government.

Stendel, M., Christensen, J. H. and Petersen, D. (2008). Arctic climate and climate change with a focus on Greenland. *Advances in Ecological Research*, **40**, 13–43.

Stern, L. N. (2009). *Plenary presentation at the International Scientific Congress on Climate Change 2009*. Available online at: http://climatecongress.ku.dk/presentations/congresspresentations.

Stott, P. A., Stone, D. A. and Allen, M. R. (2004). Human contribution to the European heatwave of 2003. *Nature*, **432**, 610–14.

Stott, P. A., Jones, G. S., Lowe, J. A. *et al.* (2006). Transient climate simulations with the HadGEM1 model: Causes of past warming and future climate change. *Journal of Climate*, **19**, 3055–69.

Suliman, M. (1999). *Ecology, Politics and Violent Conflict*. London: Zed Books.

Sun, Y., Solomon, S., Dai, A. and Portmann, R. W. (2007). How often will it rain? *Journal of Climate*, **20**, 4801–4818.

Theisen, O. M. and Brandsegg, K. B. (2007). The environment and non-state conflicts in Sub-Saharan Africa. *International Studies Association 48th Annual Convention*, 28 February–3 March, Chicago, IL, http://www.allacademic.com/meta/p179143_index.html.

The Lancet (2009). Set of six papers on the direct health co-benefits of tackling climate change. *The Lancet*, 27 November, 374.

Towner, R. H. and Dean, J. S. (1996). Questions and problems in pre-Ft. Sumner Navajo archaeology. In *The Archaeology of Navajo Origins*, ed. R. H. Towner. Salt Lake City, UT: University of Utah Press, pp. 1–18.

Trenberth, K. E., Jones, P. D., Ambenje, P. *et al.* (2007). Observations: Atmospheric surface and climate change. In *Climate Change 2007: The Physical Science Basis. Contribution of Working Group I to the Fourth Assessment Report of the Intergovernmental Panel on Climate Change*, eds. S. Solomon, D. Qin, M. Manning, Z. Chen, M. Marquis, K. B. Averyt, M. Tignor and H. L. Miller. Cambridge, UK and New York, NY: Cambridge University Press, pp. 235–336.

Trotman, A., Gordon, R. M., Hutchinson, S. D., Singh, R. and McRae-Smith, D. (2009). Policy responses to GEC impacts on food availability and affordability in the Caribbean community. *Environmental Science & Policy*, **12**, 529–41.

UCDP (2008). Uppsala Conflict Data Program. UCDP Database, http://www.ucdp.uu.se/database. Uppsala, Sweden: Uppsala University.

United Nations International Strategy for Disaster Reduction (UNISDR) (2007). http://www.unisdr.org/disaster-statistics/impact-economic.htm.

von Grebmer, K., Nestorova, B., Quisumbing, A. *et al.* (2009). *Global Hunger Index: The Challenge of Hunger: Focus on Financial Crisis and Gender Inequality*. Bonn, Washington, D.C., Dublin: Welthungerhilfe, International Food Policy Research Institute and Concern Worldwide.

Weinstein, J. M. (2007). *Inside Rebellion: The Politics of Insurgent Violence*. New York, NY: Cambridge University Press.

Westerling, L., Hidalgo, H. G., Cayan, D. R. and Swetnam T. W. (2006). Warming and earlier spring increase western U.S. forest wildfire activity. *Science*, **18**, 940–43.

WorldFish (2007). *The Threat to Fisheries and Aquaculture from Climate Change*. Penang, Malaysia: The WorldFish Center.

Wright, S. (2009a). Emerging military responses to climate change: The new technopolitics of exclusion. *IOP Conference Series: Earth and Environmental Sciences*, **6**, 562001.

Wright, S. (2009b). Climate change & the new techno-politics of border exclusion & zone denial. Presentation at *Climate/Security, conference organised by Centre for Advanced Security Theory, Copenhagen, 9 March 2009*.

Xu, J., Grumbine, R. E., Shrestha, A., Eriksson, M., Yang, X., Wang, Y. and Wilkes, A. (2009). The melting Himalayas: Cascading effects of climate change on water, biodiversity, and livelihoods. *Conservation Biology*, **23**, 520–30.

Xu, J., Shrestha, A. B., Vaidya, R., Eriksson, M. and Hewitt, K. (2007). *The Melting Himalayas: Regional Challenges and Local Impacts of Climate Change on Mountain Ecosystems and Livelihoods*. Kathmandu: International Centre for Integrated Mountain Development (ICIMOD).

Yao, T., Wang, Y., Liu, S. *et al.* (2004). Recent glacial retreat in High Asia and its impact on water resource in Northwest China. *Science in China Series D: Earth Sciences*, **47**, 1065–75.

Zhang, X., Zwiers, F. W., Hegerl, G. C. *et al.* (2007). Detection of human influence on twentieth-century precipitation trends. *Nature*, **448**, 461–65.

Zhao, M., Held, I. M., Shian-Jiann, L. and Vecchi, G. A. (2009). Simulations of global hurricane climatology, interannual variability, and response to global warming using a 50-km resolution GCM. *Journal of Climate*, **22**, 6653–78.

Zhou, X.-N., Yang, G.-J., Yang, K. *et al.* (2008). Potential impact of climate change on schistosomiasis transmission in China. *The American Society of Tropical Medicine and Hygiene*, **78**, 188–94.

Zomer, R. J. (2008). Biodiversity goods and services – Increasing benefits for mountain communities. In Proceedings of the International Mountain Biodiversity Conference. Biodiversity Conservation and Management for Enhanced Ecosystem Services: Responding to the Challenges of Global Change. 16–18 November. Kathmandu: ICIMOD, http://www.icimod.org/imbc.

6

Impacts of climate change on the biotic fabric of the planet

'The world that is bequeathed to us is, in our hands and in our time, being unmade ... the future complexity of life – our potential gift to the future – is being eradicated'[1]

Biodiversity represents the fabric of life itself. It is comprised not only of numbers of species, but also includes the variety of all life forms and their genes as well as the communities and ecosystems of which they are a part. The consequences of climate change for biodiversity are potentially profound.

Many acknowledge the importance of biodiversity in its own right, without any consideration of utilitarian value. However, biodiversity also provides the underpinning of the ecosystem services on which human societies are ultimately dependent. Without well-functioning ecosystems we humans, as a biological species, could not exist.

Even before anthropogenic climate change became a significant issue, biodiversity has been in decline in many parts of the world. The reasons are almost entirely linked to human numbers, economic activity and resource use; the proximate causes include landscape modification and conversion, direct predation (i.e. hunting and fishing), introduction of alien species, extraction of water resources, and application of excess nutrients. Climate change now represents an additional stressor, usually interacting with the already existing stressors in complex ways. The acceleration of biodiversity decline with climate change, and the consequent further degradation and loss of ecosystem services, provides important information to inform the discussion of what constitutes dangerous climate change.

6.1 Observed impacts of climate change on biodiversity

In many cases, it is difficult to attribute changes in biodiversity directly to the impacts of climate change because so many other stressors are already affecting biodiversity. For example, 45% of European bird species are experiencing population declines (Birdlife

---

[1] Deborah Bird Rose, scholar of the ecological humanities, Macquarie University, Australia.

Climate Change: Global Risks, Challenges and Decisions, Katherine Richardson, Will Steffen and Diana Liverman *et al.* Published by Cambridge University Press. © Katherine Richardson, Will Steffen and Diana Liverman 2011.

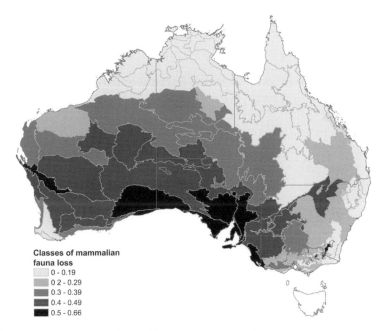

Figure 6.1 Continental-scale decline in biodiversity due to existing stressors; proportion of loss of mammalian fauna in 76 bioregions of mainland Australia. The proportion is calculated by adding the number of extinct species to one half of the species in decline. Shown are five classes: 0–0.19, 0.2–0.29, 0.3–0.39, 0.4–0.49, 0.5–0.66. Source: McKenzie *et al.* (2007).

International, 2008), and 48% of bird species in the United States are of conservation concern (http://www.stateofthebirds.org). Over 20% of the 5000 mammal species world-wide are already threatened with extinction (http://www.iucnredlist.org/documents/summarystatistics/2009RL_Stats_Table_4a.pdf). Over the past two centuries on the continent of Australia, over 50% of mammalian species have gone extinct in south-central Australia and over 30% in most of the rest of the continent (Figure 6.1); 49 species of vascular plants have become extinct (compared to 27 for the whole of Europe); and many other species of plants, mammals and reptiles have reached such low numbers that they are functionally extinct; that is, they cannot play their previous functional role in ecosystems (Steffen *et al.*, 2009). Changes in marine ecosystems are just as dramatic: in coastal seas and estuaries about 85% of large whales are gone, 87% of sea turtles, 90% of oysters, 90% of manatees and dugongs, 65% of sea grass and 67% of wetlands (Jackson, 2008). About 50% of the world's existing coral reefs are on the verge of collapse (Stone, 2007).

However serious these non-climate stressors have been, climate change is a stressor like none other, as it directly influences the basic physical and chemical environment underpinning all life. It affects basic evolutionary biology and ecology – physiology, metabolism, phenology, reproductive strategies, competition, mutualism, trophic interaction and more. Even with 'just' a 0.7 °C rise in global mean temperature, the effects of climate change are already discernible in many different ways on all continents, and in all coastal seas and ocean basins around the Earth (Box 6.1; Parmesan, 2006). More specific examples of these effects include the following.

- An upward range shift of 29 m per decade has been observed in 171 forest plant species in Western Europe (Lenoir *et al.*, 2008), as well as an upward range shift of 500 m over 50 years for half of the 28 mammalian species studied in Yosemite National Park, USA (Moritz *et al.*, 2008) and the movement of warm water fish into northern seas (Møller, 2009).
- A study of 542 plant species in Europe from 1971 to 2000 shows an advance in spring leaf unfolding of 2.5 days per decade and fruit ripening by 2.4 days per decade, with 75% of species showing advances (Cleland *et al.*, 2007). Migration of songbirds has shown a significantly earlier spring migration over a 46-year period (Van Buskirk *et al.*, 2009).
- Extreme climate events are particularly important; heavy rainfall events associated with the El Niño phase in California have significantly changed ecosystem dynamics in alpine, grassland and coastal marine ecosystems (Hobbs and Mooney, 1995).
- Shifting distributions of species have been observed as the climate shifts, such as the invasion of shrubs into herb-dominated vegetation in tundra due to warming (Mølgaard, 2009).

Box 6.1
Observed responses of biodiversity to climate change

PAUL LEADLY

There is powerful evidence that biodiversity is already responding to late twentieth-century and early twenty-first-century climate change (Parmesan and Yohe, 2003; Millennium Ecosystem Assessment, 2005; Parmesan, 2006; IPCC, 2007a). These changes include modifications of the genetic structure of populations, changes in species abundance, shifts in species and biome ranges, and species extinctions. In some cases, these changes are associated with important losses of ecosystem services.

Shifts in the abundance of plant and animal species within communities are often the first signs of climate impacts on biodiversity. Rising temperatures over the past century have favoured warm-adapted species over cold-adapted species in the majority of communities studied (Parmesan and Yohe, 2003). There is also good evidence that rising atmospheric CO_2 concentrations, changes in precipitation patterns, storms and climate-related changes in fire regimes, insect outbreaks, and other disturbance regimes are driving major changes in species abundance and ecosystem services at regional scales (IPCC, 2007a). A large number of experiments have helped to identify the mechanisms underlying community-level responses to climate change and rising CO_2 (Potvin *et al.*, 2007). Climate-driven changes in species abundance in communities are having major impacts on ecosystem services when they include large-scale declines or shifts in key species (Hoegh-Guldberg *et al.*, 2007). Two of the most important, well-documented examples of ongoing community-level changes are increased dominance of shrubs in Arctic tundra (Tape *et al.*, 2006) and large-scale bleaching and dieback of hard corals (Hoegh-Guldberg *et al.*, 2007).

Over longer time periods, species respond to climate change by shifting their ranges. There is compelling evidence that this is starting to occur for a very broad array of terrestrial and marine species in response to warming over the past century (Figure 6.2; Beaugrand *et al.*, 2002; Parmesan and Yohe, 2003; Perry *et al.*, 2005). Range shifts are also one of the best-documented responses of species and biomes to climate change in the palaeological record; species' ranges have often shifted many hundreds of kilometres during cycles of glaciation and deglaciation (Willis and Bagwat, 2009). The largest observed range shifts over the past several decades tend to be in temperature-sensitive and mobile species groups, such as butterflies, but this is almost certainly a harbinger of major range shifts of species with longer response times, such as trees that have much larger impacts on ecosystem services.

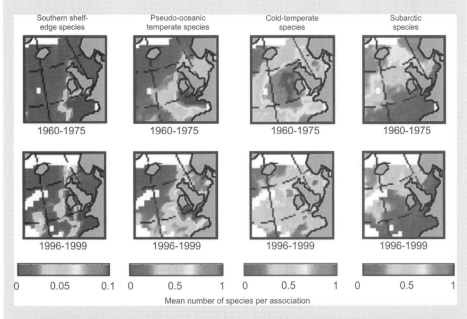

Figure 6.2 Changes in phytoplankton abundance in the Eastern North Atlantic at the end of the twentieth century based on the Continuous Phytoplankton Recorder database (Beaugrand *et al.*, 2002). Warm colours indicate high abundance and cool colours indicate low abundance. Both cold water-adapted species (cold-temperate and subarctic) and warm-water adapted species (southern shelf-edge and pseudo-oceanic temperate) moved northward. Reproduced with permission of AAAS. (In colour as Plate 17.)

Box 6.1 (*cont.*)

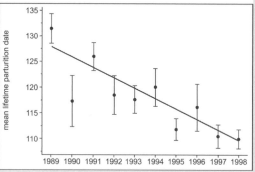

Figure 6.3 Red squirrels in northwest Canada gave birth 20 days earlier in 1998 than in 1989. Genetic analyses show that part of this response was due to changes in the genetic structure of their population. Source: Réale *et al.* (2003).

There is growing evidence for 'rapid evolution' of species in response to climate change (Reusch and Wood, 2007). The most common underlying mechanism is the selection of genotypes within a population that are best adapted to new climates, i.e. 'directional selection'. As such, species may be able to adapt more quickly to climate change than had previously been thought (Figure 6.3). However, directional selection erodes genetic diversity within populations and reduces their capacity to adapt to additional environmental stress.

Evidence for recent climate-induced extinctions is rare. One of the best documented cases is mass extinction of amphibians in the American tropics where changes in climate have been identified as one of the key culprits in combination with introduction of pathogens and habitat destruction (Pounds *et al.*, 2006). The palaeological record provides long-term insight into current trends and suggests that some groups of species, such as plants, have relatively low extinction rates even during periods of large and rapid changes in climate, while other groups may have substantially higher rates (Willis and Bagwat, 2009). This contrasts with scenarios of mass extinctions across broad taxonomic spectra projected for the twenty-first century (Thomas *et al.*, 2004; Millennium Ecosystem Assessment, 2005).

Observations of modern and palaeological responses of biodiversity to climate change provide strong qualitative support of model-based scenarios of twenty-first-century changes in species abundance and large-scale range shifts in species and biomes.

Although not strictly a climate change impact, the increasing acidity of the ocean due to the increase of atmospheric CO_2, the subsequent increase of its dissolution in the ocean and the consequent formation of carbonic acid is having demonstrable impacts on marine ecosystems (Moy *et al.*, 2009). For example, observations from Australia's Great Barrier Reef show that the rate of calcification of the coral species *Porites*, which is influenced by both ocean acidity and temperature, has declined by 14% since 1990 (De'ath *et al.*, 2009). The decline was abrupt, with no change in calcification rate from the period from 1900 to 1990. This is a classic case of threshold/abrupt change behaviour, in which a slow change in a control variable such as ocean acidity can trigger a very rapid response when a threshold is crossed.

In almost all of the cases cited above, the observations show differential effects on species; some respond to a changing climate more rapidly and with greater magnitude than others. These differential effects imply a change in community composition – often the formation of novel ecosystems – with follow-on effects for ecosystem functioning and the delivery of ecosystem services. For example, these differential shifts are often disrupting the matches or mismatches of the timing of members in trophic chains (Visser and Both, 2005). More fundamentally, changing community composition is challenging well-established views regarding biodiversity conservation, such as the preservation of existing ecosystems. In many cases, the goal in the future may need to change from protecting individual species towards the conservation of well-functioning ecosystems, even if they are undergoing inevitable shifts in species composition as the climate changes.

6.2 Projected further impacts on biodiversity

The impacts described in the previous section are likely to be just the tip of the iceberg. They are associated with a current global mean temperature rise of about 0.7 °C above pre-industrial levels. With the temperature virtually certain to rise to at least 2 °C above pre-industrial levels this century, the impacts on biodiversity will be much more severe (Warren *et al.*, 2009). Nonlinear changes of various types, especially abrupt changes and completely unanticipated events – true surprises – will become much more common, due largely to the rate of climate change. Adaptive capacity will be overwhelmed in many cases at the higher temperature ranges.

The most obvious effects of future climate change will be associated with attempts by species to self-adapt by moving to stay within their climatic envelopes. The current trends of species and biome shifts will accelerate, with many species and biomes projected to move many hundreds of kilometres by mid century. Under moderate or severe climate change scenarios (above +2 to 3 °C), the rate of climate change will outpace the capability of many species to migrate to tolerable climates. Specific examples of the many projected changes include significant turnover and likely local extinctions of bird species in sub-Saharan Africa as they are unable to migrate (Hole *et al.*, 2009); reductions in the abundance of wild relatives of agriculturally significant plants in sub-Saharan Africa (Jarvis *et al.*, 2009); and large shifts of tree species, such as in Italy, moving many of them out of protected areas (Attorre *et al.*, 2009; Figure 6.4).

Less predictable but, perhaps, more profound changes are projected to occur in association with widespread disruption of current plant and animal communities, leading to novel combinations of species and, hence, to new ecosystems. For example, experimental results suggest that increasing drought and higher temperature conditions will lead to significant reductions in flowering and recruitment from seeds in grasslands, leading to changes in community composition as the differential effects of these processes flow through the ecosystems (Pedersen *et al.*, 2009; Stampfli and Zeiter, 2009). Model simulations show reduced capacity of kelp beds to recover from storm damage in warmer marine environments (Staehr *et al.*, 2009).

The case of coral reefs demonstrates the interaction of stressors on a complex ecosystem, in this case the interaction between ocean acidity and rising sea surface temperature. Ocean acidification (Chapter 2) is a direct consequence of rising atmospheric CO_2 concentrations while the sea surface temperature increase is indirectly due to the CO_2 increase via its influence on the climate system as a whole. Figure 6.5 shows the environmental space defined by sea surface temperature (plotted as deviation from today's temperature) and carbonate ion concentration, which depends on acidity. Both temperature and acidity have varied naturally between ice ages and interglacial periods, defining an environmental space to which present-day coral reef systems are adapted. The carbonate ion concentration today is already well outside of this envelope. The CO_2 and climate trajectories are moving coral reefs towards thresholds in both temperature and carbonate ion concentration, the crossing of which – likely in the second half of this century without vigorous mitigation – would lead to the elimination of coral-dominated reefs and their conversion into algae-dominated ecosystems (Hoegh-Guldberg *et al.*, 2007).

Most impact studies – either observed impacts today or projected future impacts – focus on particular species, taxa, ecosystems or regions. Syntheses of this information (Parmesan, 2006) give important insights into general patterns of impacts and responses. What has been lacking, however, is a systematic, top-down global risk analysis for the planet's ecosystems under scenarios of future climate. Box 6.2 reports on such an analysis, using a global vegetation model to simulate shifts in ecosystem structure and in important ecosystem processes such as surface water runoff. The results provide a direct link from projected changes in ecosystem structure and functioning at the global scale to the provision of ecosystem services.

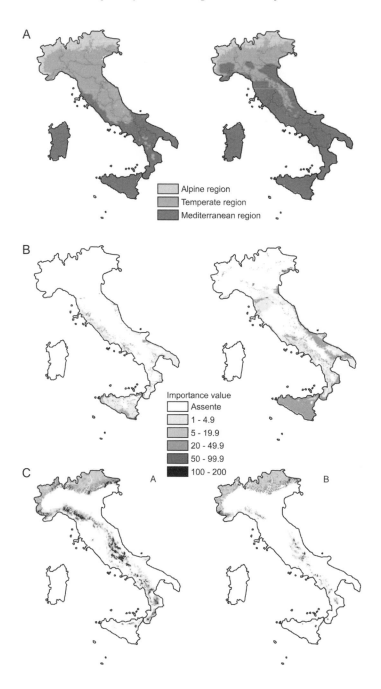

Figure 6.4 (A) Current (left) and future predicted (right) distribution of Mediterranean (dark grey), temperate (mid-grey) and alpine (light grey) regions in Italy; (B) current (left) and future predicted (right) distribution of cork oak (*Quercus suber*); and (C) current (left) and future predicted (right) distribution of beech (*Fagus sylvatica*). Source: Attorre *et al.* (2009).

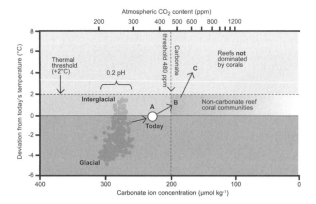

Figure 6.5 Temperature, atmospheric CO_2 concentration and carbonate ion concentrations reconstructed for the past 420 000 years. Carbonate concentrations were calculated from Vostok ice
core data. Acidity of the ocean has varied by $\pm$ 0.1 pH units over the past 420 000 years. The
thresholds for major changes to coral communities are indicated for thermal stress (+2 °C) and
carbonate ion concentration of 200 μM kg^{-1}, which corresponds to an approximate aragonite saturation Ω_{arag} of 3.3 and an atmospheric CO_2 concentration of 480 p.p.m. Black arrows pointing
towards the upper right indicate the pathway currently followed towards atmospheric CO_2 concentration of more than 500 p.p.m. The letters A, B and C refer to reef states described in the figure. Source: Hoegh-Guldberg *et al.* (2007), including details of the reconstructions. Reproduced
with permission of AAAS.

Box 6.2
A risk analysis for world ecosystems under future climate change

MARKO SCHOLZE

Our current world is under the threat of severe changes in the global environment system due
to anthropogenic intervention to the natural system. The UNFCCC commits nations to avoiding
'dangerous' climate change and 'allowing ecosystems to adapt naturally', but the concept of
'dangerous' climate change ultimately requires a normative decision based on value judgments.
However, it is not clear how likely are different amounts of climate change to have major
impacts on the world's ecosystems. So far, relatively minor climate changes that have occurred
during recent decades have already impacted local ecosystems (Walther *et al.*, 2002), and much
larger changes are projected for the twenty-first century (IPCC, 2007b).

The risks of climate-induced changes in key ecosystem processes during the twenty-first
century are quantified by forcing the Lund-Potsdam-Jena (LPJ) dynamic global vegetation
model (Sitch *et al.*, 2003) with outputs from 16 coupled atmosphere–ocean general circulation
models. The analysis is carried out by mapping the proportions of model runs showing
exceedance of natural variability in freshwater runoff and wildfire frequency, as well as the
proportions of model runs showing forest/non-forest shifts. The outputs represent four emission
scenarios: 'committed' climate change (i.e. atmospheric composition held constant from 2000),
and SRES A1B, A2 and B1.

All the climate model runs were initialised for pre-industrial conditions and run up to 2000 with radiative forcing based on observations, and then to 2100 under one of the four scenarios. To capture physiological effects of rising CO_2, LPJ was run with the time series of global mean CO_2 concentrations for the simulation period.

The analysis does not assign probabilities to scenarios, or weights to models. Instead, the distribution of outcomes within three sets of model runs grouped according to the amount of global warming they simulate (global mean surface temperature difference between 2071–2100 and 1961–1990) is considered: <2 °C (including simulations in which atmospheric composition is held constant, i.e. in which the only climate change is due to greenhouse gases already emitted, 16 runs), 2–3 °C (20 runs), and >3 °C (16 runs). Critical or dangerous change is defined based on the difference between the 2071–2100 and the 1961–1990 means; more precisely when the change in mean exceeds ±1σ of the interannual variability during 1961–1990, based on climate observations. For an extreme event occurring once every 100 years, a shift in the mean by 1σ in the direction of the extreme translates into a ten-fold increase in its frequency: thus, a small shift in the mean means that the '100-year event' becomes the '10-year event'. The analysis focuses on the risk of impacts of changes in extreme events on ecosystems. For changes in ecosystem type, a critical change is defined as a shift between forest and non-forest states.

Increased risks for more frequent wildfire are shown in Amazonia, the far north and many semiarid regions (Figure 6.6). For more runoff, high risks are displayed north of 50 °N, and in tropical Africa and north-western South America; and for less runoff in West Africa, Central America, southern Europe and the eastern USA (Figure 6.7). High risk of forest loss is shown for Eurasia, eastern China, Canada, Central America and Amazonia, with forest extensions

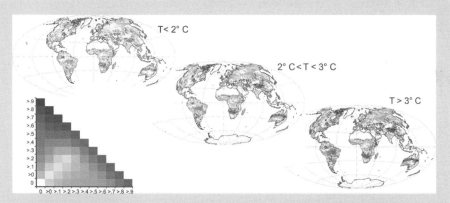

Figure 6.6 Probability of exceeding critical levels of change for wildfire frequency between 1961–1990 and 2071–2100 for three levels of global warming. Critical change is defined where the change in the mean of 2071–2100 exceeds ±1 σ of the observed (1961–1990) interannual variability (see inset for scale of probability of critical change: red for increase (vertical axis), green for decrease (horizontal axis), mixed colours for both exceeding the critical level by +1 σ as well as by −1 σ). Colours are shown only for grid cells with less than 75% cultivated and managed areas. Source: modified from Scholze *et al.* (2006). Copyright 2006 National Academy of Sciences, USA. (In colour as Plate 18.)

Box 6.2 (*cont.*)

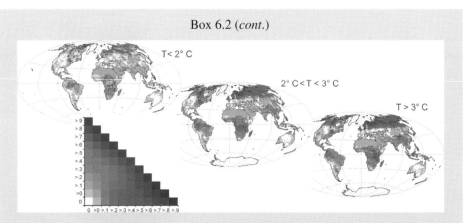

Figure 6.7 Same as Figure 6.6 but for freshwater runoff (blue for increase, red for decrease, mixed colours for both exceeding the critical level by $+1\sigma$ as well as by -1 σ). Grey areas denote grid cells with less than 10 mm yr^{-1} mean runoff for 1961–1990. Source: modified from Scholze *et al.* (2006). Copyright 2006 National Academy of Sciences, USA. (In colour as Plate 19.)

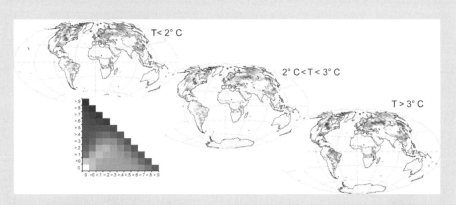

Figure 6.8 Biome change from forest to non-forest (blue) or vice versa (green). Colours are shown only for grid cells with less than 75% cultivated and managed areas. Source: modified from Scholze *et al.* (2006). Copyright 2006 National Academy of Sciences, USA. (In colour as Plate 20.)

into the Arctic and semiarid savannas (Figure 6.8). Substantially larger areas are affected for global warming >3 °C than for <2 °C; some features appear only at higher warming levels. A more detailed discussion of the results can be found in Scholze *et al.* (2006). This study clearly cannot provide an unambiguous definition of dangerous climate change. However, it may help to inform policy discussions by drawing attention to the steeply increasing risks to ecosystem services, such as for runoff associated with global climate changes beyond the range to which the climate system is already committed.

6.3 Consequences for the provision of ecosystem services

As for biodiversity itself, ecosystem services – which include provisioning services (e.g. food, fibre, water), regulating services (e.g. erosion control, pest and disease regulation, pollination) and cultural services (e.g. spiritual, aesthetics, recreation) – have already been significantly affected by human activities before anthropogenic climate change has had any appreciable effects. Over 60% of these services have been adversely affected by human activities, with most of that impact occurring in the past 50 years (Millennium Ecosystem Assessment, 2005). In effect, the increase in a few types of provisioning services – for example, the impressive increase in food production to support a growing human population – has been at the expense of many other, less direct ecosystem services that are nevertheless important for human well-being.

Despite these widespread pre-existing changes to ecosystem services, the additional impacts of climate change itself are already clearly observable on ecosystem services in some cases. The consequences for humans are distributed very unevenly; in many cases the poor and more subsistence-oriented populations are the most directly and severely affected. For example, there are reports of losses of genetic resources that could be important for agriculture in developing countries (Hodgkin *et al.*, 2009; Jarvis *et al.*, 2009), and climate-exposed subsistence agriculture provides the livelihoods for 50%–80% of Indonesia's rural and peri-urban poor (Miller *et al.*, 2009).

The impacts of climate change have also been observed on food resources through changes in marine biodiversity, an example of impacts on provisioning services. Examples include the following.

- Higher temperatures are increasing the stability and decreasing the mixing in Danish lakes and coastal waters, thereby decreasing the oxygen content in the bottom water, which in turn changes the timing of the spring bloom of phytoplankton with implications for water quality and ecosystem composition (Hansen and Bendtsen, 2009; Hansen *et al.*, 2009).
- Climate-driven changes are occurring to the species composition of intertidal invertebrates and macroalgae along the British coast, with potential flow-on effects to biomass, primary production and fish nursery habitat (Mieszkowska *et al.*, 2009).

Some impacts can be discerned already on regulating services. For example, in the northern high latitudes, changes in vegetation distribution – the movement of boreal forest northward into tundra – is changing the ability of the land surface to reflect incoming solar radiation during winter. This generates a positive (reinforcing) feedback on climate and constitutes a degradation of the regulating services that ecosystems can provide at the scale of regional climate (Chapin *et al.*, 2008). Disease control is another important regulating service. The recent re-emergence of two infectious diseases (American *Mucocutaneous leishmaniasis* and *Bartonellosis*) in southwestern Amazonia has been linked to biodiversity loss in the region, which in turn has been driven by land-use change exacerbated by a drying climate (Cesario *et al.*, 2009).

As another example, the loss of mangrove forests along many tropical coastlines has diminished the ability of coastal ecosystems to act as buffers towards extreme events such as cyclones and tsunamis. Most of the loss of mangrove forests is due to direct conversion to shrimp farms but this process is now be exacerbated by sea-level rise and rising temperature. Whatever the causes, the loss of the regulating services when mangrove forests are lost are likely to be huge. This is illustrated by the December 2004 tsunami in Asia – which killed 200 000 people – made 2 million homeless and caused economic losses of US$6 billion (Mooney, 2007). The greatest damage from the tsunami occurred in areas with the most extensive clearing of mangrove forests (Kathiresan and Rajendran, 2006). By comparison, the value of the shrimp farms that replaced the mangroves is estimated to be US$1–2 billion (Barbier and Cox, 2002). The comparison becomes even more dramatic when the value of the shrimp industry (~US$194–209 per hectare) is compared to the value (up to US$35,000 per hectare) of the complete range of timber and non-timber services of mangrove forests to local communities (Lindenmayer, 2009).

The potential disruption of a wide range of ecosystem services provided by coral reefs is another well-studied example of the consequences of climate change, interacting with other stressors (Box 6.3). Coral reefs are an interesting case study as they provide ecosystem services in all three categories: provisioning – food production via fisheries; regulating – the buffering of coasts against storm surges; and cultural – recreational diving and the tourism industry in general.

Box 6.3
Impacts of climate change on the ecosystem services provided by coral reefs

OVE HOEGH-GULDBERG

Coral reef ecosystems are common to the world's tropical coastlines. Despite occupying less than 1% of the Earth's surface, coral reefs provide important habitat to at least a million species (Reaka-Kudla, 1997), and food and livelihoods for over 500 million people (Bryant *et al.*, 1998; WRI, 2006). Coral reefs also play important roles as coastal barriers to wave energy that would otherwise damage coastal ecosystems such as mangroves and seagrass beds, and human infrastructure such as coastal towns and cities. Despite their importance, coral reefs are currently undergoing rapid contraction due to the influence of threats such as overfishing and declining water quality (Bryant *et al.*, 1998). Global surveys of the status of coral reefs have revealed that at least 30% of the coral on coral reefs has disappeared as a result of these threats (Wilkinson, 1999; Bruno and Selig, 2007; Wilkinson, 2008), and that further losses are almost certain under the rapidly growing pressure from human activities on coral reefs.

Until a decade ago, global climate change was considered to be a distant threat to coral reefs (Wilkinson and Buddemeier, 1994). This perspective has changed dramatically under our improved understanding of the impacts of increasing atmospheric CO_2 on coral reefs, as well as after a series of global impacts that underscored the seriousness of the problem. Coral bleaching occurs when the fundamentally important symbiosis between corals (simple animals) and tiny plant-like dinoflagellates breaks down (Figure 6.9). When this happens, usually as a result of thermal stress, the brown dinoflagellates leave the coral turning it a brilliant white colour (bleached). This deprives the coral of access to the organic carbon that dinoflagellates normally provide to corals as a result of their photosynthetic activities, and leads to an increased risk of starvation, disease and death. Coral bleaching first became evident on coral reefs in 1979, and has become a regular feature of the world's coral reefs since then (Hoegh-Guldberg, 1999). Mass coral bleaching events are caused by small increases (1–2 °C) in sea temperature over six to eight weeks. The more intense and long-lasting, the greater the effect temperature has in driving mass coral bleaching and mortality. In 1998, coral reefs in almost every part of the world underwent mass coral bleaching (Hoegh-Guldberg, 1999), and an estimated 16% of the world's corals disappeared in a little over 12 months (Wilkinson *et al.*, 1999).

In addition to the impact of warming seas, corals are likely to be severely impacted by ocean acidification (Kleypas and Langdon, 2006), which occurs as a result of the increased flux of CO_2 into the world's oceans (Figure 6.10 and Chapter 2). When CO_2 enters the ocean, it reacts with water to produce a dilute acid (carbonic acid). The acid then reacts with carbonate

Figure 6.9 Coral bleaching occurs when the symbiosis between corals and tiny plant-like organisms called dinoflagellates breaks down. This generally happens as a response to stress. Mass coral bleaching events that started to occur in 1979 have been traced to small increases in sea temperature above the long-term summer maxima (1–2 °C). When deprived of their source of photosynthetic energy, corals tend to experience increased levels of starvation, disease, and death. Source: O. Hoegh-Guldberg.

Box 6.3 (*cont.*)

ions, converting them into bicarbonate ions and thereby decreasing the concentration of carbonate ions. Reduced carbonate ion concentrations directly decreases the ability of corals to produce their calcium carbonate skeletons. The majority of evidence suggests that exceeding atmospheric CO_2 concentrations of 450 p.p.m. will decrease carbonate ion concentrations to a point where corals are no longer able to maintain calcium carbonate reef structures (Hoegh-Guldberg *et al.*, 2007). Recent studies suggest that the impacts of these changes to ocean temperature and acidity are already being felt. Coral populations are decreasing at the rate of 1%–2% per year in most parts of South-East Asia and the western Pacific (Bruno and Selig, 2007), while coral calcification has declined by 15% since 1990 – a situation unprecedented in the 400 years of record investigated so far in Australia (De'ath *et al.*, 2009) and Thailand (Tanzil *et al.*, 2009).

Projections of future ocean temperatures and acidities suggest that coral-dominated ecosystems will be rare even under the mildest IPCC scenarios (Figure 6.11; Hoegh-Guldberg, 1999; Done *et al.*, 2003; Donner *et al.*, 2005; Hoegh-Guldberg *et al.*, 2007). Given that corals are crucial for building and maintaining the carbonate structures of coral reefs, this change has serious implications for the million species that depend on coral reefs, and ecological services provided by coral reefs for coastal human populations (Hoegh-Guldberg

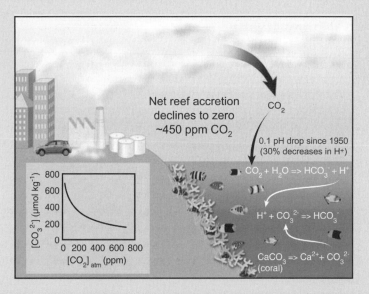

Figure 6.10 Increasing concentrations of atmospheric carbon dioxide have resulted in a greater flux of CO_2 into the ocean. This has acidified the world's oceans, driving down the concentrations of carbonate ions. The latter are important for marine calcifiers such as corals. At CO_2 concentrations of 450 p.p.m., concentrations of carbonate ions decrease to levels that are unable to sustain calcification rates to keep up with reef erosion. Source: adapted from Hoegh-Guldberg *et al.* (2007).

Figure 6.11 Extant examples of reefs from the Great Barrier Reef that are used as analogues for the ecological structures anticipated by Hoegh-Guldberg *et al.* (2007) under increasing amounts of atmospheric CO_2. (A) If concentrations of CO_2 remain at today's level, many coral dominated reefs will survive although there will be a compelling need to increase their protection from local factors such as deteriorating coastal water quality and overfishing. (B) If CO_2 concentrations continue to rise as expected, reefs will become less dominated by corals and increasingly dominated by other organisms such as seaweeds. (C) If CO_2 levels continue to rise as we burn fossil fuels, coral reefs will disappear and will be replaced by crumbling mounds of eroding coral skeletons. In concert with the progression from left to right is the expectation that much of the enormous and largely unexplored biodiversity of coral reefs will disappear. This will almost certainly have major impacts on the tourist potential of coral reefs as well as their ability to support fisheries, both indigenous and industrial. Source: modified from Hoegh-Guldberg *et al.* (2007). (In colour as Plate 21.)

et al., 2000, 2009). While there are clear linkages between climate change, ecosystem health and human well-being, our understanding of the details of how these massive changes to tropical coastal ecosystems will affect people is still in its infancy and should be a future research priority.

Charismatic species such as polar bears are often the pin-up stars of biodiversity–climate change discussions, but a systematic analysis of climate change impacts on cultural services more generally has seldom been undertaken. Box 6.4 describes the relationship between climate change, ecosystem services and social systems and, in particular, explores the consequences for human health of a loss of cultural services.

Box 6.4
Climate, ecosystem services and social systems

THOMAS ELMQVIST

Climate change is not only altering ecological systems and biodiversity, but also poses fundamental challenges for managers, policymakers and other actors in developing strategies to respond to uncertainties in projected climate change impacts on ecosystems (Scheffer *et al.*, 2001; Lobell *et al.*, 2008) and possible nonlinear responses (Carpenter *et al.*, 2009). A proactive approach to meeting these challenges could be based on the view that ecosystems provide crucial options for communities to buffer the impacts of environmental disturbances, extreme events and change. From such a perspective, well-functioning ecosystems must be maintained and degraded services restored where possible. Ecosystem services that, in this context, need to receive considerable attention are regulating services related to protection from hazards, i.e. reducing the impacts of natural forces on social systems and the managed environments. However, this capacity may easily be eroded, and the magnitude of many hazards is sensitive to human interaction with the natural environment – including flash floods due to extreme rainfall events on heavily managed ecosystems that cannot retain rainwater; landslides and avalanches on deforested slopes; and fires caused by prolonged droughts in managed landscapes (TEEB, 2008). In some cases, management may lead to ecosystem degradation, resulting in complete loss of protective and buffering capacity. Strategies to meet expected changes in ecosystem services – including building more flexible and adaptive institutions and governance systems – have to take on a wider agenda on baseline vulnerability, and acknowledge the need to build resilience as a way to buffer or steer away from abrupt ecosystem changes and loss of buffering capacity.

In what ways do the temporal and spatial dynamics of ecosystem change – under climate impacts – challenge existing governance arrangements at multiple levels of societal organisation, and why do governance systems often fail to protect vital ecosystem functions and resources? Important external factors can include a lack of political interest, administrative fragmentation and inefficiency, and the presence of free-riding behaviour. However, a lack of capacity to understand the behaviour of multi-scale ecosystems and their associated services is a fundamental but seldom elaborated factor. At worst, policymaking can unintentionally increase vulnerability to climate change.

It is in this context important to distinguish between adaptive capacity – the incremental and frequent adjustments by social actors undertaken more or less daily to deal with change in order to maintain the status quo, i.e. to sustain the current development pathways; and transformative capacity – the ability to fundamentally alter the nature of the system over the long term, when current ecological, social, or economic conditions become untenable or are undesirable.

The challenge of climate change can be used proactively to stimulate innovation and learning within society to build resilience, and produce forward-looking decisions and strategies that could contribute to a needed transformation of social–ecological systems (Folke *et al.*, 2005). Specifically, future research on climate–ecosystem services–social systems should focus on (i) analysing different trajectories in adaptation processes in linked social–ecological systems and identifying which trajectories represent undesirable lock-ins, and (ii) identifying how novel management of ecosystems may facilitate desirable transformations.

6.4 Limits to adaptive capacity

Adaptation is now unavoidable, given the observed impacts on biodiversity and those sure to come given the committed climate change already in the pipeline. Many tools and strategies for adapting biodiversity conservation to a changing climate are rapidly being developed and are beginning to be deployed (Chapter 14). Here the focus is on the limits to adaptive capacity, which is critical information for defining dangerous climate change and, thus, for informing the mitigation approaches required to avert dangerous climate change.

In exploring the limits to adaptive capacity, several important constraints to adaptation need to be considered, including the fundamental fact that many adaptation approaches may be costly and thus require widespread community support to be implemented, support that does not yet exist. Perhaps an even more fundamental constraint is the daunting lack of knowledge of how complex systems like natural ecosystems respond to multiple, interacting stresses and to particular management activities. For example, savannas are influenced by a very complex web of interactions involving elevated CO_2 concentration, temperature, precipitation, reproductive strategies, grass–shrub–tree balance, herbivory, disturbance and others (Pedersen *et al.*, 2009). In contrast, the most common research tool at present for climate impact analysis of such systems is the projection of changing distributions of individual species driven by a single climatic variable such as temperature. More sophisticated models that simulate functional responses to multiple stresses at ecosystem and landscape levels are under development. However, serious questions remain about our ultimate capacity to anticipate, even to a first-order approximation, the actual outcomes of such complex dynamics.

In the face of such high levels of uncertainty, enhancing the resilience of well-functioning ecosystems at all scales is often a central pillar of an effective adaptation strategy. Increasing resilience can indeed be effective up to a point, beyond which facilitating transformation will be a more appropriate management response when 'saving' existing ecosystems becomes futile. It appears likely that this resilience–transformation threshold may be transgressed often under continuing climate change.

The transformation of existing ecosystems will often involve the change of well-loved ecosystems towards new and different systems that may not be recognisable and hence not as valued as the ecosystems they replace. Thus, this strategy may well lead to social conflict and challenge the perspectives of those with a more static view of the natural world. An example of facilitating transformation is the proposed assisted colonisation of coral species to new locations to enable them to move fast enough to deal with climate change (Hoegh-Guldberg *et al.*, 2008), an activity sometimes called 'ecoengineering'. In addition to being controversial, these approaches are invariably costly and not guaranteed of success.

Where adaptive capacity has reached its limit and there are still strong reasons to conserve particular species, *ex situ* conservation (preservation of a species or relevant genetic material in an environment that is not its natural home, i.e. in captivity) can play an important complementary role. For example, *ex situ* conservation of crop cultivars can provide insurance against *in situ* extinctions and increase flexibility in the response of agricultural systems to climate change. *Ex situ* conservation can also play a role in assisted migration.

However, *ex situ* approaches will only be feasible for a few valuable species, and cannot substitute for the maintenance of well-functioning ecosystems in the natural world.

Ultimately, the ability of ecosystems to cope with the rates and magnitudes of climate change projected for this century will depend on their capability to self-adapt; that is, the ability of their biotic components to evolve or to move through landscapes and seascapes at the required rates to track their environmental envelopes and to reorganise into new, functioning ecosystems. While such rapid adaptation may be possible in principle, human modification of the land, the coastal zone and the coastal seas presents considerable barriers and constraints to self-adaptation. The ultimate limit to adaptation of biodiversity, then, may be our willingness and ability to significantly reconfigure coastal and landscape patterns and uses (e.g. away from maximising production towards multi-functional agricultural landscapes) to maximise the fluidity for species migration (Manning *et al.*, 2009). Such an undertaking on a global scale is unprecedented in human history.

6.5 The need for strong mitigation action

In terms of biodiversity and the provision of ecosystem services, the trade-offs between adaptation and mitigation are strikingly clear. Although adaptation to climate change is an urgent priority for biodiversity conservation, many species and ecosystems will be unable to adapt – even with assistance – to the rapid rates of climate change projected for the rest of this century and beyond if the current trajectories of greenhouse gas emissions are not turned sharply downwards. The capabilities of many species either to evolve or to migrate will simply not be sufficient to deal with such pervasive changes to their abiotic environment in such a short period of time (Box 6.5). Strong mitigation action is therefore essential if there is to be any chance of averting a mass extinction event this century.

Box 6.5
Implications of the rate and magnitude of climate change for biodiversity

WILL STEFFEN

The magnitude of the looming biodiversity crisis can perhaps be best understood by considering the severe challenge that species, communities and ecosystems face in adapting to climate change. Although the biosphere has adapted in the past to climates far different from that of the Holocene, it is probably only during the abrupt climatic shifts associated with mass extinction events that the biosphere has experienced global-scale change at the rate projected for this century without rapid and effective mitigation.

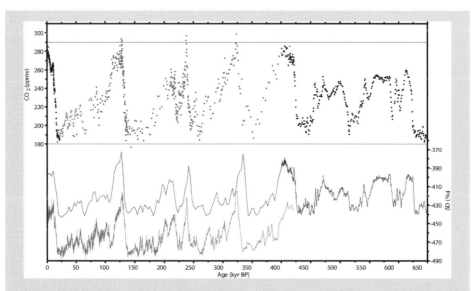

Figure 6.12 A composite CO_2 record over six and a half ice age cycles, back to 650 000 years b.p. (before present) (Siegenthaler *et al.*, 2005). The record results from the combination of CO_2 data from three Antarctic ice cores: Dome C (black), 0 to 22 k years b.p. and 390 to 650 k years b.p.; Vostok (blue), 0 to 420 k years b.p., and Taylor Dome (light green), 20 to 62 years b.p. Black line indicates δD from Dome C, 0 to 400 k years b.p. and 400 to 650 k years b.p. Blue line indicates δD from Vostok, 0 to 420 k years b.p. δD (deuterium) is a proxy for temperature. Source: references to individual ice core data are found in Siegenthaler *et al.* (2005). (In colour as Plate 25.)

Today's biosphere, which delivers the ecosystem services on which contemporary society depends, has evolved in the environmental envelope of the late Quaternary period of the past few million years. Ice core data from Antarctica (Figure 6.12; EPICA community members, 2004; Siegenthaler *et al.*, 2005) quantify the nature of this envelope (Chapter 4).

The rate of the climatic changes being observed now and projected for this century is particularly important for assessing the biosphere's capability to adapt. Figure 6.13 shows the projected increases in temperature by 2100 from the IPCC Third Assessment Report (IPCC, 2001) (the more recent projections from the Fourth Assessment Report are not significantly different) on the same timescale as that for the natural variability observed over the past millennium. The dark grey shading in the figure defines the envelope of natural variability for present-day ecosystems. Even the most optimistic scenario for 2100 implies a sharp excursion from the envelope of natural variability at a rapid rate. At the upper range of projections – a ~5 °C increase over pre-industrial levels – the biosphere would have to adapt in just 100 years to the same magnitude of climatic change as between the last ice age and the current warm period, a transition that spanned 5000–10 000 years (Lorius *et al.*, 1990).

Box 6.5 (*cont.*)

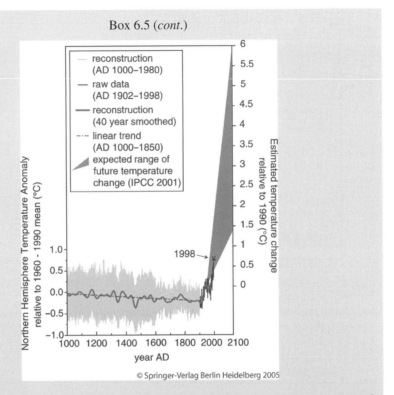

Figure 6.13 The range of future projections of temperature change compared to the envelope of natural variability for the past millennium. Source: derived from Mann *et al.* (1999) and IPCC (2001). Reproduced/modified by permission of American Geophysical Union.

A complementary way of representing the adaptation challenge that the biosphere faces is shown in Figure 6.14. A dynamic global vegetation model has been used to simulate the equilibrium biome distribution for Australia for the Last Glacial Maximum (~20 000 years ago), the present climate and the climate projected for a 600 p.p.m. CO_2 world in 2100. Although the details of the distributions are sensitive to the climate and vegetation models used, the magnitude of simulated changes are robust. The changes to Australia's biota are at least as great between now and the end of the century as during the 5000–10 000-year transition from the last ice age to the present. The selective pressure on many organisms, particularly long-lived organisms, to adapt to such rapid climatic change will be extreme.

The message from such analyses is clear. In a world where much of the land surface is highly fragmented, the ocean is acidifying and the composition of marine ecosystems has been altered by human predation, the biosphere faces an impossible task in trying to adapt to exceptionally rapid climate change. The best hope for minimising biodiversity loss this century is undoubtedly rapid and deep reductions in greenhouse gas emissions.

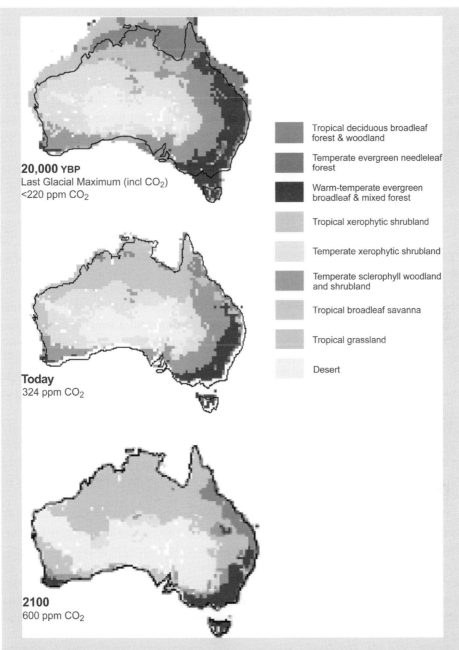

Figure 6.14 Modelled historical Australian biome distribution. Model data show distribution at the Last Glacial Maximum (approximately 200 p.p.m. CO_2), the 'present' (>324 p.p.m. CO_2) and simulated for 2100 (600 p.p.m. CO_2 scenario) using the dynamic global vegetation model BIOME 4.0. The details of the simulated biome distributions in 2100 are sensitive to the nature of the particular climate and vegetation models used for the projections; shown is a typical example of change. Source: S. Harrison, Bristol University, personal communication, modified from Steffen *et al.* (2009). (In colour as Plate 23.)

Conserving as much biodiversity as possible goes well beyond saving charismatic species for their own sake, and is critical for maintaining ecosystem functioning and the provision of ecosystem services. Impacts on provisioning services such as food and water resources are the most obvious (Chapter 5), but the many subtle roles that biodiversity plays in the provision of other services – from pest and disease control through the regulation of regional climate to cultural services that are important for many people – may ultimately have just as great an impact on human well-being. The consequences of a failed global mitigation policy could literally be catastrophic, playing out through a collapse of many critical ecosystem services and the subsequent destabilisation of the geopolitical landscape through conflict and mass migration.

6.6 Implications for conservation policy

The long-standing goals of many current biodiversity conservation strategies are to preserve species in the locations they are currently found and to maintain ecosystems with the same composition they currently have. These often strongly held goals will become increasingly counterproductive as climate change accelerates, and species and ecosystems attempt to cope with these changes to their abiotic environment. Changing existing biodiversity conservation goals and policies will not be easy, despite the obviously severe impacts of climate change. Overcoming the constraints to change will require considerable resources and a major effort in lifting the importance of biodiversity, and the consequent provision of ecosystem services, to a major societal concern equal to that of security and economic prosperity.

The sheer magnitude of what is required to achieve this transformation of attitude and action towards biodiversity and its conservation is, itself, an important consideration for the definition of dangerous climate change. The less effective mitigation is, the more costly will become attempts at adaptation, the less likely will become their success, and the more probable will become rapid deterioration of ecosystem services.

At a minimum, existing conservation goals will need to change towards maintaining good ecological functioning and diversity *in situ* everywhere, and minimising loss of biodiversity overall. This implies a major shift in values and a massive increase in investment: (i) reserve systems will need to be expanded, hopefully to encompass future 'biorefugia' as they develop; (ii) better integration of on-reserve and off-reserve conservation will need to be achieved; and (iii) policies on invasive species and quarantine will need to be reviewed and modified. Such changes to conservation goals may raise significant challenges in gaining social and cultural acceptance.

Furthermore, given the very fundamental nature of climate change, many of the policy and management tools that have been developed over the past century to implement biodiversity conservation will be rendered ineffective. The past will no longer be a reliable guide to the future. New and modified policy and management approaches will be most robust if they go back to fundamental ecological principles, requiring a significant overhaul of education and training, institutional architecture, and policy development (Steffen *et al.*, 2009).

The resource requirements to support these updated conservation goals and the adaptation activities based on them, especially for scenarios of no or little mitigation, are probably at least an order of magnitude more than is currently invested in natural capital (Steffen *et al.*, 2009). A significant part of this investment should go towards enhancing the reserve system, perhaps 2.5 to 3 times the current level of investment (Andelman *et al.*, 2009). This investment is urgently required as land-use decisions being made now are foreclosing options in many countries.

6.7 Summary and conclusions

The message from the most recent research on biodiversity and climate change is exceedingly clear.

- Anthropogenic climate change is a new and different stressor; it affects the fundamental conditions for life and its rate will likely exceed the capacity of many species and ecosystems to cope.
- The impacts of climate change on biodiversity are observable now, and are very likely to intensify through the rest of the century.
- The potential consequences for the provision of ecosystem services are severe.
- The limits to adaptive capacity will likely be reached at lower levels of temperature rise than for other sectors that need to adapt.

Without very rapid and effective mitigation, climate change will likely drive a further increase in the already high level of biodiversity loss. There will likely be significant losses of species and genetic diversity, and major shifts in species and biome distributions and ecosystem composition. The result will be significant further degradation of ecosystem services. Even with the strongest and most effective possible mitigation, active adaptive management of ecosystems is required to maintain and enhance the functional resilience of ecosystems. It is critical that this challenge is met in order to maintain 'biodiversity security', defined by the UNFCCC as 'sufficient stock of current and future biodiversity and ecosystem services to sustain human beings'.

In both this and the previous chapter, when considering the potential impacts of climate change on living organisms, it has been assumed that climate change occurs linearly, i.e. gradually and incrementally. Evidence is accumulating, however, that this might not be the case and that there is a risk of abrupt change in environmental conditions occurring in response to climate change. This evidence is examined in the next chapter.

References

Andelman, S., Midgley, G., Polasky, S. *et al.* (2009). Ensuring biodiversity security in the face of climate change. *IOP Conference Series: Earth and Environmental Science*, **6**, 312005.

Attorre, F., Vitale, M., Tomasetto, F. *et al.* (2009). Effect of climate change on tree species distribution to support the elaboration of adaptive management strategies in natural protected areas. *IOP Conference Series: Earth and Environmental Science*, **6**, 312012.

Barbier, E. and Cox, M. (2002). Economic and demographic factors affecting mangrove loss in the coastal provinces of Thailand, 1979–1996. *Ambio*, **31**, 351–57.

Beaugrand, G., Reid, P. C., Ibañez, F., Lindley, J. A. and Edwards, M. (2002). Reorganisation of North Atlantic marine copepod biodiversity and climate. *Science*, **296**, 1692–94.

Birdlife International (2008). *State of the World's Birds: Indicators for our Changing World.* Cambridge, UK: Birdlife International.

Bruno, J. F. and Selig, E. R. (2007). Regional decline of coral cover in the Indo-Pacific: Timing, extent, and subregional comparisons. *PLoS ONE*, **2**, e711.

Bryant, D., Burke, L., McManus, J. and Spalding, M. (1998). *Reefs at Risk: A Map-based Indicator of Threats to the World's Coral Reefs.* Washington, D.C.: World Resources Institute.

Carpenter, S. R., Mooney, H. A., Agard, J. *et al.* (2009). Science for managing ecosystem services: Beyond the Millennium Ecosystem Assessment. *Proceedings of the National Academy of Sciences (USA)*, **106**, 1305–12.

Cesario, M., Cesario, R. R. and Andrade-Morraye, M. (2009). The impacts of climate change on ecosystem services: The case of disrupted disease regulation in South-Western Amazonia. *IOP Conference Series: Earth and Environmental Science*, **6**, 302007.

Chapin III, F. S., Randerson, J. T., McGuire, A. D., Foley, J. A. and Field, C. B. (2008). Changing feedbacks in the climate–biosphere system. *Frontiers in Ecology and the Environment*, **6**, 313–20.

Cleland, E. E., Chuine, I., Menzel, A., Mooney, H. A. and Schwartz, M. D. (2007). Shifting plant phenology in response to global change. *Trends in Ecology & Evolution*, **22**, 357–65.

De'ath, G., Lough, J. M. and Fabricius, K. E. (2009). Declining coral calcification on the Great Barrier Reef. *Science*, **323**, 116–19.

Done, T., Whetton, P., Jones, R. *et al.* (2003). *Global Climate Change and Coral Bleaching on the Great Barrier Reef.* Final report to the State of Queensland Greenhouse Taskforce through the Department of Natural Resources and Mining. Brisbane: Queensland Government Department of Natural Resources.

Donner, S. D., Skirving, W. J., Little, C. M., Oppenheimer, M. and Hoegh-Guldberg, O. (2005). Global assessment of coral bleaching and required rates of adaptation under climate change. *Global Change Biology*, **11**, 2251–65.

EPICA community members (2004). Eight glacial cycles from an Antarctic ice core. *Nature*, **429**, 623–28.

Folke, C., Hahn, T., Olsson, P. and Norberg, J. (2005). Adaptive governance of social–ecological systems. *Annual Review of Environment and Resources*, **30**, 441–73.

Hansen, J. L. S. and Bendtsen, J. (2009). Effects of climate change on hypoxia in the North Sea–Baltic Sea transition zone. *IOP Conference Series: Earth and Environmental Science*, **6**, 302016.

Hansen, J. W., Nedergaard, M. and Skov, F. (2009). Ecological consequences of climate change and how to measure them. *IOP Conference Series: Earth and Environmental Science*, **6**, 302015.

Hobbs, R. J. and Mooney, H. A. (1995). Effects of episodic rain events on mediterranean-climate ecosystems. In *Time Scales of Biological Responses to Water Constraints*, eds. J. Roy, J. Aronson and F. di Castri. Amsterdam: SPB Academic Publishing, pp. 71–85.

Hodgkin, T., Snook, L., Bellon, M. and Frison, E. (2009). Redefining in situ and ex situ conservation of useful plant diversity in an era of climate change. *IOP Conference Series: Earth and Environmental Science*, **6**, 312003.

Hoegh-Guldberg, O. (1999). Climate change, coral bleaching and the future of the world's coral reefs. *Marine and Freshwater Research*, **50**, 839–66.

Hoegh-Guldberg, O., Hoegh-Guldberg, H., Stout, D. K., Cesar, H. and Timmerman, A. (2000). *Pacific in Peril: Biological, Economic and Social Impacts of Climate Change on Pacific Coral Reefs*. Suva, Fiji: Greenpeace.

Hoegh-Guldberg, O., Hoegh-Guldberg, H., Veron, J. E. *et al.* (2009). *The Coral Triangle and Climate Change: Ecosystems, People and Societies at Risk*. Brisbane: WWF Australia, http://www.panda.org/coraltriangle.

Hoegh-Guldberg, O., Hughes, L., McIntyre, S. *et al.* (2008). Assisted colonization and rapid climate change. *Science*, **321**, 345–46.

Hoegh-Guldberg, O., Mumby, P. J., Hooten, A. J. *et al.* (2007). Coral reefs under rapid climate change and ocean acidification. *Science*, **318**, 1737–42.

Hole, D., Turner, W., Brooks, T. *et al.* (2009). Towards an adaptive management framework for climate change across species, sites and land/seascapes. *IOP Conference Series: Earth and Environmental Science*, **6**, 312006.

Intergovernmental Panel on Climate Change (IPCC) (2001). *Climate Change 2001: The Scientific Basis. Contribution of Working Group I to the Third Assessment Report of the Intergovernmental Panel on Climate Change*, eds. J. T. Houghton, Y. Ding, D. J. Griggs, M. Noguer, P. J. van der Linden, X. Dai, K. Maskell and C. A. Johnson. Cambridge, UK and New York, NY: Cambridge University Press.

Intergovernmental Panel on Climate Change (IPCC) (2007a). Summary for policymakers. In *Climate Change 2007: Impacts, Adaptation and Vulnerability. Contribution of Working Group II to the Fourth Assessment Report of the Intergovernmental Panel on Climate Change*, eds. M. L. Parry, O.F. Canziani, J. P. Palutikof, P. J. van der Linden and C.E. Hanson. Cambridge, UK and New York, NY: Cambridge University Press, pp. 7–22.

Intergovernmental Panel on Climate Change (IPCC) (2007b). *Climate Change 2007: The Physical Science Basis. Contribution of Working Group I to the Fourth Assessment Report of the Intergovernmental Panel on Climate Change*, eds. S. Solomon, D. Qin, M. Manning, Z. Chen, M. Marquis, K. B. Averyt, M. Tignor and H. L. Miller. Cambridge, UK and New York, NY: Cambridge University Press.

Jackson, J. B. C. (2008). Ecological extinction and evolution in the brave new ocean. *Proceedings of the National Academy of Sciences (USA)*, **105**, 11458–65.

Jarvis, A., King, N., Gaiji, S. *et al.* (2009). GBIF: Infrastructure, standards and access to data and tools to forecast changes in agricultural production. In *Climate Change: Global Risks, Challenges and Decisions. International Scientific Congress*, Copenhagen, Denmark, 10–12 March. London: Institute of Physics IOP, p. 1.

Kathiresan, K. and Rajendran, N. (2006). Coastal mangrove forests mitigated tsunami. *Estuarine, Coastal and Shelf Science*, **65**, 601–06.

Kleypas, J. A. and Langdon, C. (2006). Coral reefs and changing seawater chemistry. In *Coral Reefs and Climate Change: Science and Management. AGU Monograph Series, Coastal and Estuarine Studies*, eds. J. T. Phinney, O. Hoegh-Guldberg, J. Kleypas, W. Skirving and A. Strong. Washington, D.C.: American Geophysical Union.

Lenoir, J., Gégout, J. C., Marquet, P. A., de Ruffray, P. and Brisse, H. (2008). A significant upward shift in plant species optimum elevation during the 20th century. *Science*, **320**, 1768–71.

Lindenmayer, D. (2009). Ecosystem services: An example of monetary valuation. In *Australia's Biodiversity and Climate Change*, eds. W. Steffen, A. Burbidge, L. Hughes, R. Kitching, D. Lindenmayer, W. Musgrave, M. Stafford Smith and P. Werner. Melbourne: CSIRO Publishing.

Lobell, D. B., Burke, M. B., Tebaldi, C. *et al.* (2008). Prioritizing climate change adaptation needs for food security in 2030. *Science*, **319**, 607–10.

Lorius, C., Jouzel, J., Raynaud, D., Hansen, J. and Le Treut, H. (1990). The ice-core record: Climate sensitivity and future greenhouse warming. *Nature*, **347**, 139–45.

Mann, M. E., Bradley, R. S. and Hughes, M. K. (1999). Northern hemisphere temperatures during the past millennium: inferences, uncertainties, and limitations. *Geophysical Research Letters*, **26**, 759–62.

Manning, A. D., Fischer, J., Felton, A. *et al.* (2009). Landscape fluidity – a new perspective for understanding and adapting to global change. *Journal of Biogeography*, **36**, 193–99.

McKenzie, N. L., Burbidge, A. A., Baynes, A. *et al.* (2007). Analysis of factors implicated in the recent decline of Australia's mammal fauna. *Journal of Biogeography*, **34**, 597–611.

Mieszkowska, N., Hawkins, S. and Burrows, M. (2009). Climate-driven changes in coastal marine biodiversity: trends, forecasts & management implications. *IOP Conference Series: Earth and Environmental Science*, **6**, 312021.

Millennium Ecosystem Assessment (2005). *Ecosystems and Human Well-being. General Synthesis*. Washington, D.C.: Island Press.

Miller, C., Indriani, G., Suroso, D., Jovanovic, T. and Alexander, K. (2009). Reintegrating native biodiversity in Indonesia's agricultural landscapes: Reducing the vulnerability of the rural poor to climate change; IARU International Scientific Congress on Climate Change, Copenhagen, 10–12 March 2009, Session 31 (Spoken paper).

Mølgaard, P. (2009). Herbs are hurt, shrubs will thrive in a warmer arctic climate results of ITEX – The International Tundra Experiment. *IOP Conference Series: Earth and Environmental Science*, **6**, 312022.

Møller, P. R. (2009). Is a recent increase of the Greenland fish fauna caused by climate change? *IOP Conference Series: Earth and Environmental Science*, **6**, 312023.

Mooney, H. A. (2007). The costs of losing and restoring ecosystem services. In *Managing and Designing Landscapes for Conservation*, eds. D. B. Lindenmayer and R. J. Hobbs. Oxford: Blackwell Press, pp. 365–75.

Moritz, C., Patton, J. L., Conroy, C. J. *et al.* (2008). Impact of a century of climate change on small-mammal communities in Yosemite National Park, USA. *Science*, **322**, 261–64.

Moy, A. D., Howard, W. D., Bray, S. G. and Trull, T. W. (2009). Reduced calcification in modern Southern Ocean planktonic foraminifera. *Nature Geoscience*, **2**, 276–80.

Parmesan, C. (2006). Ecological and evolutionary responses to recent climate change. *Annual Review of Ecology, Evolution, and Systematics*, **37**, 637–69.

Parmesan, C. and Yohe, G. (2003). A globally coherent fingerprint of climate change impacts across natural systems. *Nature*, **421**, 37–42.

Pedersen, J. K., Arndal, M. F. and Schmidt, I. K. (2009). Above and belowground phenology in a heathland during future climate change. *IOP Conference Series: Earth and Environmental Science*, **6**, 312011.

Perry, A. L., Low, P. J., Ellis, J. R. and Reynolds, J. D. (2005). Climate change and distribution shifts in marine fishes. *Science*, **308**, 1912–15.

Potvin, C., Chapin III, F. S., Gonzalez, A. *et al.* (2007). Plant biodiversity and responses to elevated carbon dioxide. In *Terrestrial Ecosystems in a Changing World*, eds.

J. Canadell, D. Pataki and L. Pitelka., Heidelberg, Germany: Springer-Verlag, pp. 125–34.

Pounds, J. A., Bustamante, M. R., Coloma, L. A. *et al.* (2006). Widespread amphibian extinction from epidemic disease driven by global warming. *Nature*, **439**, 161–67.

Reaka-Kudla, M. L. (1997). Global biodiversity of coral reefs: A comparison with rainforests. In *Biodiversity II: Understanding and Protecting Our Biological Resources*, eds. M. L. Reaka-Kudla, D. E. Wilson and E. O. Wilson. Washington, D.C.: Joseph Henry Press.

Réale, D., McAdam, A. G., Boutin, S. and Berteaux, D. (2003). Genetic and plastic responses of a northern mammal to climate change. *Proceedings of the Royal Society B*, **270**, 591–96.

Reusch, T. B. H. and Wood, T. E. (2007). Molecular ecology of global change. *Molecular Ecology*, **16**, 3973–92.

Scheffer, M., Straile, D., van Nes, E. H. and Hosper, H. (2001). Climatic warming causes regime shifts in lake food webs. *Limnology and Oceanography*, **46**, 1780–83.

Scholze, M., Knorr, W., Arnell, N. W. and Prentice, I. C. (2006). A climate-change risk analysis for world ecosystems. *Proceedings of the National Academy of Sciences (USA)*, **103**, 13116–20.

Siegenthaler, U., Stocker, T. F., Monnin, E. *et al.* (2005). Stable carbon cycle–climate relationship during the late Pleistocene. *Science*, **310**, 1313–17.

Sitch, S., Smith, B., Prentice, I. C., *et al.* (2003). Evaluation of ecosystem dynamics, plant geography and terrestrial carbon cycling in the LPJ dynamic global vegetation model. *Global Change Biology*, **9**, 161–85.

Staehr, P., Wernberg, T., Thomsen, M. *et al.* (2009). Global warming is eroding the resilience of kelp beds. *IOP Conference Series: Earth and Environmental Science*, **6**, 312008.

Stampfli, A. and Zeiter, M. (2009). Prediction and mitigation of undesirable change in semi-natural grassland due to extreme droughts and hay management. *IOP Conference Series: Earth and Environmental Science*, **6**, 312027.

Steffen, W., Burbidge, A., Hughes, L. *et al.* (2009). *Australia's Biodiversity and Climate Change*. Melbourne: CSIRO Publishing.

Stone, R. (2007). A world without corals? *Science*, **316**, 678–81.

Tanzil, J. T. I., Brown, B. E., Tudhope, A. W. and Dunne, R. P. (2009). Decline in skeletal growth of the coral *Porites lutea* from the Andaman Sea, South Thailand between 1984 and 2005. *Coral Reefs*, **28**, 519–28.

Tape, K., Sturm, M. and Racine, C. (2006). The evidence for shrub expansion in Northern Alaska and the Pan-Arctic. *Global Change Biology*, **12**, 686–702.

TEEB (2008). *The Economics of Ecosystems and Biodiversity*. European Communities, http://www.teebweb.org.

Thomas, C. D., Cameron, A., Green, R. E. *et al.* (2004). Extinction risk from climate change. *Nature*, **427**, 145–48.

Van Buskirk, J., Mulvihill, R. S. and Leberman, R. C. (2009). Variable shifts in spring and autumn migration phenology in North American songbirds associated with climate change. *Global Change Biology*, **15**, 760–71.

Visser, M. E. and Both, C. (2005). Shifts in phenology due to global climate change: the need for a yardstick. *Proceedings of the Royal Society B*, **272**, 2561–69.

Walther, G.-R., Post, E., Convey, P. *et al.* (2002). Ecological responses to recent climate change. *Nature*, **416**, 389–95.

Warren, R., Price, J., Fischlin, A., Midgley, G. and de la Nava Santos, S. (2009). Increasing impacts of climate change upon ecosystems with increasing global mean temperature rise. *IOP Conference Series: Earth and Environmental Science*, **6**, 302037.

Wilkinson, C. (2008). *Status of Coral Reefs of the World: 2008*. Townsville, Australia: Global Coral Reef Monitoring Network and Reef and Rainforest Research Centre.

Wilkinson, C. R. (1999). Global and local threats to coral reef functioning and existence: Review and predictions. *Marine Freshwater Research*, **50**, 867–78.

Wilkinson, C. R. and Buddemeier, R. W. (1994). *Global Climate Change and Coral Reefs: Implications for People and Reefs*. Report of the UNEP-IOC-ASPEI-IUCN Global Task Team on the Implications of Climate Change on Coral Reefs. Gland: IUCN.

Wilkinson, C., Lindén, O., Cesar, H. *et al.* (1999). Ecological and socioeconomic impacts of 1998 coral mortality in the Indian Ocean: An ENSO impact and a warning of future change? *Ambio*, **28**, 188–96.

Willis, K. J. and Bhagwat, S. A. (2009). Biodiversity and climate change. *Science*, **326**, 806–07.

World Resources Institute (WRI) (2006). The value of coastal ecosystems. *Earth Trends*, November 2006 monthly update.

7

Tipping elements: jokers in the pack

'This is an externality like no other... there is a big probability of a devastating outcome'[1]

Striking developments in the climate system in recent years have reinforced the view that anthropogenic global warming is unlikely to cause a smooth transition into the future. The record minimum area coverage of Arctic sea-ice in September 2007 drew widespread attention, as has the accelerating loss of water from the Greenland and West Antarctic ice sheets (Chapter 3). These large-scale components of the Earth System are among those that have been identified as potential 'tipping elements' – climate sub-systems that could exhibit a 'tipping point' where a small change in forcing (in particular, global temperature change) causes a qualitative change in their future state (Lenton *et al.*, 2008). The resulting transition may be either abrupt or irreversible, or in the worst cases, both. The most 'policy-relevant' tipping elements have been defined as those that (i) have a tipping point that could be crossed this century, (ii) would undergo, as a consequence, a qualitative change within this millennium, thereby (iii) affecting (if not damaging) a large number of people. For a full definition of a tipping element and its tipping point, see Box 7.1 (and Lenton *et al.*, 2008). In IPCC terms such changes are referred to as 'large-scale discontinuities' (Smith *et al.*, 2009). Should they occur, they would surely qualify as dangerous climate changes (Schellnhuber *et al.*, 2006) (although not all are equally dangerous, as we will explore further).

We use the metaphor of 'jokers in the pack' because large-scale tipping points are conventionally viewed as 'high-impact but relatively low probability' events. Recent work suggests there are rather more jokers in the climate pack of cards than the metaphor might lead one to expect: a shortlist of nine policy-relevant tipping elements in the climate system has been identified that might be triggered by human activities this century (Lenton *et al.*, 2008). The metaphor is apt in another way: When gambling with the climate, we will never know exactly when a joker is going to turn up, because the trigger of any

---

[1] Professor Lord Nicholas Stern, IARU International Scientific Congress on Climate Change, Copenhagen, 10–12 March 2009.

Climate Change: Global Risks, Challenges and Decisions, Katherine Richardson, Will Steffen and Diana Liverman *et al.* Published by Cambridge University Press. © Katherine Richardson, Will Steffen and Diana Liverman 2011.

tipping-point change is likely to be a combination of (stochastic) natural variability on top of an underlying (deterministic) forcing due to human activities. Thus, we can only talk in terms of probabilities of passing particular tipping points. However, recent expert elicitation has obtained some useful information on these probabilities for different future warming scenarios (Kriegler *et al.*, 2009). The probabilities are necessarily imprecise, but *even with the most conservative assumptions* they indicate that in a 4 °C warmer world it is more likely than not that at least one of five large-scale thresholds will be passed. The key message from recent studies is that tipping-point change now appears significantly closer, in terms of global temperature change, than it did in earlier assessments (Smith *et al.*, 2009).

This chapter has two main aims. First, we want to revisit (and slightly revise) the list of potential policy-relevant tipping elements (Lenton *et al.*, 2008), based on the IARU congress and recent literature, assessing some new candidates and reassessing others in the light of the definition (Box 7.1), and provide some updates on the science of specific tipping elements, especially where there is new insight into the mechanisms behind them, or new information about the proximity of tipping points. Second, we want to expand on the policy implications of tipping elements, and explore how we might move towards risk management of them.

Figure 7.1 shows an updated map of the potential policy-relevant tipping elements in the climate system introduced by Lenton *et al.* (2008). As there are rather a lot of 'jokers'

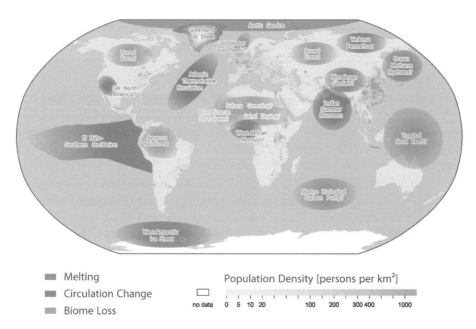

Figure 7.1 Map of potential policy-relevant tipping elements overlain on global population density – adjusted from Lenton *et al.* (2008) based on contents of this chapter. Question marks indicate systems whose status as policy-relevant tipping elements is particularly uncertain. Figure by Veronika Huber, Timothy M. Lenton and Hans Joachim Schellnhuber. (In colour as Plate 24.)

to cover, we subdivide the potential tipping elements into those involving melting of large masses of ice, those involving changes in the circulation of the atmosphere and ocean, and those involving the loss of unique biomes.

Box 7.1
Defining a tipping element and its tipping point

TIMOTHY M. LENTON

In colloquial terms, the phrase 'tipping point' captures the notion that 'little things can make a big difference' (Gladwell, 2000), in other words, at a particular moment in time, a small change can have large, long-term consequences for a system. To apply the term usefully to the climate (or in any other scientific context), it is important to be precise about what qualifies as a tipping point, and about the class of systems that can undergo such change. To this end, the term 'tipping element' has been introduced (Lenton *et al.*, 2008) to describe large-scale subsystems (or components) of the Earth system that can be switched – under certain circumstances – into a qualitatively different state by small perturbations. In this context, the tipping point is the corresponding critical point – in forcing and a feature of the system – at which the future state of the system is qualitatively altered.

   To formalise this further, it is important to define a spatial scale; only components of the Earth system associated with a specific region or collection of regions, which are at least sub-continental in scale (length scale of order ~1000 km), are considered. For such a system to qualify as a tipping element, it must be possible to identify a single control parameter (ρ), for which there exists a critical control value (ρ_{crit}), from which a small perturbation ($\delta\rho > 0$) leads to a qualitative change ($\hat{F}$) in a crucial feature of the system (F), after some observation time ($T > 0$). The actual change (ΔF) is measured with respect to a reference state of the feature at the critical value:

$$|\Delta F| = |F(\rho \geq \rho_{crit} + \delta\rho \,|T)| - F(\rho_{crit}\,|T) \geq \hat{F} > 0. \qquad (7.1)$$

In this definition, the critical threshold (ρ_{crit}) is the tipping point, beyond which a qualitative change occurs, and the change may occur immediately after the cause or much later.

   Following Lenton *et al.* (2008) we restrict ourselves to tipping elements in the climate system that are potentially relevant to current policy. The subset of 'policy-relevant' tipping elements is defined by the following (additional) conditions. (i) Human activities are interfering with the system such that decisions taken within a 'political time horizon' ($T_P \sim 100$ years) can determine whether the tipping point (ρ_{crit}) is crossed. If it is crossed, (ii) the time to observe a qualitative change (including the time to trigger it) lies within an 'ethical time horizon' ($T_E \sim 1000$ years). (iii) A significant number of people care about the fate of the system because either it contributes significantly to the overall mode of operation of the Earth system, or it contributes significantly to human welfare, or it has great value in itself as a unique feature of the biosphere.

   For a system to possess a tipping point there must be strong positive feedback in its internal dynamics. Many scientists take 'tipping point' to be synonymous with a 'bifurcation point'

Box 7.1 (*cont.*)

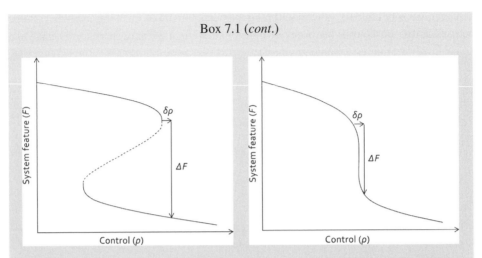

Figure 7.2 Schematic of two types of tipping element that can exhibit a tipping point where a small change in control ($\delta\rho$) results in a large change in a system feature (ΔF), illustrated here in terms of the time-independent equilibrium solutions of the system: (left) a system with bi-stability passing a bifurcation point, (right) a mono-stable system exhibiting highly non-linear change. Source: Timothy M. Lenton.

in the equilibrium solutions of a system, as schematically illustrated in Figure 7.2 (left). This implies that passing a tipping point necessarily carries some irreversibility. However, other classes of non-linear transition can meet the definition above, and one schematic example is given in Figure 7.2 (right). Again this shows the (time-independent) equilibrium solutions of a system, but here they are continuous (there is no bifurcation) and, therefore, the transition is reversible. In reality, the existence or not of a tipping point should be considered in a time-dependent fashion, and there could be several other possible types of tipping elements (for the mathematical details, see the Supplementary Information of Lenton *et al.*, 2008). Theoretically, one can construct elements that react infinitely slowly to tipping, yet do this in an entirely irreversible fashion.

7.1 Tipping elements involving melting of ice

The concept of a threshold is intuitively obvious when thinking about ice melting to liquid water – an example of a first-order phase transition. However, that happens on a relatively small scale. For major masses of ice on Earth to qualify as climate tipping elements they must exhibit a large-scale threshold due to strong positive feedbacks in their internal dynamics, coupled to the climate and, on passing this threshold, they must undergo a qualitative change in state (Box 7.1).

7.1.1 Arctic sea-ice

The summer minimum area cover of Arctic sea-ice has declined markedly in recent decades, most strikingly in 2007. Increased heating due to exposure of the dark ocean surface (the ice–albedo feedback) is playing a key role (Box 7.2). Ice cover has fallen below all IPCC model projections, despite the models having been in agreement with the observations of summer ice cover in the 1970s (Stroeve *et al.*, 2007). Winter sea-ice is also declining in area (though less rapidly) and there is an overall, progressive thinning of the ice cap. Further warming appears likely to occur and this raises the possibility that the Arctic may already be committed to a qualitative change in which the ocean becomes largely ice-free during summer months. One summer, perhaps as early as the 2030s or 2040s, will see a combination of natural variability and overall thinning causing loss of nearly all the ice. The year that the North Pole becomes ice-free in the summer will likely be seen as a 'tipping point' by non-experts, but it is unlikely to involve an irreversible bifurcation (as in Figure 7.2; Eisenman and Wettlaufer, 2009). Summer sea-ice quickly recovers once the climate turns cold again because of a stabilising feedback related to the ice growth rate (Notz, 2009). Nevertheless, it may still qualify as a tipping element because, as the ice cap gets thinner, it becomes prone to larger fluctuations in area, which can be triggered by relatively small changes in forcing (Holland *et al.*, 2006). Also, it is conceivable that loss of summer ice involves additional dynamical feedbacks that lead to a qualitative change in atmosphere or ocean circulation and heat transports. If so, the impacts are likely to be felt further afield, for example in Europe.

Loss of winter (i.e. year-round) ice is more likely to represent a bifurcation (Figure 7.2, left), where the system can switch rapidly and irreversibly from one state (with seasonal ice) to another (without any) (Eisenman and Wettlaufer, 2009). However, the threshold for year-round ice loss requires around 13 °C warming at the North Pole (Winton, 2006). Whether this is accessible this century depends on the uncertain strength of polar amplification of warming.

Box 7.2
The decline of Arctic sea-ice

DONALD PEROVICH

The Arctic Ocean is covered by floating sea-ice, with an areal extent of millions of square kilometres. While vast in extent, it is only a thin veneer that is a few metres thick. This combination makes it an excellent proxy indicator for warming in the Arctic. Stated simply, if the Arctic is warming, there should be a decrease in ice extent and thickness. The warming signal will be greatest at the end of the summer melt season when the ice cover reaches its annual minimum. Figure 7.3 shows the time-series of yearly minimum ice extent determined from satellite observations from 1979 to 2009. While year-to-year fluctuations are seen in the record, an accelerating downward trend is evident. Most notable is the dramatic retreat in the summer of 2007, when the ice extent reached a record minimum of 4.2 million square kilometres – a 46% decrease from 1980. There have been modest increases in 2008 and 2009,

Defining 'dangerous climate change'

Box 7.2 (*cont.*)

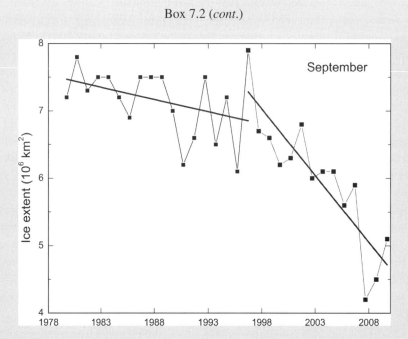

Figure 7.3 Time-series of sea-ice extent at the end-of-summer minimum in September. The straight lines are least squares linear regression for the periods 1979–1996 and 1996–2009. Source: data from National Snow and Ice Data Center, USA.

but those years still rank as the second and third smallest ice extents. Other key characteristics of the Arctic sea-ice are changing as well. There has been a decrease in the thicker, more robust, multi-year ice and an increase in the thinner first-year ice. In the past decade, there has been a loss of 1.5 million square kilometres in multi-year ice coverage (Nghiem *et al.*, 2007). Ice thickness observations made using satellite altimetry and submarine sonar show a decrease of mean winter multi-year ice thickness from 3.6 m to 1.9 m over the past three decades (Kwok and Rothrock, 2009).

The observed decline in sea-ice is directly linked to sea-ice dynamic and thermodynamic forcing, and is affected by feedback processes. Recent declines are consistent with acceleration due to the ice–albedo feedback as open ocean water causes increased absorption of solar radiation. The solar heating is warming the upper ocean and contributing significantly to melting on the bottom of the ice. From 1979 to 2007, 85% of the Arctic region received an increase in solar heat input at the surface, with an increase of 5% per year in some regions including the Beaufort Sea (Perovich *et al.*, 2007). *In situ* measurements in this region show that there was a three times greater bottom-ice melt in 2007 compared to earlier years (with relatively little change in surface melt) (Perovich *et al.*, 2008). Other factors contributing to record ice loss include patterns of atmospheric circulation (Rigor and Wallace, 2004; Maslanik *et al.*, 2007) and ocean circulation (Nghiem *et al.*, 2007), which have exported multi-year ice out of the Arctic basin through the Fram Strait, reductions in summertime cloud cover

(Kay *et al.*, 2008) and increased input of ocean heat from the Pacific (Shimada *et al.*, 2006; Woodgate *et al.*, 2006).

The declining Arctic sea-ice cover has prompted considerable discussion about the possibility of a summer ice-free Arctic Ocean. While there are many estimates, there is no definitive answer as to when, or if, this will occur. However, there is a related question that can be answered: When will the decrease of summer Arctic sea-ice affect human activity? This threshold has already occurred. In recent years, the declining ice cover has led to increased shipping, tourism and resource exploration in the Arctic Ocean, a trend that is expected to continue.

7.1.2 Greenland Ice Sheet (GIS)

The Greenland Ice Sheet (Box 3.3) is currently losing mass at a rate that has been accelerating (Rignot *et al.*, 2008). In summer 2007, there was an unprecedented increase in surface melt, mostly south of 70° N and also along the west side of Greenland, due to an up to 50-day longer melt season than average with an earlier start (Mote, 2007). This is part of a longer-term trend of increasing melt extent since the 1970s. Recent observations show that seasonal surface melt has led to accelerated glacier flow (Joughin *et al.*, 2008; van de Wal *et al.*, 2008). The surface mass balance of the GIS is still positive (there is more incoming snowfall than melt at the surface, on an annual average), but the overall mass balance of the GIS is negative due to an increased loss flux from calving of glaciers that outweighs the positive surface mass balance. The margins of the GIS are thinning at all latitudes (Pritchard *et al.*, 2009), and the rapid retreat of calving glaciers terminating in the ocean, most notably Jakobshavn Isbræ, is probably linked to warming ocean waters (Holland *et al.*, 2008).

The GIS will be committed to irreversible meltdown if the surface mass balance goes negative, because as the altitude of the surface declines it gets warmer (a strong positive feedback). Initial assessments put the temperature threshold for this to occur at around 3 °C of regional warming, based on a positive-degree-days model for the surface mass balance (Huybrechts and De Wolde, 1999). Results from an expert elicitation concur that if global warming exceeds 4 °C there is a high probability of passing the threshold (Kriegler *et al.*, 2009). However, an alternative surface energy balance model predicts a more distant threshold at 8 °C regional warming or ~6 °C global warming (J. Bamber, personal communication). The actual threshold for massive GIS shrinkage is presumably passed before the surface mass balance goes negative. A more nuanced possibility, which is emerging from some coupled climate–ice sheet model studies, is that there might be multiple stable states for GIS volume and, hence, multiple tipping points (J. Lowe, personal communication). Passing a first tipping point where the GIS retreats onto land could lead to ~15% loss of the ice sheet and about 1 m of global sea-level rise. As for the rate at which this could occur, an upper limit is that the GIS could contribute around 50 cm to global sea-level rise this century (Pfeffer *et al.*, 2008).

7.1.3 West Antarctic Ice Sheet (WAIS)

The West Antarctic Ice Sheet (Box 3.4) is also losing mass at present, and some parts – particularly draining into the Amundsen Sea – are thinning rapidly (Pritchard *et al.*, 2009). Air temperatures have recently been shown to be warming across West Antarctica (Steig *et al.*, 2009), but the shrinkage of the WAIS is more sensitive to the intrusion of warming ocean waters, and the collapse of floating ice shelves that buttress the main ice sheet. The WAIS may be vulnerable to large-scale collapse, due to retreat of the grounding line where the ice sheet is pinned to the bedrock below sea level (Weertman, 1974; Mercer, 1978). Having been strongly questioned when it was first introduced, the paradigm of a potential abrupt collapse of the WAIS has recently gained new momentum (Vaughan, 2008). Recent theory has confirmed the potential for multiple stable states of the grounding line and hence bifurcation-type tipping points (Figure 7.2, left) (Schoof, 2007). Also, new palaeo-data have shown that the WAIS collapsed repeatedly during the ~3 °C warmer world of the early Pliocene (5–3 million years ago) (Naish *et al.*, 2009). Modelling supports this and suggests further collapses during some (but not all) of the more recent Pleistocene interglacials (Pollard and DeConto, 2009). Furthermore, East Antarctic ice cores show anomalous spikes of warmth (above present) during all of the past four interglacial intervals, which might be explained by repeated WAIS collapse (P. Holden *et al.*, personal communication). Data from the last (Eemian) interglacial suggest the up-to-2 °C-warmer world of the time had peaks of sea level up to 9 m higher than present and rates of sea-level rise of 1.6 ± 1.0 m per century (Rohling *et al.*, 2008). To achieve such rates of sea-level rise probably required rapid grounding line retreat of the WAIS, and possibly parts of the periphery of the East Antarctic Ice Sheet that are also grounded below sea level. Current models put the threshold for WAIS collapse when the surrounding ocean warms by ~5 °C (Pollard and DeConto, 2009), and expert elicitation concurs that if global warming exceeds 4 °C, it is more likely than not that the WAIS will collapse (Kriegler *et al.*, 2009). Depending on the rate of dynamic ice-sheet decay, Antarctica could contribute up to about 60 cm to global sea-level rise this century (Pfeffer *et al.*, 2008).

7.1.4 Yedoma permafrost

Continuous permafrost is the perennially frozen soil, which currently covers ~10.5 million km^2 of the Arctic land surfaces, but is melting rapidly in some regions (Box 4.4). This area could be reduced to as little as 1.0 million km^2 by the year 2100, which would represent a qualitative change in state (Lawrence and Slater, 2005). However, permafrost did not make the original shortlist of tipping elements (Lenton *et al.*, 2008) because of a lack of evidence for a large-scale threshold for permafrost melt. Instead, in future projections, the local threshold of freezing temperatures is exceeded at different times in different localities. More recent work has suggested, however, that at least one large area of permafrost could exhibit coherent threshold behaviour. The frozen loess (windblown dust) of Eastern Siberia (150–168° E and 63–70° N), also known as Yedoma, is deep (~25 m) and has an extremely high carbon content (2%–5%), thus it may contain ~500 Gt C (Zimov *et al.*, 2006). Recent studies have shown the potential for this regional frozen-carbon store to undergo self-sustaining collapse, due to an internally generated source of heat released by biochemical decomposition of the carbon, triggering further melting in a runaway positive feedback (Khvorostyanov *et al.*, 2008a, 2008b). Once underway, this process could release 2.0–2.8 Gt C yr^{-1} (mostly as CO_2 but with some methane) over about a century, removing ~75% of the initial carbon stock. The collapse would be irreversible in the strongest sense that removing the forcing would not stop it continuing. To pass the tipping point requires an estimated >9 °C of regional warming (Khvorostyanov *et al.*, 2008a). However, this is a region already experiencing strongly amplified warming, partly linked to shrinkage of the Arctic sea-ice (Lawrence *et al.*, 2008). During August–October 2007, Arctic land temperatures jumped around 3°C above the mean for the preceding 30 years (from analysis of HadCRUT data). Thus, the Yedoma tipping point may be accessible this century under high emissions scenarios, so we add it to the shortlist of policy-relevant tipping elements.

7.1.5 Ocean methane hydrates

Recent model estimates suggest that up to 2000 Gt C are stored as methane hydrates beneath the ocean floor (Archer *et al.*, 2009). As the deep ocean warms, heat diffuses into the sediment layer and may destabilise this reservoir of frozen methane. Bubbles associated with the melting of methane may trigger submarine landslides (Kayen and Lee, 1991). This finding raised the concern that the destabilisation of methane hydrates could result in abrupt massive release of greenhouse-potent methane into the atmosphere. If this scenario was plausible, methane hydrates would clearly qualify as a policy-relevant tipping element (Box 7.1). However, recent palaeo-climatic evidence makes this scenario very unlikely (Archer, 2007). Instead, the most likely impact of a melting hydrate reservoir is a long-term chronic methane source (Archer *et al.*, 2009). Archer *et al.* (2009) estimated an additional warming of 0.4–0.5 °C from the hydrate response to fossil fuel CO_2 release, persisting over several millennia. This estimate is subject to large uncertainties, in particular with regard to the magnitude of temperature forcing required to trigger the destabilisation of methane hydrates. Thus, methane hydrates (Section 4.3) are unlikely to qualify as a policy-relevant tipping element (Box 7.1; as defined by Lenton *et al.*, 2008), but they can be considered a slow and, for societal purposes, irreversible tipping element in the global carbon cycle.

7.1.6 Himalayan glaciers

It has been suggested that the Hindu Kush–Himalaya–Tibetan (HKHT) glaciers should be added to the list of tipping elements because much of their mass could be lost this century (Ramanathan and Feng, 2008). The downstream human impacts of this will be profound

(Box 5.3). However, they do not meet the definition of a tipping element (Box 7.1) unless a large-scale threshold can be identified. Instead, we expect HKHT mass loss to be roughly proportional to warming. Perhaps a new category of changes is required to encompass such high-impact eventualities.

7.2 Changes in atmospheric and oceanic circulation

The circulations of the ocean and atmosphere coupled together and to the land surface, can exhibit different dynamical stable states and modes of variability (Chapter 2), with potential thresholds between them. They can also be particularly sensitive to gradients of forcing, as these are usually what drive the circulations in the first place. Monsoons are a seasonal example, driven by more rapid heating of the land than the ocean, which causes warm air to rise over the continent, creating a pressure gradient that sucks in moist air from over the ocean, which then rises, its water condenses and rain falls, releasing latent heat that reinforces the circulation (Levermann *et al.*, 2009).

7.2.1 El Niño Southern Oscillation (ENSO)

The ENSO phenomenon is the most significant natural mode of coupled ocean–atmosphere variability in the climate system (Box 2.2). Over the past century, warming has been greater in the eastern than the western equatorial Pacific, and this has been linked to El Niño events becoming more severe (e.g. in 1983 and 1998). Recently, a changing pattern of El Niño has been noted toward 'Modoki' events where the warm pool shifts from the west to the middle (rather than the east) of the equatorial Pacific (Ashok and Yamagata, 2009; Yeh *et al.*, 2009). In future projections, the first coupled model studies predicted a shift from current ENSO variability to more persistent or frequent El Niño conditions. Now that numerous models have been compared, there is no consistent trend in frequency. However, in response to a stabilised 3–6 °C warmer climate, the most realistic models simulate increased El Niño amplitude (Guilyardi, 2006). Also, a shift toward 'Modoki' events has been forecast (Yeh *et al.*, 2009). Furthermore, palaeo-data indicate different ENSO regimes under different climates of the past (Box 2.1). Despite large persisting uncertainties, the probability of ENSO either vanishing or becoming overly strong is estimated to be rather low during the twenty-first century (Latif and Keenlyside, 2009). The mechanisms and timescale of any transition are unclear, but a gradual increase in El Niño amplitude and/or a shift in location are consistent with the recent observational record and would, nevertheless, have severe impacts in many regions. In summary, it is unclear whether ENSO qualifies as a tipping element (Box 7.1) because a large-scale threshold has yet to be clearly identified.

7.2.2 Atlantic thermohaline circulation (THC)

The archetypal example of a tipping element is the THC, which is prone to collapse when sufficient freshwater enters the North Atlantic to halt density-driven deep water (NADW) formation there. State-of-the-art modelling, which minimises artefacts arising from numerical diffusion, shows that a hysteresis-type response to freshwater perturbations is a characteristic, robust feature of the THC (Hofmann and Rahmstorf, 2009). However, the shutdown of the THC may actually be one of the more distant tipping points. Expert elicitation suggests that THC collapse only becomes as likely as not with >4 °C warming this century (Kriegler *et al.*, 2009). The IPCC (2007) views the threshold as even more remote, but recent analysis suggests the AR4 models are systematically biased towards a stable THC (Box 7.3). Although a collapse of the THC may be one of the more distant tipping points, a weakening of the THC this century is robustly predicted (IPCC, 2007) due to freshening of the North Atlantic by increased precipitation in high latitudes and melting of ice. This in turn will have similar, though smaller, effects than a total collapse. A potential tipping point that occurs in some models is a shut-off of deep convection and NADW formation in the Labrador Sea region (to the west of Greenland) and switch to convection only in the Greenland–Iceland–Norwegian seas (to the east of Greenland). This would have dynamic effects on sea level, increasing it down the eastern seaboard of the USA by around 25 cm in the regions of Boston, New York and Washington, D.C. (in addition to the global steric effects of ocean warming) (Yin *et al.*, 2009).

Box 7.3
Is the THC bi-stable at present?

S. L. WEBER AND S. S. DRIJFHOUT

Many modelling studies have shown that the THC exhibits hysteresis behaviour, where it can be either in the mono-stable regime (that is, only the present active state exists) or in the bi-stable regime (both the active and the collapsed states exist). If the present THC is in the bi-stable regime, then global warming and associated changes in the hydrological cycle, combined with GIS melt, might induce a permanent collapse. If it is in the mono-stable regime, such a collapse is extremely unlikely – even in the case of considerable warming.

Hence, it is of great interest to diagnose the stability of the THC. How can we do this? Model experiments suggest that the freshwater budget of the Atlantic basin is an important indicator of stability (de Vries and Weber, 2005). Several processes contribute to the budget. First, there is the exchange of water at the air–sea interface: precipitation, evaporation and runoff from the surrounding continents. Second, there are freshwater transports by ocean currents at the southern border of the basin and, to a lesser extent, through the Bering Strait. The southern transport is due to the azonal, mainly wind-driven, circulation in the upper layers and to the THC itself. The latter term, called the overturning freshwater transport Mov, seems to play a key role. Model sensitivity experiments show that the THC is bi-stable, if the overturning freshwater transport is associated with a net loss of freshwater (Mov negative). It is mono-stable if Mov is positive. In the latter case, the overturning recovers after a collapse, because of a large-scale salinity overturning feedback, which reinforces the remnant circulation cell that exists in the collapsed state. The sign of the overturning freshwater transport is, thus, a diagnostic for THC stability.

Box 7.3 (*cont.*)

The distribution of the southern transport over the azonal component and the THC-related component depends on the relative salinities and strengths of these flows. As such, it is difficult to measure these quantities directly. Using inverse-model data, it has been suggested that the present-day ocean has Mov = -0.20 Sv (Weijer *et al.*, 1999). This implies a bi-stable state for the present THC.

We now consider the Atlantic freshwater budget, as it is obtained from data analyses and from different climate models (Figure 7.4). Three of these models are coupled atmosphere–ocean GCMs that have been presented in the IPCC Fourth Assessment Report (AR4) (IPCC, 2007) for past and future climate simulations. Two are Earth System Models of Intermediate Complexity that have also contributed to AR4 and one of these, ECBilt/CLIO, has been used in the sensitivity experiments described previously. Of the latter model, we show two versions: the standard version, which was found to be mono-stable (Mov positive); and a corrected version, which was found to be bi-stable (Mov negative). All AR4 models are seen to have serious flaws in their freshwater budget, likely due to errors in the salinity distribution and/or the circulation in the southern Atlantic. As a result, all models have positive Mov, indicating mono-stability in contrast to the observational data (Weber *et al.*, 2007). This strongly suggests that current climate models are systematically biased towards a stable THC, overestimating the distance to the tipping point. A further detailed analysis of more AR4 models (research in progress) has confirmed this view.

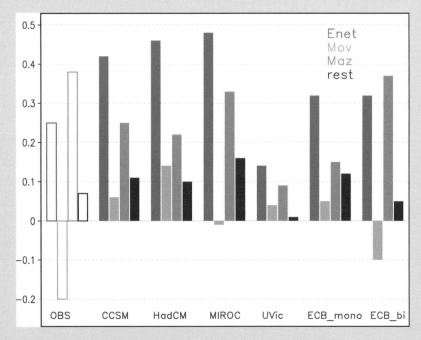

Figure 7.4 The terms of the Atlantic freshwater budget as obtained from data analysis (OBS) and different climate models that were used in AR4: CCSM3.0, HadCM3M2, MIROC3.2, UVic ESCM2.7 and two versions of ECBilt2/CLIO, which have been demonstrated to be mono-stable (ECB_mono) and bi-stable (ECB_bi). The terms (in Sv = Sverdrup = $10^6\,\mathrm{m^3\,s^{-1}}$) are net evaporation (dark grey), the overturning (light grey) and azonal (mid-grey) ocean transports and a rest term (black). AR4 models are seen to be biased as compared to observations, suggesting that THC stability is overestimated.

7.2.3 *West African Monsoon (WAM) and Sahel/Sahara precipitation*

Past intervals of severe drought in West Africa have been linked to weakening of the THC (Chang *et al.*, 2008; Shanahan *et al.*, 2009). The latter event seems to trigger a phenomenon known as the Atlantic Niño (by analogy with El Niño events in the equatorial Pacific), involving reduced stratification and warming of the sea surface in the Gulf of Guinea. This disrupts the WAM, which is usually enhanced by the development of a 'cold tongue' in the eastern equatorial Atlantic that increases the temperature contrast between the Gulf of Guinea and the land to the north. In a typical year, there is also a northward 'jump' of the monsoon into the Sahel in July, which corresponds to a rapid decrease in coastal rainfall and the establishment of the West African westerly jet in the atmosphere (Hagos and Cook, 2007). The jump is due to a tipping point in atmospheric dynamics: when the east/west wind changes sharply in the north/south direction, this instability causes the northward perturbation of an air parcel to generate additional northward flow (a strong positive feedback).

It is not clear in which direction the WAM might shift in future. The more benign alternative is a greening of the Sahel/Sahara, by a mechanism that can be related to observations of the seasonal northward shift of the WAM (Box 7.4). The more dangerous option is a southward shift of the WAM. If ocean temperatures change such that the West African westerly jet fails to form or is weakened below the tipping point needed to create inertial instability, then the rains may fail to move into the continental interior, drying the Sahel. Recent simulations suggest a tipping point if the THC weakens below ~8 Sv (a Sverdrup is equivalent to 10^6 m^3 s^{-1}), causing the subsurface North Brazil Current to reverse and an abrupt warming in the Gulf of Guinea (a persistent Atlantic Niño state). The WAM then shifts such that there is a large reduction in rainfall in the Sahel, and an increase in the Gulf of Guinea and coastal regions (Chang *et al.*, 2008). Such a transition is forecast in one of only three IPCC (2007) models that produces a realistic present climate in this region (Cook and Vizy, 2006). However, the other two models give conflicting responses; in one the Sahel gets markedly wetter despite a collapse of the WAM (Box 7.4), in the other there is little net change.

Changes in atmospheric circulation patterns and precipitation over West Africa would have profound effects on dust output from the Bodélé Depression and other major dust sources of the Sahara (Washington *et al.*, 2009). Palaeoclimatic evidence suggests that dust production in this region ceased abruptly during wetter conditions of the mid-Holocene. North African dust, transported westwards over the Atlantic, supplies minerals to coral reefs in the Caribbean region and to the Amazonian rainforests. Therefore, a shift of the WAM that triggers a strong increase or decrease in dust emissions would have direct consequences extending far beyond Africa.

Box 7.4
The potential for abrupt change over northern Africa

KERRY H. COOK

The complicated climate system of West Africa and the Sahel presents several possibilities for tipping points, and the potential for abrupt and/or irreversible changes in rainfall distributions. An intricate system of atmospheric jets accompanies the West African monsoon flow, and the region exhibits multiple sources of moisture, nonlinear dynamics and global connectivity.

A number of modelling studies have suggested that the regional vegetation–climate system in West Africa has two stable climate states, with a potential for rapid changes between them (Prentice *et al.*, 1992; Brovkin *et al.*, 1998). One is today's climate, with a dry Sahara and marginally moist Sahel. The other is the 'green Sahara' climate of African Humid Periods, the most recent of which occurred about 8000 years ago in association with stronger summer insolation compared to today. Recent work suggests that a strong positive feedback between atmospheric circulation patterns and soil moisture distributions creates these dual stable states (Patricola and Cook, 2008). Today, the mid-tropospheric African easterly jet marks the northern extent of the West African monsoon circulation, and transports moisture off the continent below the level of condensation. Sources of moisture include the south-westerly monsoon flow across the Guinean coast and the West African westerly jet, which flows from the eastern Atlantic onto the continent at about 10 °N. A shallow vertical system forms over the high temperatures of the Sahara Desert, with the Saharan high centered near 600 hPa over the continental thermal low (Figure 7.5). In the alternative state (Figure 7.5), the low-level monsoon inflow is similar to today but the African easterly jet does not form. Moist convection extends into what is now the Sahara, with moisture supplied by the intensification and deepening of the West African westerly jet. The tipping point between the states occurs when savannah vegetation extends to about 18°N due to, for example, decadal variability, sea surface temperature forcing, and/or land surface changes.

Internal atmospheric dynamics can – and do – give rise to abrupt changes in precipitation fields over West Africa every spring, suggesting a potential for abrupt climate change. The northward progression of rainfall into the Sahel in boreal spring and summer is not smooth, but involves a sudden shift of the rainfall maximum into the continental interior (Sultan and Janicot, 2003). High rainfall rates develop along the Guinean coast in May, and linger there typically until sometime in mid June to mid July. At some point in that month-long period, the rains suddenly diminish along the coast and, within a few days, the rainfall maximum reforms in the Sahel. This is the onset of the West African monsoon, also called the 'monsoon jump', and the population of the Sahel is highly dependent on these seasonal rains for agriculture and water resources. The jump is related to the development of inertially unstable flow along the Guinean coast as the continental thermal low progresses north and becomes established over the Western Sahel and Sahara. The resulting positive meridional (north–south) gradients in the zonal (east–west) wind are large enough to pass the threshold for inertial instability and, as a

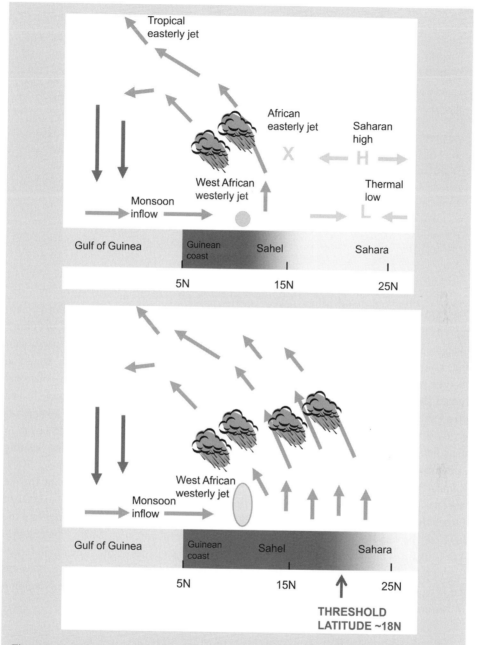

Figure 7.5 Schematic latitude/height cross-section of the summer circulation across West Africa in (upper) today's climate, and (lower) the African Humid Period. The West African westerly jet is directed out of the page, and the African easterly jet is directed into the page. Dark grey shading indicates the presence of vegetation; light grey of desert.

Box 7.4 (*cont.*)

result, low-level divergence of the flow occurs along the Guinean coast. The resulting sudden demise of rainfall along the coast promotes the northward jump of the rainfall maximum (K. H. Cook, unpublished data; Hagos and Cook, 2008). If warming oceans or other changes in future climate prevent this instability from developing, the rains may not move into the continental interior and the Sahel will become drier.

It is clear that there is a potential for abrupt climate change in northern Africa, as such rapid change is recognised in the palaeo-climate record and we can capture mechanisms of abrupt change over northern Africa in our models. But a *potential* for abrupt change is different from a proven reality, and the nature of the physical processes suggests that there will be significant regionality in climate change. Improvements in our ability to model and understand climate change and variability in northern Africa, which include sustaining and expanding observational systems, are needed for more confident prediction.

7.2.4 Indian Summer Monsoon (ISM)

The ISM system is already being influenced by aerosol and greenhouse gas forcing. Palaeo-records indicate its volatility, with flips on and off of monsoonal rainfall linked to climate changes in the North Atlantic (Burns *et al.*, 2003; Gupta *et al.*, 2003). Greenhouse warming, which is stronger over northern hemisphere land than over the Indian Ocean, would on its own be expected to strengthen the monsoonal circulation. However, the observational record shows declines in ISM rainfall, which have been linked to an 'atmospheric brown cloud' haze created by a mixture of black carbon (soot) and sulphate aerosols (Ramanathan *et al.*, 2005). The haze is more concentrated over the continent than over the ocean to the south, and it causes more sunlight to be absorbed in the atmosphere and less heating at the surface. Hence, it tends to weaken the monsoonal circulation (Meehl *et al.*, 2008; Ramanathan and Carmichael, 2008). In simple models, there is a tipping point for the regional planetary albedo (reflectivity; Box 1.2) over the continent which, if exceeded, causes the ISM to collapse altogether (Zickfeld *et al.*, 2005). Data analysis suggests the ISM may be relatively close to a critical threshold (Levermann *et al.*, 2009). The real picture is likely to be more complex, with the potential for switches in the strength and location of the monsoonal rains. Increasing aerosol forcing could further weaken the monsoon but if then removed, greenhouse warming could trigger a stronger monsoon, producing a 'roller-coaster ride' for hundreds of millions of people (Zickfeld *et al.*, 2005).

7.2.5 Southwest North America (SWNA)

Increased humidity in a warmer world causes increased moisture divergence, thus changing global atmospheric circulation – including poleward expansion of the Hadley cells and the subtropical dry zones (Held and Soden, 2006; Lu *et al.*, 2007); a development that tends to strongly reduce runoff in these regions (Milly *et al.*, 2005). One area that may be particularly affected is SWNA, defined as all land in the region 125–95° W and 25–40° N. Aridity in this domain is robustly predicted to intensify and persist in the future, and a transition is probably already underway – to something that has been described as 'unlike any climate state we have seen in the instrumental record' (Seager *et al.*, 2007). However, it does not qualify as a tipping element unless a large-scale threshold can be identified (Box 7.1). Recently, increased SWNA aridity has been linked to the potential for increased flooding in the Great Plains (Cook *et al.*, 2008). The key driver is model-projected relatively higher summer warming over land than over ocean (analogous to what drives seasonal monsoons). In simulations of future dynamics, an increased contrast between the continental low and the North Atlantic sub-tropical high strengthens the Great Plains low-level jet, which transports moisture from the Caribbean to the upper Great Plains, triggering flooding there but starving SWNA of moisture (Cook *et al.*, 2008).

7.3 Biome loss

At the ecosystem scale, there are probably many potential thresholds related to climate change, the most obvious of which is the 'extinction' (disappearance) of a unique type of ecosystem, because it has nowhere to retreat to (e.g. the Fynbos on the southern tip of Africa, or ecosystems high up mountains). Such changes are clearly a concern to policymakers, but it is not clear that they meet the definition of a tipping element, which focuses on the larger scale of what on land are called biomes, and looks for the existence of a threshold in the internal dynamics triggered by (or coupled to) the climate.

7.3.1 *Amazon rainforest*

A severe drought occurred from July to October in 2005 in western and southern parts of the Amazon basin, which led the Brazilian government to declare a state of emergency. Despite subsequent 'greening up' of large areas of forest (Saleska *et al.*, 2007), the 2005 drought made the Amazon region a significant episodic carbon source, when otherwise it has been a carbon sink (Phillips *et al.*, 2009). The 2005 drought has been linked to unusually warm sea-surface temperatures in the North Atlantic (Cox *et al.*, 2008). But lengthening of the Amazon dry season is also part of a wider trend in seasonality, associated with weakening of the zonal tropical Pacific atmospheric circulation as attributed to anthropogenic greenhouse gas forcing (Vecchi *et al.*, 2006). The trend of a lengthening dry season is forecast to continue, with one model predicting that the 2005 drought will be the norm by 2025 (Cox *et al.*, 2008). If drying continues, several model studies have shown the potential for significant dieback of up to ~70% of the Amazon rainforest by late this century, and its replacement by savannah and caatinga (mixed shrubland and grassland) (Cook and Vizy, 2008). There are positive feedbacks related to the current ways rainforests store and recycle water to the atmosphere, which could accelerate the

Amazon demise. Ecosystem disturbance processes, such as increased fire frequency and pest infestation, could also amplify a transition initially driven by drought. Experts suggest Amazon dieback is more likely than not if global warming exceeds 4 °C (Kriegler *et al.*, 2009). However, the Amazon rainforest may lag climate forcing significantly and hence it may be committed to some dieback long before it is apparent (Box 7.5). In one model, committed dieback begins at 1 °C global warming above pre-industrial levels, even though it does not begin to be observed until global warming approaches 4 °C (Jones *et al.*, 2009). The existence and extent of forecast Amazon dieback depends on the choice of climate model (Scholze *et al.*, 2006; Salazar *et al.*, 2007), because not all GCM projections give a regional, seasonal drying trend in the Amazon. A recent analysis based on 19 GCMs has indicated that many models tend to underestimate current rainfall and that, although dry-season water stress is likely to increase over the twenty-first century, the rainfall regime of eastern Amazonia is likely to shift in the direction of seasonal forests rather than savannah (Malhi *et al.*, 2009). Dieback is generally less sensitive to the choice of vegetation model, but the direct effect of CO_2 increasing the water-use efficiency of vegetation can have a strong effect of tending to shift the dieback threshold further away (P. M. Cox, personal communication).

Box 7.5
Committed ecosystem changes

CHRIS D. JONES AND JASON LOWE

Terrestrial ecosystems may respond sensitively to climate change and have been identified as possible tipping elements (Cox *et al.*, 2004; Lenton *et al.*, 2008). This is in part because local biophysical feedbacks exist between the land surface and climate (Claussen *et al.*, 1999). In Amazonia, increased (decreased) forest cover leads to increased (decreased) precipitation, and vice-versa (Betts *et al.*, 2004). Such a feedback opens the possibility of bi-stability or tipping-point behaviour as defined in this chapter.

Bi-stability is not required for a tipping element, but local feedbacks that create strong nonlinearities are (Figure 7.2). Jones *et al.* (2009) showed that loss of the Amazon forest is reversible in their model under recovery of global climate to initial conditions, but the degree of committed Amazon forest cover is still determined by these local feedbacks. Figure 7.6 shows the degree of committed forest loss as a function of global temperature for the coupled climate–vegetation simulations of Jones *et al.* (2009). The forest cover response resembles Figure 7.2, right, in that the transition is finite but narrow (rather than vertical as in Figure 7.2, right). Further, this critical transition zone – or tipping point – occurs between 1 °C and 3 °C of warming of global temperature above pre-industrial levels: a very policy-relevant temperature range starting close to present-day warming and centred on the often-quoted target of 2 °C to avoid dangerous climate change. Within this range, very small additional global warming can give rise to significant additional loss of Amazon forest (4% loss of forest for each 0.1 °C of warming).

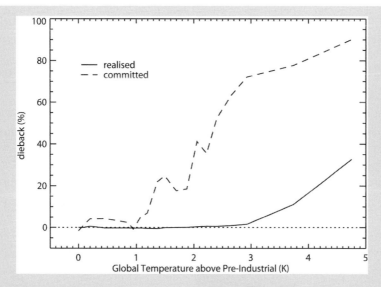

Figure 7.6 Realised and committed Amazon forest extent as a function of global mean temperature above pre-industrial levels. Percentage of complete dieback (in the region 40–70° W, 15° S–5° N as marked in Figure 7.7) as it evolves dynamically ('realised' – solid lines) through the SRES A2 simulation and the corresponding 'committed' state (dashed lines). Source: Jones *et al.* (2009). © Crown copyright 2009, the Met Office.

Additionally, components of the Earth System affected by climate change will respond on their own natural timescales. For some components, these timescales are slow compared with the rate of climate change and, hence, the system response lags behind the degree of global warming. Thus, when the climate is stabilised, these systems are not yet in equilibrium and continue to respond for perhaps long into the future. Physical components such as sea-level rise and ice-sheet melt have long timescales, and their commitments have been studied (Wigley, 1995; Gregory and Huybrechts, 2006). Ecosystems have been less studied in this respect but it is likely that they also exhibit slow timescales and, therefore, long-term commitments. Jones *et al.* (2009) showed that the global terrestrial biosphere continues to change for decades after climate stabilisation and may even be committed to long-term change before any response is observable.

Figure 7.7 shows an example from the Met Office Hadley Centre climate–vegetation model HadCM3LC. The left panel shows the simulated extent of Amazon forest cover under the SRES A2 transient scenario when it reaches a global warming of 2 °C. The right panel shows the eventual (committed) extent of forest cover if climate were stabilised at 2 °C. Significant dieback occurs after stabilisation and, in this case, takes about 100 years to fully manifest (Jones *et al.*, 2009).

Ecosystem commitments do not only include dieback. A warmer future climate may enable northward expansion of the Boreal forest into present tundra regions (Scholze *et al.*, 2006; Sitch *et al.*, 2008) and our same model also simulates significant committed increases in

Box 7.5 (*cont.*)

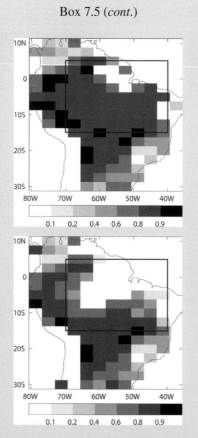

Figure 7.7 Distribution of Amazon forest tree cover at 2 °C: (upper) realised state on reaching 2 °C, and (lower) committed state after maintaining 2 °C, represented as fractional coverage of broadleaf tree simulated by the model. The black box shows the region used for calculating mean forest cover plotted in Figure 7.6. Source: Jones *et al.* (2009). © Crown copyright 2009, the Met Office.

high-latitude tree cover following climate stabilisation at 2 °C. In this case, the realised changes on reaching 2 °C are small and the timescale of forest response is many centuries.

Given that large-scale changes in ecosystem extent inevitably have a long timescale to fully adjust to a changed climate, the concept of ecosystem commitments is robust, although the magnitude and even sign of commitment will vary regionally and between models. This has significant implications for both ecosystem services and terrestrial carbon storage.

There is no requirement for such systems with long-term commitments to also have a tipping point, but ecosystems are likely to exhibit both. The local biophysical feedbacks between ecosystems and climate are responsible for the tipping-point behaviour and also influence the degree of commitment. In such a system, a further consideration arises. Not only must we understand where thresholds in the system lie, but also we must understand the timescales of the system response, as we may have passed a threshold without knowing it. The full extent of commitment will be determined by the time spent beyond a threshold as well as the threshold itself. Such considerations must be made when assessing the implications of overshoot, or 'peak and decline' scenarios. Committed ecosystem changes, in addition to realised changes, must be considered when defining dangerous climate change and forming policy to avoid it.

7.3.2 Boreal forest

The boreal forest in western Canada is currently suffering from an invasion of mountain pine beetle that has caused widespread tree mortality (Kurz *et al.*, 2008). Fire frequencies have also been increasing across the boreal forest zone. In the future, widespread dieback has been predicted in at least one model, when regional temperatures reach around 7 °C above present, corresponding to around 3 °C global warming. Limited expert elicitation concurs that above 4 °C global warming dieback becomes more likely than not (Supplementary Information of Kriegler *et al.*, 2009). The causes are complex such as increasingly warm summers that become too hot for the currently dominant tree species, increased vulnerability to disease, more frequent fires causing significantly higher mortality and decreased reproduction rates. The contemporary forest would be replaced over large areas by open woodlands or grasslands that tolerate increased fire frequency.

7.3.3 Coral reefs

Coral bleaching events linked to ocean warming have become much more widespread and detrimental in recent decades, and marine biologists are talking about being at a 'point of no return' for tropical coral reefs (Box 6.4; Veron *et al.*, 2009). Ocean acidification (Chapter 2), as a direct consequence of rising atmospheric CO_2 emissions, may also contribute to threshold-like changes in coral reef ecosystems (Riebesell *et al.*, 2009). Cold-water corals that grow down to 3000 m depth will be particularly vulnerable to acidification. They will be first affected as the saturation horizon of aragonite (a crystalline form of calcium carbonate) shallows because of ocean acidification. Once bathed in corrosive waters under-saturated in aragonite, the skeletons and shells may dissolve and the reefs collapse. It has been estimated that with unabated CO_2 emissions, 70% of the presently known deep-sea coral reef locations will be in corrosive waters by the end of this century (Guinotte *et al.*, 2006). Whether large areas, such as the Great Barrier Reef or the cold-water coral reef systems extending from Northern Norway to the West coast of Africa, might qualify as tipping elements warrants further research.

7.3.4 Marine biological carbon pump

The global oceans currently absorb approximately 2 Gt C yr^{-1}, which is equivalent to around 20% of annual anthropogenic carbon emissions (Le Quéré *et al.*, 2009).

Sequestering of carbon into the oceans is largely mediated by the so-called biological carbon pump, i.e. fixation of carbon by phytoplankton photosynthesis and the export of organic carbon to deeper ocean layers. Increasing temperatures and acidifying waters may affect the biological pump in a multitude of non-linear ways (Box 4.2). Hence a decline in the ocean carbon sink has been discussed as a potential tipping element (Kriegler *et al.*, 2009). Recently, it has been suggested that tipping-point behaviour may arise from the detrimental effects of ocean acidification on marine calcifying organisms (Riebesell *et al.*, 2009). Reduced biogenic calcification critically lowers the supply of calcium carbonate, which serves as a carrier of organic carbon and forms a component of the mineral ballast transport to the deeper ocean. Weakening of the biological pump would represent a positive feedback on atmospheric CO_2 concentrations and, in turn, on ocean acidification, but the attenuation of biogenic calcification also creates a negative, stabilising feedback on atmospheric CO_2 levels, which may outweigh it (Hofmann and Schellnhuber, 2009). The loss of calcium carbonate ballast could also be at least partially compensated by increased aggregation of organic matter and non calcium carbonate minerals in carbon-enriched waters (Riebesell *et al.*, 2009). As an additional factor, de-oxygenation could trigger a qualitative change in the functioning of the biological carbon pump. Widespread hypoxia may result from warming waters and decreased ventilation of the oceans (Keeling *et al.*, 2010), but has also been linked to ocean acidification (Hofmann and Schellnhuber, 2009). Besides potentially harmful consequences for a variety of marine ecosystems, 'oxygen holes' may provoke the loss of bio-available nitrogen by favouring denitrification and ammonium oxidation. In parts of the oceans, where primary productivity is limited by nitrogen, spreading of anoxic conditions could reduce the efficiency of the biological pump (Riebesell *et al.*, 2009). However, the marine biological carbon pump will not qualify as a tipping element unless a large-scale threshold can be identified (Box 7.1).

7.4 Policy implications – towards tipping-point risk management

Having detailed the potential tipping elements in the climate system, the overriding question becomes: How should (climate) policy respond? In human endeavours, the prospect of having to deal with jokers in the pack – high-impact, relatively low probability events, including a strong element of unpredictability – is not new. Think of earthquakes or hurricanes making landfall. We have systems (albeit flawed ones) for dealing with such events, and they hinge around a risk management approach. Although these are relatively short-timescale 'events', some of the risk management principles may be usefully mapped over to climate tipping-point timescales.

Risk, in the formal sense, is the product of the likelihood (or probability) of something happening and its (negative) impact. So a meaningful risk assessment of tipping elements would demand careful assessment of the impacts of passing different tipping points as well as of the associated likelihoods (under different forcing scenarios).

There already exists some information about the likelihood of passing different tipping points, as a function of global temperature change. Results of a workshop and literature review covered nine tipping elements (Lenton and Schellnhuber, 2007; Lenton *et al.*, 2008), and a process of expert elicitation considered six of these under three different future climate trajectories, and involved eliciting imprecise probability statements from 52 experts (Kriegler *et al.*, 2009). Useful results were obtained for five of these: the Greenland Ice Sheet, West Antarctic Ice Sheet, ENSO, Amazon rainforest and the THC. The imprecise probability statements were then formally combined to give lower-bound probabilities. These reveal that the likelihood of passing at least one of five tipping points rises from >16% under a mid-range (2–4 °C) global warming corridor to >56% (i.e. more likely than not) under a high-warming (>4 °C) corridor. In Figure 7.8, we update the diagram of Lenton and Schellnhuber (2007) to summarise likelihoods as a function of global warming, based on the expert elicitation results and recent literature.

There also exists some information on the impacts of tipping the different elements, but the gaps are larger and this area needs more detailed work. A recent study has articulated the implications of four different tipping-point scenarios for the insurance sector (Lenton *et al.*, 2009a), considering Amazon rainforest dieback, ISM disruption (coupled with melt of HKHT glaciers), a shift to a more arid climate in south-west North America (including loss of mountain snowpacks), and high-end sea-level rise from melting ice sheets with additional regional sea-level rise along the north-eastern seaboard of the USA related to weakening of the THC. However, tipping-point impacts will depend on human responses and are, thus, a more epistemologically contested area than assigning likelihoods to events. The resulting ambiguity needs to be reduced if risk assessment is to be usefully pursued (Stirling, 2003).

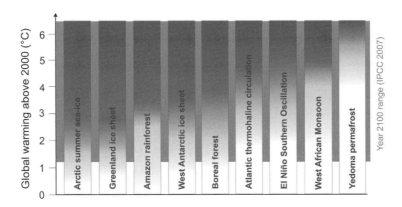

Figure 7.8 A summary of current information on the likelihood of tipping different elements under different degrees of global warming – updated from Lenton and Schellnhuber (2007) based on expert elicitation results (Kriegler *et al.*, 2009) and recent literature discussed herein. The darkness of the grey shading indicates the likelihood of tipping. Figure by Timothy M. Lenton and Hans Joachim Schellnhuber.

Table 7.1. *A simple 'straw man' example of tipping element risk assessment, by Timothy M. Lenton*

Tipping element	Likelihood of passing a tipping point (by 2100)	Relative impact** of change in state (by 3000)	Risk score (likelihood x impact)	Risk ranking
Arctic summer sea-ice	High	Low	3	4
Greenland ice sheet	Medium–High*	High	7.5	1 (highest)
West Antarctic ice sheet	Medium*	High	6	2
Atlantic THC	Low*	Medium–High	2.5	6
ENSO	Low*	Medium–High	2.5	6
West African monsoon	Low	High	3	4
Amazon rainforest	Medium*	Medium	4	3
Boreal forest	Low	Low–Medium	1.5	8 (lowest)

* Likelihoods informed by expert elicitation

** Initial judgment of relative impacts is the subjective assessment of T.M.L.

With these caveats, let us offer an initial 'straw man' illustration of how a tipping-point risk assessment might look (Table 7.1). Here we assume a scenario of *partial* mitigation of greenhouse gas emissions leading to roughly 3°C global warming by 2100. We concentrate on tipping elements from the original shortlist where a threshold can be meaningfully linked to global temperature change (thus excluding the ISM). We start with likelihoods and relative impacts on a five-point scale: low, low–medium, medium, medium–high and high. Information on likelihood is taken from review of the literature (Lenton and Schellnhuber, 2007; Lenton *et al.*, 2008) and where available, expert elicitation (Kriegler *et al.*, 2009). Impacts are considered in relative terms based on an initial subjective judgment (noting that most tipping-point impacts, if placed on an absolute scale compared to other climate eventualities, would be high). Impacts depend on timescale and here we consider the full 'ethical time horizon' introduced elsewhere (Lenton *et al.*, 2008) and assume minimal discounting of impacts on future generations. Likelihood and impacts are simply multiplied together to give a measure of risk, and a ranking emerges (Table 7.1).

What this simple straw man readily illustrates are some familiar dilemmas for the would-be risk manager – relatively high-impact, low-probability events, such as WAM collapse, come out with a similar risk to relatively lower-impact, high-probability events, such as Arctic summer sea-ice loss. However, what stand out are the *high-impact, high-probability scenarios* as a priority for risk management effort, in this case: melt of the Greenland Ice Sheet, followed by collapse of the West Antarctic Ice Sheet. These risks will be best managed by restricting the extent of future warming of these systems and their surrounding ocean. Of course, this exercise would be better conducted with a wider team of experts and relevant stakeholders to get a more scientifically credible and socially legitimate assessment. We simply hope our straw man encourages some thought and activity in this area.

7.5 Prospects for early warning

Perhaps the most useful information that science could provide to help society manage global warming is some early warning of an approaching tipping point. There are several degrees of early warning, from simply identifying possible threats, to being able to forecast that a tipping point is imminent. For other high-impact events, such as hurricanes or tsunamis, there are quite sophisticated early-warning systems in place, from which we could potentially learn.

Existing work has probably not identified all possible tipping elements in the climate system, so more research is definitely warranted to systematically search for them. The latest methods for detecting threshold behaviour could be applied to a wide range of palaeo-data as well as to the instrumental record; and to the output of existing climate model runs, such as from the IPCC Assessment Reports. A useful theoretical starting point would be to try to identify all the potentially strong positive feedbacks in components of the Earth System that could manifest at large spatial scales – because these are a necessary condition for tipping-point behaviour. Historically, climate science has been good at identifying global-scale positive feedbacks on temperature.

However, we are talking here about an intermediate (but still large) spatial scale, and about feedbacks that are internal to the dynamics of a part of the system, and sometimes have little or no effect on global temperature (although they may be triggered by it changing). In recent years, at least one such feedback has been discovered (the potential for runaway breakdown of Yedoma permafrost), and others have been better formalised (the multiple stable states of ice sheet grounding lines). Once identified, such feedbacks should then be included in Earth System models, and the phase space of the models systematically searched for multiple stable states at the regional scale, and other signs of strong non-linearity. There are recent examples of the successful detection of multiple states in complex models, for example in the Amazon basin (Oyama and Nobre, 2003). We encourage further effort in this area, although with state-of-the-art models it will challenge current computing resources.

Where a potential tipping-point threat has been convincingly identified, the challenge becomes: Can we detect any early warning signs before the threshold is reached? Recent progress has been made in identifying and testing generic potential early-warning indicators of an approaching tipping point (Livina and Lenton, 2007; Dakos *et al.*, 2008; Lenton *et al.*, 2008, 2009b) (Box 7.6). Slowing down in response to perturbation is a nearly universal property of systems approaching various types of tipping points (Scheffer *et al.*, 2009). It manifests as increasing autocorrelation in time-series data (in simple terms; each data point becomes more like the surrounding ones). This has been successfully detected in past climate records approaching different transitions (Livina and Lenton, 2007; Dakos *et al.*, 2008), and in model experiments (Livina and Lenton, 2007; Dakos *et al.*, 2008; Lenton *et al.*, 2009b).

Flickering between states may also occur prior to a more permanent transition (Bakke *et al.*, 2009). Other early-warning indicators are being explored for ecological tipping points, which could potentially be applied to climate. These include increasing variance (Biggs *et al.*, 2009), skewness (Guttal and Jayaprakash, 2008; Biggs *et al.*, 2009) and their spatial equivalents (Guttal and Jayaprakash, 2009).

It is encouraging that there is some theoretical potential for early warning of an approaching threshold, but there are considerable practical limitations on whether an effective early-warning system could be deployed for specific systems. A key consideration is: What is the longest internal timescale of the system in question? It is changes in this that the 'critical slowing down' method is trying to detect (Box 7.6). In the case of the ocean circulation or ice-sheet dynamics, these timescales are long (in the thousands of years). Therefore, one needs a long and relatively high-resolution (palaeo) time-series record for the system in question, in order to get an accurate picture of its natural state of variability from which to detect changes. Often such records are lacking. However, all hope is not lost: some potential tipping points have much faster dynamics and relatively little internal memory, for example the monsoons. For such systems, existing observational time-series data may be sufficient. Also, for specific tipping elements, such as the THC, there may be other leading indicators of vulnerability that are deducible from observational data (Box 7.3).

Box 7.6
Tipping point early-warning methods

TIMOTHY M. LENTON AND VALERIE N. LIVINA

Physical systems that are approaching bifurcation points show a nearly universal property of becoming more sluggish in response to perturbation. This is referred to as 'critical slowing down' in dynamical systems theory. To visualise this, picture the present state of a system as a ball in a curved potential well (attractor) that is being nudged around by some stochastic noise process, e.g. weather (Figure 7.9). The ball continually tends to roll back towards the bottom of the well – its lowest potential state – and the rate at which it rolls back is determined by the curvature of the potential well. As the system is forced towards a bifurcation point, the potential well becomes flatter. Hence the ball will roll back ever more sluggishly. At the bifurcation point, the potential becomes flat and the ball is destined to roll off into some other state (alternative potential well). Mathematically speaking, the leading eigenvalue, which characterises the rates of change around the present equilibrium state, tends to zero as the bifurcation point is approached.

So, for those tipping elements that exhibit true bifurcation points (e.g. Figure 7.2, left), we should be able to look for slowing down in time-series data as a basis for early warning.

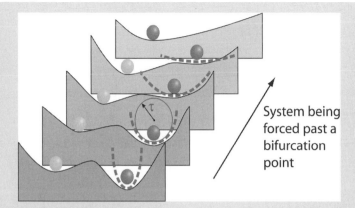

System being
forced past a
bifurcation
point

Figure 7.9 Schematic representation of a system being forced past a bifurcation point. The system's response time to small perturbations, τ, is related to the growing radius of the potential well. Source: Hermann Held, from Lenton *et al.* (2008).

Following this rationale, Held and Kleinen (2004) developed a method of examining decay rate to perturbations using a simple lag-1 autocorrelation function (ACF). They averaged over short-term variability in order to isolate the dynamics of the longest imminent timescale of a system (Held and Kleinen, 2004). The approach was subsequently modified by using detrended fluctuation analysis (DFA) to assess the proximity to a threshold from the power law exponent describing correlations in time-series data (Livina and Lenton, 2007). At a critical threshold, the data become highly correlated across short- and middle-range timescales, and the time-series behaves as a random walk with uncorrelated steps. The DFA approach does not demand aggregation of the data and, remarkably, the ACF method has also been found to work without averaging (Dakos *et al.*, 2008). However, both methods need to span a sufficient time interval to capture what can be a very slow decay rate.

   Model tests have shown that both ACF and DFA methods work in principle in simple (Dakos *et al.*, 2008), intermediate complexity (Held and Kleinen, 2004; Livina and Lenton, 2007), and fully three-dimensional (Lenton *et al.*, 2009b) models. The challenge is to get the methods to work in practice, in the complex and noisy climate system. Initial tests found that the ending of the last ice age recorded in ice core data is detected as a critical transition using the DFA method (Livina and Lenton, 2007). Subsequent work showed increasing autocorrelation in eight palaeo-climate time-series approaching transitions, using the ACF method (Dakos *et al.*, 2008).

   In Figure 7.10, we compare existing DFA analysis of an ice core record (Livina and Lenton, 2007) with the ACF method (with or without detrending, i.e. removing the underlying trend). We do not interpolate the sparse data because this introduces a high degree of correlation and removes any clear signal in the propagators. Both approaches detect critical behaviour during the last deglaciation (the propagators approach or exceed a critical value of 1). However, using only the more sparse data prior to the transition (to the left of the dotted vertical lines) the upward trends in the propagators, indicative of slowing down, are rather weak.

Box 7.6 (*cont.*)

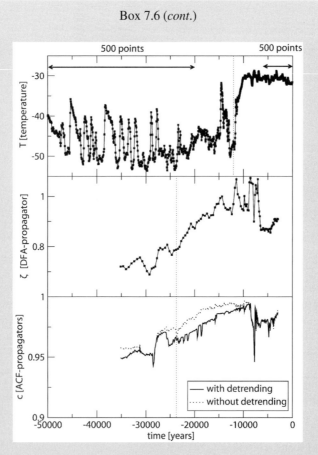

Figure 7.10 A test of tipping point early-warning methods on Greenland ice core temperature proxy data (Livina and Lenton, 2007), with the addition of ACF analysis. A sliding window of 500 data points is used, and the data are non-uniform, as indicated in the top panel. Results are plotted in the middle of the windows, as would be the case when trying to forecast a tipping point. The vertical dotted line in the bottom two panels indicates where the analyses include the 500 points before the transition into the Holocene marked in the top panel. The ACF method was applied with and without the detrending method described in Dakos *et al.* (2008). Source: Valerie N. Livina.

These results, using real palaeo time-series data, are somewhat encouraging, but there remain serious practical constraints to developing a workable real-time tipping point early-warning system.

The lack-of-data problem is that long, high-resolution and preferably evenly spaced time-series are required for successful forewarning of thresholds. For slow-transition systems, notably ocean and ice sheets, observation records will need to be extended further back in time (e.g. for the THC beyond the ~150-year sea-surface temperature record). Real-time observation systems also need to be carefully targeted at key variables that reveal the underlying dynamics of a system.

The lag problem is that if a tipping element is forced slowly (keeping it in quasi-equilibrium), proximity to a threshold may be inferred in a model-independent way. However, humans are forcing the climate system relatively rapidly, so inherently 'slow' tipping elements, such as ice sheets and the THC, will be well out of equilibrium with the forcing. This means that a dynamical model will also be needed to establish proximity to a threshold.

The noise problem is that, for a method of anticipating a threshold to be useful, the time it takes to find out proximity to a threshold must be shorter than the time in which noise would be expected to cause the system to change state – the 'mean first exit time' (Kleinen *et al.*, 2003). The level of noise in the climate system can be relatively high, meaning that a tipping element could get 'knocked' out of its present state well before a bifurcation point is reached. To accommodate this, a sophisticated early-warning system should take account of the noise level for a particular tipping element and adjust its estimates accordingly.

For those tipping elements that do not exhibit a true bifurcation (e.g. Figure 7.2, right), the methods described may still yield some useful information, because transitions that are not strictly bifurcations are expected to resemble bifurcation-type behaviour to a certain degree. In particular, deducing the longest imminent timescale in a system determines its linear response characteristics to external forcing.

7.6 Response strategies

Assuming science could provide society with reliable early warning of an approaching tipping point – and that this information was viewed as sufficiently credible, salient and legitimate to warrant action – the question becomes what could (or should) we do about it? Obvious courses of action are to try and avoid reaching the tipping point, or to try and 'pre-adapt' (build resilience) to better cope with the changes that are due to occur. Which is feasible or most appropriate will depend somewhat on the tipping element in question, and the forcing factors responsible, but let us start with some fairly general observations. The natural response to warning of the prospect of a high-impact event is to try to avoid it. However, our options for effective intervention ('steering') in the climate system are actually somewhat limited. That is because there are several sources of inertia in the Earth System – one might liken heading towards a climate tipping point as like the *Titanic* heading towards the fateful iceberg: it is a big ship and it is hard to change its trajectory quickly. An analogous problem of avoiding an approaching tipping point in an ecological system (a fishery) shows that once there is a reliable early warning of an approaching tipping point, it is probably too late for slow intervention methods to avoid it (Biggs *et al.*, 2009).

The conventional avoidance strategy is to mitigate the emissions of greenhouse gases, especially carbon dioxide (Chapter 8), but this is a slow intervention. The climate is responding to the concentration of these agents (and the resulting radiative forcing), not their emissions. We know that the prime mitigation target, CO_2, is a very long-lived pollutant; and the current policy challenge is still framed as stopping it rising, and holding a stabilisation level – not bringing its concentration down. Even if radiative forcing is stabilised, the climate will continue to warm for some considerable time because heat will still be entering the system (primarily the deep ocean) and slow, positive feedbacks will be operating (such as loss of methane from hydrates).

Potentially, radiative forcing could be reduced by mitigating emissions of short-lived greenhouse gases (e.g. tropospheric ozone) and black carbon aerosols, whose concentrations will decline fairly rapidly. Ozone in the lower atmosphere currently contributes to global warming 20% as much as CO_2. Rigorous global implementation of air pollution regulations and available technologies could drive down emissions of ozone precursors (especially carbon monoxide and nitrogen oxides) quickly, producing a climate response within decades (Molina *et al.*, 2009). Black carbon (Chapter 1) is estimated to be the second or third largest global warming agent, although large uncertainties exist about its exact radiative forcing. Due to its deposition on snow and resulting positive feedbacks, black carbon may be responsible for as much as ~0.5–1.4 °C of the 1.9 °C warming observed in the Arctic from 1890 to 2007 (Shindell and Faluvegi, 2009), and for approximately 0.6 °C of the 1 °C warming in the Tibetan Himalayas since 1950 (Ramanathan and Carmichael, 2008). Halving black carbon emissions could be achieved by 2030 with full application of existing technologies (Cofala *et al.*, 2007). Lowering black carbon emissions would have the 'double benefit' of reducing global warming and raising the temperature level at which tipping points involving melting of ice and snow are triggered (Hansen and Nazarenko, 2004). A further 'fast-action' strategy to consider is phasing down the production and consumption of hydrofluorocarbon with high global warming potential (Molina *et al.*, 2009).

Nevertheless, the bulk of mitigation will always be a relatively slow way of turning the climate ship, and we have not even mentioned the inertia in society when it comes to replacing energy and existing infrastructure. We might speed up the steering by starting to actively remove CO_2 (or other positive radiative forcing agents) from the atmosphere (Lenton and Vaughan, 2009), but we are still talking about timescales of half a century before we can stabilise the CO_2 concentration and at least a century before temperature stops rising. A comprehensive analysis of the complex control problem involved is given in Schellnhuber *et al.* (2009), which is based on the limitation of cumulative CO_2 emissions worldwide.

So, are there any faster avoidance strategies available? Here we come to the controversial topic of geoengineering the amount of sunlight absorbed by the Earth, with the aim of deliberately counteracting the positive radiative forcing coming primarily from accumulated greenhouse gases (Chapter 17; Lenton and Vaughan, 2009). Some proponents of this type of geoengineering are arguing that it should be developed as a potential emergency response to the prospect of an approaching climate tipping point. For their argument to hold up, we would need to be confident that such intervention could work fast enough to avoid the transgression of a threshold. The currently much-discussed method of injecting aerosols (probably of sulphate) globally into the stratosphere (Crutzen, 2006), could conceivably get the Earth back in approximate radiative energy balance, and could begin to cool the climate within a year, if it mimicked the Mt Pinatubo volcanic eruption. However, it would take several years of repeated injection to feel the full climate cooling effect (as the temperature of the ocean mixed layer adjusted) (Wigley, 2006). To this must be added the time to develop the technology and the infrastructure needed to deploy it, which is probably at least a decade at present. So, we are probably talking about one to two decades for this type of intervention to take effect. In principle, this could be helpful in avoiding an approaching temperature threshold in a relatively sluggish system such as the Greenland Ice Sheet. Perhaps a tailor-made regional version of radiation management – 'regio-engineering' – would be the preferable option under these circumstances. However, geoengineering is not going to be much use for fast-response systems or ones where the threshold is not clearly linked to global temperature, such as the West African or Indian monsoons. Indeed, past volcanic aerosol injections are known to have slowed down the hydrological cycle and reduced rainfall in such regions (Trenberth and Dai, 2007; Robock *et al.*, 2008) (and the theory behind this is robust). So, aerosol geoengineering interventions might pose a greater risk to some tipping elements than the reduction in risk they achieved for others. Once again, a careful risk management approach is needed to weigh up the odds.

We take the view that in a rational world, any early warning would be useful, even if it turns out to be impossible to avoid an approaching tipping point, because it would give societies some time to prepare themselves. Effective adaptation can certainly happen faster than mitigation can alter the climate trajectory, and some types of adaptation can probably happen faster than geoengineering could alter the climate trajectory. However, these may be types of adaptation, such as mass migration, that carry their own considerable risks of triggering undesirable social 'tipping points', i.e. conflicts rather than cooperative responses (Chapters 5 and 13). Research on 'social tipping elements' is urgently needed to better anticipate this type of dynamics. An example of a geographically localised social tipping element are the agro-cultures of South-East Asia that, fed by glacial melt-water and monsoon precipitation, might collapse under the regional impacts of unabated global warming (Schellnhuber, 2009). Furthermore, there is the problem that humans are not perfectly rational actors. Hence, receiving an alarm signal carries the danger of triggering maladaptive responses, especially when our fallible, internal, human methods of risk assessment are at play. Historically, for example, tsunami warnings have caused numerous people to head to the coast to see the incoming wave, thus greatly increasing their exposure to risk. We cannot rule out such maladaptive responses to early warning of approaching climate tipping points. Indeed, despite the known incidence of hurricane landfalls, there has been a demographic trend of people moving *to* Florida. Ironically, US citizens have also historically been leaving the central Great Plains and moving to the southwest, which is one of the regions we have highlighted as undergoing a transition to a new climate state less able to support agriculture and people.

7.7 Recovery prospects

If we pass tipping points, and the corresponding tipping elements change state, is there any prospect for recovering their original state? Again the answer will depend on the system in question and the timescale. The definition of a tipping element includes systems that can exhibit reversible or irreversible changes in state (Figure 7.2). A reversible transition means that if the forcing is returned below the tipping point, the system will recover its original state (either abruptly or gradually). An irreversible transition means that it will not – it takes a larger change in forcing to recover (and hence there is some hysteresis in the trajectory of the system in phase space). An example of a transition that should be reversible in principle is the loss of Arctic summer sea-ice. Transitions that could exhibit some irreversibility are the loss of large ice sheets, changes in the Atlantic THC or the loss of major biomes.

Even if a transition is reversible in principle, it does not mean that the changes will be reversible in practice. A key problem (discussed previously) is that it is rather difficult to reduce radiative forcing of the climate system, unless one indulges in geoengineering, and the climate further lags the forcing (the problem of 'committed climate change'). In other words, global warming is hard to reverse. Even loss of Arctic summer sea-ice, which seems like a highly reversible system because the ice regrows each winter, has some longer-term 'memory'. It cannot be completely recovered in one season because some of the ice that has been lost consisted of thick, multi-year strata that take several winters to accumulate.

Irreversible transitions vary in what is needed to recover from them, and even with respect to whether recovery is possible at all. The 'strongest' form of irreversibility is extinction – the loss of species and hence genetic information and diversity. Extinctions would surely accompany major biome transitions such as dieback of the Amazon rainforest or boreal forests, should they occur. Thus, although something resembling the current rainforest or boreal forest might eventually grow back, they would never be the same. Ice sheets such as those on Greenland and West Antarctica, if lost, might eventually be recovered with appropriate forcing, but we are talking about timescales of tens of thousands of years to regrow a major ice sheet. Alternatively, human activities might fundamentally alter the dynamics of the climate system – switching it out of its recent mode of roughly 100,000-year glacial cycles and back into an earlier (Pliocene-like) state in which Greenland lives up to its name and remains un-glaciated, West Antarctica only sporadically has an ice sheet, East Antarctica holds less ice and sea levels are, in the long term, around 25 m higher than those experienced today (Rohling *et al.*, 2009).

7.8 Conclusion

Although we have used the metaphor of 'jokers in the pack', recent work suggests we should no longer be talking about climate tipping points as 'high-impact, low-probability events'. If the business-as-usual approach continues, in some cases we are talking about 'high-impact, high-probability' events. We may already be at (or very close to) a tipping point for some large-scale systems in the Arctic. There is no obvious clustering of thresholds to support, for example, the 2 °C policy target adopted in various political agreements, including the Copenhagen Accord (Chapter 13). However, 2 °C is clearly a better target than 3 °C or 4 °C, because it is evident that increasing global warming increases the risk from climate tipping points. Even in a 2 °C world, adaptation will probably be required for some unavoidable transitions that we may already be committed to, even if they have yet to become apparent because of internal lags in the systems in question.

References

Archer, D. (2007). Methane hydrate stability and anthropogenic climate change. *Biogeosciences*, **4**, 521–44.

Archer, D., Buffett, B. and Brovkin, V. (2009). Ocean methane hydrates as a slow tipping point in the global carbon cycle. *Proceedings of the National Academy of Sciences (USA)*, **106**, 20596–601.

Ashok, K. and Yamagata, T. (2009). Climate change: The El Niño with a difference. *Nature*, **461**, 481–84.

Bakke, J., Lie, O., Heegaard, E. *et al.* (2009). Rapid oceanic and atmospheric changes during the Younger Dryas cold period. *Nature Geoscience*, **2**, 202–05.

Betts, R. A., Cox, P. M., Collins, M. *et al.* (2004). The role of ecosystem-atmosphere interactions in simulated Amazonian precipitation decrease and forest dieback under global climate warming. *Theoretical and Applied Climatology*, **78**, 157–75.

Biggs, R., Carpenter, S. R. and Brock, W. A. (2009). Turning back from the brink: Detecting an impending regime shift in time to avert it. *Proceedings of the National Academy of Sciences (USA)*, **106**, 826–31.

Brovkin, V., Claussen, M., Petoukhov, V. and Ganopolski, A. (1998). On the stability of the atmosphere-vegetation system in the Sahara/Sahel region. *Journal of Geophysical Research*, **103**, 31613–24.

Burns, S. J., Fleitmann, D., Matter, A., Kramers, J. and Al-Subbary, A. A. (2003). Indian Ocean climate and an absolute chronology over Dansgaard/Oeschger events 9 to 13. *Science*, **301**, 1365–67.

Chang, P., Zhang, R., Hazeleger, W. *et al.* (2008). Oceanic link between abrupt change in the North Atlantic Ocean and the African monsoon. *Nature Geoscience*, **1**, 444–48.

Claussen, M., Kubatzki, C., Brovkin, V. *et al.* (1999). Simulation of an abrupt change in Saharan vegetation in the mid-Holocene. *Geophysical Research Letters*, **26**, 2037–40.

Cofala, J., Amman, M., Klimont, Z., Kupiainen, K. and Höglund-Isaksson, L. (2007). Scenarios of global anthropogenic emissions of air pollutants and methane until 2030. *Atmospheric Environment*, **41**, 8486–99.

Cook, K. H. and Vizy, E. K. (2006). Coupled model simulations of the west African Monsoon System: Twentieth- and twenty-first-century simulations. *Journal of Climate*, **19**, 3681–703.

Cook, K. H. and Vizy, E. K. (2008). Effects of twenty-first-century climate change on the Amazon rain forest. *Journal of Climate*, **21**, 542–60.

Cook, K. H., Vizy, E. K., Launer, Z. S. and Patricola, C. M. (2008). Springtime intensification of the Great Plains low-level jet and Midwest precipitation in GCM simulations of the twenty-first century. *Journal of Climate*, **21**, 6321–40.

Cox, P. M., Betts, R. A., Collins, M. *et al.* (2004). Amazonian forest dieback under climate-carbon cycle projections for the 21st century. *Theoretical and Applied Climatology*, **78**, 137–56.

Cox, P. M., Harris, P. P., Huntingford, C. *et al.* (2008). Increasing risk of Amazonian drought due to decreasing aerosol pollution. *Nature*, **453**, 212–15.

Crutzen, P. J. (2006). Albedo enhancement by stratospheric sulfur injections: A contribution to resolve a policy dilemma? *Climatic Change*, **77**, 211–19.

Dakos, V., Scheffer, M., van Nes, E. H. *et al.* (2008). Slowing down as an early warning signal for abrupt climate change. *Proceedings of the National Academy of Sciences (USA)*, **105**, 14308–12.

de Vries, P. and Weber, S. L. (2005). The Atlantic freshwater budget as a diagnostic for the existence of a stable shut-down of the meridional overturning circulation. *Geophysical Research Letters*, **32**, L09606.

Eisenman, I. and Wettlaufer, J. S. (2009). Nonlinear threshold behavior during the loss of Arctic sea ice. *Proceedings of the National Academy of Sciences (USA)*, **106**, 28–32.

Gladwell, M. (2000). *The Tipping Point: How Little Things Can Make a Big Difference*. New York, NY: Little Brown.

Gregory, J. M. and Huybrechts, P. (2006). Ice-sheet contributions to future sea-level change. *Philosophical Transactions of the Royal Society A – Mathematical, Physical & Engineering Sciences*, **364**, 1709–31.

Guilyardi, E. (2006). El Niño-mean state-seasonal cycle interactions in a multi-model ensemble. *Climate Dynamics*, **26**, 329–48.

Guinotte, J. M., Orr, J., Cairns, S. *et al.* (2006). Will human-induced changes in seawater chemistry alter the distribution of deep-sea scleractinian corals? *Frontiers in Ecology and the Environment*, **4**, 141–46.

Gupta, A. K., Anderson, D. M. and Overpeck, J. T. (2003). Abrupt changes in the Asian southwest monsoon during the Holocene and their links to the North Atlantic Ocean. *Nature*, **431**, 354–57.

Guttal, V. and Jayaprakash, C. (2008). Changing skewness: An early warning signal of regime shifts in ecosystems. *Ecology Letters*, **11**, 450–60.

Guttal, V., and Jayaprakash, C. (2009). Spatial variance and spatial skewness: Leading indicators of regime shifts in spatial ecological systems. *Theoretical Ecology*, **2**, 3–12.

Hagos, S. M. and Cook, K. H. (2007). Dynamics of the west African monsoon jump. *Journal of Climate*, **20**, 5264–84.

Hagos, S. M. and Cook, K. H. (2008). Ocean warming and late-twentieth-century Sahel drought and recovery. *Journal of Climate*, **21**, 3797–814.

Hansen, J. and Nazarenko, L. (2004). Soot climate forcing via snow and ice albedos. *Proceedings of the National Academy of Sciences (USA)*, **101**, 423–28.

Held, H. and Kleinen, T. (2004). Detection of climate system bifurcations by degenerate fingerprinting. *Geophysical Research Letters*, **31**, L23207.

Held, I. M. and Soden, B. J. (2006). Robust responses of the hydrological cycle to global warming. *Journal of Climate*, **19**, 5686–99.

Hofmann, M. and Rahmstorf, S. (2009). On the stability of the Atlantic meridional overturning circulation. *Proceedings of the National Academy of Sciences (USA)*, **106**, 20584–89.

Hofmann, M. and Schellnhuber, H. J. (2009). Ocean acidification affects marine carbon pump and triggers extended marine oxygen holes. *Proceedings of the National Academy of Sciences (USA)*, **106**, 3017–22.

Holland, D. M., Thomas, R. H., de Young, B., Ribergaard, M. H. and Lyberth, B. (2008). Acceleration of Jakobshavn Isbræ triggered by warm subsurface ocean waters. *Nature Geoscience*, **1**, 659–64.

Holland, M. M., Bitz, C. M. and Tremblay, B. (2006). Future abrupt reductions in the summer Arctic sea ice. *Geophysical Research Letters*, **33**, L23503.

Huybrechts, P. and De Wolde, J. (1999). The dynamic response of the Greenland and Antarctic ice sheets to multiple-century climatic warming. *Journal of Climate*, **12**, 2169–88.

Intergovernmental Panel on Climate Change (IPCC) (2007). *Climate Change 2007: The Physical Science Basis. Contribution of Working Group I to the Fourth Assessment Report of the Intergovernmental Panel on Climate Change*, eds. S. Solomon, D. Qin, M. Manning, Z. Chen, M. Marquis, K. B. Averyt, M. Tignor and H. L. Miller. Cambridge, UK and New York, NY: Cambridge University Press.

Jones, C., Lowe, J., Liddicoat, S. and Betts, R. (2009). Committed ecosystem change due to climate change. *Nature Geoscience*, **2**, 484–87.

Joughin, I., Das, S. B., King, M. A. *et al.* (2008). Seasonal speedup along the western flank of the Greenland Ice Sheet. *Science*, **320**, 781–83.

Kay, J. E., L'Ecuyer, T., Gettelman, A., Stephens, G. and O'Dell, C. (2008). The contribution of cloud and radiation anomalies to the 2007 Arctic sea ice extent minimum. *Geophysical Research Letters*, **35**, L08503.

Kayen, R. E. and Lee, H. J. (1991). Pleistocene slope instability of gas hydrate-laden sediment of Beaufort Sea margin. *Marine Geotechnology*, **10**, 125–41.

Keeling, R. F., Körtzinger, A. and Gruber, N. (2010). Ocean deoxygenation in a warming world. *Annual Review of Marine Science*, **2**, 199–29.

Khvorostyanov, D. V., Ciais, P., Krinner, G. and Zimov, S. A. (2008a). Vulnerability of east Siberia's frozen carbon stores to future warming. *Geophysical Research Letters*, **35**, L10703.

Khvorostyanov, D. V., Krinner, G., Ciais, P., Heimann, M. and Zimov, S. A. (2008b). Vulnerability of permafrost carbon to global warming. Part I: model description and the role of heat generated by organic matter decomposition. *Tellus B*, **60**, 250–64.

Kleinen, T., Held, H. and Petschel-Held, G. (2003). The potential role of spectral properties in detecting thresholds in the Earth system: Application to the thermohaline circulation. *Ocean Dynamics*, **53**, 53–63.

Kriegler, E., Hall, J. W., Held, H., Dawson, R. and Schellnhuber, H.-J. (2009). Imprecise probability assessment of tipping points in the climate system. *Proceedings of the National Academy of Sciences (USA)*, **106**, 5041–46.

Kurz, W. A., Dymond, C. C., Stinson, G. *et al.* (2008). Mountain pine beetle and forest carbon feedback to climate change. *Nature*, **452**, 987–90.

Kwok, R. and Rothrock, D. A. (2009). Decline in Arctic sea ice thickness from submarine and ICESat records: 1958–2008. *Geophysical Research Letters*, **36**, L15501.

Latif, M. and Keenlyside, N. S. (2009). El Niño/Southern Oscillation response to global warming. *Proceedings of the National Academy of Sciences (USA)*, **106**, 20578–83.

Lawrence, D. M. and Slater, A. G. (2005). A projection of severe near-surface permafrost degradation during the 21st century. *Geophysical Research Letters*, **32**, L24401.

Lawrence, D. M., Slater, A. G., Tomas, R. A., Holland, M. M. and Deser, C. (2008). Accelerated Arctic land warming and permafrost degradation during rapid sea ice loss. *Geophysical Research Letters*, **35**, L11506.

Le Quéré, C., Raupach, M. R., Canadell, J. G. and Marland, G. (2009). Trends in the sources and sinks of carbon dioxide. *Nature Geosciences*, **2**, 831–36.

Lenton, T. M. and Schellnhuber, H.-J. (2007). Tipping the scales. *Nature Reports Climate Change*, **1**, 97–98.

Lenton, T. M., Held, H., Kriegler, E. *et al.* (2008). Tipping elements in the Earth's climate system. *Proceedings of the National Academy of Sciences (USA)*, **105**, 1786–93.

Lenton, T. M., Footitt, A. and Dlugolecki, A. (2009a). *Major Tipping Points in the Earth's Climate System and Consequences for the Insurance Sector*. n.p.: Tyndall Centre for Climate Change Research.

Lenton, T. M., Myerscough, R. J., Marsh, R. *et al.* (2009b). Using GENIE to study a tipping point in the climate system. *Philosophical Transactions of the Royal Society A – Mathematical, Physical & Engineering Sciences*, **367**, 871–84.

Lenton, T. M. and Vaughan, N. E. (2009). The radiative forcing potential of different climate geoengineering options. *Atmospheric Chemistry and Physics Discussions*, **9**, 2559–608.

Levermann, A., Schewe, J., Petoukhov, V. and Held, H. (2009). Basic mechanism for abrupt monsoon transitions. *Proceedings of the National Academy of Sciences (USA)*, **106**, 20572–77.

Livina, V. N. and Lenton, T. M. (2007). A modified method for detecting incipient bifurcations in a dynamical system. *Geophysical Research Letters*, **34**, L03712.

Lu, J., Vecchi, G. A. and Reichler, T. (2007). Expansion of the Hadley cell under global warming. *Geophysical Research Letters*, **34**, L06805.

Malhi, Y., Aragão, L. E. O. C., Galbraith, D. *et al.* (2009). Exploring the likelihood and mechanism of a climate-change-induced dieback of the Amazon rainforest. *Proceedings of the National Academy of Sciences (USA)*, **106**, 20610–15.

Maslanik, J., Drobot, S., Fowler, C., Emery, W. and Barry, R. (2007). On the Arctic climate paradox and the continuing role of atmospheric circulation in affecting sea ice conditions. *Geophysical Research Letters*, **34**, L03711.

Meehl, G. A., Arblaster, J. M. and Collins, W. D. (2008). Effects of black carbon aerosols on the Indian monsoon. *Journal of Climate*, **21**, 2869–82.

Mercer, J. H. (1978). West Antarctic ice sheet and CO_2 greenhouse effect: A threat of disaster. *Nature*, **271**, 321–25.

Milly, P. C. D., Dunne, K. A. and Vecchia, A. V. (2005). Global pattern of trends in streamflow and water availability in a changing climate. *Nature*, **438**, 347–50.

Molina, M., Zaelke, D., Sarma, K. M. *et al.* (2009). Reducing abrupt climate change risk using the Montreal Protocol and other regulatory actions to complement cuts in CO_2 emissions. *Proceedings of the National Academy of Sciences (USA)*, **106**, 20616–21.

Mote, T. L. (2007). Greenland surface melt trends 1973–2007: Evidence of a large increase in 2007. *Geophysical Research Letters*, **34**, L22507.

Naish, T., Powell, R., Levy, R. *et al.* (2009). Obliquity-paced Pliocene West Antarctic ice sheet oscillations. *Nature*, **458**, 322–28.

Nghiem, S. V., Rigor, I. G., Perovich, D. K. *et al.* (2007). Rapid reduction of Arctic perennial sea ice. *Geophysical Research Letters*, **34**, L19504.

Notz, D. (2009). The future of ice sheets and sea ice: Between reversible retreat and unstoppable loss. *Proceedings of the National Academy of Sciences (USA)*, **106**, 20590–95.

Oyama, M. D. and Nobre, C. A. (2003). A new climate-vegetation equilibrium state for tropical South America. *Geophysical Research Letters*, **30**, CLM5.1–CLM5.4.

Patricola, C. M. and Cook, K. H. (2008). Atmosphere/vegetation feedbacks: A mechanism for abrupt climate change over northern Africa. *Journal of Geophysical Research*, **113**, D18102.

Perovich, D. K., Light, B., Eicken, H. *et al.* (2007). Increasing solar heating of the Arctic Ocean and adjacent seas, 1979–2005: Attribution and role in the ice-albedo feedback. *Geophysical Research Letters*, **34**, L19505.

Perovich, D. K., Richter-Menge, J. A., Jones, K. F. and Light, B. (2008). Sunlight, water, and ice: Extreme Arctic sea ice melt during the summer of 2007. *Geophysical Research Letters*, **35**, L11501.

Pfeffer, W. T., Harper, J. T. and O'Neel, S. (2008). Kinematic constraints on glacier contributions to 21st-century sea-level rise. *Science*, **321**, 1340–43.

Phillips, O. L., Aragão, L. E. O. C., Lewis, S. L. *et al.* (2009). Drought sensitivity of the Amazon rainforest. *Science*, **323**, 1344–47.

Pollard, D. and DeConto, R. M. (2009). Modelling West Antarctic ice sheet growth and collapse through the past five million years. *Nature*, **458**, 329–32.

Prentice, I. C., Cramer, W., Harrison, S. P. *et al.* (1992). A global biome model based on plant physiology and dominance, soil properties and climate. *Journal of Biogeography*, **19**, 117–34.

Pritchard, H. D., Arthern, R. J., Vaughan, D. G. and Edwards, L. A. (2009). Extensive dynamic thinning on the margins of the Greenland and Antarctic ice sheets. *Nature*, **461**, 971–75.

Ramanathan, V., Chung, C., Kim, D. *et al.* (2005). Atmospheric brown clouds: Impacts on South Asian climate and hydrological cycle. *Proceedings of the National Academy of Sciences (USA)*, **102**, 5326–33.

Ramanathan, V. and Carmichael, G. (2008). Global and regional climate changes due to black carbon. *Nature Geoscience*, **1**, 221–27.

Ramanathan, V. and Feng, Y. (2008). On avoiding dangerous anthropogenic interference with the climate system: formidable challenges ahead. *Proceedings of the National Academy of Sciences (USA)*, **105**, 14245–50.

Riebesell, U., Körtzinger, A. and Oschlies, A. (2009). Sensitivities of marine carbon fluxes to ocean change. *Proceedings of the National Academy of Sciences (USA)*, **106**, 20602–09.

Rignot, E., Box, J. E., Burgess, E. and Hanna, E. (2008). Mass balance of the Greenland ice sheet from 1958 to 2007. *Geophysical Research Letters*, **35**, L20502.

Rigor, I. G. and Wallace, J. M. (2004). Variations in the age of Arctic sea-ice and summer sea-ice extent. *Geophysical Research Letters*, **31**, L09401.

Robock, A., Oman, L. and Stenchikov, G. L. (2008). Regional climate responses to geoengineering with tropical and Arctic SO$_2$ injections. *Journal of Geophysical Research*, **113**, D16101.

Rohling, E. J., Grant, K., Hemleben, Ch. *et al.* (2008). High rates of sea-level rise during the last interglacial period. *Nature Geoscience*, **1**, 38–42.

Rohling, E. J., Grant, K., Bolshaw, M. *et al.* (2009). Antarctic temperature and global sea level closely coupled over the past five glacial cycles. *Nature Geoscience*, **2**, 500–04.

Salazar, L. F., Nobre, C. A. and Oyama, M. D. (2007). Climate change consequences on the biome distribution in tropical South America. *Geophysical Research Letters*, **34**, L09708.

Saleska, S. R., Didan, K., Huete, A. R. and da Rocha, H. R. (2007). Amazon forests green-up during 2005 drought. *Science*, **318**, 612.

Scheffer, M., Bacompte, J., Brock, W. A. *et al.* (2009). Early warning signals for critical transitions. *Nature*, **461**, 53–59.

Schellnhuber, H. J. (2009). Tipping elements in the Earth System. *Proceedings of the National Academy of Sciences (USA)*, **106**, 20561–63.

Schellnhuber, H. J., Cramer, W., Nakicenovic, N., Wigley, T. and Yohe, G. (2006). *Avoiding Dangerous Climate Change*. Cambridge, UK: Cambridge University Press.

Schellnhuber, H. J. *et al.* (2009). Solving the climate dilemma: the budget approach. WBGU Special Report, WBGU, Berlin.

Scholze, M., Knorr, W., Arnell, N. W. and Prentice, I. C. (2006). A climate-change risk analysis for world ecosystems. *Proceedings of the National Academy of Sciences (USA)*, **103**, 13116–20.

Schoof, C. (2007). Ice sheet grounding line dynamics: Steady states, stability, and hysteresis. *Journal of Geophysical Research*, **112**, F03S28.

Seager, R., Ting, M., Held, I. *et al.* (2007). Model projections of an imminent transition to a more arid climate in southwestern North America. *Science*, **316**, 1181–84.

Shanahan, T. M., Overpeck, J. T., Anchukaitis, K. J. *et al.* (2009). Atlantic forcing of persistent drought in west Africa. *Science*, **324**, 377–80.

Shimada, K., Kamoshida, T., Itoh, M. *et al.* (2006). Pacific ocean inflow: Influence on catastrophic reduction of sea ice cover in the Arctic Ocean. *Geophysical Research Letters*, **33**, L08605.

Shindell, D. and Faluvegi, G. (2009). Climate response to regional radiative forcing during the twentieth century. *Nature Geoscience*, **2**, 294–300.

Sitch, S., Huntingford, C., Gedney, N. *et al.* (2008). Evaluation of the terrestrial carbon cycle, future plant geography and climate-carbon cycle feedbacks using five Dynamic Global Vegetation Models (DGVMs). *Global Change Biology*, **14**, 2015–39.

Smith, J. B., Schneider, S. H., Oppenheimer, M. *et al.* (2009). Assessing dangerous climate change through an update of the Intergovernmental Panel on Climate Change (IPCC) 'reasons for concern'. *Proceedings of the National Academy of Sciences (USA)*, **106**, 4133–37.

Steig, E. J., Schneider, D. P., Rutherford, S. D. *et al.* (2009). Warming of the Antarctic ice-sheet surface since the 1957 International Geophysical Year. *Nature*, **457**, 459–62.

Stirling, A. (2003). Risk, uncertainty and precaution: Some instrumental implications from the social sciences. In *Negotiating Environmental Change: New Perspectives from Social Science*, eds. F. Berkhout, M. Leach and I. Scoones. London: Edward Elgar, pp. 33–76.

Stroeve, J., Holland, M. M., Meier, W., Scambos, T. and Serreze, M. (2007). Arctic sea ice decline: Faster than forecast. *Geophysical Research Letters*, **34**, L09501.

Sultan, B. and Janicot, S. (2003). The West African monsoon dynamics Part II: The preon-set and onset of the summer monsoon. *Journal of Climate*, **16**, 3407–27.

Trenberth, K. E. and Dai, A. (2007). Effects of Mount Pinatubo volcanic eruption on the hydrological cycle as an analog of geoengineering. *Geophysical Research Letters*, **34**, L15702.

van de Wal, R. S. W., Boot, W., van den Broeke, M. R. *et al.* (2008). Large and rapid melt-induced velocity changes in the ablation zone of the Greenland Ice Sheet. *Science*, **321**, 111–13.

Vaughan, D. G. (2008). West Antarctic Ice Sheet collapse: The fall and rise of a paradigm. *Climatic Change*, **91**, 65–79.

Vecchi, G. A., Soden, B. J., Wittenberg, A. T. *et al.* (2006). Weakening of tropical Pacific atmospheric circulation due to anthropogenic forcing. *Nature*, **441**, 73–76.

Veron, J. E. N., Hoegh-Guldberg, O., Lenton, T. M. *et al.* (2009). The coral reef crisis: The critical importance of <350 p.p.m. CO$_2$. *Marine Pollution Bulletin*, **58**, 1428–36.

Washington, R., Bouet, C., Cautenet, G. *et al.* (2009). Dust as a tipping element: The Bodélé Depression, Chad. *Proceedings of the National Academy of Sciences (USA)*, **106**, 20564–71.

Weber, S. L., Drijfhout, S. S., Abe-Ouchi, A. *et al.* (2007). The modern and glacial overturning circulation in the Atlantic ocean in PMIP coupled model simulations. *Climate of the Past*, **3**, 51–64.

Weertman, J. (1974). Stability of the junction of an ice sheet and an ice shelf. *Journal of Glaciology*, **13**, 3–13.

Weijer, W., de Ruijter, W. P. M., Dijkstra, H. A. and van Leeuwen, P. J. (1999). Impact of interbasin exchange on the Atlantic overturning circulation. *Journal of Physical Oceanography*, **29**, 2266–84.

Wigley, T. M. L. (1995). Global mean-temperature and sea level consequences of greenhouse gas concentration stabilization. *Geophysical Research Letters*, **22**, 45–48.

Wigley, T. M. L. (2006). A combined mitigation/geoengineering approach to climate stabilization. *Science*, **314**, 452–54.

Winton, M. (2006). Does the Arctic sea ice have a tipping point? *Geophysical Research Letters*, **33**, L23504.

Woodgate, R. A., Aagaard, K. and Weingartner, T. J. (2006). Interannual changes in the Bering Strait fluxes of volume, heat and freshwater between 1991 and 2004. *Geophysical Research Letters*, **33**, L15609.

Yeh, S.-W., Kug, J.-S., Dewitte, B. *et al.* (2009). El Niño in a changing climate. *Nature*, **461**, 511–14.

Yin, J., Schlesinger, M. E. and Stouffer, R. J. (2009). Model projections of rapid sea-level rise on the northeast coast of the United States. *Nature Geoscience*, **2**, 262–66.

Zickfeld, K., Knopf, B., Petoukhov, V. and Schellnhuber, H. J. (2005). Is the Indian summer monsoon stable against global change? *Geophysical Research Letters*, **32**, L15707.

Zimov, S. A., Schuur, E. A. G. and Chapin, F. S. (2006). Permafrost and the global carbon budget. *Science*, **312**, 1612–13.

8

Linking science and action: targets, timetables and emission budgets

'There are times in the history of humanity when fateful decisions are made. The decision ... on whether to enter a comprehensive global agreement for strong action on climate change is one of them. ...On a balance of probabilities, the failure of our generation would lead to consequences that would haunt humanity until the end of time.'[1]

Political leaders of 194 countries (as of January 2010) have signed the United Nations Framework Convention on Climate Change (UNFCCC), which states in Article 2 that the Convention's 'ultimate objective ... is to achieve... stabilization of greenhouse gas concentrations in the atmosphere at a level that would prevent dangerous anthropogenic interference with the climate system ...'. As we have seen in the preceding chapters, scientists are able to document with a very high degree of certainty that human activities exert measurable influence on several components of the climate system. In addition, most scientists agree that, on the basis of observed changes in the climate system, there is a very high probability (>90% according to the IPCC) that these changes are primarily the result of anthropogenic influences on the climate system. Furthermore, many impacts on human societies and nature have already been recorded (Chapters 5 and 6). These impacts can only be expected to increase in the future.

With Article 2 of the UNFCCC as the point of departure, the next step is to determine what actually constitutes 'dangerous anthropogenic interference with the climate system', or 'dangerous climate change' as it is often called in shorthand. Scientists can identify the nature of human influences on the climate system, and are increasingly able to quantify impacts of climate change resulting from these influences. What they cannot do, however, is define what constitutes 'dangerous anthropogenic interference with the climate system'. Defining 'dangerous' is a value judgement that must be made by society as a whole, but informed by the observed and projected impacts of climate change on a wide range of social and natural systems that ultimately affect human well-being (Chapters 5 and 6). In

---

[1] Ross Garnaut, in *The Garnaut Climate Change Review*, Cambridge University Press, 2008, pp. 591, 597.

addition, the concept of 'dangerous' is increasingly informed by the risks of abrupt, large-scale shifts in the climate system associated with tipping elements (Chapter 7).

Ultimately, then, defining dangerous climate change is a judgement of risk. How much risk are societies willing to accept as the impacts and consequences of climate change escalate with rising temperatures? Based on the evidence to date, well over 100 countries around the world have now agreed that anthropogenic climate change should be limited to a temperature rise of no more than 2 °C above pre-industrial levels – the so-called 2 °C guardrail (Council of the European Union, 2005; IPCC, 2007a; Copenhagen Accord, 2009). As noted in Chapters 5–7, the 2 °C guardrail clearly does not guarantee a 'safe' level of climate change, but limits the risk of unmanageable climate change. Once agreed, the 2 °C guardrail can then be translated into trajectories of greenhouse gas emissions that would be required to confine the global temperature increase to a maximum of 2 °C. The chain of logic typically flows from (i) linking temperature to the atmospheric concentration of greenhouse gases and then to (ii) linking the atmospheric concentration of greenhouse gases to the magnitude and rate of greenhouse gas emissions from human activities, or, alternatively, linking temperature directly to the cumulative emission of greenhouse gases. These simple chains of logic – which form the basis for global efforts to curb greenhouse gas emissions – nevertheless contain significant uncertainties and complexities that make a probabilistic, rather than a deterministic, approach to the linking of science to action most appropriate.

8.1 Linking temperature and greenhouse gas concentrations

The linking of atmospheric greenhouse gas concentration to global average temperature is a scientific problem, and is normally addressed through the concept of 'climate sensitivity' (Box 8.1). In the context of contemporary climate change, climate sensitivity is defined as the long-term temperature increase that would result if the atmospheric CO_2 concentration is doubled relative to pre-industrial levels. That is, climate sensitivity refers to the increase in temperature in response to around 560 p.p.m. CO_2 over the long term, when the climate system has reached equilibrium.

Box 8.1

Estimating climate sensitivity

THOMAS SCHNEIDER VON DEIMLING

Climate sensitivity – the equilibrium change in global mean surface temperature after a doubling of the atmospheric CO_2 concentration – is a key parameter for describing how responsive the climate system is to changes in its radiative balance. It is a common measure to compare differences in the simulated climate response and gives insight into the overall feedback strength of a specific model. In simple climate models it is used as a tuning parameter, while for complex climate models climate sensitivity is not known *a priori* and depends on a model's description of many physical processes and feedbacks. However, any

Box 8.1 (*cont.*)

long timescale feedbacks, such as ice-sheet and vegetation change, are usually not included in model-based estimates of climate sensitivity (Box 1.5).

Thirty years ago, the now famous 1.5–4.5 °C climate sensitivity range was first postulated based on the modelling results of two global climate models (GCMs) (Charney, 1979). Since then, this estimate had been broadly corroborated by successive IPCC reports, which reported climate sensitivity ranges based on the most recent GCM results available at the time (3.8 °C ± 0.8 °C in 1995 (17 models), 3.5 °C ± 0.9 °C in 2001 (15 models)). The latest generation of Fourth Assessment Report (AR4) models – which show highly improved model skill for simulating present-day climatology (Randall *et al.*, 2007) – span a range of 2.1–4.4 °C (18 models), with most models clustering around a sensitivity of just above 3 °C.

Is the clustering seen in these IPCC multimodel ensembles strong support for a consensus estimate of 3 °C? Given that the sampling of the IPCC models is rather arbitrary and not designed to cover the full range of possible model configurations (Tebaldi and Knutti, 2007), various approaches of testing the robustness of this best-guess estimate have been explored. A key question is to what extent alternative models could be constructed (based on different parameterisations or on different parameter settings), which simulate present-day climatology with similar accuracy but which will reveal different climate sensitivities.

The large increase in computer power in recent years and the use of distributed computing sources made it possible only recently to test very large sets of alternative model assumptions. Stainforth *et al.* (2005) have generated a broad set of model versions by perturbing key parameters in a single GCM (e.g. those affecting the description of clouds, convection, and precipitation). This deliberate detuning of the model parameters (the *perturbed physics approach*) resulted in a broad spread in simulated climate sensitivities. Model versions with sensitivities exceeding 4.5 °C could not be rejected based on annual mean climatology, but the majority of ensemble members yielded a sensitivity close to 3 °C. Sensitivities below 2 °C were very rare in this ensemble.

When subjecting the same model ensemble to tighter constraints by using statistical measures of the correspondence between simulated and observed fields of present-day climate, models with rather low or high sensitivity proved to be less consistent with the data (Piani *et al.*, 2005).[2] This conclusion is supported when observational evidence about the seasonal cycle of present-day climate is used to constrain the model ensemble (Knutti *et al.*, 2006). The best agreement is achieved for models with sensitivity near 3 °C. Yet it should be noted that these two studies are based on a version of the same climate model and may share common biases in model representation. A further study, which analysed a million-member ensemble constructed from emulations of the same model, but which considered additional climatology data for testing model performance, found a best agreement between the models and the data for slightly higher sensitivity values (Sanderson *et al.*, 2008).

Further support for sensitivity within the IPCC range of 2.0–4.5 °C comes from the analysis of huge perturbed physics ensembles constructed from reduced-complexity models. By accounting for uncertainty in model sensitivity, simulated oceanic heat uptake and twentieth-century radiative forcing, individual model members can be tested to reproduce

---

[2] A similar finding is seen in a perturbed-physics ensemble from Murphy *et al.* (2004) based on the same model but with a different parameter choice.

observed twentieth-century warming. A recent study assembled the findings of these analyses and showed that models with a climate sensitivity between 2 °C and 3.5 °C prove to be closer to observations than models with a sensitivity parameter outside this range (Meinshausen *et al.*, 2009).

Using, and especially combining, additional observational evidence (Annan and Hargreaves, 2006) – such as the cooling following large volcanic eruptions or the pronounced decline in global mean temperature between today and the Last Glacial Maximum (Schneider von Deimling *et al.*, 2006) – allows for independent testing of how well climate feedbacks are represented in a model. The comparison of studies that have used various additional data information favours a most likely estimate of climate sensitivity around 3 °C (Knutti and Hegerl, 2008). Future perturbed physics ensembles based on structurally different models will help clarify to what extent this estimate is robust when alternative physical descriptions and different parameter choices are considered (Collins, 2007). Given the set of models available today and the situation of limited computational resources, 3 °C is currently our best answer for a most likely estimate of the magnitude of climate sensitivity.

As future models will describe the climate system in more detail (e.g. by modelling so far neglected geochemical processes of the Earth System such as greenhouse gas release from thawing permafrost regions), sensitivity estimates will become possible that better account for the full complexity of the Earth climate system. Compared to climate sensitivity estimates derived from current models, these new estimates then should provide us with a further improved measure for quantifying the effects of anthropogenic greenhouse gas emissions on global temperatures.

One obvious approach to estimate the climate sensitivity is to run climate models to equilibrium with a CO_2 concentration of 560 p.p.m. and to note the calculated temperature increase. These model simulations include, in addition to the direct radiative effect of increased CO_2 concentration, the so-called 'fast feedbacks' in the climate system – primarily the changes in type and amount of clouds, the increased water vapour in the atmosphere, and the reduced extent of sea ice. When many different models are each run many times, an ensemble of results is obtained. The probability distribution derived from such an ensemble of model runs shows a strong clustering of results around a climate sensitivity of 3 °C (Figure 8.1; IPCC, 2007b). As the increase in radiative forcing associated with a doubling of CO_2 concentration from pre-industrial levels is about 3.7 W m^{-2} (IPCC, 2007b) the climate sensitivity is about 0.8 °C per W m^{-2}.

Nature has already done the experiment of naturally raising greenhouse gas concentrations, most recently in the transition from the last ice age to the Holocene; CO_2 concentrations, for example, rose from about 180 p.p.m. to 280 p.p.m. during this transition (Petit *et al.*, 1999). The observed temperature increase during this transition should thus provide an independent estimate of climate sensitivity (Hansen *et al.*, 2008). The analysis, however, must also include the so-called 'slow feedbacks' in the climate system that operate on multi-century or millennial timescales – the changes in (i) the extent of the large polar ice sheets, and (ii) the distribution of terrestrial vegetation, both of which significantly affect the Earth's

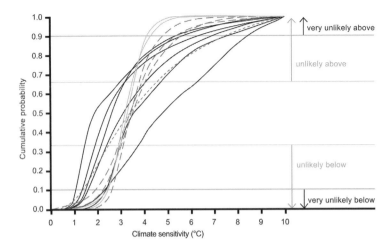

Figure 8.1 Individual cumulative distributions of climate sensitivity from the observed twentieth century warming (solid black lines), model climatology (grey dashed lines) and proxy evidence (grey dotted line). Horizontal lines and arrows mark the edges of the likelihood estimates. Source: IPCC (2007b), with further information on the methodology.

albedo (Box 1.2). Such an analysis yields a climate sensitivity of about 0.75 °C per W m^{-2}, which is identical, within the range of uncertainty, to the model-based estimate (Box 8.1).

A further check on the estimate of climate sensitivity can be made by comparing the observed post-industrial climatic changes with those expected from the change in radiative forcing. With a net change in radiative forcing of about 1.6 W m^{-2} since 1750 (IPCC, 2007b), we should now experience a temperature rise of approximately 1.6 x 0.8, or 1.3 °C. The observed temperature rise of about 0.75 °C above pre-industrial levels, coupled with the further committed rise of about 0.5 °C due to greenhouse gases already in the atmosphere (IPCC, 2007b), yields an observed equilibrium temperature rise of 1.25 °C, in excellent agreement with that expected from the estimated climate sensitivity.

Although our understanding of climate sensitivity has improved markedly over the past decade, there are still some uncertainties surrounding the current quantitative estimates. In particular, the model-based estimates show a long tail of low-probability, but much higher, sensitivities (Figure 8.1). There is a 5% probability that the real sensitivity could be as high as 5 or 6 °C. Nevertheless, the convergence of several lines of evidence towards a sensitivity of 3 °C gives us confidence to base our mitigation approaches on that value (Box 8.1).

8.2 Linking greenhouse gas concentrations and emissions: targets and timetables

We can now work back from climate sensitivity to estimate the limit on further emissions of greenhouse gases that will be required to observe the 2 °C guardrail. The most common approach is to set a global target for the stabilisation of greenhouse gas concentrations and then work back to the emission trajectory required to reach the desired stabilisation target. The approach implicitly assumes that the current fraction of carbon emissions that actually remains in the atmosphere will be maintained indefinitely into the future (but see Chapter 4 and later discussion in this chapter).

Most analyses of emission targets and timetables are carried out for both CO_2 and CO_2-equivalents (CO_2-e), in which the latter accounts for the warming effects of non-CO_2 gases (excluding water vapour, which is considered as a feedback in the climate system) and the cooling/warming effects of aerosols (Chapter 1). The approach is to calculate, based on the relative warming/cooling potential of the particle gas or type of aerosol and the lifetime of the gas/aerosol in the atmosphere, the equivalent amount of CO_2 it would take to yield the same warming/cooling effect on climate. Because the calculation is also based on the lifetime of the gas/aerosol in the atmosphere, CO_2-e is always determined for a specified period of time.

As an example, the CO_2 concentration in 2008 was about 385 p.p.m. (Le Quéré *et al.*, 2009). When the warming potential of the additional amounts of the other greenhouse gases that have accumulated since the beginning of the industrial revolution is included, the CO_2-e concentration becomes 463 p.p.m. (European Environment Agency, 2009). Inclusion of the net cooling effect of aerosols brings CO_2-e back down to 396 p.p.m., not much different from the CO_2 concentration itself. This fortuitous near-cancellation of the effects of non-CO_2 greenhouse gases and aerosols allows one to currently estimate the overall human influence on the climate system by considering the influence of CO_2 alone. However, it is unlikely that this fortuitous relationship will continue into the future as countries make efforts to reduce aerosol loading, which, if unabated, diminishes air quality and causes impacts on health at local and regional scales.

Table 8.1 shows the characteristics of various emissions trajectories designed to stabilise CO_2 and CO_2-e concentrations at particular levels (IPCC, 2007a). Note that in these scenarios the CO_2 and CO_2-e concentrations are assumed to have diverged significantly by 2050 and at stabilisation. The table presents a daunting picture of humanity's challenge to stay within the 2 °C guardrail. To keep the temperature rise within the 2.0–2.4 °C range, CO_2 concentration will need to be stabilised at 350–400 p.p.m. and CO_2-e at 445–490 p.p.m. We are already within that range for CO_2, and only the counteracting cooling effect of aerosols is keeping the CO_2-e from also lying within the target range. Thus, it is only the aerosol cooling that gives us a chance to eventually stabilise CO_2-e at no more than 450 p.p.m. To keep the CO_2 concentration from rising above 400 p.p.m., CO_2 emissions will need to peak no later than 2015, and global emissions in 2050 will need to have been reduced by 50%–85% from their 2000 levels (Table 8.1, last column). This is a Herculean task.

An obvious question is how the efforts of countries around the world to reduce their emissions, when aggregated, compare to the global targets required to have a reasonable chance of staying within the 2 °C guardrail. This question is surprisingly difficult to answer with a reasonable degree of confidence, as countries use various, incompatible ways to calculate their own emission reduction targets, and then often use a variety of approaches to implement the desired emission reductions. The latter can render the outcome of implementation strategies even more difficult to predict and interpret.

Table 8.1. *Characteristics of post-TAR stabilisation scenarios and resulting long-term equilibrium global average temperature rise*

Temperature rise	CO$_2$	CO$_2$-eq.	Year of peak emissions	% change in global emissions
Global average temperature increase above pre-industrial at equilibrium, using 'best estimate' climate sensitivity	CO$_2$ concentration at stabilisation (2005 = 379 p.p.m.)	CO$_2$-eq concentration at stabilisation including GHGs and aerosols (2005 = 375 p.p.m.)	Peaking year for CO$_2$ emissions	Change in CO$_2$ emissions in 2050 (percent of 2000 emission)
(°C)	(p.p.m.)	(p.p.m.)	(year)	(%)
2.0–2.4	350–400	445–490	2000–2015	−85 to −50
2.4–2.8	400–440	490–535	2000–2020	−60 to −30
2.8–3.2	440–485	535–590	2010–2030	−30 to +5
3.2–4.0	485–570	590–710	2020–2060	+10 to +60
4.0–4.9	570–660	710–855	2050–2080	+25 to +85
4.9–6.1	660–790	855–1130	2060–2090	+90 to +140

Source: IPCC (2007a).

Table 8.2. *Hypothetical changes in emissions as a result of using different base years*

Base year	Actual emissions (Mt)	Emissions in 2020 with 25% reduction (Mt)	% change in 2020 based on 1990	Actual change in emissions in 2020 from 1990 (Mt)
1990	500.0	375.0	−25.0	−125.0
2000	609.5	457.1	−8.6	−42.9
2005	672.9	504.7	+0.9	+4.7
2009	728.4	546.3	+8.5	+46.3

Source: W. Steffen.

For example, using different base years can yield markedly different outcomes for actual emissions. Table 8.2 shows the case of a hypothetical country with emissions of 500 Mt of carbon in 1990, and an annual growth rate in emissions of 2%. A 25% reduction in emissions by 2020, which might be considered the minimum required to keep on track to meet the 2 °C guardrail, gives strikingly different results depending on the base year for the calculation. If the reduction is applied to 1990, then the emissions in 2020 are indeed 125 Mt less than they were in 1990. However, if the 25% reduction is applied to 2005 as a base year, actual emissions in 2020 grow by about 5 Mt compared to their 1990 values. The atmosphere, and hence the climate, experience the actual change in the net flow of carbon to the atmosphere, and not the percentage changes calculated from arbitrary base years.

Another approach to calculating emission reductions, especially with respect to rapidly developing countries, is to reduce emissions against a 'business-as-usual' trajectory rather than against the observed emissions of a particular base year. While designed to recognise additional effort towards climate protection than would otherwise have occurred, this emission reduction calculation is fraught with considerable uncertainties. The most important of these is the use of a sliding future baseline, which is subject to several critical assumptions about what constitutes 'business as usual', assumptions that can be changed through time. Again, the bottom line for the climate system is the actual amount of CO_2 and other greenhouse gases that are emitted into the atmosphere.

Even more complexities can enter into the actual implementation of emission reduction targets (German Advisory Council on Global Change, 2009). For example, some would argue that countries should be allowed to transfer unused emission allowances from the first Kyoto commitment period into subsequent periods. Various land-use change sinks, including the proposed new mechanism of avoided deforestation (Reducing Emissions From Deforestation and Forest Degradation – REDD), could additionally be used as emission reduction strategies, despite the risks associated with these (Section 8.5). On top of these, the Clean Development Mechanism (CDM) allows industrialised countries to invest in emission-offsetting projects in the developing world that are often questionable as to their 'additionality' in the sense of whether or not they would have been carried out in the absence of any policy. When such implementation strategies are aggregated, it is a likely outcome that commitments by industrialised countries to reduce their emissions by, say, 30% by 2020 against a 1990 baseline could actually result in a 25% or greater *increase* in domestic emissions by 2020 (German Advisory Council on Global Change, 2009). Even more importantly, such an outcome would leave the fossil fuel-based sectors in industrial countries largely untouched, and would thus slow the transition to a low- or no-carbon economy (Chapter 11) that will be required to stabilise the climate.

In summary, the targets and timetables approach and the associated implementation strategies, although they are the preferred approach in the current negotiation arena, present a bewildering array of complexities that make it difficult to link actions to science. In fact, there are even more uncertainties and risks associated with the linkage between emission reductions and atmospheric concentration using the targets and timetables approach than with the linkage between greenhouse gas concentrations and temperature change. Because of this, the scientific community is rapidly adopting the budget approach as a more direct, simpler and intuitively understandable way to link emissions directly to temperature and associated climatic changes.

8.3 Linking temperature and emissions: the budget approach

The budget, or cumulative emissions, approach is based on a simple scientific fact: the degree of anthropogenic climate change we are observing now and can expect in the future is related primarily to the sum of additional greenhouse gases in the atmosphere (Allen *et al.*, 2009; England *et al.*, 2009; Matthews *et al.*, 2009; Meinshausen *et al.*, 2009; Zickfeld *et al.*, 2009). Thus, the best way to analyse the level of effort required by humanity to limit climate change to acceptable levels is simply through the total, temporally integrated amount of greenhouse gases – especially CO_2 – that can be emitted into the atmosphere in the future. Based on the global aggregate amount, individual emission budgets can then be allocated to various countries on the basis of equity and other considerations – which in itself is a difficult task.

This budget approach starts at the same point as the targets and timetable approach – observing the 2 °C guardrail. Then, for a given probability of meeting this goal, the total amount of CO_2 that can be released into the atmosphere by humanity is calculated. This budget can be calculated from the beginning of the industrial revolution until emissions must revert to zero (Box 8.2; Allen *et al.*, 2009), or from 2000 to 2050, assuming a small residual global budget after 2050 (Meinshausen *et al.*, 2009; German Advisory Council on Global Change, 2009). Both approaches give essentially the same answer.

Box 8.2
The importance of cumulative carbon emissions

MYLES R. ALLEN AND NIEL H. A. BOWERMAN

The single most important factor in the risk of dangerous long-term anthropogenic inter- ference in the climate system is the accumulation of carbon dioxide emissions over time, not the rate of emission in any given year or decade. This has potentially profound implications for mitigation strategy.

Unlike most other greenhouse gases, carbon dioxide does not have a well-defined atmospheric lifetime: fossil carbon emissions are mixed rapidly through the 'active carbon cycle', which comprises the atmosphere, land biosphere and near-surface ocean. Natural removal, by transport into the deep ocean and ultimately the formation of ocean sediments, takes place only over millennia. About 15%–20% of carbon emissions remain in the atmosphere, in effect, indefinitely (Solomon *et al.*, 2009). Hence even if anthropogenic emissions were reduced to zero, atmospheric concentrations would not fall back to pre- industrial levels for thousands of years. Feedbacks further complicate the picture: in some Earth System models, once temperatures reach a certain threshold, concentrations could even continue to rise after a complete cessation of emissions due to the continued release of carbon from the biosphere (Chapter 4; Lowe *et al.*, 2009).

Hence we should expect peak warming caused by carbon dioxide to be determined primarily by cumulative anthropogenic emissions up to the date at which temperatures peak, with only negligible net emissions being possible thereafter. Emissions from fossil fuel use, cement manufacturing and land use change to 2006 comprise approximately half a trillion tonnes of carbon (500 Gtc, or 1.8 trillion tonnes of CO_2) and are estimated to have committed us to ~1°C warming; that is, if anthropogenic CO_2 emissions were to cease immediately, the warming they cause – which is currently similar to total anthropogenic warming because the warming due to other greenhouse gases is approximately balanced by cooling due to anthropogenic aerosols (see Chapter 1) – would peak close to 1° C above pre-industrial temperatures.

At the simplest level, the cumulative impact of CO_2 emissions suggests that emitting a further 500 Gt C, or one trillion tonnes of carbon in total, would commit us to a peak warming of around 2 °C. Model studies confirm this: Allen *et al.* (2009) estimated the peak warming caused by a total emission of one trillion tonnes of carbon, or cumulative warming commitment (CWC), to be 2 °C with a one-standard-error uncertainty range of 1.6–2.6 °C and 5%–95% confidence interval of 1.3–3.9 °C. Meinshausen *et al.* (2009) suggested a similar warming commitment for CO_2 alone, but note that the impact of other forcing agents is likely to reduce the total cumulative CO_2 budget consistent with a given peak anthropogenic warming: they found that cumulative CO_2 emissions over all time of 900 Gt C gives a 50% chance of anthropogenic warming exceeding 2 °C, while keeping the risk of warming exceeding 2 °C to 25% implies a cumulative budget of 710 Gt C.

Matthews *et al.* (2009) estimated the CWC due to 1000 Gt C to be 1.5 °C, with a 5%–95% confidence interval of 1.0–2.1 °C, while Zickfeld *et al.* (2009) proposed a total budget of approximately 1000 Gt C to give a 66% chance of total anthropogenic warming remaining below 2 °C. These lower estimates of CWC result primarily from lower estimates of carbon dioxide-induced warming to date; such estimates are being refined continuously as the signal of anthropogenic warming strengthens, so uncertainty in CWC is likely to be resolved rapidly.

The importance of cumulative carbon dioxide emissions is shown in Figure 8.2, which shows three scenarios for future carbon dioxide emissions each corresponding to a cumulative total carbon emission of 1000 Gt C. Despite very different emission rates in 2020, the temperature responses are very similar (right panel). Hence it is not true, as is often suggested, that if

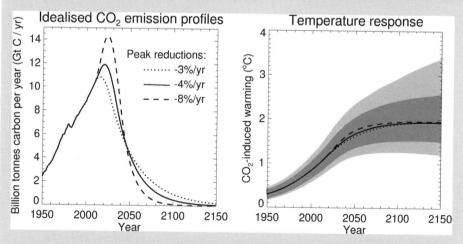

Figure 8.2. Three scenarios for carbon dioxide emissions (left panel) – each corresponding to a cumulative total emissions of one trillion tonnes of carbon – and most likely temperature response (right panel). The impact of emitting a trillion tonnes faster or slower is dwarfed by uncertainty in the response (grey shading). Source: Allen *et al.* (2009).

Box 8.2 (*cont.*)

emissions fail to peak by 2020 then we are committed to 2 °C of warming. What is true is that if emissions fail to peak by 2020 then we are committed to potentially unfeasible rates of emission reductions after 2020 if we are still to avoid 2 °C of warming (German Advisory Council on Global Change, 2009). If we assume constant per cent-per-year emission reductions after the peak, then the required rate of decline is given simply by $E_{peak}/(C_{total}-C_{peak})$ x 100%, where E_{peak} is the peak rate of emission in tonnes of carbon per year, C_{total} is the total cumulative budget and C_{peak} is cumulative emissions prior to the peak, both in tonnes of carbon. If we aim to limit cumulative carbon emissions to one trillion tonnes, then emissions would have to fall by over 2% per year from now on even if emissions were to peak immediately, and the required rate of decline is rising as emissions continue to rise (increasing E_{peak}) and accumulate (increasing C_{peak}). Extrapolating average per cent-per-year changes in fossil fuel and land-use change emissions over 1987–2006, the required rate of decline would exceed 4% per year if emissions fail to peak before the 2020s, and the trillionth tonne itself would be released in the mid 2040s.

Even if they do not become an explicit part of climate change mitigation policy (German Advisory Council on Global Change, 2009), cumulative targets provide a powerful means of monitoring the likely effectiveness of any mitigation regime. For example, we can monitor the projected date on which the trillionth tonne will be released if recent emission trends continue. Aside from deliberate geoengineering, the net impact of other anthropogenic forcing is unlikely to be negative, so if a policy regime is to have a better-than-even chance of preventing more than 2 °C of warming, then it must prevent the release of the trillionth tonne of carbon; hence, instead of advancing, the projected date of release must recede continuously into the future. An approximate real-time estimate of current cumulative carbon emissions and many of the numbers calculated in this box can be found at http:/trillionthtonne.org.

Box 8.2 summarises several approaches for estimating the probability of observing the 2 °C guardrail for a given carbon or CO_2 cumulative emission budget. For example, to achieve a 75% probability of constraining temperature rise to 2 °C or less, humanity can emit no more than 1000 Gt (one trillion tonnes) of CO_2 in the period from 2000 to 2050, or no more than about 1500 Gt CO_2-e (Meinshausen *et al.*, 2009). If we want to take a 50:50 chance of observing the 2 °C guardrail, then we can emit 1440 Gt of CO_2 (2000 Gt CO_2-e) in the period from 2000 to 2050. These estimates, which include emissions from both fossil fuel combustion and land-use change, are based on a number of assumptions regarding the behaviour of the climate system during this period, including climate sensitivity and carbon cycle–climate feedbacks.

How are we tracking towards this overall global emissions budget, and what are the implications of our current situation regarding the 2050 timeframe? Observed emissions from 2000 through 2008 were about 305 Gt CO_2 (Le Quéré *et al.*, 2009; Meinshausen *et al.*, 2009), leaving just under 700 Gt CO_2 in the budget out to 2050 if a 75% probability of observing the 2 °C guardrail is to be maintained. Thus, we have already consumed over

30% of the total global emissions budget out to mid-century in just the first nine years of the 50-year period. This emphasises the magnitude of the challenge ahead, and the need to achieve significant reductions in emission rates as quickly as possible.

The figure in Box 8.2 shows the magnitude of the challenge in a visual format, emphasising the critical importance of the peak year of emissions. Delaying the emissions peak by less than a decade beyond 2020 means that the annual rate of emission reductions in subsequent years will need to reach up to 8% compared to 3% for the earlier peak year. The coming decade is thus crucial in turning the global emissions curve downwards. Every year counts.

In addition to being much easier to understand and interpret than the targets and timetable approach, the global budget approach has some other attractive features (German Advisory Council on Global Change, 2009). The most important of these are (i) interim global budgets that can readily be defined and understood, as well as their implications for the overall budget to 2050 (e.g. Figure 8.2, in Box 8.2), (ii) national budgets that can readily be derived from the global budget (and which must aggregate to the overall global budget) and strategies for management of the budget, (iii) flexibility in terms of the temporal pathways of emissions, and (iv) flexible implementation mechanisms, which could include emissions trading within and between countries and regions.

A crucial challenge to the budget approach is the distribution of national emission budgets within the overall constraint of the global budget. One of the most thorough analyses of the budget approach (German Advisory Council on Global Change, 2009) has suggested that this distribution could be based on an underpinning, fundamental ethical principle: an equal per capita emission allowance for every human being, regardless of the country in which he or she resides. If this principle is accepted, national budgets can be calculated based only on start and end years (e.g. 2000–50), the level of ambition (e.g. 75% probability of observing the 2 °C guardrail), and the country's population in a reference year.

Determination of the start year is important. One policy option for determining national budgets could be based on a start year of 1990, which emphasises historical responsibility of emissions over future responsibility (German Advisory Council on Global Change, 2009). In this approach, the actual CO_2 emissions from fossil fuel combustion from 1990 through 2009 are included in the budget, allowing the residual left in the budget out to 2050 to be calculated. Figure 8.3 shows the results for selected countries. Three of the countries – Germany, the USA and Russia – have already consumed more than their entire budget. Setting the start year at 2010, which emphasises future responsibility rather than past, produces a somewhat different set of national budgets (Figure 8.4). Nevertheless, both approaches give much larger average annual emission budgets to developing countries.

A set of national emission budgets based on equity principles can then be aggregated up to the global scale to give a corresponding set of emission trajectories. Figure 8.5 shows a differentiation of countries into three broad groups based on per capita CO_2 emissions, with dark grey representing the highest emitters (wealthiest countries), mid-grey representing the rapidly growing economies (rapidly growing emitters) and light grey the world's poorest countries (very low emitters). Figure 8.6 shows how this differentiation plays out in terms of emission trajectories designed to yield a 67% probability of observing the 2 °C guardrail, starting

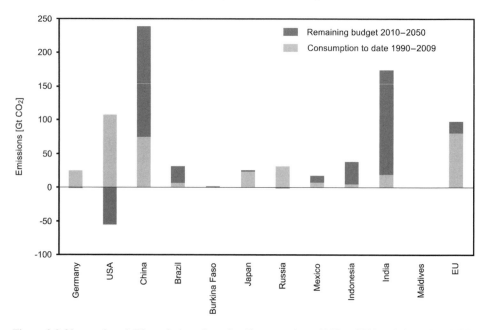

Figure 8.3 Observed total CO_2 emissions from fossil sources from 1990 to 2009 and the residual CO_2 budget to 2050 based on a start year of 1990, which emphasises historical responsibility for emissions. Source: German Advisory Council on Global Change (2009).

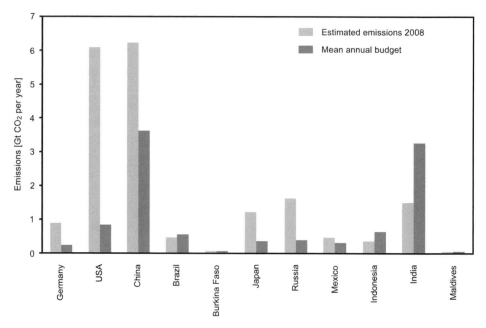

Figure 8.4 Estimated fossil CO_2 emissions in 2008 and average annual budgets to 2050 based on a start year of 2010, which emphasises future responsibility for emissions. Source: German Advisory Council on Global Change (2009).

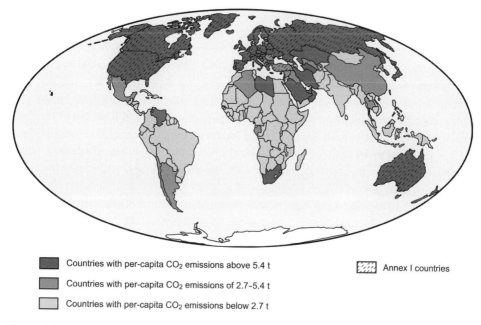

Countries with per-capita CO_2 emissions above 5.4 t

Countries with per-capita CO_2 emissions of 2.7–5.4 t

Countries with per-capita CO_2 emissions below 2.7 t

Annex I countries

Figure 8.5 Per capita fossil CO_2 emissions in 2005, organised by emission level and country. Source: German Advisory Council on Global Change (2009), based on data from WRI (2009).

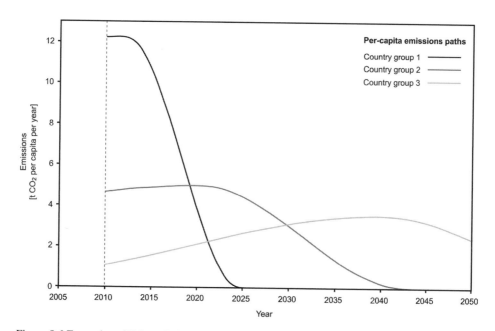

Figure 8.6 Examples of CO_2 emissions pathways for the three sets of countries defined in Figure 8.5 (German Advisory Council on Global Change, 2009). The pathways are for the 2010–50 period, and when aggregated, give a 67% probability of observing the 2 °C guardrail.

from observed emissions in 2010. The trajectories strongly reflect equity considerations. The world's largest emitters will need to begin a sharp reduction in per capita emissions immediately towards complete decarbonisation by 2025, while the world's poorest countries can raise their per capita emissions for several decades to peak in 2040. Emissions trading could modify these curves somewhat to give wealthy countries a little more time to completely decarbonise.

A proposal for another, finer-grained approach to determining emission budgets on equity considerations is based not on differentiation at the national level but at the individual level. In this approach the concept of 'common but differentiated responsibilities' has been applied to the emissions of individuals, based on the income distribution within a country as a proxy for the distribution of CO_2 emissions among the country's residents. Thus, individuals who are 'high emitters' – that is, those whose emissions exceed a universal individual emissions cap – are treated the same, regardless of whether they live in a developed or a developing country. National targets are determined by summing the excess emissions of all 'high emitters' residing in the country. Based on this methodology, the US would have to reduce its emissions by 55% as of 2030, the rest of the OECD by 28%, China by 29% and the rest of the world by 16% (Chakravarty *et al.*, 2009).

How feasible are the drastic emission reduction trajectories required to have a good chance of observing the 2 °C guardrail? Box 8.3 compares the technological options at humanity's disposal to meet a range of emission reduction targets by 2050, the most stringent of which would give a 75% probability of observing the 2 °C guardrail. Perhaps surprisingly to many people, the study shows that not only do we have the technologies we need to achieve the most ambitious levels of emission reductions, but also that we can do so at manageable cost – especially when compared to the risks avoided through emissions abatement.

Box 8.3
Achieving low CO_2 stabilisation levels: technological and economic feasibility

BRIGITTE KNOPF

In order to underpin political commitment of stakeholders and decision-makers to tackle the 2 °C target, it needs to be shown that the goal is not only technically feasible but also economically viable. The EU project ADAM[3] explored the feasibility of the 2 °C target in terms of technologies and economic costs for three different CO_2 concentration pathways (550 p.p.m., 450 p.p.m. and 400 p.p.m. CO_2-e) that roughly have a 15%, 50% and 75% chance of achieving the 2 °C goal. The latter scenario assumes negative emissions from the energy sector by the end of this century. To obtain a robust picture of mitigation costs and technological options, the analysis is based on a model comparison with five state-of-the-art energy–environment–economy models (Knopf *et al.*, 2009, Edenhofer *et al.*, 2010).

All models achieved even the low stabilisation scenario of 400 p.p.m. at moderate costs. Overall, global mitigation costs, expressed as aggregated GDP losses until 2100, are reported

---

[3] The model comparison presented here has been performed as part of the ADAM project (Adaptation and mitigation strategies: Supporting European climate policy), http://www.adamproject.eu.

to be below 0.8% for the 550 p.p.m. scenario, and below 2.5% for the 400 p.p.m. scenario. Costs increase with the stringency of the target but only linearly with the probability to reach the 2 °C goal. One of the five models reports overall gains for all stabilisation pathways.

In all mitigation scenarios, the energy mix strongly depends on the various models. This suggests that there is not just one way to achieve emission reductions but that several alternatives exist: extension of renewable energy, use of carbon capture and storage (CCS), increase of nuclear power, increase in biomass energy use or energy efficiency improvements – and different combinations of these options. But because the feasibility of CCS as well as the massive expansion of nuclear energy or renewable energy is not yet resolved, the question arises by how much costs will increase if some of these options are not available at all or only to a limited extent. This uncertainty analysis, which was performed for three of the models, conveys that achieving the 550 p.p.m. target allows for high flexibility in the deployment of different technologies; one technology can easily be replaced with another, albeit with an increase of costs in some cases (Figure 8.7, left). When the more ambitious target of 400 p.p.m. is to be achieved, this flexibility is lost in some cases (Figure 8.7, right). Interestingly, the models agree on the ranking of importance of individual energy technologies.

The availability of CCS and renewables turns out to be of pivotal importance for achieving low stabilisation levels. The analysis shows that without CCS the 400 p.p.m. CO_2-e target is not achievable ('No CCS' scenario in Figure 8.7) and costs increase for the 550 p.p.m. target.

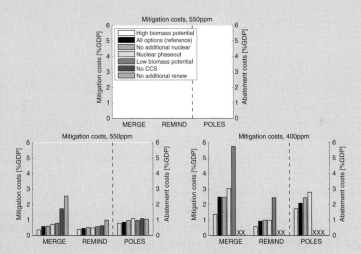

Figure 8.7 Global mitigation costs as aggregated GDP losses (for the models MERGE and REMIND) or abatement costs (model POLES) until 2100 relative to the baseline: (left) shows mitigation costs at 550 p.p.m., (right) shows mitigation costs at 400 p.p.m. An 'X' indicates a scenario where the target is not achieved. Source: modified from Edenhofer *et al.* (2010). This figure copyrighted and reprinted by permission from the International Association for Energy Economics. The figure first appeared in *The Energy Journal* (Vol. 31, Special Issue 1 on The Economics of Low Stabilization). Visit *The Energy Journal* online at http://www.iaee.org/en/publications/journal.aspx.

Box 8.3 (*cont.*)

Without the extension of renewable energy due to climate policy beyond its use in the baseline ('No additional renew'), the required emission reductions become very expensive for 550 p.p.m. CO_2-e and can no longer be achieved in case of low stabilisation at 400 p.p.m. CO_2-e. Moreover, assumptions regarding the biomass potentials considerably determine the costs of mitigation for low stabilisation. With a biomass use of less than 100 EJ per year ('Low biomass potential') compared to 200 EJ per year in the reference case, the losses are more than doubled for two models in the 400 p.p.m. case, and the third model cannot achieve the policy objective.

In contrast, the effect of nuclear energy as a mitigation option is limited, as nuclear already plays an important role in the baseline energy mix and in most models the increase induced by climate policy has nearly no effect on overall costs ('No additional nuclear'). Actually, the models show that a global phase-out of nuclear energy is possible at only slightly increased costs for both targets ('Nuclear phaseout').

In summary, the models illustrate techno-economic scenarios in which a low stabilisation target of 400 p.p.m. CO_2-e for atmospheric GHG concentrations can be achieved at moderate cost. The models provide a number of technology pathways that have a high likelihood of achieving the 2 °C objective. As a caveat, substantial dependence on particular technologies is observed in such a low stabilisation scenario, as compared to a less ambitious target of 550 p.p.m. that is more robust against the failure of particular technologies. All models assume global participation in climate policy in the near term and unconstrained technology transfer across regions. It is an enormous challenge to international climate policy to create these conditions.

8.4 Uncertainties in the natural carbon cycle

One of the assumptions common to any of the approaches to emission reductions is that the fraction of emissions that remains in the atmosphere is maintained into the future. The budget approach described in the previous section explicitly includes carbon cycle dynamics in its analysis of the probability of exceeding an average global temperature increase of 2 °C above pre-industrial levels for any given budget (Meinshausen *et al.*, 2009). Any significant change in the behaviour of the natural carbon cycle, which absorbs over half of the human emissions of CO_2, would have profound implications for the level of emission reductions required to constrain climate change to any desired level.

As discussed in more detail in Chapter 4, the most recent synthesis of knowledge on the carbon cycle (Le Quéré *et al.*, 2009) shows that although the strength of the natural carbon sinks in the ocean and on land have continued to increase as human emissions of CO_2 have increased, the ocean sink may be slowly failing to keep pace in a relative sense. The fraction of CO_2 emissions remaining in the atmosphere has likely increased from 40% to 45% over the past 50 years. There is some evidence that the land sink could weaken and possibly even become a net source of carbon to the atmosphere in the second half of the century, and possible new sources of carbon emissions to the atmosphere (e.g. from melting permafrost soils and tropical peatlands) could be activated at higher levels of global average temperature (Sections 4.2 and 4.3, and references therein).

Box 8.4 pulls together our current understanding of carbon cycle dynamics, as well as the effect of release of the 'aerosol brake', and casts this knowledge in the framework of emission reduction trajectories based on the cumulative emissions approach (note that in the box, cumulative emissions are calculated from 1750 to the far future and are represented as Gt or Pg of carbon rather than of CO_2). Two major conclusions emerge from the analysis. First, nearly all of the feedbacks and vulnerabilities operate in the same direction – to increase the warming. Second, the risks that these feedbacks will be activated rise sharply as global average temperature and cumulative emissions increase. The analysis reiterates the importance of observing the 2 °C guardrail (corresponding to cumulative emissions of approximately 1000 Gt C from 1750 (Allen *et al.*, 2009) or, equivalently, 1000 Gt CO_2 from 2000 to 2050 (Meinshausen *et al.*, 2009)); within this guardrail, the probability of activating the most serious of the carbon cycle feedbacks remains low.

Box 8.4
Relative effects of forcings and feedbacks on warming

MICHAEL R. RAUPACH AND JOSEP G. CANADELL

Uncertainties in the magnitude and rate of future climate change arise from both forcing and response. Radiative forcing is uncertain because we do not know the future trajectories of anthropogenic greenhouse gas emissions and other forcing agents. Climate response is uncertain because of lack of knowledge of physical and biogeochemical feedbacks in the Earth System, which can dampen or reinforce climate change.

Interactions between forcings and responses can be assessed through the relationship between peak warming above pre-industrial temperatures (T_p) and cumulative anthropogenic CO_2 emissions (Q) from fossil fuel combustion and net land-use change since the start of the industrial revolution around 1750. Recent papers (Allen *et al.*, 2009; Meinshausen *et al.*, 2009; Matthews *et al.*, 2009; Zickfield *et al.*, 2009) have shown that the relationship $T_p(Q)$ is robust, within quantifiable uncertainty bands. To have a 50% chance of keeping peak warming (T_p) to less than 2 K, cumulative emissions over all time must be less than about $Q = 1000$ Gt C (1 Gt C is 1 billion tonnes of elemental carbon in CO_2). Cumulative emissions to the end of 2008 were about 530 Gt C, rising at nearly 10 Gt C per year (Le Quéré *et al.*, 2009), so more than half of the 1000 Gt C quota has been used already.

Figure 8.8 shows the relationship between cumulative emissions Q (forcing) and peak warming T_p (response). The curves give the predicted warming T_p exceeded with 50% (median) probability, from Allen *et al.* (2009), Zickfield *et al.* (2009) and Matthews *et al.* (2009); black points show cumulative emissions and warmings to 2100 from IPCC scenarios (IPCC, 2007a). Forcing is measured by how far along the horizontal (Q) axis humanity chooses to proceed before CO_2 emissions are reduced to near zero. Response is measured on the vertical (T_p) axis, with several factors contributing to its uncertainties.

(1) *Uncertainty in climate sensitivity.* The response of climate to a given radiative forcing is uncertain because of poorly known physical and biogeochemical feedbacks in the

Box 8.4 (*cont.*)

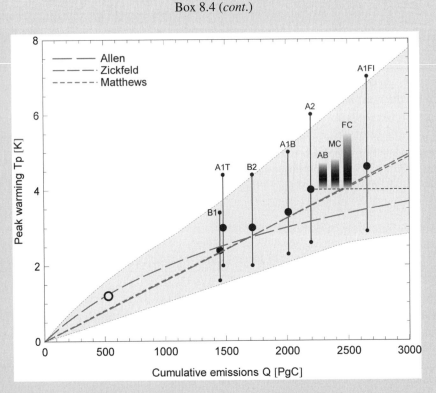

Figure 8.8 Peak warming from pre-industrial times (T_p) as a function of cumulative emissions (Q) from both fossil fuels and land-use change, from 1750 to the far future. Curves are from Allen *et al.* (2009), Zickfeld *et al.* (2009) and Matthews *et al.* (2009); shaded area represents the outer limit of the 17%–83% uncertainty ranges from these studies. Solid points show IPCC scenarios for 2100 (IPCC, 2007b), with uncertainty bars giving *likely* ranges in IPCC terminology (probability 67%–90% of an outcome within this range). Grey bars show ranges of possible effects (relative to the warming in 2100 from the A2 scenario as a base) from: (FC) coupled carbon–climate feedbacks on C sinks, from C4MIP (Friedlingstein *et al.*, 2006); (MC) mobilisation of previously immobile C pools; (AB) release of aerosol brake. All ranges are indicative only and include high uncertainty. Open circle shows cumulative emissions Q to 2008, with peak warming T_p, including 0.7 K of warming observed to date plus 0.5 °K of committed warming with radiative forcing stabilised at 2008 levels (IPCC, 2007b).

Earth System. Reinforcing feedbacks are particularly important because they lead to 'long-tailed' probability distributions for future climate change (Knutti and Hegerl, 2008; Roe and Baker, 2007). In the figure, much of the uncertainty in IPCC projections reflects uncertainty in climate sensitivity.

(2) *Carbon–climate feedbacks on land and ocean CO_2 sinks.* The influence of climate change on land and ocean CO_2 sinks occurs through both changes in atmospheric composition, including CO_2 and other gases; and also changes in climate, particularly the averages and

distributions of temperature and precipitation (Raupach and Canadell, 2008). The effects have been studied in several carbon–climate model inter-comparisons, notably C⁴MIP (Friedlingstein *et al.*, 2006), which found that carbon–climate feedbacks engendered by coupling of climate and carbon-cycle models increased warming (under an A2 scenario) by 0.1 to 1.5 K. This range is indicated by the 'FC' bar in the figure.

(3) *Releases of carbon from disturbed pools*. Several hitherto largely immobile carbon pools can be disturbed by climate change, leading to release to the atmosphere. A major potential pool is the organic carbon in frozen soils, estimated at nearly 1700 Gt C in total (Tarnocai *et al.*, 2009), of which around 100 Gt C may be vulnerable to release by thawing over the next century (Schuur *et al.*, 2009). There is also a significant pool of carbon in tropical peatland soils, mainly in the South-East Asian archipelago, of which around 30 Gt C may be vulnerable to decomposition and fire following drainage (Hooijer *et al.*, 2009). Net releases of carbon in other forest ecosystems are also likely through fire, insect attack and ecological transitions (Kurz *et al.*, 2008a, 2008b). Release of methane hydrates from reservoirs on the ocean floor and beneath permafrost is a high-impact, highly uncertain risk but is presently thought likely to occur over longer timescales than the other releases noted above (Bohannon, 2008); the biggest short-term vulnerability for this pool may well be through its use as a fuel source. The 'MC' bar shows a conservative range for the overall warming consequences of these vulnerabilities over timescales of the order of a century.

(4) *Release of the 'aerosol brake'*. There are significant vulnerabilities associated with anthropogenic influences on the Earth System other than through carbon–climate feedbacks. One concern is additional radiative forcing from release of the 'aerosol brake' on warming, which may occur as anthropogenic aerosol loads in the atmosphere decrease through efforts to improve air quality. This would reduce the present significant (though highly uncertain) net negative contribution from aerosol radiative forcing (Ramanathan and Feng, 2008). An indicative range for the resulting additional warming is shown by the 'AB' bar.

These vulnerabilities act mainly to increase warming. Increases may be higher than indicated if the climate system crosses thresholds leading to further positive feedbacks. The various contributions augment one another, but the heights of the bars cannot simply be added, as they represent indicative ranges for the effects of interactive processes.

8.5 Land-use change and fossil fuel emissions: A carbon cycle perspective

Human-driven emissions of CO_2 to the atmosphere are derived primarily from the combustion of fossil fuels and from land-use change, predominantly deforestation. Historically, emissions from fossil fuel combustion and cement production account for about 80% of total human emissions, although there is considerable variability from year to year in the land-use change component. For example, in 2008, land-use change accounted for only about 12% of total human CO_2 emissions, compared to the longer-term average of 20%, which had been the case for the 1990–2000 period (Le Quéré *et al.*, 2009).

Because the additional CO_2 in the atmosphere comes from both fossil fuel combustion and land-use change, and because it is the total amount of additional CO_2 in the atmosphere that perturbs the climate system, there is an obvious temptation to equate emissions from fossil fuel combustion and those from land-use change in emission reduction strategies. That is, reduction in emissions from the land-use change sector is often used or proposed as an offset to emissions from the fossil fuel sector.

However, although molecules of CO_2 from these two sources are chemically and radiatively equivalent (but have different isotopic signatures), they are quite different from the perspective of the natural carbon cycle. Carbon associated with land-use change and with the land sink is part of the dynamic carbon cycle that continually interchanges and redistributes carbon among the land, ocean and atmospheric reservoirs. Thus, land-use change represents a human modification of an active part of the natural carbon cycle. Combustion of fossil fuel represents an injection of new carbon into the atmosphere–ocean–land system; this is carbon that has been locked away from the active carbon cycle at the Earth's surface for millions of years. As this fossil carbon is activated and released into the atmosphere, it is redistributed to the land, ocean and atmospheric reservoirs.

It is certainly important, for a whole range of reasons from reversing land degradation to conserving biodiversity, that deforestation is quickly slowed and stopped, and that the historic 20% of additional CO_2 in the atmosphere is returned to the soil and to vegetation. However, from the perspective of the natural carbon cycle, reduction of one tonne of CO_2 emissions from land systems is not equivalent to the reduction of one tonne of emissions from fossil fuel combustion.

Consider the following scenario. Many countries decide to offset a large part of their emissions with land system uptake over the next few decades. More carbon is stored in the land but fossil emissions continue to rise. More importantly, new fossil fuel-intensive energy infrastructure and low-efficiency built infrastructure as well as transport systems are put into place. While global CO_2 emissions slow for a while, the land sinks eventually saturate and overall emissions surge again as fossil emissions begin to dominate. The temperature rises towards, and then past, 3 °C above pre-industrial levels. The land sink begins to reverse with the higher temperatures as feedback processes and new sources are activated. Most of the additional carbon that was absorbed into land systems earlier in the century is returned to the atmosphere. Fossil emissions continue to grow. The climate system is now beyond human control.

Thus, emission reduction by land uptake or by reduced fossil fuel use is not a question of 'this can offset that', but rather a '*both* this and that' imperative. In terms of the science of the carbon cycle and the climate system, if the 2 °C guardrail is to be observed and dangerous climate change is to be avoided, it is critical that the land-use source of CO_2 emissions is eliminated, and that uptake of CO_2 into soils and vegetation be enhanced *simultaneous*ly with vigorous, ongoing reductions of emissions from fossil fuel combustion.

8.6 Summary and conclusions

The first seven chapters of this book have put forward the case for humanity to take decisive action on the climate change challenge, and to do so with a sense of urgency. The climate system is changing faster than expected, creating serious and growing risks associated with sea-level rise, extreme events and changes in the hydrological cycle. Many impacts of climate change are now observable both in the natural world and in socio-economic sectors of importance to humans. Unabated climate change will raise the risk of abrupt and/or irreversible changes as tipping elements in the climate system are triggered. In response to these threats, over 100 countries are converging on the 2 °C guardrail (Copenhagen Accord, 2009); that is, on constraining the rise in global average temperature to no more than 2 °C above pre-industrial levels.

With an already recorded global average temperature rise of approximately 0.75 °C above pre-industrial levels and the already committed further increase of around 0.5°C (IPCC, 2007b), humanity has little time to act decisively. Although human actions influence the climate system in many ways, the single most important influencing factor is the emission of CO_2 into the atmosphere. The most common approach for reducing CO_2 emissions is through the adoption of a set of legally binding targets and timetables. However, the many complexities and uncertainties associated with the targets and timetables approach renders them difficult to understand, and hard to implement and verify. More recently, the research community has described a budget approach to emission reduction, based on the cumulative emissions that are allowable globally if the 2 °C guardrail is to be observed. This approach can be translated directly into a set of national CO_2 emission budgets that are easy to understand and interpret but very challenging to implement.

The remainder of this book explores the responses of humanity to the challenge of climate change, both in terms of emission reductions to avoid dangerous climate change and the adaptation measures required to deal with climate change that is already unavoidable. We examine the technologies, economic instruments and institutions that we need to turn around the emissions trajectories and implement the ambitious emission reductions outlined in this chapter. In addition, we outline the strategies that can be used to deal with the risks of climate change for food, water, health and security as well as for the natural world. With many countries still needing to develop further to bring their populations out of poverty, the integration of climate change adaptation and mitigation with development becomes of overriding importance. Finally, we explore the importance of values, ethics, world views, and religious and spiritual beliefs in mobilising the world's population to meet the challenge of human-induced climate change.

References

Allen, M. R., Frame, D. J., Huntingford, C. *et al.* (2009). Warming caused by cumulative carbon emissions: Towards the trillionth tonne. *Nature*, **458**, 116366.

Annan, J. D. and Hargreaves, J. C. (2006). Using multiple observationally-based constraints to estimate climate sensitivity. *Geophysical Research Letters*, **33**, L06704.

Bohannon, J. (2008). Weighing the climate risks of an untapped fossil fuel. *Science*, **319**, 1753.

Chakravarty, S., Chikkatur, A., de Coninck, H. *et al.* (2009). Climate policy based on individual emissions. *IOP Conference Series: Earth and Environmental Science*, **6**, 102005.

Charney, J. G. (1979). *Carbon Dioxide and Climate: A Scientific Assessment*. Washington, D.C.: National Academies Press.

Collins, M. (2007). Ensembles and probabilities: a new era in the prediction of climate change. *Philosophical Transactions of the Royal Society A – Mathematical, Physical & Engineering Sciences*, **365**, 1957–70.

Copenhagen Accord (2009). Draft decision -/CP.15, http://unfccc.int/resource/docs/2009/cop15/eng/l07.pdf

Council of the European Union (2005). *Presidency Conclusions*. Brussels, 22–23 March. European Commission.

Edenhofer, O., Knopf, B., Leimbach, M. and Bauer, N. (2010). The economics of low stabilization: Model comparison of mitigation strategies and costs. *The Energy Journal*, **31**, 11–48.

England, M. H., Sen Gupta, A. and Pitman, A. J. (2009). Constraining future greenhouse gas emissions by a cumulative target. *Proceedings of the National Academy of Sciences (USA)*, **106**, 16539–40.

European Environment Agency (2009). CSI 013 – Atmospheric greenhouse gas concentrations–Assessment published Mar 2009. http://themes.eea.europa.eu/IMS/IMS/ISpecs/ISpecification20041007131717/IAssessment1234255180259/view_content.

Friedlingstein, P., Cox, P., Betts, R. *et al.* (2006). Climate-carbon cycle feedback analysis: Results from the C4MIP model intercomparison. *Journal of Climate*, **19**, 3337–53.

German Advisory Council on Global Change (WBGU) (2009). *Solving the Climate Dilemma: The Budget Approach*. Special Report. Berlin: WBGU Secretariat.

Hansen, J., Sato, M., Kharecha, P. *et al.* (2008). Target atmospheric CO_2: Where should humanity aim? *Open Atmospheric Science Journal*, **2**, 217–31.

Hooijer, A., Page, S., Canadell, J. G. *et al.* (2009). Current and future CO_2 emissions from drained peatlands in Southeast Asia. *Biogeosciences Discussions*, **6**, 7207–30.

Intergovernmental Panel on Climate Change (IPCC) (2007a). *Climate Change 2007. Synthesis Report*, eds. The Core Writing Team, R. K. Pachauri and A. Reisinger. Cambridge, UK and New York, NY: Cambridge University Press.

Intergovernmental Panel on Climate Change (IPCC) (2007b). *Climate Change 2007: The Physical Science Basis. Contribution of Working Group I to the Fourth Assessment Report of the Intergovernmental Panel on Climate Change*, eds. S. Solomon, D. Qin, M. Manning, M. Marquis, K. Averyt, M. M. B. Tignor, H. L. Miller Jr and Z. Chen. Cambridge, UK and New York, NY: Cambridge University Press.

Knopf, B., Edenhofer, O., Barker, T. *et al.* (2009). The economics of low stabilisation: Implications for technological change and policy. In *Making Climate Change Work for Us – ADAM Synthesis Book*, eds. M. Hulme and H. Neufeldt. Cambridge, UK: Cambridge University Press.

Knutti, R. and Hegerl, G. C. (2008). The equilibrium sensitivity of the Earth's temperature to radiation changes. *Nature Geoscience*, **1**, 735–43.

Knutti, R., Meehl, G. A., Allen, M. R. and Stainforth, D. A. (2006). Constraining climate sensitivity from the seasonal cycle in surface temperature. *Journal of Climate*, **19**, 4224–33.

Kurz, W. A., Dymond, C. C., Stinson, G. *et al.* (2008a). Mountain pine beetle and forest carbon feedback to climate change. *Nature*, **452**, 987–90.

Kurz, W. A., Stinson, G., Rampley, G. J., Dymond, C. C. and Neilson, E. T. (2008b). Risk of natural disturbances makes future contribution of Canada's forests to the global carbon cycle highly uncertain. *Proceedings of the National Academy of Sciences (USA)*, **105**, 1551–55.

Le Quéré, C., Raupach, M. R., Canadell, J. G. *et al.* (2009). Trends in the sources and sinks of carbon dioxide. *Nature Geoscience*, **2**, 831–36.

Lowe, J. A., Huntingford, C., Raper, S. C. B. *et al.* (2009). How difficult is it to recover from dangerous levels of global warming? *Environmental Research Letters*, **4**, 014012.

Matthews, H. D., Gillett, N. P., Stott, P. A. and Zickfeld, K. (2009). The proportionality of global warming to cumulative carbon emissions. *Nature*, **459**, 829–32.

Meinshausen, M., Meinshausen, N., Hare, W. *et al.* (2009). Greenhouse gas emission targets for limiting global warming to 2 °C. *Nature*, **458**, 1158–62.

Murphy, J. M., Sexton, D. M. H., Barnett, D. N. *et al.* (2004). Quantification of modelling uncertainties in a large ensemble of climate change simulations. *Nature*, **430**, 768–72.

Petit, J. R., Jouzel, J., Raynaud, D. *et al.* (1999). Climate and atmospheric history of the past 420,000 years from the Vostok ice core, Antarctica. *Nature*, **399**, 429–36.

Piani, C., Frame, D. J., Stainforth, D. A. and Allen, M. R. (2005). Constraints on climate change from a multi-thousand member ensemble of simulations. *Geophysical Research Letters*, **32**, L23825.

Ramanathan, V. and Feng, Y. (2008). On avoiding dangerous anthropogenic interference with the climate system: Formidable challenges ahead. *Proceedings of the National Academy of Sciences (USA)*, **105**, 14245–50.

Randall, D. A., Wood, R. A., Bony, S. *et al.* (2007). Climate models and their evaluation. In *Climate Change 2007: The Physical Science Basis. Contribution of Working Group I to the Fourth Assessment Report of the Intergovernmental Panel on Climate Change*, eds. S. Solomon, D. Qin, M. Manning, M. Marquis, K. Averyt, M. M. B. Tignor, H. L. Miller Jr and Z. Chen. Cambridge, UK and New York, NY: Cambridge University Press, pp. 589–662.

Raupach, M. R. and Canadell, J. G. (2008). Observing a vulnerable carbon cycle. In *The Continental-scale Greenhouse Gas Balance of Europe*, eds. A. J. Dolman, R. Valentini and A. Freibauer. New York, NY: Springer, pp. 5–32.

Roe, G. H. and Baker, M. B. (2007). Why is climate sensitivity so unpredictable? *Science*, **318**, 629–32.

Sanderson, B. M., Knutti, R., Aina, T. *et al.* (2008). Constraints on model response to greenhouse gas forcing and the role of subgrid-scale processes. *Journal of Climate*, **21**, 2384–2400.

Schneider von Deimling, T., Held, H., Ganopolski, A. and Rahmstorf, S. (2006). Climate sensitivity estimated from ensemble simulations of glacial climate. *Climate Dynamics*, **27**, 149–63.

Schuur, E. A. G., Vogel, J. G., Crummer, K. G. *et al.* (2009). The effect of permafrost thaw on old carbon release and net carbon exchange from tundra. *Nature*, **459**, 556–59.

Solomon, S., Plattner, G.-K., Knutti, R. and Friedlingstein, P. (2009). Irreversible climate change due to carbon dioxide emissions. *Proceedings of the National Academy of Sciences (USA)*, **106**, 1704–09.

Stainforth, D. A., Aina, T., Christensen, C. *et al.* (2005). Uncertainty in predictions of the climate response to rising levels of greenhouse gases. *Nature*, **433**, 403–06.

Tarnocai, C., Canadell, J. G., Schuur, E. A. G. *et al.* (2009). Soil organic carbon pools in the northern circumpolar permafrost region. *Global Biogeochemical Cycles*, **23**, GB2023.

Tebaldi, C. and Knutti, R. (2007). The use of the multi-model ensemble in probabilistic climate projections. *Philosophical Transactions of the Royal Society A – Mathematical, Physical & Engineering Sciences*, **365**, 2053–75.

World Resources Institute (WRI) (2009). Climate Analysis Indicators Tool Version 6.0, http://cait.wri.org.

Zickfeld, K., Eby, M., Matthews, H. D. and Weaver, A. J. (2009). Setting cumulative emissions targets to reduce the risk of dangerous climate change. *Proceedings of the National Academy of Sciences (USA)*, **106**, 16129–34.

Part III
Equity issues

Climate change is having, and will have, strongly differential effects on people within and between countries and regions, on this generation and future generations, and on human societies and the natural world...

9

The equity challenge and climate policy: responsibilities, vulnerabilities and inequality in the response to climate change

'Our climate is owned by no one and yet needed by everyone, rich and poor. As we struggle to act together to save it, we stumble across one devastating idea. Our survival depends on equity.'[1]

The discussions about responses to climate change are dominated by concerns about equity – between rich and poor, North and South, present and future generations, investment in mitigation and adaptation, and between humans and the rest of nature. These discussions contrast with the global biophysical nature of climate change, wherein greenhouse gases are well mixed globally and quickly lose any biophysical connection with the place where they were emitted and in which the complexities of the Earth System produce impacts in regions far from those that emit most of the greenhouse gases.

We are dealing with a tragedy of the global commons and with the enormous challenges of a collective action problem. As the Stern Review argues: 'no two countries will face exactly the same situation in terms of impacts or the costs and benefits of action, and no country can take effective action to control the risks that they face alone. International collective action to tackle the problem is required because climate change is a global public good – countries can free-ride on each others' efforts – and because co-operative action will greatly reduce the costs of both mitigation and adaptation. The international collective response to the climate change problem required is therefore unique, both in terms of its complexity and depth' (Stern, 2007). However, as noted at several points in Chapter 8, the response strategies to deal with climate change invariably raise equity issues when we confront the complexity of allocating emissions rights or emission reduction obligations by country, by groups of people or by individuals. Similarly, the responsibility for funding adaptation and the allocation of adaptation funds are also fraught with considerations of equity and justice. In fact, all aspects of climate change – causes, impacts and response options – are exceedingly complex. Climate change and climate policy are already having, and will have, a variety of strongly differential outcomes for peoples within a country and

---

[1] Christian Aid (1999). *Who owes who? Climate change, debt, equity and survival.* http://www.ecologicaldebt.org/Who-owes-Who/Who-owes-who-Climate-change-dept-equity-and-survival.html.

Climate Change: Global Risks, Challenges and Decisions, Katherine Richardson, Will Steffen and Diana Liverman *et al.* Published by Cambridge University Press. © Katherine Richardson, Will Steffen and Diana Liverman 2011.

between countries, for current and future generations, and will include a range of social and biophysical consequences that are highly variable in space and time (Adger *et al.*, 2006; Richardson *et al.*, 2009).

Many elements of the highly complex and contentious equity dimensions of climate change were laid bare at the COP15 meeting in Copenhagen in December 2009 where, at various stages in the negotiations, there were very notable disagreements about who should reduce emissions and by how much, who should pay for adaptation and be eligible for the funds, and who had the right to make decisions and monitor international commitments (see also Chapter 13).

9.1 General principles

Dealing with the equity aspects of climate change intersects closely with larger questions of relationships between countries, peoples and the natural environment. First, the cross-scale nature of the climate change challenge defies simplistic solutions focused either on 'saving' the global environment regardless of the human condition, or on development and poverty alleviation without concern for our global life support system. Bringing large numbers of people out of poverty and bridging the gap between wealthy and poor must continue to be a very high priority, but can ultimately be counterproductive if it significantly damages the global life support system. Protecting this life support system for humans is obviously non-negotiable for the future of humanity, but a sufficient level of protection cannot be achieved without accounting for the vast differences in institutions and governance, resource access, and human well-being across the world's societies. The resistance of many developing nations to binding limits on their greenhouse gas emissions can be traced in part to their not having achieved the basic levels of human development, and to their frustration in the difficulties they face in gaining even their most basic goals in many types of global negotiations, including on trade and development (Najam, 1995, 2004; Müller, 1999, 2001; Roberts and Parks, 2007).

Equity concerns also intersect closely with discussions about ethics and rights, including those about ethical obligations to future generations and non-human species (Chapter 10), and about basic rights to a healthy life or to food and water (Caney, 2005). The Greenhouse Development Rights framework proposes that total emissions (capped to protect the atmosphere) be allocated according to principles that protect those below an income threshold (e.g. \$7500) and take account of historical responsibility (per capita emissions from fossil fuels since 1990) (Baer *et al.*, 2008).

Equity is enshrined in the international climate regime through two key principles in the United Nations Framework Convention on Climate Change (UNFCCC, 1992) – the precautionary principle and the principle of common but differentiated responsibility. The precautionary principle assumes that where there are threats of serious or irreversible damage, uncertainty shall not be used as a reason for inaction. The principle of common but differentiated responsibility recognises that while all countries should protect the climate system for present and future generations, the developed countries should take the lead in combating climate change as they bear the greatest responsibility for historical emissions and have the most capacity to respond.

Equity issues also emerge in struggles over the process of climate policymaking with calls for greater procedural justice that would include all nations and affected peoples, provide opportunities for participation, and build scientific and other types of capacity for different countries to engage in assessment and negotiations. In many cases, the arguments over process are linked to questions of power including who is controlling the negotiations, which private interests are influencing the process and outcomes, and how less powerful actors can claim a role in the governance of climate policies. Women are often marginalised in the process of policymaking, although they are already offering much in terms of insights and action on both mitigation and adaptation (Buckingham and Kulcur, 2009).

Careful analysis of claims about equity, and the distribution of climate impacts and responsibilities, start to break down simple cause–effect logic and isolated understanding, perception and categorisation of climate change. The intersection of climate change and poverty tends to be focused at a simplistic aggregate level, for example, between richer and poorer nations; whereas inequality within individual nations can mean that the rich in the developing world are less vulnerable or have greater greenhouse gas emissions than the poor in the developed world. Equity questions require approaches that are dynamic, holistic, multi-dimensional, multi-scale and differentiated (Tanner and Mitchell, 2008). For example, an analysis of historical emissions in Brazil shows how care needs to be taken to distinguish land-use emissions from livestock emissions, and to weight emissions according to land area (de Araujo *et al.*, 2007).

Equity issues are, perhaps, most evident in work on the impacts of climate change and the capacity to adapt to them. Climate impacts may differentially greatly impact the poor when compared to possible impacts for the wealthy, although the range, speed, and magnitude of impacts may be overwhelming for both. Careful identification of the areas, sectors and groups that are most vulnerable to climate change is needed, as vulnerability differs across households, communities and regions (Smit and Wandel, 2006). Changing climatic conditions impact local livelihoods and adaptive capacities usually quite differently (Kasperson and Kasperson, 2001; Adger *et al.*, 2006, 2007; Kirsch-Wood *et al.*, 2008). Adaptation cannot, thus, be an afterthought, but must rather be integral to any truly viable global climate regime (Chapter 15), and as such must also be brought under the overarching umbrella of 'common but differentiated responsibilities' (Vashist *et al.*, 2009).

9.2 Mitigation approaches and changing equity considerations

Historical responsibility for the anthropogenic burden of greenhouse gases that has accumulated in the atmosphere since the advent of the industrial revolution lies predominantly with the world's wealthier countries (Figure 9.1). However, the nature of this inequality is rapidly changing as several large countries – particularly China, India and Brazil – have

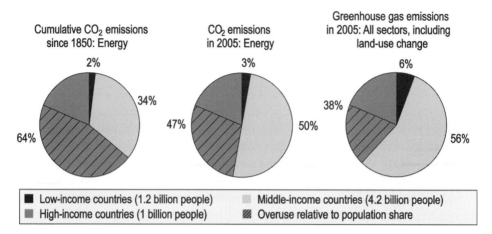

Figure 9.1 Share of global greenhouse gas emissions from the energy sector for low-, middle- and high-income countries for 2005 and historically (from 1850). The right-hand chart includes emissions from land-use change for 2005. Source: World Bank (2010). Overuse relative to population share is that share of emissions in high income countries that is a result of higher than average per capita greenhouse gas emissions.

begun rapid trajectories towards 'developed' status. The implications for climate policy are enormous and exceedingly complex, particularly at this time – in the midst of the great economic, social and geopolitical transitions of the twenty-first century.

Attempts to confront the rapidly changing geopolitical landscape, and associated ethical minefield, in the design of climate mitigation strategies have produced varying results. As described in Chapter 8, on the one hand, the emission budget approach yields significantly different results when the baseline is taken to emphasise historical responsibility than when it is skewed more towards current and future responsibility or to per capita emissions (Figure 9.2). On the other hand, others (den Elzen and Dellink, 2009; Voigt, 2009) have found fairly marginal differences between burden sharing proposals based on current per capita emissions and historical responsibility approaches. These differences are fairly stable even when different periods of responsibility are considered – back to 1990, 1900 or even 1750.

Individual mitigation activities will also have strongly differential consequences for various groups within societies and between countries. Some countries are far more generously endowed with fossil fuel energy resources than others – including developing countries such as Nigeria and Venezuela – and thus have more to lose in an energy transition. Others are planning to develop based on manufacturing, which will increase their carbon emissions – such as Greenland, which may expand resource extraction and processing (such as smelting) as it loses ice cover and traditional lifestyles. Some societies have far greater access to new technologies than others. Education levels, income and gender equality, and social cohesiveness vary significantly across countries and between groups within countries. All are relevant for climate mitigation approaches.

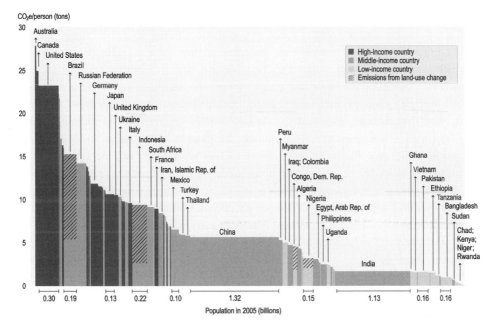

Figure 9.2 Ranking of countries by per capita emissions, as measured in tonnes of CO_2-e. The width of the bars is scaled to the population of the country. Source: World Bank (2010).

There is also disagreement about how to allocate emissions associated with the import and export of goods between countries, as a large proportion of some countries' emissions are associated with their exports. Would it be more equitable to assign emissions to the country where goods are consumed, or to regulate or tax imported goods in relation to their embodied emissions (Wobst *et al.*, 2007)?

One thing is clear– the equity conundrum is front and centre in the climate change challenge. Ignoring equity issues, or wishing that they would go away, will doom any potential mitigation approach to failure.

9.2.1 Energy and human development

Equity debates over emissions are integrally linked to those over energy consumption and fuel poverty. In the least developed parts of the world, increasing the production and consumption of energy is an urgent requirement. Energy is key to development because of its linkages to poverty alleviation, food security, human health, environmental quality, investments and national security. Energy-related emissions have tended to grow with economic development where energy supply is from sources that emit greenhouse gases. Recognising the role of energy in a wide range of development-related activities, a sustainable energy strategy is obviously central for achieving integrated development goals while simultaneously dealing with climate change.

When addressing the equity dimensions of climate and energy policies at the household level, country characteristics – such as population density, available energy sources and district and more local heating – are important as they serve as indicators for the way in which household CO_2 emissions vary between and within countries. Specific patterns of energy consumption are important in considering the impact of various climate policy options on households and individuals. Energy consumption patterns, and thus CO_2 emissions, vary with income, nature of the national energy supply and distribution system, urban or rural location, gender, and age, among other key factors (Khanna and Rao, 2009). For example, in some European countries (e.g. The Netherlands and the UK) emission intensities are lower for higher income households; whereas for others (e.g. Norway and Sweden) the intensity increases (Kerkhof *et al.*, 2009). Given the less-than-adequate level of energy services available to very poor people, climate policies that are sensitive to equity considerations should positively address energy poverty in addition to focusing on price-based instruments to reduce emissions. There are more than two billion people who use traditional forms of energy – especially bioenergy – with low per capita emissions, but where policies for efficient stoves, community renewables and reforestation could provide cheaper and more sustainable energy sources (Sagar and Kartha, 2007). There are also important connections between energy access and use, human health, the provision of ecosystem services, access to potable water, and transport solutions.

Given the close nexus between energy and most aspects of human development, many proposed solutions to reduce the emissions of greenhouse gases or to enhance the uptake of CO_2 by land systems from the atmosphere intersect strongly with equity considerations. The increasing use of land systems for climate mitigation (e.g. biofuels, see Chapter 11) can have major effects on economic growth and poverty that can be beneficial or adverse for poor and smaller-scale farmers depending on how projects are implemented (Arndt *et al.*, 2009). For example, many of the actions taken to reduce greenhouse gas emissions via increasing carbon bio-sequestration in forests and soils, if carefully planned, will likely enhance population health (Haines *et al.*, 2009; Kollanus *et al.*, 2009; Smith *et al.*, 2009), but can also result in the conversion of cropland and a reduction in food security.

9.2.2 Equity implications of Clean Development Mechanism projects

Although the developing world was not expected to commit to emission reductions under the first phase of the Kyoto Protocol, the Clean Development Mechanism (CDM) provides a mechanism for greenhouse gas reduction projects in non-Annex I (developing) countries. The CDM allows for such projects to obtain certified emission reductions (CERs) that can be purchased by developed countries and firms to offset their emissions and meet their emission reduction commitments under Kyoto, the EU trading scheme or voluntarily. For developing countries, CDM projects are seen as an 'additional financial mechanism' (as outlined in the UNFCCC) for achieving sustainable development while simultaneously reducing greenhouse gas emissions.

Contribution to sustainable development is one of the most important criteria for the selection of a CDM project, as it aims at orienting the future development of non-Annex I Parties along a more sustainable path and preparing developing countries for eventually assuming some type of emission reduction targets in the future. Host (developing) country governments are asked to certify that CDM projects meet sustainable development criteria.

The CDM has had some successes, including ancillary benefits such as air pollution abatement of urban smog (Pittel and Rübbelke, 2008; Rübbelke and Rive, 2008). However, some argue that it is largely failing to provide the kinds of sustainable development and equity benefits that were expected and that projects are often implemented without local input. Research suggests that as a market-based mechanism, the CDM has flowed very unevenly across regions and sectors, and is creating fewer jobs than expected (Hultman *et al.*, 2009). The projects were also initially biased to certain technologies (such as industrial gas reductions) rather than to those that benefit the poor (such as efficient woodstoves or community renewable energy; Hultman *et al.*, 2009). Voigt (2008), for example, suggested that it may be that the goals of the CDM cannot be met by every project; some projects do better on cost effectiveness or environmental integrity, while some deliver more sustainable development benefits. It is still not clear, however, how CDM includes multiple considerations and addresses dimensions of environmental integrity, sustainable development and economic efficiency in an integrated way. An important equity-related question remains: what role, if any, might the CDM have in really addressing poverty reduction?

The CDM is one instrument that contributes to the broad goal of technology transfer. Given the very large differences in technical capability between the industrialised and developing countries, particularly the least developed of the latter, there is an urgent need to greatly enhance the transfer and diffusion of low-emission energy generation technologies, as well as technologies in other parts of the economy – especially built infrastructure and transport – that significantly reduce greenhouse gas emissions. Box 9.1 discusses the critical issue of technology transfer in more detail.

Box 9.1

Equity issues in technology transfer

ARNE JACOBSON

The successful transfer of renewable energy, energy efficiency and other GHG mitigation technologies to middle- and low-income countries can play a key role in addressing global climate change. This transfer also has *potential* to help address equity concerns related to energy access and economic development in a carbon-constrained world. The equity issues that arise at the intersection of climate change, development and international relations are many and complex, however, and technology transfer offers no guarantee for improved equity. As discussed below, equity issues are linked to the structure of technology transfer programs.

Box 9.1 (*cont.*)

Existing economic, social and cultural factors can also strongly influence the distribution of benefits and costs associated with a program. This box includes discussion of both international and intra-national equity dimensions of technology transfer, using examples to highlight key issues.

The Clean Development Mechanism (CDM) is the world's largest climate change-related framework for technology transfer. The stated goals of the CDM are to provide Annex I polluters with low-cost GHG reduction investment opportunities and, in so doing, to support technology transfer that contributes to sustainable development (UNFCCC, 1992). While the CDM's GHG reduction track record is controversial (Wara and Victor, 2008; Schneider, 2009), Annex I polluters have benefited from low-cost GHG mitigation investment opportunities (Capoor and Ambrosi, 2009). The CDM has also stimulated over US\$20 billion in GHG reduction investments in non-Annex I countries, including an increasing number of renewable energy and energy efficiency projects (Capoor and Ambrosi, 2007, 2008, 2009; McNish *et al.*, 2009). This 'North–South' investment flow arguably represents a modest initial step toward addressing some of the international equity issues raised in climate change negotiations. Whether this investment constitutes a contribution to sustainable development is debatable, however, and an examination of CDM investments to date reveals numerous equity concerns.

Most CDM investment has been captured by a few countries, with, for example, China, India and Brazil accounting for over 75% of year 2012 certified emission reductions (CERs) under CDM (this includes all CERs issued to date as well as those expected by 2012). In contrast, all African countries in aggregate account for only 3.2% of these CERs, and least developed countries worldwide account for only 1% (UNEP, 2010). Thus, while the CDM has supported the transfer of GHG mitigation technologies to non-Annex I countries, the distribution of projects and investment among non-Annex I countries has been highly uneven. These trends are an arguably predictable outcome of the structure of the CDM, which provides incentives for Annex I polluters to seek least-cost GHG reduction investments. Given this incentive structure, it is unsurprising that they would choose to channel most funds into a few fast-growing, profitable markets with substantial emissions reduction opportunities.

The structure for verifying emissions reductions under the CDM also raises concerns about equity. Monitoring and verification are unquestionably necessary to ensure that projected GHG reductions are realised, but current rules represent a barrier for small and distributed projects (Sutter, 2001; UNEP, 2010). This is true because the verification cost is generally much lower per tonne of avoided CO_2-e for large, centralised projects than for small projects or even larger aggregated projects involving distributed reductions. Certain small-scale and distributed projects offer the most direct routes for low-income people to benefit from CDM projects. Recent efforts to develop new methodologies for small and distributed projects may help reduce this divide but, to date, poor people in developing countries have received few benefits from CDM technology transfer efforts.

Looking beyond program structure, existing conditions 'on the ground' that influence people's ability to access and use technologies and services also strongly affect the equity dimensions of technology transfer. Here, the term 'ability to access' is defined broadly to include factors such as people's legal standing to access program benefits, purchasing power, relevant capabilities and knowledge, community and household social relations, national and local politics, and others (Ribot and Peluso, 2003).

Rural electrification with solar energy offers an example of how a delivery approach interacts with contextual factors to influence outcomes. Solar home systems can be a relatively low-cost alternative to grid extension for basic electricity in service rural areas. In Kenya, over 200 000 small solar home systems have been sold through unsubsidised channels to rural households in off-grid areas. These systems represent an important expansion of access relative to the status quo, but most rural Kenyans cannot afford them (Jacobson, 2007). In market-oriented delivery models, access is strongly determined by purchasing power, so those with more money have better electricity access. For those who can afford a solar home system, equity issues do not end with the purchase. For instance, although many solar products perform well, some do not. Those unfortunate enough to buy low-quality goods can suffer significant financial losses. Unequal access to product performance information links equity to the effectiveness of government and international quality assurance efforts (Duke *et al.*, 2002; Jacobson and Kammen, 2007). Additional equity issues emerge in the context of household social dynamics. Small solar systems provide tiny amounts of electricity that must be allocated among competing possible uses within the home (common appliances include televisions, lights, radios and mobile phones). Intra-household dynamics, including gender and elder–junior relations, shape how and by whom electricity is used in each household. In other words, solar electricity can facilitate activities like children's studying, but the energy is frequently allocated in other ways (Jacobson, 2007). Thus, while solar electrification may result in increased energy access for some, a number of factors shape the distribution of beneficiaries. Structural changes to the delivery approach, including possibly targeted subsidies, could improve equity – but each alternative brings additional challenges that must also be considered.

Technology transfer is a critical tool for climate change mitigation, and it can also help address some equity concerns raised in international negotiations. However, even the best technology transfer programs raise equity concerns in some form. Policymakers should consider how program structure influences outcomes at both the international and intra-national scales, as well as how a variety of additional factors shape the equity dimensions of technology transfer efforts.

9.2.3 Equity and the forest deal: Implications of REDD for indigenous people

Deforestation is a major source of global greenhouse gas emissions, but was not covered under the international climate regime and carbon trading where only new or replanted forests could generate carbon credits. New proposals for reducing emissions from deforestation and forest degradation (REDD) would provide funding and carbon credits to countries that control forest loss, and thus protect and enhance land-based carbon sinks. As with the CDM, REDD promises a flow of funds to the developing world for carbon mitigation but has raised considerable concern about equity, especially in the risks and opportunities the approach presents for forest-dwelling communities and those whose livelihoods depend on the forests (Box 9.2; Okereke and Dooley, 2010).

Box 9.2

REDD (Reducing Emissions from Deforestation and Forest Degradation)
and equity: realising synergies and managing trade-offs

FRANCES SEYMOUR

Reducing Emissions from Deforestation and Forest Degradation (REDD) is an approach for
mitigating global greenhouse gas emissions that was included in the Bali Action Plan (2007)
(see also UNFCCC, 2007) and referenced in the Copenhagen Accord (2009). Under proposed
REDD schemes currently being negotiated, industrialised countries would provide finance to
developing countries in return for reductions in emissions that result from conversion of forests
to other land uses (deforestation), and from decreases in forest carbon stocks (degradation)
that result from such processes as destructive logging, unsustainable fuelwood extraction or
repeated burning. The proposed scope of REDD has recently been expanded (connoted by
'REDD+' or 'REDD plus') to include enhancement of forest carbon stocks (for example,
through restoration of degraded forests).

REDD presents a mitigation option that is potentially attractive to both industrialised and
developing countries. International financial transfers – likely mediated by a combination of
fund- and market-based mechanisms – would compensate developing countries for reductions
in forest-based emissions below an agreed reference level. Even after taking into account
the significant institutional constraints, REDD is likely to be less expensive than most other
mitigation options. REDD may thus make it possible to achieve deeper global emission
reductions than would otherwise be possible (Lubowski, 2008).

If a REDD mechanism is combined with sufficiently large emission reduction commitments
on the part of industrialised and middle-income countries, REDD could be a key element of
a global climate protection strategy that could enhance North–South equity by prompting
significant financial transfers. Establishment of the reference level will be a key factor in
determining how much a particular country stands to gain. However, overly generous reference
levels, i.e. those that exceed the true business-as-usual scenario, would come at the expense of
climate protection objectives (Angelsen, 2008b).

The design of the global architecture of REDD will influence its equity implications by
determining how well different countries will be positioned to benefit from REDD finance.
To maximise emission reductions in the short term, REDD funds would target countries
such as Indonesia and Brazil where current forest-related emissions are highest. However,
due to the problem of international 'leakage' – the potential displacement of deforestation
and degradation to other countries – participation in REDD must also include countries
with significant remaining forest, such as those in the Congo Basin, even if current rates of
deforestation and degradation are low. Expanding the scope of REDD to include activities
(such as reforestation) relevant to countries with little remaining forest would increase the
number of countries that could benefit, but at the likely expense of efficiency in meeting
climate protection objectives (Brown *et al.*, 2008).

A REDD design that relies on a market mechanism to allocate funds would likely
concentrate activities in countries seen to have more favourable investment climates, at

the expense of those (mostly poorer) countries seen as higher risk. Significant public investment will be needed to build capacity for monitoring, reporting and verification of forest-based emission reductions, as well as to remedy various governance deficiencies before significant REDD finance is likely to flow to the poorest countries (Dutschke and Wertz-Kanounnikoff, 2008).

The potential equity implications of REDD within societies is highly uncertain due to the poor governance conditions that characterise many countries and areas within countries with significant remaining forest area. In principle, REDD funds could be invested in ways that reduce forest-based carbon emissions, and also result in co-benefits in the form of rural poverty reduction and reduced vulnerability to climate change. For example, a modest REDD revenue stream could provide sufficient incentives for cocoa farmers in west and central Africa to maintain high-carbon agroforestry systems that, while of lower productivity than more intensive monoculture cultivation systems, provide more rural employment and better maintenance of ecosystem services (Sonwa *et al.*, 2009). However, research shows that few forested areas meet the necessary criteria for such direct payment schemes: a recent analysis of threatened forest in the Brazilian Amazon estimated that while more than half of the area could be economically viable for payment schemes to achieve REDD objectives, only about 25% have the land tenure status necessary to implement such schemes (Börner *et al.*, 2010).

A second constraint on the potential of REDD finance to improve equity within countries is the lack of capacity for transparent and accountable management of revenue flows. In the past, funds earmarked for reforestation and benefit-sharing with forest communities have been subject to corruption and misallocation to politically well-connected parties (Barr *et al.*, 2009). For REDD funds to benefit disadvantaged rural communities, new financial management capacities and accountability mechanisms (such as systems of independent audit) will need to be developed.

There is also a risk that REDD could result in the unintended negative consequence of making some of the world's most vulnerable people even worse off (Seymour, 2010). Owing in part to their remoteness and the ethnic diversity of forest communities, forests are often home to the poorest and most politically marginalised communities within countries. To the extent that such communities are currently being disadvantaged as forests are degraded or converted to other uses, REDD could potentially make maintaining forests a viable alternative. But if forests are given new value where land rights are insecure, such communities could see their access to forest resources restricted – or even be driven off their land – as more powerful actors position themselves to benefit from a new global market for forest carbon. How to design and implement safeguards at national and international levels, and allocate the costs and benefits of changing forest-related policies and practices within countries, are among the many contentious issues surrounding the prospect of REDD.

In summary, REDD has the potential to improve equity within and between countries, but the degree to which it will do so depends on the design of the global architecture, governance conditions within countries, and how trade-offs with the principal objective of reducing emissions are managed.

Strong arguments have been made that REDD agreements should protect local rights and livelihoods for at least two reasons: (i) it is ethically and morally appropriate to do so, and (ii) millions of forest-dependent people living in or adjacent to forests can play a key role in enhancing the effectivess of the overall program (Amacher *et al.*, 2009; Carlson and Curran, 2009). That is, REDD must be equitable to be effective because poor people will need alternatives to deforestation in order to participate in REDD. While it will often not be practical to provide direct payments to those who might otherwise degrade or cut the forests (owing to the potentially high transaction costs), policies associated with REDD can be designed to ensure that farmers and local communities gain increased revenue from sustainable timber production and forest conservation (Karsenty, 2008).

Thus, there is no doubt that REDD programs present potential vast benefits for for-est-dependent communities but, like any global sustainable development initiative, they must be carefully designed with indigenous and traditional people's interests in mind for these benefits to be realised. Currently, some advocacy organisations are concerned that a new global forestry initiative, if not carefully designed and implemented, will pose some inherent dangers to indigenous communities (Lovera, 2008; Adhikari, 2009). For example, many REDD schemes propose that recurrent payments should be made to the state, includ-ing some upfront payments to establish a forest cover baseline and a reliable monitoring system. Others propose to give credit for commercial forest plantations. These important features, however, are not enough on their own to ensure that REDD schemes are equitable. A number of key questions will need to be addressed to ensure that local communities can take advantage of the benefits REDD offers and that revenues from REDD indeed find their way to these communities.

To realise the potential benefits for local communities, REDD efforts will need to avoid the mistakes of previous efforts to control deforestation, and to utilise improved governance approaches and policy tools that are sensitive to the political and economic drivers of forest loss. Policy reforms could also be justified by social and biodiversity co-benefits, as well as by the inclusion of payments for ecosystem services more broadly. Critical questions remain regarding the choice and measurement of baselines (which greatly affect the amount of payment for a given activity), and how implementation will take into account wide variations in ecosystem type and disparities in capacity (Carlson and Curran, 2009). Regarding the latter, capacity-building activities in areas such as reserve management and fire control could form an important component of a REDD package (Angelsen, 2008a).

Modelling results indicate that REDD has the potential to be an efficient and effective emissions reduction strategy under a range of design options (Figure 9.3; Busch *et al.*, 2009). It is particularly important to ensure that the REDD program, with its suite of financial incentives, is extended to countries with historically low deforestation rates; this is to prevent leakage of deforestation activities from large countries with currently high deforestation rates, such as Brazil and Indonesia (this initiative has been termed REDD+), to those countries with low deforestation rates. In addition, in many countries, clearing of forests is required to release land for agricultural production, so the success

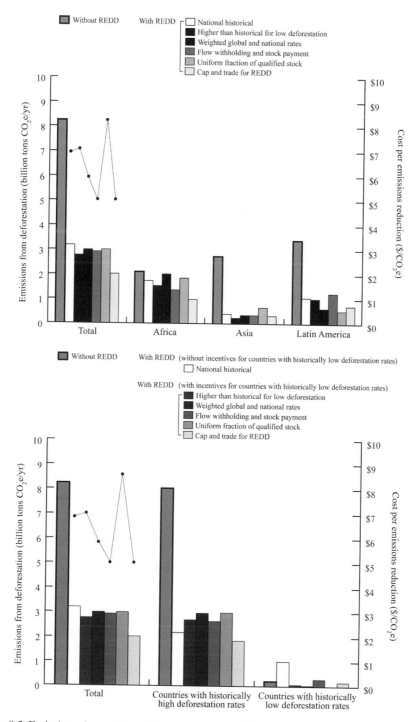

Figure 9.3 Emissions from deforestation under seven REDD (reducing emissions from deforestation and forest degradation) design options: (top) by region, (bottom) by historical deforestation rate and (next page) at varying elasticity of demand for frontier land agricultural output. Source: Busch *et al.* (2009).

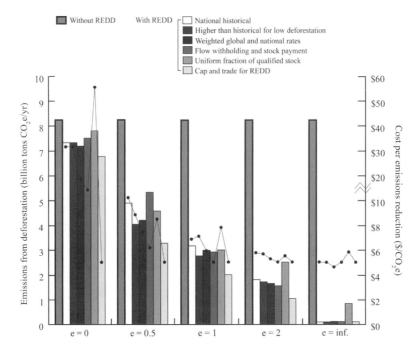

Figure 9.3 (cont.)

of REDD will depend on the ability to meet food demands outside the tropical forest frontier (Busch *et al.*, 2009).

9.3 Vulnerability and adaptation to climate change

Equitable solutions to climate change must be underpinned by an understanding of differential vulnerability to climate change (Adger *et al.*, 2006). Issues requiring attention and serious consideration include vulnerability and equity between nations, how vulnerability and adaptation interact with poverty, regional imbalances in adaptive capacity, and differential absorptive capacity to undertake adaptation projects and programs (Smit *et al.*, 2001; Adger *et al.*, 2006). If climate policy is to respond to equity concerns, then inequities within countries and across different social and gender groups also need to be acknowledged as well as identifying winners and losers, opportunities and challenges (Leichenko and O'Brien, 2008). It is also important to be able to clearly and effectively identify vulnerability 'hotspots', and to consistently and rigorously identify where past, present and expected future vulnerabilities and impacts have been and are likely to occur.

In moving from the assessment of vulnerability towards building effective adaptation approaches, the complexity of the challenge stretches from the types of adaptive responses required (e.g. proactive, reactive and inaction) to contested notions of a number of ethical and justice dimensions (Dow *et al.*, 2006). In the context of climate change, equity issues

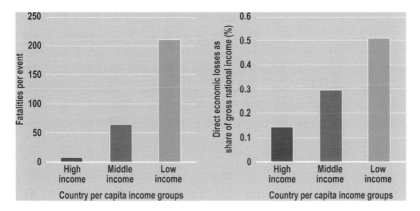

Figure 9.4 A disproportionate share of the human and economic burdens from natural disasters (1980–2004) falls on low-income and lower middle-income countries. Source: IIASA (2009). Reproduced with permission from the International Institute for Applied Systems Analysis (IIASA).

should not only be seen as simply making efforts to ensure that the vulnerable are equally targeted and fairly buffered from disproportionally shouldering the burdens of impacts. A broader range of pertinent issues includes decision-making processes, response measures, policies and institutional contexts that facilitate or hinder actions, as well as ensuring inclusion and 'fairness' (Adger *et al.*, 2006).

Local and regional vulnerabilities to climate change are a result of inequities that exist within and across nations, and may be attributed to factors such as unequal economic relationships, unequal access to entitlements, differential social capital, power relationships and institutional factors; the outcome of these inequalities is evident, for example, in the human and economic burdens from natural disasters (Figure 9.4).

Some of the complex and unequal vulnerabilities to climate change – intensified by multiple interacting drivers of change – are outlined below for food, water, health and overall human security. These are not the only sectors that require attention. Those selected here give some perspectives on the interacting and exacerbating stresses and opportunities associated with climate change.

9.3.1 Climate change and food security

Numerous studies have documented how political economy, natural resources and other factors interact to create unequal access to food within and between countries, regions and households. Food is one of the most critical sectors already being impacted by climate variability (Chapter 5) and climate change, particularly in poorer, rural contexts. In the 1990s, early first-order climate impact models suggested that climate change would result in a range of agricultural winners and losers, with developing countries being more vulnerable (Rosenzweig and Parry, 1994). More recent assessments note that this possible inequality may worsen (Parry *et al.*, 2004; Fischer *et al.*, 2005). Cline (2007), for example, applied agricultural impact models combined with climate model projections to develop estimates for agricultural impacts in more than 100 countries with and without CO_2 fertilisation effects. Agricultural productivity for the entire world is shown to possibly decline from levels that would otherwise be reached by between 3% and 16% by the 2080s. Some developing countries, many of which have average temperatures that are already near or above crop tolerance levels, for example, are also shown to experience an average 10%–25% decline in agricultural productivity by the 2080s. Some of the largest negative impacts to agricultural productivity are expected in regions where the poor derive a large share of their livelihoods from agriculture.

Other more recent assessments of climate change impacts (Alcamo, 2009; IFPRI, 2009; World Bank, 2009) move beyond crop production only to analyse a number of 'combined impacts' (e.g. increased food prices, weakened health status, etc) that could be particularly severe for various developing regions, including South Asia, Latin America and Sub-Saharan Africa. Using a global agricultural supply-and-demand model linked to a biophysical crop model (Design Support System for Agrotechnology Transfer, DSSAT), mixed results are shown for parts of Sub-Saharan Africa, and the Latin America and Caribbean regions (Rosengrant *et al.*, 2008 see, for example, Table 10).

Several assessments of the adaptation and mitigation costs required for the agricultural sector have also recently been undertaken (e.g. Pan African Climate Justice Alliance (PACJA), 2009; Parry *et al.*, 2009; see also several websites, e.g. http://www.gecafs.org; http://www.fanrpan.org; http://www.cgiar.org). Estimated costs range from 10% of Africa's GDP by the end of the century with a 4.1 °C warming (PACJA, 2009) to estimates of at least $7 billion per year in additional funding required to finance the research, rural infrastructure and irrigation investments needed to offset negative impacts of climate change across several developing regions. The needs are strongly differentiated across regions, with, as potential examples, Sub-Saharan Africa requiring substantial investment in infrastructure such as roads, a strong need for agricultural research in Latin America and a requirement for irrigation efficiency in Asia (IFPRI, 2009). Pro-growth and pro-poor development strategies that support food security and agricultural sustainability may also be appropriate options for supporting climate change adaptation (Tanner and Mitchell, 2008; IFPRI, 2009).

9.3.2 Climate change and water security

Inequities and differences between levels of development and resource use are very clear in the water sector. Water access, demand and effective usage are already issues of concern, for example, between large industrial users and smaller holder users, between public utility users and private users, and between sub-national and trans-boundary water usage. Several assessments have been undertaken on how climate variability and change can aggravate such interactions. Assessments of climate change and hydrological impacts, of flooding and drought situations, and of the role of climate change and variability in extreme events and their consequences have been undertaken (IPCC, 2007; Bates *et al.*, 2008; Parry *et al.*, 2009).

Changes in both water quality and quantity, exacerbated by climate change, are expected to have a range of knock-on impacts on societies (e.g. for food availability, stability, access and utilisation; Bates *et al.*, 2008). As Bates *et al.* (2008) state: 'Many semi-arid and arid areas (e.g. the Mediterranean Basin, western USA, southern Africa and north-eastern Brazil) are particularly exposed to the impacts of climate change and are projected to suffer decreases of water resources due to climate change'. Adaptation costs in the water sector, however, are not limited to those linked directly to water service and provision. Costs coupled to increased flood risks and flood management have also been estimated, in some isolated cases under different climate and urbanisation scenarios, for example, to be several hundred millions annually in the Sacramento Basin, California (Zhu *et al.*, 2007). Such costs would be outside the capacities of many developing countries, again reinforcing the point that climate change will have strongly differential impacts around the world.

Recent assessments show that adaptation options in the water sector for various regions and sub-regions will vary not only on financial resources, but also on availability of existing infrastructure, and institutional and governance capability (Bates *et al.*, 2008; Ludwig *et al.*, 2009). For example, for most European and Mediterranean countries, buffers can usually be developed to soften the impacts of periods of water stress. By contrast, many countries in Africa (e.g. those in western and eastern Africa) have a limited ability to buffer against periods of induced water stress and can be hampered by a lack of storage infrastructure. Even if these countries could invest 5% of their GDP in water infrastructure, it would still take several decades to create sufficient storage (e.g. about 40–70 years in parts of eastern Africa).

Adapting to the impacts of climate change on water resources will take much creative ingenuity and political will. Current water management practices, therefore, may not be sufficient and robust enough to cope with the impacts of climate change (Bates *et al.*, 2008), particularly in poor, developing countries that have intrinsically low adaptive capacity. In these countries, reduced technological capability and financial resources may drive adaptation more oriented towards demand management. In Morocco, for example, a recent study (Khebiza *et al.*, 2009) showed that water-saving practices may be better options for adaptation to water stress than technical and high risk and high cost engineering solutions (e.g. pipeline irrigation).

9.3.3 Climate change and human health

Impacts on food security and access to water resources are two ways in which climate change impinges indirectly on the health and well-being of populations. In fact, human health can be considered as an integrator, or an ultimate 'bottom line', of the various consequences of a changing climate for humans. The risks to health and survival highlight the profound moral dimension of climate change. Those risks impinge, now and in the future, unequally between and within populations (McMichael *et al.*, 2008). The poor, the marginal, the uneducated and the geographically vulnerable are at greatest risk of climate-related under-nutrition, disease, injury and premature death. Meanwhile, the countries that have been the largest (per capita) greenhouse gas emitters generally incur lower risks to health. The further moral dilemma spans generations: today's high-emitting world population faces lesser risks to health than will future generations. The spectrum of risks to human health is broader than previously thought, including impacts of severe droughts on mental health (Kjellstrom *et al.*, 2009) and the release of toxins such as mercury by melting ice sheets (Grandjean *et al.*, 2008). These escalating risks to human health impinge in highly unequal ways between and within populations.

In general, climate change will amplify the existing health inequalities around the world. Based on the original estimates by the World Health Organization for the years 2000 and 2030 (McMichael *et al.*, 2004), it is likely that, in low-income countries, climate change currently causes an additional 200 000 premature deaths annually from a subset of climate-sensitive health outcomes: from crop failure and under-nutrition, diarrhoeal disease, malaria, and flooding. Around 85% of those deaths are in children. Furthermore, the health risks faced by future generations from climate change will be far greater than those faced by people living today and, if progress is not made rapidly to alleviate poverty in many parts of the world, the poor and most vulnerable of the future will suffer proportionally even greater consequences. A more detailed discussion of the equity dimensions of climate change impacts on human health is given in Box 9.3.

Box 9.3
Differential effects of climate change on health both between and within populations

ROBERTO BERTOLLINI, DIARMID CAMPBELL-LENDRUM AND
ANTHONY J. MCMICHAEL

All regions of the world will be affected by changing climate, but the resulting health risks to human populations vary greatly, depending on where and how people live, on their age and pre-existing disease status, and perhaps most importantly on poverty and inadequate coverage of health services (Confalonieri *et al.*, 2007). A general tendency will be for climate change to amplify existing health inequities in the world.

   Geography is an important determinant of vulnerability to health impacts of climate change. Small island developing states and other low-lying regions are particularly vulnerable to death and injury, and destruction of their public health infrastructure, from increasingly severe tropical storms (WHO, 2006); salinisation of freshwater and agricultural land from sea-level rise; as well as the effects of droughts, floods and high temperatures on nutritional deficiencies and diarrhoea (Singh *et al.*, 2001; Hashizume *et al.*, 2008). In cities – particularly tropical megacities – the effects of long-term climatic change are compounded by the urban heat island effect and elevated levels of air pollution (increasing vulnerability to heatwaves); extensive coverage of impervious surfaces and precarious housing (increasing the risks of flash floods); and high population densities, inadequate water and sanitation services (raising vulnerability to climate-sensitive infectious diseases such as diarrhoea and dengue) (Campbell-Lendrum and Corvalan, 2007). Mountain populations are also vulnerable to a range of health risks, such as glacier lake outburst floods and reduction of summer water availability as glaciers retreat (WHO/SEARO, 2006), as well as exposure to malaria and other vector-borne diseases as warmer temperatures enable transmission at higher altitudes (Pascual *et al.*, 2006).

Individual characteristics such as age and gender greatly influence health vulnerability to weather and climate. In the 1991 cyclone disasters that killed 140 000 people in Bangladesh, death rates among women were almost four times greater than those among men. However, the effect of age is probably more important; in the same events, death rates among children under 10 years of age were more than six times greater than those of older children (Bern *et al.*, 1993). The most important climate-sensitive diseases globally are also those of poor children. About 90% of the burden of malaria and diarrhoea, and almost all of the burden of diseases associated with under-nutrition, are borne by children aged five years or less, mostly in developing countries (WHO, 2008). Climate change threatens to intensify these burdens further. Studies project that rising temperatures and changing rainfall patterns are likely to increase the population at risk of malaria in Africa by over 170 million in 2030 (Hay *et al.*, 2006), as well as increasing the number of months of exposure to transmission (Tanser *et al.*, 2003). The resulting disease burden can be expected to affect mainly children.

The most important determinant of health vulnerability at the global scale is probably poverty. The poorest populations already suffer from high burdens of climate-sensitive disease and lack effective public health systems to protect them from the increased risks associated with climate change. For example, the per capita mortality rate from vector-borne diseases is almost 300 times greater in developing nations than in developed regions (WHO, 2008). In simple arithmetic terms, and noting that climate change tends to act as a multiplier of existing risks or problems, the greater the background rate of (say) child diarrhoeal disease, the greater the absolute increment in that rate due to climate change. Climate-related health risks are also often greater for poor individuals within any population. In developing countries, diseases transmitted by water, soil and vectors – such as schistosomiasis (Ximenes *et al.*, 2003), hookworm (Hotez, 2008) and filariasis (Ottesen *et al.*, 1997) – are often many times more common among people with the lowest socioeconomic status in any one site. The phenomenon also occurs in rich countries; in the wake of Hurricane Katrina in the USA, children from lower-income groups were at increased risk of developing severe mental health symptoms (McLaughlin *et al.*, 2009).

The strong association between poverty and climate-sensitive diseases means that climate change and associated development patterns threaten to widen existing health inequalities. A World Health Organization assessment of the burden of disease caused by climate change suggested that the modest warming that has occurred since the 1970s was already causing over 140 000 excess deaths annually by the year 2004 (WHO, 2009). The estimated per capita impacts were many times greater in regions that already had the greatest disease burden (McMichael *et al.*, 2004). The disproportion of populations that have contributed the least to climate change and are the most vulnerable to health risks is graphically presented in Figure 9.5.

Box 9.3 (*cont.*)

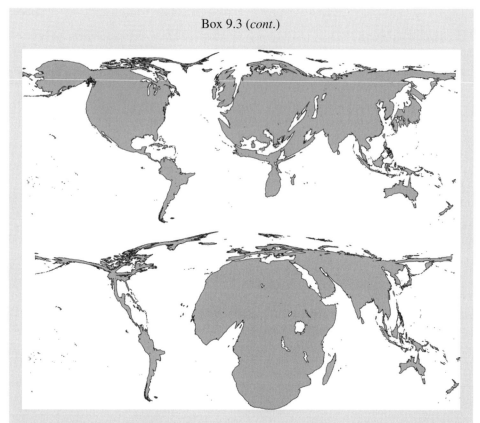

Figure 9.5 Poorer countries contribute little to greenhouse gas emissions, but are most vulnerable to health impacts. The top image shows countries scaled by total emissions of CO_2 to 2002; the bottom image shows countries scaled by WHO regional estimates of per capita mortality from climate change in 2000. Sources: McMichael *et al.* (2004); Patz *et al.* (2007).

Conversely, a series of new studies (Haines *et al.*, 2009) shows that climate mitigation policies in a range of sectors can have significant direct beneficial effects on health, in some cases fully offsetting the costs of mitigation. They also suggest that these benefits may disproportionately accrue to poorer populations. For example, more sustainable transport systems would be expected to bring some 7400 additional years of healthy life per million population annually in London, UK, versus approximately 13 000 in New Delhi, India (Woodcock *et al.*, 2007). Feasible improvements in UK housing energy efficiency would bring about 850 additional years of healthy life per million population, whereas cleaner-burning cook stoves in India could bring about 12 500 additional years; roughly equivalent to halving the cancer burden of the country (Markandya *et al.*, 2009).

Overall, climate change threatens to increase health inequities within and between populations – and, ominously, between generations. In contrast, effective and fair adaptation and mitigation policy has the potential to safeguard and improve health, and to reduce disparities between populations.

9.3.4 *Climate change and human security*

Climate change is often viewed as a 'threat multiplier', exacerbating security threats caused by persistent poverty, weak institutions for resource management and conflict resolution, a history of mistrust between communities and nations, and inadequate access to information or resources. The growing understanding of climate change as an issue of equity that disproportionally undermines human security of some regions and groups is essential to addressing the underlying causes that drive security-related processes and influence outcomes and response measures. Sustainable development, with its key elements of equity, long-term management of resources and mainstreaming environmental considerations into policy and decision-making, could potentially help to address the human security implications resulting from climate change impacts (Barnett, 2003).

Human security is defined as a state that is achieved when and where individuals and communities have the options necessary to end, mitigate, or adapt to threats to their human, environmental and social rights; have the capacity and freedom to exercise these options; and actively participate in pursuing these options (GECHS, 2009).

Considerable research has already been carried out to identify a number of potential human security implications that may arise from differential climate change impacts, in some instances aggravating already existing stresses (Gleditsch, 1998; Adger *et al.*, 2006; Grasso, 2010). For example, Hurricane Katrina exposed how climate variability, poverty and policy failures can undermine human security even in rich countries with well-established disaster management and information systems. It shows the multiple risks to human security (e.g. flooding, wind, pollution) when large populations are located on storm-prone coasts (http://www.gechs.org/aviso/14.pdf). In the Pacific Islands, climate change in the form of sea-level rise is starting to undermine human security as salt water contaminates ground water, waves damage infrastructure, and traditional livelihoods are undermined by cultural and economic changes (GECHS Science Plan, http://www.ihdp.uni-bonn.de/html/publications/reports/report11/gehssp.htm#chapterII22).

Human security should be seen as an issue that cuts across national boundaries, and needs to be addressed at different scales and units of analysis. This complex arena, however, is also further complicated by the lack of reliable and easily accessible data, and the ability to effectively monitor change (Chapter 1).

Identifying those most vulnerable and at risk to climate change, using various approaches including national indicators of vulnerability, may guide adaptation funding and other risk reduction efforts that ultimately may reduce security risks (Symons and Harris, 2009; Vajjhala, 2009). Such indicators, for example, may include overall counts of human suffering from climate disasters (e.g. numbers made homeless, disaster loss estimations associated with hurricanes, floods, droughts, heatwaves; Kirsch-Wood *et al.*, 2008). Biophysical indicators of areas of heightened vulnerability may also be explored, perhaps focusing on areas of current or predicted ecosystem damage and thus loss of ecosystem services.

9.4 Equity and adaptation

The scale of the adaptation challenge, especially at the global level, is enormous. Monetary compensation can probably never repay or repair damages from climate change; perhaps a great majority of all damage cannot be adapted to, regardless of how much funding were available. Assessment of adaptation costs (including benefits) should therefore consider not only economic criteria, but also indicators of social welfare and equity. These considerations make it difficult to estimate the amounts of funds needed for meeting the 'adaptation gap' to allow developing nations to avoid the worst impacts of climate change. Adaptation deficits, in many cases, are closely linked to development, particularly for those countries where there are already deficits to coping and adapting to present climate (Mitchell and Tanner, 2006; Parry *et al.*, 2009).

There is now a strong focus at all scales, from global to local, on the effective sourcing and distributing of resources to help people and ecosystems reduce their vulnerability to climate change and to build resilience to climate impacts. Opportunities arising from climate change, and not only challenges, need to be identified and seized (Müller, 2008). A major frontier in research is thus estimating the types of adaptation interventions that will be needed, based on different climate scenarios, on different social contexts, in both the short and long term, and on different understanding of how 'adaptation' is being used and applied (Hof *et al.*, 2009; Parry *et al.*, 2009). For example, payments for environmental services, such as forest protection, can generate income that can help communities adapt to climate change (Kalame *et al.*, 2009).

Notwithstanding the difficulties of identifying and describing appropriate and effective adaptation activities, the cost of adaptation has inevitably received much attention (Parry *et al.*, 2009), particularly during and since COP15. Estimates of the overall global costs for adaptation range from US$8 billion to over US$100 billion a year (World Bank, 2005; Müller and Hepburn, 2006; Oxfam, 2007; UNDP, 2007). Differences in these figures are due to different definitions of what constitutes adaptation, of whether one only addresses current investments in development or also improves older infrastructure, and so on. On a sector basis, a range of various adaptation costing scenarios have recently been made (Parry *et al.*, 2009). The costs, for example, of treating the extra medical cases and associated illnesses coupled to climate change are estimated to be about US$3–10 billion a year by 2030 (Ebi and Meehl, 2007; Ebi, 2009). In their careful analysis of adaptation costing under various scenarios and constraints, Parry *et al.* (2009) provided some top-down estimations (e.g. UNFCCC) in the range of US$11.3 billion–US$12.6 billion for the year 2030 for the agriculture, forestry and fisheries sector. They conclude, however, that this estimation is on the low side, particularly if other assessments are considered (e.g. bottom-up assessments). By contrast to these various figures, the worldwide stimulus packages in response to the 2008–09 financial crisis exceeded US$4 trillion.

Estimating the potential costs of adaptation is difficult enough. Raising adequate and predictable funding is, however, proving to be even more difficult, as voluntary funds established under the Kyoto Protocol and the UNFCCC are grossly inadequate by any measure

(Roberts *et al.*, 2008; Müller, 2008; Müller and Gomez-Echeverri, 2009; Vidal, 2009). In most developing nations, insurance and relief are *far* less available than they are in wealthy countries, so most of the costs of adapting to climate change will usually be borne by individuals and governments in these countries.

A key stumbling point at the COP15 was the issue of differentiated responsibilities in both adaptation and mitigation funding. Most of the attention on justice and climate adaptation funding, and the need for a 'just transition' to low-carbon economies, has focused on raising funds and that this funding must be predictable, adequate and appropriate (Müller and Gomez-Echeverri, 2009). Rather than view such funding as voluntary or charity, Müller and Gomez-Echeverri (2009) argued that adaptation and transition funding must be considered restitution and compensation for damages inflicted by those countries most responsible for the additional greenhouse gases in the atmosphere. Therefore, the existing model of foreign aid, with donors deciding who gets the money and under what conditions, is not generally appropriate for climate finance (Byrd and Peskett, 2008) as it is critical to know how the allocation of adaptation funding incorporates national issues such as governance, transparency and effectiveness.

Distributing the funds adds powerful ethical and equity components to the above-mentioned challenges (Grasso, 2006). Beyond how much money is raised, additional questions include: *How* is it distributed? And more importantly: *Who* calls the shots? *Who* controls the money? (Müller, 2008; Müller and Gomez-Echeverri, 2009; Grasso, 2010). The answers to these questions are critical. Additional inequities could well be created if we make poor decisions in funding efforts to address climate change.

9.5 Summary and conclusions

Climate change is inherently unfair. Its origins, impacts and solutions all have strongly differential elements, intersecting with inequalities between countries and regions, between groups of people, and between individuals. The complexity is heightened by the highly dynamic situation in the early twenty-first century. The historically high emitting countries are being joined by others on rapid development pathways, the impacts of climate change are already discernible beyond the influences of natural variability and are mounting rapidly, and the capacity of the most vulnerable people and regions to cope is already being stretched.

The principle of a 'common but differentiated' responsibility for reducing human emissions of greenhouse gases is widely agreed around the world, but the practical implementation of the principle in terms of emission reduction targets and timetables is foundering, largely because of complex equity issues. The need to adapt to the unavoidable consequences of climate change is now widely recognised, as is the need for an adaptation safety net for the poorest and most vulnerable people and societies. However, the funding and administration of adaptation activities – especially the sources and amount of the funding – the rules by which and by whom it will be allocated, and the ways in which accountability for expenditure will be ensured, are contentious, with points of view often following the traditional fractures of global society that align with equity issues.

These difficulties can easily lead to an atmosphere of pessimism but the same challenges posed by the intersection of climate change and equity concerns can be viewed as opportunities. Climate change offers an excellent opportunity, not only to stabilise and secure the climate for the future, but also to tackle the deep inequalities that confront contemporary society and to turn it onto a more sustainable and equitable pathway. The challenge is daunting but it is also achievable; the only way to guarantee failure is to ignore equity issues.

References

Adger, W. N., Agrawala, S., Mirza, M. M. Q. *et al.* (2007). Assessment of adaptation practices, options, constraints and capacity. In *Climate Change 2007: Impacts, Adaptation and Vulnerability. Contribution of Working Group II to the Fourth Assessment Report of the Intergovernmental Panel on Climate Change.* Cambridge, UK and New York, NY: Cambridge University Press, pp. 717–43.

Adger, W. N., Paavola, J., Huq, S. and Mace, M. J. (eds.) (2006). *Fairness in Adaptation to Climate Change.* Cambridge, MA and London, England: MIT Press.

Adhikari, B. (2009). Reduced emissions from deforestation and degradation: Some issues and considerations. *Journal of Forest and Livelihood,* **8**, 14–24.

Alcamo, J. (2009). Simulating the interactions between climate change, food, and water in future Africa. *IOP Conference Series: Earth and Environmental Science,* **6**, 512002.

Amacher, G., Koskela, E. and Ollikainen, M. (2009). Deforestation and land use under insecure property rights. *Environment and Development Economics,* **14**, 281–303.

Angelsen, A. (ed.) (2008a). *Moving Ahead with REDD: Issues, Options and Implications.* Bogor, Indonesia: CIFOR, http://www.cifor.cgiar.org/publications/pdf_files/Books/BAngelsen0801.pdf.

Angelsen, A. (2008b). How do we set the reference levels for REDD payments? In *Moving Ahead with REDD: Issues, Options and Implications,* ed. A. Angelsen. Bogor, Indonesia: CIFOR, pp. 53–64, http://www.cifor.cgiar.org/publications/pdf_files/Books/BAngelsen0801.pdf.

Arndt, C., Benfica, R., Tarp, F., Thurlow, J. and Uaiene, R. (2009). Biofuels, poverty, and growth: A computable general equilibrium analysis of Mozambique. *Environment and Development Economics,* **15**, 81–105.

Baer, P., Fieldman, G., Athanasiou, T. and Kartha, S. (2008). Greenhouse development rights: Towards an equitable framework for global climate policy. *Cambridge Review of International Affairs,* **21**, 649–69.

Bali Action Plan (2007). Decision 1/CP.13. FCCC/CP/2007/6/Add.1 and Reducing emissions from deforestation in developing countries: approaches to stimulate action, Decision 2/CP.13. FCCC/CP/2007/6/Add.1, http://unfccc.int/resource/docs/2007/cop13/eng/06a01.pdf#page=3.

Barnett J. (2003). Security and climate change. *Global Environmental Change,* **13**, 7–17.

Barr, C., Dermawan, A., Purnomo, H. and Komarudin, H. (2009). *Readiness for REDD: Financial Governance and Lessons from Indonesia's Reforestation Fund (RF).* CIFOR Infobrief. Bogor, Indonesia: CIFOR, http://www.cifor.cgiar.org/publications/pdf_files/Infobrief/020-infobrief.pdf.

Bates, B.C., Kundzewicz, Z. W., Wu, S. and Palutikof, J. P. (eds.) (2008). *Climate Change and Water*. Technical Paper of the Intergovernmental Panel on Climate Change. Geneva: IPCC Secretariat.

Bern, C., Sniezek, J., Mathbor, G. M. *et al.* (1993). Risk factors for mortality in the Bangladesh cyclone of 1991. *Bulletin of the World Health Organization*, **71**, 73–78.

Brown, D., Seymour, F. and Peskett, L. (2008). How do we achieve REDD co-benefits and avoid doing harm? In *Moving Ahead with REDD: Issues, Options and Implications*, ed. A. Angelsen. Bogor, Indonesia: CIFOR, pp. 107–18, http://www.cifor.cgiar.org/publications/pdf_files/Books/BAngelsen0801.pdf.

Buckingham, S. and Kulcur, R. (2009). Gendered geographies of environmental injustice. *Antipode*, **41**, 659–83.

Busch, J., Strassburg, B., Cattaneo, A. *et al.* (2009). Comparing climate and cost impacts of reference levels for reducing emissions from deforestation. *Environmental Research Letters*, **4**, 044006.

Byrd, N. and Peskett, L. (2008). Recent bilateral initiatives for climate financing: Are they moving in the right direction? Overseas Development Institute. *Opinion*, **112**.

Börner, J., Wunder, S., Wertz-Kanounnikoff, S. *et al.* (2010). Direct conservation payments in the Brazilian Amazon: Scope and equity implications. *Ecological Economics*, **69**(6), 1272–82.

Campbell-Lendrum, D. and Corvalán, C. (2007). Climate change and developing-country cities: Implications for environmental health and equity. *Journal of Urban Health*, **84**, 109–17.

Caney, S. (2005). Cosmopolitan justice, responsibility, and global climate change. *Leiden Journal of International Law*, **18**, 747–75.

Capoor, K. and Ambrosi, P. (2007). *State and Trends of the Carbon Market 2007*. Washington, D.C.: World Bank.

Capoor, K. and Ambrosi, P. (2008). *State and Trends of the Carbon Market 2008*. Washington, D.C.: World Bank.

Capoor, K. and Ambrosi, P. (2009). *State and Trends of the Carbon Market 2009*. Washington, D.C.: World Bank.

Carlson, K. M. and Curran, L. M. (2009). REDD pilot project scenarios: Are costs and benefits altered by spatial scale? *Environmental Research Letters*, **4**, 031003.

Cline, W. R. (2007). *Global Warming and Agriculture: Impact Estimates by Country*. Washington, D.C.: Center for Global Development, Peterson Institute for International Economics.

Confalonieri, U., Menne, B., Akhtar, R. *et al.* (2007). Human health. In *Climate Change 2007: Impacts, Adaptation and Vulnerability. Contribution of Working Group II to the Fourth Assessment Report of the Intergovernmental Panel on Climate Change*, eds. M. L. Parry, O. F. Canziani, J. P. Palutikof, P. J. van der Linden and C. E. Hanson. Cambridge, UK and New York, NY: Cambridge University Press, pp. 391–431.

Copenhagen Accord (2009). Decision -/CP.15, http://unfccc.int/files/meetings/cop_15/application/pdf/cop15_cph_auv.pdf.

de Araujo, M. S. M., de Campos, C. P. and Rosa, L. P. (2007). GHG historical contribution by sectors, sustainable development and equity. *Renewable and Sustainable Energy Reviews*, **11**, 988–97.

Dellink, R., den Elzen, M., Aiking, H. *et al.* (2009). *Sharing the Burden of Adaptation Financing: An Assessment of the Contributions of Countries*. Working Papers 2009.59. Fondazione Eni Enrico Mattei.

Dow, K., Kasperson, R. E. and Bohn, M. (2006). Exploring the social justice implications of adaptation and vulnerability. In *Fairness in Adaptation to Climate Change*, eds. Adger, W. N., Paavola, J., Huq, S. and Mace, M. J. The MIT Press, pp. 79–96.

Duke, R., Jacobson, A. and Kammen, D. (2002). Product quality in the Kenyan solar home systems market. *Energy Policy*, **30**, 477–99.

Dutschke, M. and Wertz-Kanounnikoff, S. (2008). How do we match country needs with financing sources? In *Moving Ahead with REDD: Issues, Options and Implications*, ed. A. Angelsen. Bogor, Indonesia: CIFOR, pp. 41–52, http://www.cifor.cgiar.org/publications/pdf_files/Books/BAngelsen0801.pdf.

Ebi, K. (2009). Health adaptation choices at local and regional scales. *IOP Conference Series: Earth and Environmental Science*, **6**, 142019.

Ebi, K. L. and Meehl, G. A. (2007). The heat is on: Climate change & heatwaves in the midwest. In *Regional Impacts of Climate Change: Four Case Studies in the United States*, eds. K. L. Ebi, G. A. Meehl, D. Bachelet, R. R. Twilley and D. F. Boesch. Arlington, VA: Pew Center on Global Climate Change.

Fischer, G., Shah, M., Tubiello, F. N. and van Velhuizen, H. (2005). Socio-economic and climate change impacts on agriculture: an integrated assessment, 1990–2080. *Philosophical Transactions of the Royal Society of London Biological Sciences*, **360**, 2067–83.

GECHS (2009). *Human Security in an Era of Global Change*. GECHS Synthesis Conference. University of Oslo, Oslo, 22–24 June 2009, http://www.gechs.org/human-security/.

Gleditsch, N. P. (1998). Armed conflict and the environment: A critique of the literature. *Journal of Peace Research*, **35**, 381–400.

Grandjean, P., Bellinger, D., Bergman, A. *et al.* (2008). The Faroes statement: Human health effects of developmental exposure to chemicals in our environment. *Basic & Clinical Pharmacology & Toxicology*, **102**, 73–75.

Grasso, M. (2006). An ethics-based climate agreement for the South Pacific region. *International Environmental Agreements: Politics, Law and Economics*, **6**, 249–70.

Grasso, M. (2010). An ethical approach to climate adaptation finance. *Global Environmental Change*, **20**, 74–81.

Haines, A., McMichael, A. J., Smith, K. R. *et al.* (2009). Public health benefits of strategies to reduce greenhouse-gas emissions: Overview and implications for policy makers. *The Lancet*, **374**, 2104–14.

Hashizume, M., Wagatsuma, Y., Faruque, A. S. *et al.* (2008). Factors determining vulnerability to diarrhoea during and after severe floods in Bangladesh. *Journal of Water Health*, **6**, 323–32.

Hay, S. I., Tatem, A. J., Guerra, C. A. and Snow, R. W. (2006). *Foresight on Population at Malaria Risk in Africa: 2005, 2015 and 2030*. Foresight Project Infectious Diseases: Preparing for the Future. London: Office of Science and Innovation.

Hof, A. F., de Bruin, K. C., Dellink, R. B. *et al.* (2009). The effect of different mitigation strategies on international financing of adaptation. *Environmental Science and Policy*, **12**, 832–43.

Hotez, P. (2008). Hookworm and poverty. *Annals of the New York Academy of Sciences*, **1136**, 38–44.

Hultman, N., Boyd, E., Roberts, J. T. *et al.* (2009). How can the clean development mechanism better contribute to sustainable development? *Ambio*, **38**, 120–22.

IFPRI (International Food Policy Research Institute) (2009). *Climate Change: Impact on Agriculture and Costs of Adaptation*. Washington, D.C.: IFPRI.

IIASA (2009). *Climate Change and Extreme Events: What Role for Insurance?* IIASA Policy Brief. International Institute for Applied Systems Analysis, http://www.iiasa.ac.at/Admin/PUB/policy-briefs/pb04.html.

Intergovernmental Panel on Climate Change (IPCC) (2007). *Climate Change 2007. Impacts, Adaptation and Vulnerability. Contribution of Working Group II to the Fourth Assessment Report of the Intergovernmental Panel on Climate Change*, eds. M. Parry, O. Canziani, J. Palutikov, P. van der Linden and C. Hanson. Cambridge, UK and New York, NY: Cambridge University Press.

Jacobson, A. (2007). Connective power: Solar electrification and social change in Kenya. *World Development*, **35**, 144–62.

Jacobson, A. and Kammen, D. M. (2007). Engineering, institutions, and the public interest: Evaluating product quality in the Kenyan solar photovoltaics industry. *Energy Policy*, **35**, 2960–68.

Kalame, F. B., Nkem, J., Idinoba, M. and Kanninen, M. (2009). Matching national forest policies and management practices for climate change adaptation in Burkina Faso and Ghana. *Mitigation and Adaptation Strategies for Global Change*, **14**, 135–51.

Karsenty, A. (2008). The architecture of proposed REDD schemes after Bali: Facing critical choices. *International Forestry Review*, **10**, 443–57.

Kasperson, R. E. and Kasperson, J. X. (2001). *Climate Change, Vulnerability, and Social Justice*. Stockholm: Stockholm Environment Institute.

Kerkhof, A. C., Benders, R. M. J. and Moll, H. C. (2009). Determinants of variation in household CO_2 emissions between and within countries. *Energy Policy*, **37**, 1509–17.

Khanna, M. and Rao, N. D. (2009). Supply and demand of electricity in the developing world. *Annual Review of Resource Economics*, **1**, 567–96.

Khebiza, M. Y., Messouli, M., Hammadi, F. and Ghallabi, B. (2009). Adaptation actions to reduce water system vulnerability to climate change in Tensift River basin (Morocco). *IOP Conference Series: Earth and Environmental Science*, **6**, 292006.

Kirsch-Wood, J., Korreborg, J. and Linde, A.-M. (2008). What humanitarians need to do. *Forced Migration Review*, **31**, 41–42.

Kjellstrom, T., Holmer, I. and Lemke, B. (2009). Workplace heat stress, health and productivity – An increasing challenge for low and middle-income countries during climate change. *Global Health Action*, doi: 10.3402/gha.v2i0.2047.

Kollanus, V., Jantunen, M., Pohjola, M. V., Ahtoniemi, P. and Tuomist, J. T. (2009). Health impacts of climate change mitigation. *IOP Conference Series: Earth and Environmental Science*, **6**, 142022.

Leichenko, R. M. and O'Brien, K. (2008). *Environmental Change and Globalization: Double Exposures*. New York, NY: Oxford University Press.

Lovera, S. (2008). *The Hottest REDD Issues: Rights, Equity, Development, Deforestation and Governance by Indigenous Peoples and Local Communities*. Gland: CEESP Commission on Environmental, Economic and Social Policies and Global Forest Coalition.

Lubowski, R. (2008). What are the costs and potentials of REDD? In *Moving Ahead with REDD: Issues, Options and Implications*, ed. A. Angelsen. Bogor, Indonesia: CIFOR, pp. 23–30, http://www.cifor.cgiar.org/publications/pdf_files/Books/BAngelsen0801.pdf.

Ludwig, F., Kabat, P., Hagemann, S. and Dortlandt, M. (2009). Impacts of climate variability and change on development and water security in Sub-Saharan Africa. *IOP Conference Series: Earth and Environmental Science*, **6**, 292002.

Markandya, A., Armstrong, B. G., Hales, S. *et al.* (2009). Public health benefits of strategies to reduce greenhouse-gas emissions: low-carbon electricity generation. *The Lancet*, **374**, 2006–15.

McLaughlin, K.A., Fairbank, J. A., Gruber, M. J. *et al.* (2009). Serious emotional disturbance among youths exposed to Hurricane Katrina 2 years postdisaster. *Journal of the American Academy of Child and Adolescent Psychiatry*, **48**, 1069–78.

McMichael, A. J., Campbell-Lendrum, D., Kovats, S. *et al.* (2004). Global climate change. In *Comparative Quantification of Health Risks: Global and Regional Burden of Disease Due to Selected Major Risk Factors*, eds. M. Ezzati, Lopez, A.D., Rodgers, A., Murray, C.J.L. Geneva: World Health Organization, pp. 1543–1650, http://www.who.int/publications/cra/chapters/volume2/1543–1650.pdf.

McMichael, A. J., Friel, S., Nyong, A. and Corvalan, C. (2008). Global environmental change and health: Impacts, inequalities, and the health sector. *British Medical Journal*, **336**, 191–94.

McNish, T., Jacobson, A., Kammen, D., Gopal, A. and Deshmukh, R. (2009). Sweet carbon: An analysis of sugar industry carbon market opportunities under the Clean Development Mechanism. *Energy Policy*, **37**, 5459–68.

Mitchell, T. and Tanner, T. (2006). *Adapting to Climate Change: Challenges and Opportunities for the Development Community*. Institute of Development Studies and Tear Fund.

Müller, B. (1999). *Justice in Global Warming Negotiations: How to Achieve a Procedurally Fair Compromise*. Oxford, UK: Oxford Institute for Energy Studies.

Müller, B. (2001). Varieties of distributive justice in climate change. *Climatic Change*, **48**, 273–88.

Müller, B. (2008). *International Adaptation Finance: The Need for an Innovative and Strategic Approach*. Oxford, UK: Oxford Institute for Energy Studies.

Müller, B. and Gomez-Echeverri, L. (2009). *The Reformed Financial Mechanism of the UNFCCC. Part I: Architecture and Governance*. Oxford, UK: Oxford Institute for Energy Studies.

Müller, B. and Hepburn, C. (2006). *IATAL – An Outline Proposal for an International Air Travel Adaptation Levy*. Oxford, UK: Oxford Institute for Energy Studies.

Najam, A. (1995). An environmental negotiation strategy for the South. *International Environmental Affairs*, **7**, 249–87.

Najam, A. (2004). The view from the South: developing countries in global environmental politics. In *The Global Environment: Institutions, Law, and Policy*, eds. R. S. Axelrod, D. L. Downie and N. J. Vig. Washington, D.C.: CQ Press, pp. 225–43.

Okereke, C. and Dooley, K. (2010). Principles of justice in proposals and policy approaches to avoided deforestation: towards a post-Kyoto climate agreement. *Global Environmental Change*, **20**, 82–95.

Ottesen, E. A., Duke, B. O., Karam, M. and Behbehani, K. (1997). Strategies and tools for the control/elimination of lymphatic filariasis. *Bulletin of the World Health Organization*, **75**, 491–503.

Oxfam (2007). *Adapting to Climate Change: What's Needed in Poor Countries and Who Should Pay*. Oxfam Briefing Paper No. 104. Oxfam International, http://www.oxfam.org/sites/www.oxfam.org/files/adapting%20to%20climate%20change.pdf.

Pan African Climate Justice Alliance (2009). *The Economic Cost of Climate Change in Africa*. Nairobi: Pan African Climate Justice Alliance.

Parry M. L., Rosenzweig, C., Iglesias, A., Livermore, M. and Fischer, G. (2004). Effects of climate change on global food production under SRES emissions and socio-economic scenarios. *Global Environmental Change*, **14**, 53–67.

Parry, M. L., Arnell, N., Berry, P. *et al.* (2009). *Assessing the Costs of Adaptation to Climate Change: A Review of the UNFCC and Other Recent Assessments*. IIED, Imperial College London, Gratham Institute for Climate Change.

Pascual, M., Ahumada, J. A., Chaves, L. F., Rodó, X. and Bouma, M. (2006). Malaria resurgence in the East African highlands: temperature trends revisited. *Proceedings of the National Academy of Sciences (USA)*, **103**, 5829–34.

Patz, J. A., Gibbs, H. K., Foley, J. A., Rogers, J. V. and Smith, K. R. (2007). Climate change and global health: quantifying a growing ethical crisis. *Ecohealth*, **4**, 397–405.

Pittel, K. and Rübbelke, D. T. G. (2008). Climate policy and ancillary benefits: A survey and integration into the modelling of international negotiations on climate change. *Ecological Economics*, **68**, 210–20.

Ribot, J. C. and Peluso, N. L. (2003). A theory of access. *Rural Sociology*, **68**, 153–81.

Richardson, K., Steffen, W., Schellnhuber, H. J. *et al.* (2009). *Synthesis Report. Climate Change: Global Risks, Challenges and Decisions Conference*, Copenhagen, 10–12 March.

Roberts, J. T., Starr, K., Jones, T. and Abdel-Fattah, D. (2008). *The Reality of Official Climate Aid. Oxford Energy and Environment Comment*, November. Oxford, UK: Oxford Institute for Energy Studies.

Roberts, J. T. and Parks, B. C. (2007). *A Climate of Injustice: Global Inequality, North-South Politics, and Climate Policy*. Cambridge, MA: MIT Press.

Rosengrant, M. W., Ringler, C., Msangi, S. *et al.* (2008). *International Model for Policy Analysis of Agricultural Commodities and Trade (IMPACT): Model Description*. Washington, D.C.: International Food Policy Research Institute.

Rosenzweig, C. and Parry, M. L. (1994). Potential impact of climate change on world food supply. *Nature*, **367**, 133–38.

Rübbelke, D. T. G. and Rive, N. (2008). *Effects of the CDM on Poverty Eradication and Global Climate Protection*. Working Papers 2008.93, Fondazione Eni Enrico Mattei.

Sagar, A. D. and Kartha, S. (2007). Bioenergy and sustainable development? *Annual Review of Environment and Resources*, **32**, 131–67.

Schneider, L. (2009). Assessing the additionality of CDM projects: Practical experiences and lessons learned. *Climate Policy*, **9**, 242–54.

Seymour, F. (2010). Forests, climate change and human rights: managing risks and trade-offs. In *Human Rights and Climate Change*, ed. S. Humphreys. Cambridge, UK: Cambridge University Press, pp. 207–37.

Singh, R. B. K., Hales, S., de Wet, N. *et al.* (2001). The influence of climate variation and change on diarrheal disease in the Pacific Islands. *Environmental Health Perspectives*, **109**, 155–59.

Smit, B. and Wandel, J. (2006). Adaptation, adaptive capacity and vulnerability. *Global Environmental Change*, **16**, 282–92.

Smit, B., Pilifosova, O., Burton, I. *et al.* (2001). Adaptation to climate change in the context of sustainable development and equity. In *Climate Change 2001: Impacts, Adaptation and Vulnerability*, eds. J. McCarthy, O. Canziani, N. A. Leary, D. J. Dokken and K. S. White. Cambridge, UK: Cambridge University Press, pp. 877–912.

Smith, H.J., Fahrenkamp-Uppenbrink, J. and Coontz, R. (2009). Carbon capture and sequestration. Clearing the air. Introduction. *Science*, **325**, 1641.

Sonwa, D. J., Weise, S. F., Nkongmeneck, B. A., Tchatat, M. and Janssens, M. J. J. (2009). Carbon stock in smallholder chocolate forest in southern Cameroon and potential role in climate change mitigation. *IOP Conference Series: Earth and Environmental Science*, **6**, 252008.

Stern, N. (2007). *The Economics of Climate Change: The Stern Review*. Cambridge, UK: Cambridge University Press.

Sutter, C. (2001). *Small-scale CDM Projects: Opportunities and Obstacles*. Zurich: Swiss Agency for Development and Co-operation, http://www.up.ethz.ch/publications/documents/Sutter_2001_Small-Scale_CDM_Vol1.pdf.

Symons, J. and Harris, P. G. (2009). *Justice and Adaptation to Climate Change in the Asia Pacific Region: Designing International Institutions*. Lingnan University Centre for Asian Pacific Studies Working Paper Series, No.195 (August). CAPS.

Tanner, T. and Mitchell, T. (2008). Introduction: Building the case for pro-poor adaptation. *IDS Bulletin, Poverty in a Changing Climate*, **39**, 1–5.

Tanser, F.C., Sharp, B. and le Sueur, D. (2003). Potential effect of climate change on malaria transmission in Africa. *The Lancet*, **362**, 1792–98.

UNDP (2007). *Human Development Report 2007/2008: Fighting Climate Change – Human Solidarity in a Divided World*. Ed. K. Watkins. Palgrave MacMillan.

UNFCCC (1992). *United Nations Framework Convention on Climate Change*. New York, NY: United Nations.

UNFCCC (2007). Decision 1/CP.13. FCCC/CP/2007/6/Add.1 Bali Action Plan, http://unfccc.int/resource/docs/2007/cop13/eng/06a01.pdf#page=3.

United Nations Environment Program (UNEP) (2010). UNEP Risoe CDM/JI Pipeline Analysis and Database, http://cdmpipeline.org/.

United Nations Framework Convention on Climate Change (UNFCCC) (1992). *Kyoto Protocol to the United Nations Framework Convention on Climate Change*. Bonn: UNFCCC, http://unfccc.int/essential_background/kyoto_protocol/items/1678.php.

Vajjhala, S. (2009). Building a 'global adaptation atlas': Setting geographic priorities for funding adaptation to climate change. Paper read at IOP Conference Series.

Vashist, S., Alam, M., Krishnaswamy, S. *et al.* (2009). Common but differentiated responsibilities' defined on the basis of equity for future climate change regime. *IOP Conference Series: Earth and Environmental Science*, **6**, 112001.

Vidal, J. (2009). Rich nations failing to meet climate aid pledges. *Guardian*, 20 February.

Voigt, C. (2008). Is the Clean Development Mechanism sustainable? Some critical aspects. *Sustainable Development Law and Policy*, **7**, 15–21.

Voigt, C. (2009). Responsibility for the environmental integrity of the CDM: Judicial review of executive board decisions. In *Legal Aspects of Carbon Trading: Kyoto, Copenhagen and Beyond*, eds. D. Freestone and C. Streck. Oxford, UK: Oxford University Press, pp. 272–94.

Wara, M. W. and Victor, D. G. (2008). *A Realistic Policy on International Carbon Offsets*. Stanford, CA: Stanford Program on Environmental Sustainability and the Environment, http://iis-db.stanford.edu/pubs/22157/WP74_final_final.pdf.

WHO (2006). *Climate Variability and Change and Their Health Effects in Small Island States: Information for Adaptation Planning in the Health Sector*. Geneva: World Health Organization.

WHO (2008). *The Global Burden of Disease: 2004 Update*. Geneva: World Health Organization.

WHO (2009). *Global Health Risks: Mortality and Burden of Disease Attributable to Selected Major Risks*. Geneva: World Health Organization.

WHO/SEARO (2006). *Human Health Impacts of Climate Variability and Climate Change in the Hindu Kush-Himalaya Region: Report of a Regional Workshop.* Mukteshwar, India: WHO Regional Office for South-East Asia.

Wobst, P., Anger, N., Veenendaal, P. *et al.* (2007). *Competitiveness Effects of Trading Emissions and Fostering Technologies to Meet the EU Kyoto Targets: A Quantitative Economic Assessment.* Industrial Policy and Economic Reforms Papers No. 4. Brussels: European Commission Enterprise and Industry Directorate-General.

Woodcock, J., Banister, D., Edwards, P., Prentice, A. M. and Roberts, I. (2007). Energy and transport. *The Lancet*, **370**, 1078–88.

World Bank (2005). *World Development Report 2005.* Washington, D.C.: The World Bank.

World Bank (2009). *World Development Report 2009.* Washington, D.C.: The World Bank.

World Bank (2010). *World Development Report 2010.* Washington, D.C.: The World Bank, p. 3, http://go.worldbank.org/ZXULQ9SCC0.

Ximenes, R., Southgate, B., Smith, P. G. and Neto, L. G. (2003). Socioeconomic determinants of schistosomiasis in an urban area in the northeast of Brazil. *Rev Panam Salud Publica*, **14**, 409–21.

Zhu, T., Lund, J. R., Jenkins, M. W., Marques, G. F. and Ritzema, R. S. (2007). Climate change, urbanization, and optimal long-term floodplain protection. *Water Resources Research*, **43**, W06421.

10

A long-term perspective on climate change: values and ethics

'There is no getting away from the fact that making policy towards climate change unavoidably requires one to take stance on ethical questions...'[1]

The equity dimensions of the climate change challenge discussed in the previous chapter are central to the current debate and, if not solved, will frustrate any attempts to find truly global solutions. The strong focus of the policy dialogue on the present and near future is appropriate and necessary to turn around the trajectory of greenhouse gas emissions and, hopefully, avert what many consider to be dangerous climate change. But just as longer-term studies of the dynamics of the Earth System in the past provide insights into the behaviour of the climate system today, a much longer gaze into the future can provide insights into the implications of the choices and decisions that we are making today.

Our ancestors have clearly shown an ability to think about their long-term futures, and to organise their institutions, economies and resource base to deliver desired outcomes hundreds of years into the future. If they had not had this capability, we would today be unable to marvel at the Great Wall of China, the exquisite ruins of the Mayan civilisation in meso-America, the great cathedrals of medieval Europe or the hauntingly beautiful old Buddhist city of Bagan in northern Myanmar (Figure 10.1). Our ancestors also differed from contemporary society in their capability to integrate utilitarian and spiritual views of the rest of nature. While evidence shows unequivocal environmental impacts of earlier societies at local and regional scales, oral traditions, prehistoric art and written records also point to a much deeper respect and reverence for the natural world than is commonly expressed in contemporary society. For them, nature was much more than a resource base.

This chapter addresses equity issues of a long-term nature. The focus is on non-economic, non-contemporary consequences of climate change, those that are not considered in the

---

[1] Nicholas Stern (2009). *The Global Deal*. Public Affairs, p. 77.

Climate Change: Global Risks, Challenges and Decisions, Katherine Richardson, Will Steffen and Diana Liverman *et al*. Published by Cambridge University Press. © Katherine Richardson, Will Steffen and Diana Liverman 2011.

Figure 10.1 Great human monuments of the past demonstrate a capability to take a long-term perspective and to organise societies to achieve long-term goals. (Left to right) The Great Wall of China (image used under license from Big Stock Photo); Salisbury Cathedral, UK (image copyright FrankfurtDave, 2010. Used under license from Shutterstock.com); the ancient city of Bagan, Myanmar (image used under license from Big Stock Photo). (In colour as Plate 22.)

benefit-cost analyses that are often carried out to inform the current policy dialogue on climate change. Here we emphasise (i) the long-term nature of climate change, particularly the implications for future generations of humans, and (ii) the consequences, from an ethical perspective, of our inadvertent experiment on the planet's climate for the rest of nature.

10.1 The long-term implications of climate change

Nearly all of the emphasis in climate change science to date has been focused strongly on the contemporary period – the recent past of 100 years or so, for which we have instrumental records of many climatic parameters, and the next approximately 100 years out to the end of this century, which is the timeframe of many model simulations of possible trajectories of the climate system. The vigorous policy dialogue is set in this timeframe, or shorter, with emission reduction targets often proposed for 2050 with an interim target of 2020. These are timeframes with which contemporary society is comfortable.

In addition, there is usually an unspoken assumption throughout society that the climate will return to its pre-industrial state soon after the emission reduction targets are achieved, and certainly by the end of the century. Scientific research provides a much different picture of the timescales over which anthropogenic forcings will continue to have a major influence on the Earth's climate. From a scientific perspective, anthropogenic climate change is not a decadal problem, and not even a century-scale one. Its fingerprint will almost certainly be discernible for millennia into the future.

Model simulations of the effects of anthropogenic emissions of CO_2 on climate out to the year 3000 show that human-induced climate change is essentially irreversible over that period (Solomon *et al.*, 2009). Once emissions peak, the concentration of atmospheric CO_2 will slowly decrease as it is absorbed by the oceans but the amount of CO_2 remaining in the atmosphere at the year 3000 is still about 40% of peak concentration enhancement over the pre-industrial value. Atmospheric surface temperature, however, does not drop significantly over the period to 3000. The reason is that the decrease in radiative forcing due to decreasing CO_2 concentration is largely compensated by slower transfer of heat from the atmosphere to the ocean.

Other features of the climate system also do not return to their pre-industrial values for at least a millennium, according to the model simulations. For example, in many parts of the world, dry-season rainfall reductions may become locked in if CO_2 concentrations reach a peak of 450–600 p.p.m. (Solomon *et al.*, 2009) (note that the current concentration is about 387 p.p.m.). Areas that experience such irreversible rainfall reductions – effectively

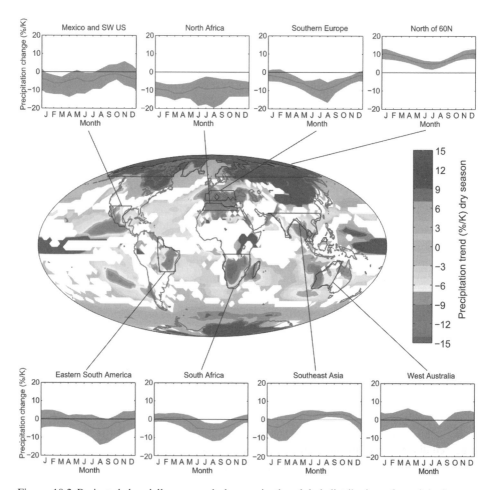

Figure 10.2 Projected decadally averaged changes in the global distribution of precipitation per degree of warming in the dry period, expressed as a percentage of change from a 1900–50 baseline period. Source: Solomon *et al.* (2009), which includes more details on the methods. (In colour as Plate 27.)

never-ending droughts – include southern Europe, North Africa, Mexico and the southwest USA, South Africa and western Australia (Figure 10.2).

As noted in Chapter 3, the time needed for anthropogenically driven sea-level rise to finally reach equilibrium is about a millennium, or probably longer. This is confirmed by the Solomon *et al.* (2009) analysis. Thermal expansion alone would guarantee an irreversible sea-level rise of 0.4 to 1.0 m if CO_2 concentration reaches 600 p.p.m. this century, and a rise of 0.6 to 1.9 m if CO_2 exceeds 1000 p.p.m. Loss of significant portions of the large polar ice sheets, which would be irreversible on the timescale of many millennia, would add much more to the rise in sea level. Figure 3.4 suggests if a temperature rise of 3 °C above pre-industrial levels, for example, is maintained for a millennium or longer, it is plausible that sea level could rise by about 25 m when equilibrium is finally achieved.

10.2 The consequences of these long-term changes for human societies

The essentially irreversible nature of many aspects of climate change raises strongly normative questions that are often expressed in terms of inter-generational equity. As Australian economist, Ross Garnaut, concluded in his comprehensive review of the climate change problem: '... the failure of our generation would lead to consequences that would haunt humanity until the end of time' (Garnaut, 2008). This statement points towards a dialogue on climate change that goes well beyond the utilitarian framework that dominates the present-day discourse. A key question is: Should those of us alive today, and our children and grandchildren, be concerned about the fate of humans in 2300? Or 2500? Or 3000?

The standard method of accounting for the welfare of future generations in economic analysis is through use of a discount rate (Chapter 12). The value of the discount rate chosen can be a strong determinant of the outcome of the analysis in terms of the rate and magnitude of mitigation that is considered to maximise welfare. However, it is clear that these standard economic approaches do not reflect the diversity of perspectives on our obligations to future generations. The enormity of the risks associated with climate change, and the very long timescales over which it operates, inevitably engages difficult ethical issues as well as economic ones. The relationship between ethics and economics can be complex; consideration of ethical issues can inform the types of economic questions asked, while economic analyses can inform ethical discussions by exploring the consequences of particular ethical perspectives (Dietz *et al.*, 2009).

The publication of economic analyses of climate change, such as the Stern Review, has triggered discussion of the normative assumptions that lie behind standard economic analysis, and have generated alternative approaches for dealing with the rights of future generations. One such approach, described in Box 10.1 (Roser, 2009), is based on the principle that future generations have a right to a certain threshold level of utility or well-being. The approach describes how this principle can be formalised, and it inevitably leads to risk-averse behaviour in the present generation.

Box 10.1
A non-consequentialist alternative to economics

DOMINIC ROSER

Do the benefits of a given climate policy exceed its costs? The answer to this question is a key criterion for many economic assessments of climate policy. However, as benefits and costs of climate policy do not fall on the same persons/generations, the relevance of this criterion is put into question by non-consequentialist theories. Non-consequentialist theories resist adding up good and bad consequences of a policy across persons/generations as if these constituted one single large agent. The non-consequentialist's distinctive focus on rights, duties and justice makes the concerns of future generations relevant for present-day policy choice in an entirely different manner than the present generation's own concerns. As far as future generations are concerned, the primary issue for the present generation is its duty to lift or keep them above a certain morally relevant threshold. The threshold owed to future generations could be specified in various ways, for example in terms of human rights or basic needs, or based on, say, an egalitarian theory of intergenerational distributive justice. The primacy of this duty towards future generations makes the outlook of non-consequentialism both more and less demanding than the consequentialist approach of economics. It is more demanding because measures to ensure that future generations achieve the morally relevant threshold must be pursued even if the costs of doing so exceed the benefits. It is also less demanding because the present generation has no duty to pursue measures that benefit future generations beyond the owed threshold even if the potential benefits should far outweigh the corresponding costs.

Note that in a non-consequentialist framework, the discount rate loses much of its relevance. When economics trades off present against future costs and benefits it must assign a certain weight – represented by a discount rate – to future costs and benefits (Chapter 12). It is a hotly debated issue whether to ascribe equal or lesser weight to values accruing in the future. A non-consequentialist framework, in contrast, argues that the very idea of weighing up present and future values (which makes the determination of a discount rate necessary in the first place) is inappropriate. Costs to the present generation for engaging in climate policy are not justified by being outweighed by discounted benefits for future generations but rather by their necessity for achieving the relevant threshold.

Assessing climate policy according to non-consequentialist standards has implications for our perspective on scientific uncertainty. Scientific uncertainty is relevant because many people seem to share a certain precautionary intuition that under conditions of uncertainty it is 'better to be safe than sorry' or, to give a more precise benchmark in terms of an example, that a forecast for a temperature increase of 1–6 °C is more worrisome than an (imaginary) precise forecast of a 3.5 °C increase. It is unclear, however, what kind of rationale could justify this precautionary intuition. One important suggestion – the rationale on which the consequentialism of economics relies – is the idea that non-linear relationships between instrumental and intrinsic values yield a reason for risk-aversion. If, for example, well-being losses (i.e. an intrinsic value) rise disproportionately with temperature increase (i.e. an instrumental value) and if we ought to ultimately be concerned with *expected* well-being, then there is a sound rationale for being risk-averse with respect to temperature increase, i.e. for being more worried about a 1–6 °C forecast than about a 3.5 °C forecast.

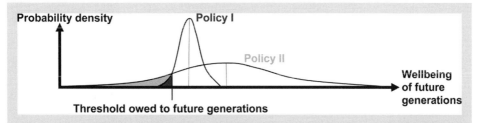

Figure 10.3 Comparison of policies with different degrees of uncertainty. Source: D. Roser.

This rationale, however, does not do full justice to our precautionary intuition, not least because it can only justify the relevance of uncertainty in instrumental values. Non-consequentialism can go further at this point. If the present generation has a duty to lift future generations above a certain threshold, then the main relevance of larger uncertainty lies in the potentially larger probability of future generations falling below this threshold. If we assume, for example, that the relevant threshold is a well-being level owed to future generations, then non-consequentialism evaluates a given climate policy not by the expected well-being it yields for future generations but primarily by its downside risk; that is, by the probability of falling below the threshold and by the expected (and possibly also maximal) well-being shortfall in case of actually falling below the threshold. Graphically, Policy II (Figure 10.3) is evaluated worse than Policy I because the light grey plus dark grey area is larger and extends further below the threshold than the dark grey area alone.

Climate inaction is sometimes justified on the basis of the (plausible) expectation that future generations might be better off than we are. Expected future well-being, however, is not the decisive issue for policy choice. Policy II is judged worse than Policy I even though it exhibits a larger expected well-being for future generations than Policy I. What matters is the size of the risk of falling below the owed threshold.

Other lines of argument, however, maintain that traditional economic analyses, no matter what discount rate is used, cannot capture ethical and justice concerns as they are based on crude economic utilitarianism (Barker *et al.*, 2009). The fundamental problem is that climate change is inherently unjust, as the more affluent people and societies can avoid many of the impacts of climate change through means such as private protection, while the poor and the vulnerable will suffer many serious and unavoidable impacts. This is particularly true for human health, and will be even more so for future generations. Today's high-emitting societies face lesser health risks than will future generations (Box 5.3). Thus, climate change will likely widen the gap between the well-being of the wealthy and the poor into the future. The concept of justice, therefore, asserts that the need for urgent and effective mitigation is not just a technological or economic issue, but is also a profoundly ethical issue that touches on the fundamental values that societies hold.

While there have been many theoretical and philosophical discourses on the obligations of our generation to future generations with respect to climate change, and even more economic analyses of the costs of mitigation compared to the benefits, there have been fewer empirical studies. One of the most extensive of these (Box 10.2; Helgesen *et al.*, 2009) has analysed the level of risk aversion of individuals to climate change impacts, incorporating a long-time perspective and aspects of inequality. The results strongly suggest that the current suite of economic analyses, which rely heavily on the discount rate used, are not nearly sophisticated enough to capture individuals' level of aversion to risk and inequality.

Box 10.2

Siblings, not triplets: a stated-preference survey into risk, inequality and time in climate change

GILES ATKINSON, SIMON DIETZ, JENNIFER HELGESON,
CAMERON HEPBURN AND HÅKON SÆLEN

Nowhere are the theoretical and empirical challenges of economic discounting more evident than in the context of public policies with very long-run consequences, such as climate change. The starting point of analysis remains the classic work of Ramsey (1928), who showed in a now standard model that the optimal consumption discount rate is given by

$$r_t = \delta + \eta g_t, \qquad\qquad (10.1)$$

where r_t is the consumption discount rate at time t, δ is the pure rate of time preference (sometimes also known as the utility discount rate), η is the elasticity of the marginal utility of consumption (how much extra money is worth to people at lower income levels) and g is the growth rate of consumption. Simply put, discounting assumes that people usually value benefits and losses differently in the present than in the future, with many economists assuming a lower value for the future.

Much of the recent discounting debate, stimulated by the Stern Review (Stern, 2007), has focused on the ethics of δ – the rate at which values change over time. Yet while δ is important to the economics of climate change, so too are η and g. In particular, the costs and benefits of mitigating climate change are very sensitive to the specific parameter choice for η which, in the simplest terms, describes how much an extra dollar, euro or rupee is worth when one is poor versus when one is rich.

The debate following the Stern Review has shown that there is little agreement on precisely what value η should take. This is in large part because, in the 'workhorse' model of welfare economics as applied to climate change, η simultaneously represents preferences over three significant dimensions of the policy issue: namely how people view risk, inequality within a generation (i.e. spatial inequality) and inequality in consumption between generations.

• Risk: uncertainties about the impacts of climate change are large and may in part be irresolvable (as emphasised by Stern, 2007). Furthermore, the worst-case scenarios imply very large damages on a global scale (Weitzman, 2007).

Table 10.1. *Correlations between dimensions of η (the elasticity of the marginal utility of consumption)*

	Relative risk aversion	Elasticity of inter-temporal substitution (midpoint)
Relative inequality aversion	0.133(***)	−0.123(***)
Relative risk aversion		−0.069(***)

*** Correlation is significant at the 0.001 level (one-tailed).
More details on the approach and the results are given in Atkinson *et al.* (2009).

- Time: the marginal impact of a tonne of greenhouse gas persists long after it has been emitted.
- Space: there are large spatial disparities in the relative impacts of climate change worldwide, with tropical and sub-tropical low-income regions widely expected to experience some of the greatest relative damages (IPCC, 2007).

Evidence on risk preferences has been taken to support values of η in the range 2–4. By contrast, evidence from the actual distribution of income within countries might suggest a value of unity or less. Yet another approach is to back out the appropriate value of η from evidence on the appropriate consumption discount rate. This approach generally leads to the recommendation that η be greater than unity.

This divergence suggests it would be worthwhile to examine whether public attitudes to climate change support a model in which preferences over risk, inequality and time are identical, or whether they in fact support a model in which these preferences are distinct. This was the objective of our study, which was based around the Climate Ethics Survey – a stated-preference survey delivered online to over 3000 respondents.

The survey posed direct questions about hypothetical consumption choices under each of the three defining characteristics of climate change. The results show that the correlations between attitudes to risk, inequality and inter-temporal substitution are weak.

In the dimension of risk, we found that η was in the interval 3–5. These results are comparable to stated-preference surveys that estimate aversion to personal risk (Barsky *et al.*, 1997; Carlsson *et al.*, 2005), and to some revealed-preference studies (Gollier, 2006). In the dimension of spatial inequality aversion, the median value of η was 2–3, while the modal group was in the upper extreme, $\eta > 7.5$. There was also a higher frequency of responses in the lowest category, $\eta < 0.5$, than observed in the risk dimension. Most respondents displayed a very low elasticity of inter-temporal substitution, corresponding to a very high aversion to inequality in income across time. For the median respondent, the midpoint of this range was 0.11, which corresponds to $\eta = 8.8$. This is in line with the results of Barsky *et al.* (1997).

Correlation between each pair of variables was significant, but weak. All the correlation coefficients were around 0.1 (Table 10.1). The standard model implies that these correlation coefficients should equal unity. Our results, therefore, suggest that the standard model is not well-suited for incorporating the attitudes of the public into the analysis of climate change. Rather, a new framework that is rich enough to treat these three issues as distinct is needed

> ## Box 10.2 (*cont.*)
>
> (Kreps and Porteus, 1978; Selden, 1978). Once models are used that disentangle the three dimensions, it becomes clearer that analysts have greater freedom to explore the consequences of different valuations of η for optimal climate policy. Otherwise, the policy advice produced by climate change economics will not reflect what is optimal, given the preferences of the public.

Climate change can be considered as part of the broader sustainability agenda (Gerlagh and Vollebergh, 2009). Sustainability is often treated, in ethical terms, as an overarching societal goal that emphasises the sharing of resources with future generations, or 'equal treatment' of present and future generations (Pezzey, 1997; Asheim *et al.*, 2001). In economic terms, this implies that welfare should not decrease over time. The Gerlagh and Vollebergh approach has developed this definition of sustainability further, including the possibility that the present generation may actually act to increase the welfare of future generations by achieving an outcome that goes beyond the non-decreasing welfare criterion. In the climate change context, this situation may occur if, for example, the climate change and development challenges are met simultaneously, with synergistic outcomes between them (Chapter 15).

Climate change can also be considered from a human rights perspective (Toft, 2009). Here the principle is straightforward – humans who will live in the future have the right not to suffer from the damages caused by human-driven climate change (Caney, 2005). Such a simple formulation, however, has direct implications for the current debate. The burden of ensuring that this basic human right is observed would fall strongly on the present generation – despite the significant contribution of past generations to climate change – and preferentially on the wealthy countries of this generation.

In summary, a variety of perspectives bring us to the same bottom line – a business-as-usual approach to climate change, in which rapid and deep cuts in greenhouse gas emissions are not achieved, is profoundly unjust to future generations, who have a fundamental right to an environment in which they can live. The ultimate argument for strong and rapid mitigation, however, may be a deeply existential one. Bearing and raising children in a trans-generational chain goes beyond all ethnic and cultural diversity (Cerutti, 2009). It is fundamental to being a biological species.

10.3 Investment in natural capital

The emerging research area of ecological economics offers some useful insights into the long-term consequences of climate change for the natural world. The concept of 'natural capital' is particularly important; that is, the stock of natural ecosystems that yields a flow of valuable ecosystem services into the future. From this perspective current generations, as custodians of Earth's natural capital, are leaving large debts for future generations to repay. That is, we are not only deriving services from the flows of energy and materials that ecosystems provide, but also we are degrading many of the ecosystems themselves – nature's basic capital – to provide some of these services (Millennium Ecosystem Assessment, 2005). This situation is obviously unsustainable in the long term. Climate change is both a manifestation of the consumption of natural capital and a contribution to its further erosion.

It is very difficult to estimate to a high degree of accuracy the level of reinvestment that contemporary societies need to make in natural capital to restore it to a more sustainable condition. An assessment of the vulnerability of Australia's biodiversity to climate change has argued that the level of reinvestment in natural capital that is required is at least an order of magnitude more than is currently invested, and probably even more (Steffen *et al.*, 2009). A case study focusing on four regions of the world suggests that 2.5 to 3.0 times the current investment levels in reserve systems is required to enhance the resilience of these systems towards climate change (Chapter 6; Andelman *et al.*, 2009). While these may seem like very large sums of money, they seem less daunting when compared, for example, to defense budgets or to the size of the economic stimulus packages that have recently been allocated in response to the global financial crisis. In addition, a substantial part of the increased reinvestment may come from private individuals and firms, a trend that has already begun in some countries.

The reinvestment in natural capital is urgent, as changes to some aspects of this type of capital – most importantly biodiversity loss – are irreversible. Once a species becomes extinct, it cannot be retrieved unless new technologies are developed that can reconstitute them from remnant DNA or other material. Biodiversity conservation is, therefore, of high priority in terms of maintaining and restoring natural capital. One of the most effective ways of improving the prospects for biodiversity is to enhance reserve systems. This challenge should be met as rapidly as possible, as land-use change is foreclosing options in many countries. Comprehensive and ongoing investment in monitoring of the natural environment is also needed, especially in terms of supporting active adaptive management.

10.4 Ethical aspects of biodiversity loss

With a business-as-usual approach to climate change – and even considering substantial increases in investment in natural capital – the Earth is facing a looming biodiversity catastrophe. Extinction rates are already at least 100 and possibly 1000 times background (natural) rates due to a number of stresses originating in human activities (Millennium Ecosystem Assessment, 2005). As temperature rises above $+2$ °C and ocean acidification increases, the extinction rate is predicted to rise by another order of magnitude by the end of this century. Between 10% and 30% of all mammals, birds and amphibians are now threatened with extinction (Millennium Ecosystem Assessment, 2005). The Earth is undoubtedly in the midst of its sixth great extinction event, but the first caused by a single biological species (Steffen *et al.*, 2004). The implications of these vast changes in the biotic fabric of the planet for the delivery of ecosystem services that support human well-being

Figure 10.4 Climate change challenges us to think about our relationship to the rest of nature. Do we have the right to eradicate other living species with whom we share the planet? (Left to right): Polar bear (image used under licence from Big Stock Photo); Hooded grebe (image copyright James C. Lowen; reproduced with permission); Mountain pygmy possum (image copyright Glen Johnson; reproduced with permission). (In colour as Plate 26.)

are discussed in Chapter 6. Here, we move from a utilitarian to a philosophical perspective on the looming biodiversity catastrophe. Do humans have the ethical right to allow their actions to eradicate other living organisms (Figure 10.4)?

Based on a non-anthropocentric approach to ethics, living beings in the natural world – such as plants and animals – also have moral status (Box 10.3; Melin, 2009). The vast world of living organisms can be considered in a hierarchical framework, from sentient beings like ourselves through biocentric to ecocentric perspectives. The last category places value on collections of organisms that together make up ecosystems, and offers an ethical connection to the more utilitarian perspective of ecosystem services (Chapter 6). From the non-anthropocentric perspective, the increased stress that many organisms are now experiencing – not only from climate change, but also from other human activities – should be a matter of concern. While there is always a certain level of harm as species live, die and go extinct naturally, it is clear that human activities have vastly increased the level of harm that the individuals of many other species are now experiencing (Van Houtan, 2009).

Box 10.3
Climate change and its effects on natural entities: a non-anthropocentric perspective

ANDERS MELIN

In the political debate about climate change it is for obvious reasons mostly the effects on humans that are considered. However, from an ethical point of view we should also ask ourselves if we ought to care about some natural entities for their own sake ('natural entities' here denotes non-human objects such as individual animals and plants; inanimate objects such as rivers and mountains; and collective entities such as species of animals and plants, and ecosystems). Within the disciplines of animal and environmental ethics, the anthropocentric moral standpoint – that is, the idea that only humans should be taken into account for their own sake – has been challenged and different non-anthropocentric positions have been put forward. The most important are sentientism (sometimes also labelled as zoocentrism), biocentrism and ecocentrism. A common way of defining these positions is as follows: sentientism means that we should care also for animals with a certain degree of sentience, for instance mammals, birds, reptiles and fishes, for their own sake; biocentrism that we should care for all individual living beings, including insects and plants, for their own sake; and ecocentrism that we should care also for collective entities such as species and ecosystems for their own sake. This expert box briefly discusses to what extent these positions can be rationally defended and what practical implications this has for climate politics.

As for sentientism, Singer is the most well-known representative of this position and his argumentation is here used as my point of departure. He argues from a utilitarian point of view and claims that we should show equal consideration to all interests, regardless of whether they belong to an animal or a human being. In the same way as humans, sentient animals have a morally relevant interest in not being exposed to pain (Singer, 1979). Singer has been most concerned with our moral relationship to domestic animals, but in his writings he also mentions our relationship to wild animals. He argues that we should take account of all sentient beings for their own sake when performing actions that affect the lives of other living beings. We should give as much weight to the infliction of suffering on other sentient creatures as we would give to the infliction of suffering on human beings (Singer, 2002).

The moral belief that we should take account of sentient animals for their own sake has become increasingly accepted, both within the discipline of philosophy and in society at large. However, Singer's viewpoint that we should show equal consideration to the interests of humans and non-human animals is more controversial. I think that the idea that we should take account of the suffering of non-human animals – but to a lesser degree than the suffering of humans – is in many ways easier to defend than Singer's belief, as his viewpoint seems to demand too much of humans. Although we have a unique moral capacity because of our high degree of self-consciousness, we are still in some respects an animal among others, and therefore we should be permitted to prioritise the interests of our fellow humans.

Biocentric ethicists want to go further than sentientists such as Singer. As stated above, they claim that we should take all living beings into consideration for their own sake. Varner, for instance, argues that also non-sentient beings such as trees have biological interests, which are morally relevant. They need some things, for instance water, in order to survive and flourish (Varner, 1998). However, it is questionable whether merely biological interests are morally relevant. The biocentric position has a weaker rational support than sentientism, as it is difficult to defend the idea that also living beings without subjective interests should be included in the moral community.

The ecocentric position is even more problematic than the biocentric. Some ecocentric ethicists, such as Rolston, claim that species have a form of objective reality and that they have morally relevant interests that differ from the interests of the individuals that belong to the species (Rolston, 1989). However, it is questionable whether this claim can be rationally supported. First, a lot of different species concepts exist within contemporary biology and this fact seems to indicate that species are human constructions. Second, even if we assume that species are separate entities with an objective existence, it is still questionable whether species can have morally relevant interests.

Box 10.3 (*cont.*)

Thus, the conclusion is that the sentientist position is the non-anthropocentric position that has the strongest rational support. The belief that we should take account of the suffering and deaths of sentient (non-human) animals – although to a lesser degree than the suffering and deaths of humans – would certainly have practical implications for climate politics. A significant rise in temperature would most likely lead to suffering and premature death for a great number of wild animals. Even though some individual animals may gain from it, the overall effects on animals would probably be very negative. When deciding what measures we should take in order to limit global warming, this fact should be acknowledged. However, some human actions that lead to global warming could still be morally justified if their overall effects on humans are positive and if these effects outweigh the negative effects on animals.

Climate change presents severe challenges to the ways in which we currently manage natural ecosystems. As noted in Chapter 6, the goals of biodiversity conservation will need to be changed, but that discussion masks deeper philosophical questions (Box 10.4; Averill, 2009a). What is 'natural' when the climate is changing due to human activities? How far should we go towards proactive interventions such as eco-engineering to save species or communities that may clearly be headed for extinction?

Box 10.4
Framing natural resource management for climate resilience

MARILYN AVERILL

Virtually all countries set aside areas for special protection. Designations such as parks, refuges and heritage sites identify areas with special ecological, aesthetic, recreational or other values that deserve governmental protection. Climate change provides new incentives to re-examine natural resource management practices, especially those intended to increase ecosystem resilience to change. Management decisions will depend upon the way resource challenges are framed, the stakeholders consulted and the values the designated area is perceived to protect (Averill, 2009b).

Climate change re-emphasises existing challenges, such as planning for droughts and floods, and managing human uses. Climate change also challenges managers to think beyond conditions as they have known them and to cope with a wider range of temperature and precipitation, more extreme weather events, shifts in timing of water availability, new invasive species, and sea rise.

Resource managers will need to rethink their objectives as the climate changes. Are they trying to prevent change? To ease transitions? To protect vulnerable species or areas of interest? To interpret changes as they occur? To protect resources for current and future generations? Managers seek to build ecosystem resilience, but resilience with respect to what? To any change? To changes that affect ecosystem services, or only to those that affect human appreciation and uses? When should a system be allowed to flip into a new state? Can changes due to climate be separated from other challenges? Climate change will increase stresses to human as well as to natural systems, and resource managers may find they need to build resistance to political and economic shocks, as well as to climate change itself.

The way an issue is framed sets boundaries around a problem and affects the way it is understood. Framing highlights some values and interests while excluding others, and can trigger support and opposition. Framing affects the expertise and information considered relevant to an issue and restricts the alternatives considered. Frames can be manipulated, and different groups will frame problems in a way that highlights their particular interests. Competing frames can help people to understand different approaches to a problem, but eventually one frame will need to be adopted by decision-makers, at least as a starting point.

The manner in which resource managers frame issues and define objectives depends in part on the predominant views of nature in the surrounding culture. Viewing an area as a complex system made up of ecosystems and the species that depend on them produces different management objectives from viewing an area as a resource for human uses such as food production or recreation. Non-use views grounded in spirituality or aesthetic appreciation may inspire other objectives.

The term 'natural' has many meanings. Some think of natural as pristine, untouched by human intervention. Others see a natural area as it is now, before climate changes it. Still others may see anthropogenic climate change as just one more natural process leaving its mark. Managers coping with climate change must decide how much human intervention can be tolerated before a resource is no longer regarded as natural, and possibly no longer entitled to protection.

Stakeholders include anyone or anything that will be affected by management decisions. Perhaps the most important stakeholder is the resource itself, the reason that management is needed. Managing institutions have a stake in making effective decisions. Surrounding communities may depend on the resource for food, fuel, livelihoods or spiritual values. Corporations may seek access to resources of commercial value. Tourists seek recreational and aesthetic experiences. People at a distance may feel a sense of ownership, and may value places of particular beauty or that support species of interest.

Managers must decide which stakeholders should be involved in planning, and at what stages, in what ways and for what purposes. Including a variety of stakeholders increases the chances that multiple interests will be protected but also can slow down the planning process. Sometimes laws require participation by particular groups. Others may claim a moral right to participate in decisions that will affect them. Inviting stakeholders to the table does not guarantee that they will be heard. Some communities lack the wealth, knowledge or power to participate effectively in planning activities. Natural resources require a human voice to speak for them. These soft voices must be heard to give participation meaning.

Protected areas are situated within larger natural and human contexts that must be considered in resource management. These areas are embedded within landscapes and are connected to a variety of communities both near and far. Relationships with these other communities may be affected by climate change, and those changes should be considered in management planning.

Adaptive management allows managers to learn from experiments and adjust decisions accordingly. Experimentation entails risks, however, and managers will need to decide how

Box 10.4 (*cont.*)

much risk can be tolerated and where extra protection is needed. Climate change itself presents a kind of experiment, and management responses are bound to have unanticipated outcomes. Flexibility will be needed to respond as new changes occur.

No formula exists to guide resource managers as they face the challenges of climate change. Each area will require a different approach, grounded in the values, needs, interests, objectives and institutions associated with that area. The stakeholders involved, and the way the problem is framed, will shape the way resource management decisions are made.

A perspective focusing on human welfare or human happiness offers a contrasting ethical approach to connecting human well-being with the rest of nature (Yrjönsuuri, 2009). According to this approach, the ultimate cause of climate change – and of other environmental impacts that are driving the biodiversity crisis – has been the emphasis on modern technological advance and the drive for increasing material growth, which underpins the quest for welfare in contemporary societies. Although new technologies are certainly needed, their development and deployment needs to be guided by a different ultimate goal. The key may be to abandon the notion that happiness and well-being can be delivered by ever-increasing consumption of material goods in response to individual goals and desires. Rather, the focus should shift to a more community oriented and environmentally oriented understanding of what it means to be happy. Thus, happiness and well-being are associated with a more holistic perspective on life, emphasising connections with others and with the rest of nature in a more balanced world.

The concept of justice, usually applied to relationships among humans, provides another approach to considering the human responsibility to the rest of nature in the context of climate change (Larrère, 2009). From a human perspective, justice as applied to the environment could be defined as the distribution of the benefits and costs of extracting environmental resources. Extending considerations of justice to other organisms in the natural world becomes more complex. For example, the long-term perspective discussed in Section 10.2 above plays a role here too. The current level of climate change impacts is largely the result of the emissions from several generations of people, and the responses will need to be spread over several generations into the future. How should responsibility be distributed over time? Furthermore, the question of justice must also consider the origins of impacts: Which impacts are the consequence of human actions and which are the outcomes of natural processes? In reality, the two are often intertwined in complex ways, complicating the consideration of justice.

Case studies of observed impacts can help to connect many of these conceptual and theoretical treatments of the ethical aspects of biodiversity loss to the on-the-ground reality of climate change. Box 10.5 describes the interactive effects of climate change and fire regimes on southern Australia's little penguin. This case study is an example of an 'ecological surprise', which is virtually impossible to predict *a priori*. If non-human living beings like penguins also have value or moral status, then the best – or perhaps the only – way to protect them from further harm due to climate change is through rapid and effective emission reductions.

Box 10.5
Climate, fire and the little penguin

LYNDA CHAMBERS, LEANNE RENWICK AND PETER DANN

In coastal regions, misty rain or fog following long spells of hot, dry and dusty weather can result in the ignition of powerpole cross-arms, due to a build-up of salt and dust on the insulators. The red-hot salt crust can fall from the pole and ignite vegetation at its base. In recent years, a number of such fires have occurred on Phillip Island, Australia, home to a large colony of little penguins *Eudyptula minor*.

These fires, as well as a lightning-initiated fire late in 2005 on Seal Island in Victoria, Australia, have caused the death or injury of many little penguins. In each case, the penguins did not avoid the fire, suggesting that their responses to fire are surprisingly inappropriate and maladaptive. In many cases, dead penguins were found either in their burrows (often collapsed) or within metres of burrows (Figure 10.5). Birds nesting under vegetation appeared to remain until they were severely burnt or killed. Penguins were also observed standing beside flames preening singed feathers, rather than moving away. Most live penguins suffered debilitating injuries including burns to their feet and legs, scorched feathers and blistered skin, swollen eyes, and many had difficulty breathing.

The synchronised breeding of seabirds such as penguins, when large numbers are present in a colony, makes them particularly vulnerable to such fires during their nesting seasons. This is particularly true for burrow-nesting species, such as the little penguin, which are disinclined to abandon eggs or chicks, or emerge from their burrows during daytime.

Increased occurrence of hot, dry and dusty weather is projected for the future and may result in increased fire-related risk of little penguin death and injury on Phillip Island if these

Figure 10.5 A dead little penguin and eggs after fire. Source: photo by Peter Dann.

Box 10.5 (*cont.*)

periods are followed by rain or fog. As coastal development encroaches on little penguin colonies throughout south-eastern Australia, this risk is heightened. Risk reduction options include running power underground, more regular pole inspections, improved insulator design and cleaning of the insulators. The risk of fire can be reduced further by appropriate habitat management, such as the planting of fire-retardant, native vegetation and quick response by agencies when a fire does occur.

10.5 Summary and conclusions

Much of the discussion about the consequences of climate change has been dominated by economic analyses, and by comparisons of the costs of mitigation, the costs of adaptation, and the costs of the more serious impacts resulting from lack of mitigation – and attempting to find a balance among these. These analyses are based on a strongly utilitarian analysis. However, climate change raises profoundly ethical questions, often oriented around issues of fairness, equity or justice. The two most prominent of these are (i) the implications of climate change for the well-being of future generations of humans, and (ii) the consequences of climate change for the rest of nature (non-human species) and the resulting obligations on humans towards other species. There are no consensus views on these non-utilitarian perspectives. Considering these deeply ethical issues often leads the climate change discourse into cultural, philosophical or religious considerations and the deep value sets or world perspectives derived from them. The most fundamental of these is the place of humanity in the rest of the living world that inhabits planet Earth.

References

Andelman, S., Midgley, G., Polasky, S. *et al.* (2009). Ensuring biodiversity security in the face of climate change. *IOP Conference Series: Earth and Environmental Science*, **6**, 312005.

Asheim, G., Buchholz, W. and Tungodden, B. (2001). Justifying sustainability. *Journal of Environmental Economics and Management*, **41**, 252–68.

Atkinson, G., Dietz, S., Helgeson, J., Hepburn, C. and Sælen, H. (2009). Siblings, not triplets: social preferences for risk, inequality, and time in discounting climate change. *Economics: The Open-Access, Open-Assessment E-Journal*, **3**, 1–28.

Averill, M. (2009a). Framing natural resource management for climate resilience. *IOP Conference Series: Earth and Environmental Science*, **6**, 132003.

Averill, M. (2009b). Introduction: Resilience, law, and natural resource management. *Nebraska Law Review*, **87**, 821–32.

Barker, T., Scrieciu, S. S. and Taylor, D. (2009). Climate change, social justice and development. *IOP Conference Series: Earth and Environmental Science*, **6**, 122001.

Barsky, R. B., Kimball, M. S., Juster, F. T. and Shapiro, M. D. (1997). Preference parameters and behavioral heterogeneity: An experimental approach in the health and retirement survey. *Quarterly Journal of Economics*, **112**, 537–79.

Caney, S. (2005). Cosmopolitan justice, responsibility, and global climate change. *Leiden Journal of International Law*, **18**, 747–75.

Carlsson, F., Daruvala, D. and Johansson-Stenman, O. (2005). Are people inequality-averse, or just risk-averse? *Economica*, **72**, 375–96.

Cerutti, F. (2009). Why should we care for future generations? *IOP Conference Series: Earth and Environmental Science*, **6**, 132002.

Dietz, S., Hepburn, C. and Stern, N. (2009). Economics, ethics and climate change. *IOP Conference Series: Earth and Environmental Science*, **6**, 122003.

Garnaut, R. (2008). *The Garnaut Climate Change Review: Final Report*. Cambridge, UK: Cambridge University Press.

Gerlagh, R. and Vollebergh, H. (2009). Sustainability as generosity. *IOP Conference Series: Earth and Environmental Science*, **6**, 122007.

Gollier, C. (2006). *Institute Outlook: Climate Change and Insurance: An Evaluation of the Stern Report on the Economics of Climate Change*. Barbon Institute.

Helgesen, J., Sælen, H., Atkinson, G., Dietz, S. and Hepburn, C. (2009). Individual preferences for aversion to public and private risk, inequality, and time – an empirical study. *IOP Conference Series: Earth and Environmental Science*, **6**, 122006.

Intergovernmental Panel on Climate Change (IPCC) (2007). *Climate Change 2007. Impacts, Adaptation and Vulnerability. Contribution of Working Group II to the Fourth Assessment Report of the Intergovernmental Panel on Climate Change*, eds. M. Parry, O. Canziani, J. Palutikovf, P. van der Linden and C. Hanson. Cambridge, UK and New York, NY: Cambridge University Press.

Kreps, D. M. and Porteus, E. L. (1978). Temporal resolution of uncertainty and dynamic choice. *Econometrica*, **46**, 185–200.

Larrère, C. (2009). Environmental justice, human responsibility and equity between humans and nature. *IOP Conference Series: Earth and Environmental Science*, **6**, 132007.

Melin, A. (2009). Climate change and biodiversity preservation: A non-anthropocentric perspective. *IOP Conference Series: Earth and Environmental Science*, **6**, 132006.

Millennium Ecosystem Assessment (2005). *Ecosystems and Human Well-being. Synthesis Report*. Washington, D.C.: Island Press.

Pezzey, J. C. V. (1997). Sustainability constraints versus 'optimality' versus intertemporal concern, and axioms versus data. *Land Economics*, **73**, 448–66.

Ramsey, F. P. (1928). A mathematical theory of saving. *Economic Journal*, **38**, 543–59.

Rolston III, H. (1989). *Environmental Ethics. Duties to and Values in the Natural World*. Philadelphia: Temple.

Roser, D. (2009). Economic models and the rights of future generations. *IOP Conference Series: Earth and Environmental Science*, **6**, 122002.

Selden, L. (1978). A new representation of preference over 'certain x uncertain' consumption pairs: The 'ordinal certainty equivalent' hypothesis. *Econometrica*, **46**, 1045–60.

Singer, P. (1979). *Practical Ethics*. Cambridge, UK: Cambridge University Press.

Singer, P. (2002). *Unsanctifying Human Life*. Oxford: Wiley-Blackwell, pp. 315–16.

Solomon, S., Plattner, G.-K., Knutti, R. and Friedlingstein, P. (2009). Irreversible climate change due to carbon dioxide emissions. *Proceedings of the National Academy of Sciences (USA)*, **106**, 1704–09.

Steffen, W., Burbidge, A. A., Hughes, L. *et al.* (2009). *Australia's Biodiversity and Climate Change*. Melbourne: CSIRO Publishing.

Steffen, W., Sanderson, A., Tyson, P. D. *et al.* (2004). *Global Change and the Earth System: A Planet Under Pressure*. Berlin: Springer-Verlag.

Stern, N. H. (2007). *The Economics of Climate Change: The Stern Review*. Cambridge, UK: Cambridge University Press.

Toft, K. H. (2009). Global justice and climate change. *IOP Conference Series: Earth and Environmental Science*, **6**, 122008.

Van Houtan, K. (2009). Extinction, suffering, and the cruciformity of the cosmos. *IOP Conference Series: Earth and Environmental Science*, **6**, 132005.

Varner, G. E. (1998). *In Nature's Interests? Interests, Animal Rights, and Environmental Ethics*. New York, NY: Oxford University Press.

Weitzman, M. L. (2007). A review of the Stern Review on the Economics of Climate Change. *Journal of Economic Literature*, **45**, 703–24.

Yrjönsuuri, M. (2009). Human happiness – friend or foe? *IOP Conference Series: Earth and Environmental Science*, **6**, 132004.

Part IV

Mitigation and adaptation approaches

Society already has many tools and approaches – technological, economic, behavioural,
and managerial – to deal effectively with the climate change challenge...

11

Low-carbon energy technologies as mitigation approaches

'The only engine big enough to impact Mother Nature is Father Greed: the Market. Only a market, shaped by regulations and incentives to stimulate massive innovation in clean, emission-free power sources can make a dent in global warming.'[1]

11.1 Introduction

The joint climate and development imperatives described in the previous chapters demand a strong, immediate and sustained response. As outlined in Chapter 8, the global community has approximately four decades to implement a reduction in greenhouse gas emissions in the order of 90%.

The transition to a low-carbon/low-emissions economy will likely take many forms and be played out on local, national and regional levels as well as on the global stage (Chapter 13). While little has so far been done to initiate this transition, a diverse suite of existing and near-term technologies exist that can be components of regionally tailored low-carbon energy mixes, and which can underpin an aggressive expansion of energy efficiency worldwide. The challenge is, in effect, one of inventing and implementing a new energy economy in around four decades; i.e. in a fraction of the time it took to build the entire current industrial energy infrastructure.

A new 'triple bottom line' is therefore needed: one that is low-carbon, high-growth and job-creating. To further increase this challenge, this must be achieved in not only a few small segments of the global population, but also widely across industrialised and developing nations – from household and village scales, to national and regional economies.

This unprecedented goal and challenge creates a demand for both transformative basic science and engineering that, in combination, can provide novel energy systems. In addition, there needs to be a focus on the practical implementation of technologies, economic

---

[1] Thomas Friedman, writing in the *New York Times*, 19 December 2009.

Climate Change: Global Risks, Challenges and Decisions, Katherine Richardson, Will Steffen and Diana Liverman *et al*. Published by Cambridge University Press. © Katherine Richardson, Will Steffen and Diana Liverman 2011.

systems and policies that move clean, job-creating technologies to scale. This dichotomy was wonderfully captured by Stokes (1997) in his theory of *use-inspired, basic research*. To accomplish the goal, a 'systems science' of sustainable energy needs to be developed immediately; a systems science that envisions and implements functioning economies, and that moves far more rapidly than simply via gradual increments of energy efficient technology deployment. Far more aggressive changes are needed than those seen today, where simple incremental shifts in market share for clean energy and services are superimposed on a backdrop of continued growth in the demand for energy. Instead, a material substitution will be required where technology evolution and policy measures underpin growth in the overall energy economy while, at the same time, generating a complete reversal of the current fossil fuel–renewable energy balance. The decarbonisation of the global economy will need to begin rapidly and be sustained for decades.

This chapter presents an overview of the interplay between existing and near-term technology, and other options (e.g. energy efficiency, policy levers) that can be employed to address climate issues and decarbonise economies. Energy options that address a second set of issues – improved local air quality, energy security, and synergies with poverty alleviation and local sustainable development – are also introduced (Box 11.1).

Box 11.1
Mitigating carbon emissions while advancing sustainable development goals: small-scale bioenergy systems

OMAR MASERA

Linking climate mitigation to sustainable development priorities is key to successful interventions, particularly within developing countries. Small-scale bioenergy systems, such as improved cookstoves, present sizable and cost-effective mitigation opportunities with very large health and overall improvements in living conditions of local people as co-benefits (Figure 11.1).

The use of solid biomass – in the form of fuelwood, charcoal and agricultural residues – in households and small industries in developing countries accounts for 71% of global biomass energy use and for 60% of total wood harvesting from forests (International Energy Agency, 2007). This use is growing in absolute terms, particularly within Africa but also in the poorest regions of Asia and Latin America. The biomass is burned in traditional devices such as open fires, which are very inefficient and polluting, and which have deleterious consequences for health and the local environment. Indoor air pollution from biomass burning alone affects more than 2.4 billion people and has been estimated to cause 1.6 million excess deaths per year – including 900 000 children under five – and the loss of 38.6 millon disability-adjusted life-years (DALY) per year, similar in magnitude to the burden of disease from malaria and tuberculosis (WHO, 2009).

The extensive use of fuelwood and charcoal in these devices is also associated with significant emissions of GHG – including CO_2 from deforestation and forest degradation – and many short-lived pollutants, i.e. CH_4, CO, black carbon, etc. Residential biomass may account for 1%–3% of net global emissions of CO_2, 20% of total black carbon emissions and 30% of CO emissions. Contributions can be much larger at the country level.

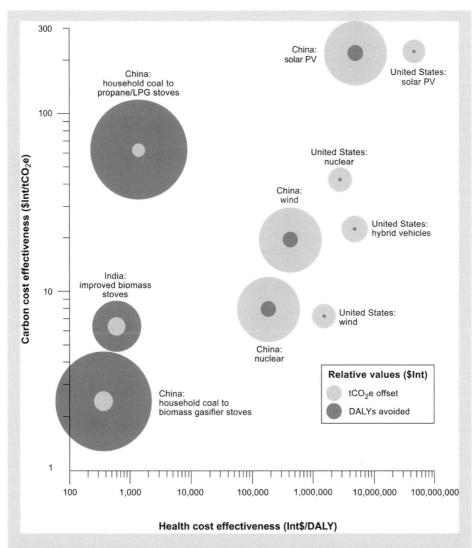

Figure 11.1 Cost-effectiveness (in terms of climate change mitigation and health) of improved cookstoves with regard to other interventions. Both axes are logarithmically scaled. The size of the circle indicates the total impact of the intervention in terms of health (dark grey) and of climate (light grey). Improved biomass cookstoves are very cost-effective in terms of both indicators. Source: Smith and Haigler (2008).

A revolution in the field of small-scale biomass systems is taking place worldwide, and a whole new generation of efficient devices such as advanced biomass cookstoves, domestic biogas systems, improved charcoal kilns, etc., is entering the market. In particular, innovative low-cost cookstoves are being distributed, which are based on small-scale gasification, clean combustion, and/or proccessed biomass fuels. These new material designs can reduce pollution and fuel use by up to 90% and 60%, respectively (Bailis *et al.*, 2005; Johnson *et al.*, 2009a). Improved marketing strategies, product design, user feedback and facilities oriented to mass

Box 11.1 (*cont.*)

production are fostering this change. Cookstoves that use the thermal-electric effect to produce both heat for cooking and power for lighting are also about to enter the market at an estimated cost of US$40/stove (Biolite, 2009).

The new generation of improved cookstoves reduces short-lived gases as well as fuel use, resulting in dramatic mitigation of greenhouse gases. In Mexico, for example, the new generation of stoves mitigates from 3–9 tCO_2-e/year (Johnson *et al.*, 2009a) depending on the renewability of fuelwood use and this does not include benefits from black carbon reduction. This makes the intervention extremely cost-effective, with benefit/cost ratios of 10 to 25:1 when co-benefits are factored in (García-Frapolli *et al.*, forthcoming). Mitigation costs reach US$1–5/$tCO_2$ and, even in some of the more industrialised developing countries such as Mexico (Johnson *et al.*, 2009a), the total mitigation reached by cookstoves alone represents 5% of the country's total from 2009 to 2030 (219 $MtCO_2$) (Johnson *et al.*, 2009b).

A major barrier for more successful programs has been the lack of long-term monitoring and feedback to the programs to assure sustained cookstove use and also to increase adoption rates. Nevertheless, some of these barriers may be partially overcome by carbon offset projects, as the key performance indicator – annual mitigation – is linked to proper and continuous stove use, and not only to cookstove installation.

11.2 Snapshots of decarbonisation pathways

The needed trajectory of net carbon emissions is clear: a four-decade, 80% or more decrease in global emissions. In Chapter 8, possible emissions trajectories for achieving this reduction are considered at the global level. Here, we illustrate what an 80% reduction in emissions from the 1990 baseline in emissions would mean for a highly industrialised country by using the example of the USA. The 1990 baseline is chosen, as this is the baseline often used in the Kyoto Protocol (Chapter 13) as well as a number of other international discussions concerning emissions reductions. A path of steady emissions growth during the period 1990–2010 in the USA will need to be altered from the business-as-usual path to a steadily accelerating path of emissions reduction – leading from the current emissions level of approximately 30% higher than the 1990 baseline, to an 80% reduction from the 1990 baseline levels by 2050.

Goals for emissions reductions by 2050 have become standard in municipal, city, state, regional and national agendas on climate. More difficult to formulate, however, have been realistic mid-term reduction scenarios, especially for the short term, i.e. by 2020 or 2030. This is largely because – at a time when global reductions in emissions have not begun – reasonable forecasts that show reductions emissions on the path to 2050 are a challenge. One such scenario, for the USA, is presented by the Natural Resources Defense Council (NRDC) (2009). A particularly important scenario for the EU finds that if cost decline trends for renewable energy can be maintained, and if the electricity grids can be expanded in scope and in the degree to which it is 'smart', dramatic decarbonisation is possible (European Climate Foundation, 2010) that can meet or achieve the 80% decarbonisation goal by mid-century.

In the NRDC blueprint for the mid-point, 2030, the envisioned transition takes a familiar form (Figure 11.3).

(1) Significant attention to energy efficiency and demand-side management provides immediate and then sustained reductions in demand – saving money at once, reducing peak demand and, thus, the cost of new technology deployment.

(2) Growth in 'new' renewable energy sectors – notably solar and wind – through specific market support policies that evolve such that each sector contributes *a fifth or more* of total electricity supply by 2030. These 'new fifths' are combined with the aggressive efficiency growth, current or somewhat expanded levels of hydro and nuclear energy (Box 11.2; note that a number of scenarios posit growth of the other sectors to the extent that this controversial sector is not materially needed, while some cases feature nuclear as the largest low-carbon sector of the entire system for a number of nations).

(3) Macroeconomic shifts in the energy markets, facilitated by a price on carbon emissions that grows to the point where it impacts significantly not only the electricity and industrial sectors, but also the notoriously hard-to-change world of transportation.

Box 11.2
Future roles for nuclear energy

PER F. PETERSON, EDWARD BLANDFORD AND LANCE KIM

Following the discovery of nuclear fission in the late 1930s, the implications of a neutron-induced fission chain reaction were quickly recognised both as a source of peaceful energy and for its destructive potential in nuclear weapons. Rapid progress on both fronts soon followed, driven by the exigencies of global conflict and growing appetites for energy. On the civilian side, the expansion of nuclear energy stalled following the energy crisis of the 1970s, when expectations of growing demand for energy failed to materialise in general, and construction cost overruns and highly varied operating performance stymied the further expansion of nuclear power.

Today, the greatly improved performance of existing nuclear power plants and nuclear energy's role as the largest source of non-carbon electricity has rekindled interest in constructing new reactors. With a few dozen nuclear power plants under construction around the globe (IAEA, 2008), several reactors in the early stages of the US Nuclear Regulatory Commission's streamlined licensing process (Nuclear Energy Institute, 2008), and bipartisan support in the US Congress, the second era of nuclear energy appears imminent.

Nuclear in a nutshell

Nuclear power plants harness the energy generated by the neutron-induced fission of heavy fissile isotopes such as uranium and plutonium. Designers of reactor cores arrange material in a critical configuration to sustain a fission chain reaction. By moving up the curve of binding

Box 11.2 (*cont.*)

energy, the fission of heavy elements liberates an enormous amount of energy per unit mass in comparison to typical chemical reactions. Per day of operation, a typical nuclear power plant with a rated capacity of 1000 megawatts of electric power (MW_e) provides enough power for roughly 1 million average US homes while consuming only 3.2 kg of uranium fuel. By comparison, a 1000 MW_e coal-fired power plant demands over 7 million kg of coal per day, and releases comparable quantities of CO_2, NO_x, SO_x and ash to the environment.

The fate of the neutrons, fission products and neutron activation products produced by a nuclear fission reaction present a number of challenges to system designers and policymakers. The production of enriched uranium to fuel reactors and the breeding of plutonium 239 within reactors raise proliferation concerns, given the utility of these materials in nuclear weapons. Radioactive byproducts of fission continue to produce decay heat following the shutdown of the reactor, necessitating highly reliable safety systems to prevent core-melt accidents. Additionally, long-lived radionuclides in spent nuclear fuel require long-term sequestration to limit impacts on the environment and public health.

Managing this broad range of issues is a high priority in the design of nuclear reactors and nuclear energy policy. The system of safeguards administered by the International Atomic Energy Agency has been effective in deterring the diversion of declared nuclear material, and continues to be improved to detect both declared and undeclared activities. National measures to provide physical protection of nuclear facilities to guard against theft and sabotage have been effective, but merit further strengthening in some countries. The safety of nuclear reactors is achieved with redundant and diverse systems performing safety functions with high reliability, a multi-barrier approach to containing radionuclides, and inherent negative feedback effects that tend to shut down the reactor if something goes awry. For new reactors, a shift in design philosophy towards simpler passive systems that operate without human intervention potentially offers a more robust and less expensive approach to reactor safety, security and proliferation resistance.

Nuclear waste management

The management of nuclear waste continues to be one of the dominant issues influencing the acceptability of nuclear energy. In contrast to the challenge of sequestering the byproducts of the combustion of fossil fuels and accommodating the health and environmental impacts of the pollutants discharged into the biosphere, the high energy density of nuclear fuel makes managing the correspondingly small quantities of spent nuclear fuel a far more manageable task. To this end, international research and development efforts have improved the understanding of the management of nuclear waste to minimise impacts on the environment and public health. A number of researchers maintain that geological isolation can facilitate the management of nuclear waste safely and reversibly over long periods of time (Ahn and Apted, 2010).

Renewed interest in nuclear energy

The performance of existing nuclear power plants is one of the drivers for the renewed interest in nuclear energy. Nuclear production costs (fuel plus operations and maintenance) are competitive with coal and less volatile in price in comparison to natural gas. As a consequence of their cost structure (high capital cost, low production cost), existing nuclear power plants have quietly achieved unprecedented levels of performance through innovations in operations and maintenance processes (including advances in human performance and risk management). On average, US nuclear power plants currently operate approximately 90% of the year, going offline at scheduled times to refuel and perform maintenance. In comparison, the fleet average during the 1980s was a mere 60%. The frequency of forced reactor outages caused by unanticipated events has also declined, resulting in a forced capability loss of less than 2% (median value) – an additional indicator of competent management.

Expectations of operating in a carbon-constrained economy are also contributing to the renewed interest in nuclear energy. With life-cycle CO_2 emissions comparable to renewable technologies (Kammen and Pacca, 2004), nuclear's role as the largest source of low-carbon electricity and the most realistic alternative to coal-generated base-load electricity cannot be ignored. With some in the European Union considering a return to coal-fired generation (Rosenthal, 2008), society is presented with stark policy alternatives between coal and nuclear for base-load generation. The French present one model for energy development. Following a 20-year construction effort, nuclear energy now constitutes the vast majority of domestic primary energy production and the French were able to close their last coal mine in early 2004.

Near-term prospects

The major near-term question to address as we consider an expansion of nuclear generation is whether new nuclear power plants can be built on schedule and at a reasonable cost. Capital costs for US reactors increased dramatically for reactors coming online through the 1980s and 1990s due to the high interest rates, public opposition, delays from poor project management, a burdensome licensing process and safety upgrades following the Three Mile Island accident. For new reactors, the overnight capital costs for new nuclear power plants vary widely from optimistic vendor estimates of $1500 per kW_e of installed capacity to a recent estimate of $2950 per kW_e (Keystone Center, 2007).

Where actual costs will fall is debatable. The high degree of optimisation for new nuclear infrastructure may offer significant cost savings for many of the control systems. Advanced light water reactors featuring passive safety systems provide substantial improvements over earlier designs, with reductions in equipment requirements and building volumes contributing to lower cost. Streamlined licensing and modular construction processes call for a high degree of standardisation, thus decreasing costly construction delays and moving more rapidly down the learning curve. On the other hand, the hiatus on nuclear plant construction and uncertain expansion plans are creating bottlenecks in supply, placing upward pressure on prices for nuclear components. The scarcity of a qualified nuclear workforce compounds the problem. The degree to which these growing pains will persist with the expansion of the nuclear energy sector remains to be seen.

In the long term, resource inputs provide an indication of future costs of energy technologies. The inflation of nuclear construction cost estimates has been attributed in part to the near doubling of construction commodity costs (for example, steel reached prices of some $600 per tonne in March 2008). But, the total cost of commodities (steel, concrete, copper, etc.)

Box 11.2 (*cont.*)

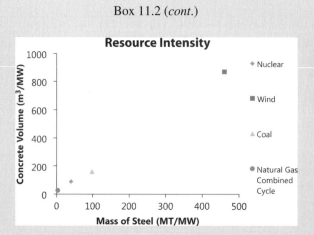

Figure 11.2 Concrete and steel intensity for selected sources of energy per average megawatt of capacity (e.g. corrected for capacity factor). Source: calculations by Kammen, unpublished.

used in nuclear construction amounts to only a small fraction of total overnight capital cost ($36 per kW_e). Thus the cost of nuclear construction is not sensitive to the cost of the material inputs, but instead depends upon the costs for manufacturing, construction, and project management and financing. As with reactor operations costs, these costs will depend primarily on the competence of the companies that build nuclear power plants.

Comparing resource inputs also provides an indication of the environmental impacts and expected competitiveness of energy technologies. Concrete and steel represent over 95% of the carbon emissions associated with construction inputs and are strongly correlated to capital cost. The relatively low requirements for these two inputs demonstrate that construction costs for nuclear energy are less sensitive to commodity costs in comparison to wind and, to a lesser extent, coal (Figure 11.2) while avoiding the high production costs and volatility of natural gas prices.

Ultimately, capturing the external costs of fossil fuel combustion can be expected to shift economic incentives in favour of nuclear for base load generation capacity. This can be expected to occur in the near term in the US as carbon control programs are implemented to reduce emissions. Under potential carbon control programs, the profitability for existing nuclear power plants is expected to increase dramatically (Smith and Haigler, 2008). Furthermore, new nuclear power plants are estimated to become the least cost source of base-load electricity, with a carbon tax of approximately $45 (Congressional Budget Office, 2008) to $100 per ton of carbon (Sailor *et al.*, 2000).

Advanced nuclear energy systems

In the longer term, the development of advanced nuclear reactor and fuel cycle technologies promise to achieve higher levels of economic performance, safety, sustainability, physical security and proliferation resistance. With a multiplicity of design options providing a suite of technological options (e.g. ranging from thermal to fast spectrum reactors, water to molten salt coolants, uranium/plutonium to uranium/thorium fuels, open to closed fuel cycles), the winner or winners of this process is difficult to predict.

Some designs, variously denoted small-medium reactors or appropriately sized reactors (less than 700 MW$_e$), have attracted the attention of venture capitalists (Riddel, 2009). These reactors are deliberately small to improve performance with respect to logistics, safety, operation and economics. While initially designed for areas with limited infrastructure inadequate to support a large reactor (e.g. small and developing countries, remote sites), sequentially deploying modular units may offer a less financially risky approach to nuclear expansion that offsets their higher levelised construction costs. Factory production of multiple modules may further improve economic performance by leading to more predictable construction schedules and enabling more opportunities for technological learning that drives down cost (Ingersoll, 2009).

Thorium-fueled reactors have attracted renewed attention recently, due to thorium's greater abundance – particularly from states such as India, with limited uranium resources. Operating on the same principle as a plutonium breeder reactor, fissile uranium 233 is produced via a neutron capture reaction in fertile thorium 222. By effectively creating more fuel than it consumes, a breeder reactor greatly expands the resource base for nuclear energy by utilising large reserves of fertile uranium 238 or thorium 222. Some claim nonproliferation provides benefits to the fissile material produced from thorium fuel, based principally on the reductions in plutonium inventories, the less attractive characteristics of the material with respect to the performance of a weapon and the reduction in incentives for reprocessing (Galperin and Raizes, 1997). However, many of the other determinants of proliferation risk (e.g. access to fissile material, a source of neutrons, nuclear technology in general) remain unchanged.

In any event, significant technological hurdles remain and progress is uncertain for advanced designs. Key challenges include the demonstration of the performance of passive safety systems under a challenging set of normal and off-normal operating conditions, as well as qualifying new fuel and structural materials for use in the harsh chemical and radiation environment of advanced nuclear reactor cores. Developing technology-neutral regulatory processes to license and manage these new designs must also confront a number of hurdles including the demonstration of a licensing approach for small modular reactors to account for multiple unit sites and factory fabrication of modules. Additionally, the US Nuclear Regulatory Commission must re-examine requirements for the Emergency Planning Zone. Existing requirements for the Emergency Planning Zone originally developed for large reactors with correspondingly large source terms (i.e. radionuclide inventories) may be unnecessarily burdensome for small reactors with passive safety features and robust containments or confinements (Ingersoll, 2009).

Conclusions

Activity in the nuclear energy sector has been substantial. Improvements in performance at existing nuclear power plants, and demand for low-carbon energy, have resulted in a dramatic turnaround in the prospects for nuclear energy. Nevertheless, the future remains uncertain and expansion could be derailed. In the US, fully implementing the provisions of the Energy Policy

Box 11.2 (*cont.*)

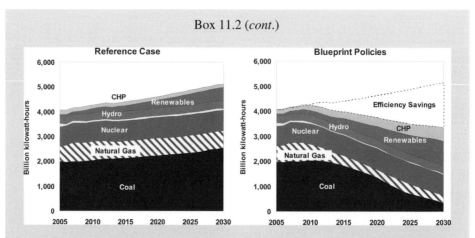

Figure 11.3 Natural Resources Defense Council (2009) reference case (left) and 'Blueprint' case (right) for energy technology development in the USA until 2030. The implementation of the 'Blueprint policies' case consists of a suite of sectoral targets for deployment of low-carbon energy supply and efficiency efforts, and the introduction of sufficiently strong carbon emissions pricing so that first coal then even oil use is reduced dramatically, and is replaced by low carbon supply and efficient use policies. Note: CHP – combined heat and power. Source: http://www. nrdc.org/globalwarming/blueprint/methodology.asp.

Act of 2005 to overcome the disincentives faced by first-movers will support near-term expansion of advanced light water reactors. Resolving the waste disposal problem will require a consistent and comprehensive approach to regulate hazardous wastes of all forms, nuclear and not. Key research and development priorities include demonstrating the cogeneration of electricity and process heat to expand demand for nuclear energy. In the longer term, the development of advanced nuclear reactor and fuel cycle technologies promises to achieve higher levels of economic performance, safety, sustainability, physical security and proliferation resistance.

Four issues initially need to be considered in evaluating the steps necessary to make these types of scenarios a reality. First, what is the track record and the prospect of a sufficient innovative research and development base for this low-carbon economy? Second, can energy efficiency be the job-creating, cost-effective and near-term 'lever of change' to reduce the needed supply of new energy? Third, can the suite of low-carbon supply options grow to the needed sectoral size – namely a fifth or more of energy supply? And finally, can a set of policies be designed and implemented that make this new economy both grow and become a force for long-term sustainable economic activity? Paramount in current discussion is the design of technology-specific policies to spur growth of each low-carbon supply-and-demand reduction option that can be implemented and sustained. This is necessary to build markets and to buy time while the larger macroeconomic change of implementing a price on pollution is not only put into place, but also so that this price can rise sufficiently to impact the entire economy.

11.3 The research and development supply

Globally, the USA contributes about one quarter of the total energy research and development (R&D) portfolio. And, until the stimulus funding infusion of 2009, the USA invested about US$1 billion less per year in energy R&D in 2007 than 20 years before (Nemet and Kammen, 2007).

This trend is remarkable, firstly because the levels in the mid 1990s had already been identified as dangerously low (Margolis and Kammen, 1999), and secondly because the decline is pervasive – across almost every energy technology category, in both the public and private sectors in the USA and at multiple stages in the innovation process, investment has been either stagnant or declining. In fact, between 1980 and 1990 only Japan and Switzerland of OECD nations saw an increase in energy R&D (Margolis and Kammen, 1999). Moreover, the decline in investment in energy has occurred while overall US R&D has grown by 6% per year, and federal R&D investments in health and defence have grown by 10 to 15% per year (Nemet and Kammen, 2007). As a result, the percentage of all US R&D invested in the energy sector has declined from 10% in the 1980s to 2% currently. Private-sector investment activity is a key area for concern. While in the 1980s and 1990s the private and public sectors each accounted for approximately half of the USA's investment in energy R&D, presently the private sector makes up only 24%.

The recent decline in private-sector funding for energy R&D is particularly troubling because it has historically exhibited less volatility than public funding – private funding rose only moderately in the 1970s and was stable in the 1980s; periods during which federal funding increased by a factor of three and then dropped by half. During the 1990s industry investment in each technology area strongly correlated with federal spending (Nemet and Kammen, 2007). The lack of industry investment in each technology area suggests that the public sector needs to play a role in not only increasing investment directly, but also in addressing market and regulatory obstacles that discourage investment in new technology. The reduced inventive activity in energy reaches back even to the earliest stages of the innovation process, in universities where fundamental research and training of new scientists occurs. For example, a study of federal support for university research raised concerns about funding for energy and the environment, as it was found that funding to universities was increasingly concentrated in the life sciences (PCAST, 1997).

A glimpse at the drivers behind investment trends in three segments of the energy economy indicates that a variety of mechanisms are at work. First, the market for fossil fuel electricity generation has been growing by 2%–3% per year, yet R&D has declined by half in the past 10 years, from $1.5 billion to $0.7 billion. In this case, the shift to a deregulated market has been an influential factor, reducing incentives for collaboration and generating persistent regulatory uncertainty. The industry research consortium, the Electric Power Research Institute (EPRI), has had its budget decline by a factor of three. Rather than shifting their EPRI contributions to their own proprietary research programs, investor-owned utilities and equipment makers have reduced both their EPRI dues and their own research programs. The data on private-sector fossil R&D validate the President's Commission on Science and Technology's (PCAST, 1997) prescient warnings in the mid-1990s about the effect of electricity sector deregulation on technology investment. Second, the decline in private sector nuclear R&D corresponds with diminishing expectations about the future construction of new plants.

Over 90% of nuclear energy R&D is now federally funded. This lack of 'demand pull' has persisted for so long that it even affects interest by the next-generation nuclear work-force; enrolment in graduate-level nuclear engineering programs has declined by 26% in the past decade. Recent interest in new nuclear construction has, thus far, not translated into renewed private-sector technology investment. Third, policy intermittency and uncertainty plays a role in discouraging R&D investments in the solar and wind energy sectors, which have been growing by 20%–35% per year for more than a decade. Improvements in technology have made wind power competitive with natural gas (Jacobson and Masters, 2001) and helped the global photovoltaic industry to expand by 50% in 2004. Nevertheless, investment by large companies in developing these rapidly expanding technologies has actually declined. By contrast, European and Japanese firms are investing and capturing a growing market share in this rapidly growing sector, thus making the USA increasingly an importer of renewable energy technologies.

Venture capital investment in energy provides a potentially promising exception to the trends in private and public R&D. While venture capital groups are often criticised in the press for taking a short-term view on the returns to investment, their search for promising investments has meant that from time to time they have provided funding where public investments were not sufficient or not sustained (Nemet and Kammen, 2007). Energy investments funded by venture capital firms in the US exceeded one billion dollars in 2000 and, despite their subsequent cyclical decline to $520 million in 2004, are still of the same scale as private R&D by large companies. Recent announcements – such as California's plan to devote up to $450 million of its public pension fund investments to environmental technology companies, and Pacific Gas and Electric's $30 million California Clean Energy Fund for funding new ventures – suggest that a new investment cycle may be starting. The emergence of this new funding mechanism is especially important because a number of studies have found that, in general, venture capital investment is 3–4 times more effective than R&D at stimulating patenting (PCAST, 1997; Kammen and Nemet, 2005). While it does not offset the declining investment by the federal government and large companies, the venture capital sector is now a significant component of the US energy innovation system, raising the importance of monitoring its activity level, composition of portfolio firms and effectiveness in bringing nascent technologies to the commercial market.

Finally, R&D investment in the drugs and biotechnology industry provides a revealing contrast to the trends seen in energy. Innovation in the medical sector has been broad, rapid and consistent. The 5000 firms in the industry signed 10 000 technology agreements during the 1990s, and the sector added over 100 000 new jobs in the past 15 years. Expectations of future benefits are high – the typical biotech firm spends more on R&D ($8.4 million) than it receives in revenues ($2.5 million), with the difference generally funded by larger firms and venture capital. Although energy R&D exceeded that of the biotechnology industry 20 years ago, today R&D investment by biotechnology firms is an order of magnitude larger than that of energy firms (Nemet and Kammen, 2007). In the mid 1980s, US companies in the energy sector were investing more in R&D ($4.0 billion) than were drug and biotechnology firms ($3.4 billion). By 2000, drug and biotech companies had increased their investment by almost a factor of 4 to $13 billion. Meanwhile, energy companies had cut their investments by more than half to $1.6 billion. From 1980 to 2000, the energy sector invested $64 billion in R&D while the drug and biotech sector invested $173 billion. Prior to the Obama Administration stimulus push, total private-sector energy R&D was less than the R&D budgets of individual biotech companies such as Amgen and Genentech. A reasonable conclusion from this comparison is that if government investment dominates a sector, there is both less sustained industry support and innovation, and the fate of an entire sector is more volatile or even at risk due to changes in singular support decisions.

11.4 Reductions in patenting intensity

Divergence in investment levels between the energy and other sectors of the economy is only one of several indicators of under-performance in the energy economy. In this section, results are presented using three methods developed to assess patenting activity, as earlier work (see Nemet and Kammen, 2007) has shown that patenting activity provides an indication of the outcomes of the innovation process.

First, records of successful US patent applications as a proxy for the intensity of inventive activity are used to demonstrate strong correlations between public R&D and patenting across a variety of energy technologies (Margolis and Kammen, 1999). Since the early 1980s, all three indicators – public sector R&D, private sector R&D, and patenting – have exhibited consistently negative trends (Nemet and Kammen, 2007). Public R&D and patenting are highly correlated for wind, photovoltaic cells, fuel cells and nuclear fusion (Figure 11.4). Nuclear fission is the one category that is not well correlated to federal R&D funding, but the story here is complicated because of investment from the Department of Defense that is difficult to track. Comparing patenting against *private* sector R&D for the more aggregated technology categories also reveals concurrent negative trends. The long-term decline in patenting across technology categories and their correlation with R&D funding levels provides further evidence that the technical improvements upon which performance-improving and cost-reducing innovations are based are occurring with decreasing frequency.

Second, in the same way that studies measure scientific importance using journal citations, patent citation data can be used to identify 'high-value' patents (Nemet and Kammen, 2007). For each patent, the number of times it is cited by subsequent patents is identified using the NBER Patent Citations Datafile. For each year and technology category, the probability of a patent being cited is estimated by recording the number of patents in that technology category in the next 15 years (Figure 11.4).

The adjusted patent citations for each year are then calculated using a base year. 'High-value' patents are those that received twice as many citations as the average

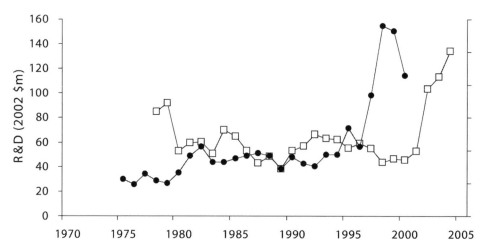

Figure 11.4 US federal R&D and patents in a range of energy technology areas. Open squares represent investment in millions of dollars and filled circles represent patents. Note the strong correlation between funding levels and innovation, measured in patents granted, for *each* technology. The one exception is arguably nuclear fission, where (significantly subsidised) commercial viability has already been achieved. Source: Kammen and Nemet (2007). With kind permission of Springer Science+Business Media.

patent in that technology category. Between 5% and 10% of the patents examined were defined as high value. The Department of Energy accounts for a large fraction of the most highly cited patents – with a direct interest in 24% (6 of the 25) of the most frequently referenced US energy patents – while only associated with 7% of total US energy patents. In the energy sector, valuable patents do not occur randomly – they cluster in specific periods of productive innovation (Marburger, 2004). The drivers behind these clusters of valuable patents include R&D investment, growth in demand and exploitation of technical opportunities. These clusters both reflect successful innovations and productive public policies, while marking opportunities to further energise emerging technologies and industries.

Finally, patent citations can be used to measure both the return on R&D investment and the health of the technology commercialisation process, as patents from government research provide the basis for subsequent patents related to technology development and marketable products. The difference between the US federal energy patent portfolio and all other US patents is striking, with energy patents earning on average only 68% as many citations as the overall US average from 1970 to 1997 (Nemet and Kammen, 2007). This lack of development of government-sponsored inventions should not be surprising, given the declining emphasis on innovation among private energy companies.

In contrast to the rest of the energy sector, investment and innovation in fuel cells have grown. Despite a 17% drop in federal funding, patenting activity intensified by nearly

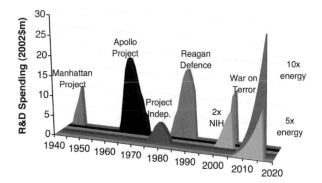

Figure 11.5 Federal R&D investments in a range of projects, including the atomic bomb (Manhattan project), 'Star Wars' (Reagan defense), the doubling of the US National Institute of Health budget, and two forecasts for federal energy research assuming a five-fold and a ten-fold increase. In real terms, the Obama Stimulus increase will grow the US federal energy research budget by somewhat more than a factor of five. Sources: Kammen and Nemet (2007); Nemet, unpublished.

an order of magnitude, from 47 in 1994 to 349 in 2001. Trends in patenting and the stock prices of the major firms in the industry reveal a strong correlation between access to capital and the rate of innovation (PCAST, 1997). The relationship between fuel cell company stock prices and patenting is stronger than that between patenting and public R&D. In a study of the fuel cell industry, only five firms accounted for 24% of patents from 1999 to 2004, while almost 300 firms received fuel cell patents from 1999 to 2004; a diverse participation by both small and large firms (Nemet and Kammen, 2007). This combination of increasing investment and innovation is unique within the energy sector. While investments have decreased as venture funding overall has receded since the late 1990s, the rapid innovation in this period has provided a large, new stock of knowledge on which new designs, new products and cost-reducing improvements can build. The industry structure even resembles that of the biotechnology industry. A large number of entrepreneurial firms and a few large firms collaborate through partnerships and intellectual property licensing to develop this earlier stage technology. The federal government, of course, need not be the only driver of innovation in the energy sector if the private sector and recent major investments by a range of multinational companies (e.g. BP, Shell, Chevron, Berkshire Hathaway, and a number of ultra-rich individuals) demonstrate the defining role that they can play in this area.

While studies of returns on R&D only indicate correlation, not causation, the case is strong that these trends indicate that when investments in energy research and development are either large enough, or sustained, very significant returns can be realised. This places the 2008–10 round of economic stimulus packages in the EU, USA and China into an important and potentially transformative category – *if they are sustained and combined with demand-pull policies*. As an example, the stimulus investment in the US will increase energy R&D spending by between a factor of five and ten (Figure 11.5), i.e. a level earlier identified as being appropriate for stimulating energy technology development (Nemet and Kammen, 2007).

11.5 The role of energy efficiency

Active research, investment and deployment of energy efficiency technologies, practices, management strategies and market mechanisms have been exercised over more than three decades. The overwhelming conclusion from these activities is that the most cost-effective form of energy is the energy that we did not need to design, build, deploy or manage. In the USA, this knowledge exists as a major resource of shared experience and expertise that is currently in the hands of investor-owned utilities, municipal utility districts, state public utility commissions, the US Department of Energy, the Environmental Protection Agency (EPA), and a number of other institutions and individuals.

The list of individual energy efficiency innovations is tremendously long, and includes: efficient water heaters; improved refrigerators and freezers; advanced building control technologies and advances in heating, ventilation, and cooling (HVAC); smart windows that can adjust to maintain a comfortable interior environment; a steady stream of new building codes to reduce needless energy use; compact fluorescent lights; and the emerging wave of even lower energy, solid-state light-emitting diode (LED) lights. Improvements in buildings alone, including insulation and energy-efficient roofing materials – where we use over 60% of all energy – have come at savings of *tens of billions of dollars* each year.

In the USA, several states – including California, New York, Rhode Island and Wisconsin – have consistently deployed energy efficiency innovations, as have a number of European nations and Japan (Figure 11.6). State planners, officials, citizens and industry leaders have found these to be tremendously cost-effective, often providing greater service at *lower personal and social cost* than the 'conventional' route of simply adding more fossil-fuel based supply technologies. This is the case for several reasons. First, energy-efficient technologies often represent upgrades in service through superior performance (e.g. higher-quality lighting, heating and cooling with greater controls, or improved reliability of service through greater ability of utilities to respond to time of peak demand). These innovations can provide better and less expensive service.

Second, a wide range of energy-efficient technologies have ancillary benefits of improved quality of life, such as advanced windows that not only save on heating and cooling expenses, but also make the workplace or home more comfortable. More efficient vehicles, for example, not only save immediately on fuel purchases, but also emit less pollutants, thus improving health and savings in medical costs to the individual and to society.

The integrated benefits of energy efficiency have been so striking that those states and nations that have invested significantly in these technologies have made considerable savings on energy costs and GHG emissions (Richter *et al.*, 2008).

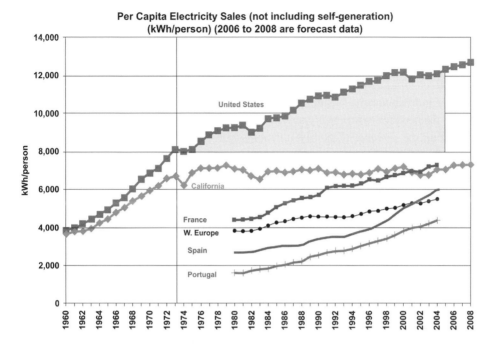

Figure 11.6 Per capita energy consumption in California, the United States, France, Spain, Portugal and Western Europe overall. Energy consumption in the most efficient states is fully 40% lower than the US national average. A number of nations, such as Denmark, are far more efficient than even the leading US states, indicating strongly that a wealth of further innovations are possible if investment and deployment of energy-efficient technologies are adopted. Source: California Energy Commission.

The adoption rate of energy-efficient technologies and energy management practices varies widely in different regions of the USA, with New York and several other states *more than 40% more efficient than the national average* on a per capita basis (Figure 11.6). According to statistics provided by the California Utilities Commission and the California Energy Commission, savings on energy in California in 2007, relative to the national average, amounted to more than $400 per person per year and are due to documented savings in energy (Richter *et al*., 2008).

The US, through the long-standing efforts of the EPA and the Department of Energy, have developed and facilitated the adoption of a number of energy efficiency practices. The 'Green Light, EnergyStar' programs and many of the US housing codes are derived from this experience. In Germany, a similar effort, GreenFreeze, is credited with similar savings both financially and in terms of GHG emissions. In total, energy efficiency investments save the USA over $170 billion annually, an amount that is rising with increasing energy costs (Richter *et al*., 2008). While these savings are impressive, a wealth of data indicates that far greater savings could be realised if these programs were expanded to a greater number of appliances and lighting systems, and if current standards were to be made more stringent. The most efficient technologies consistently provide remarkable levels of savings, often repaying their added cost in mere weeks or months, and then providing those savings year after year (Richter *et al.*, 2008).

These energy savings are *not* simply of local benefit, providing much-needed savings for individuals and businesses. In regions with aggressive energy efficiency programs, the need for new power plants has been reduced significantly – in some locations permitting the *removal* of poorly functioning or expensive fossil fuel supply options (Richter *et al.*, 2008).

Energy efficiency has provided, and continues to be, a tremendously cost-effective opportunity to reduce the need for new power generation and GHG generation. In many cases, investments in energy efficiency (Kammen *et al.*, 2008) can be made at near-zero or even *negative* cost when health – added to worker productivity, or security or other 'co-benefits' – are taken into account (Kammen and Pacca, 2004).

Opportunities for energy efficiency come in a great many technologies and practices. California has been a consistent leader in developing and deploying energy efficiency and, in the early 1980s, took the innovative step to *decouple* revenues and total sales from its investor-owned utilities. As a result in a decoupled system (Granade *et al.*, 2009), revenues are determined by a process of matching predicted and observed energy sales, with the effective price of electricity adjusted to meet an expected revenue target. This innovation has put energy efficiency and conservation on an 'even footing' with new generation, and has, in fact, institutionalised energy conservation and efficiency. The reason is that the value of efficiency is now equivalent to a new generation on a kilowatt-to-kilowatt comparison and, in fact, energy savings are generally superior due to the avoided costs of added power generation, operation and maintenance.

New energy efficiency innovations are taking place all the time, and should be featured prominently in the technical and economic assessments conducted by the EPA. On 27 September 2007, for example, the California Public Utilities Commission voted to enhance energy efficiency performance standards by adding a new incentive program. This program returns a portion of the financial savings from deploying energy efficiency (e.g. compact fluorescent lighting, improved efficiency water heating and space conditioning) innovations as a monetary incentive to the investor-owned utilities based on the level of end-user (ratepayer) efficiency.

The opportunities to continue and expand the deployment of energy efficient technologies are vast. The evolution of solid-state lighting, for example, has been sufficiently promising that the Office of Energy Efficiency and Renewable at the US Department of Energy has now set a technology-based goal of lights that are fully 50% more efficient than what we have today, and result in a decrease in total electricity consumption of 10%. Sandia National Laboratory projects that advances in solid-state lighting will save the USA over 70 GW of supply capacity, more than 100 million tons of carbon emissions annually, and save more than $42 billion per year.

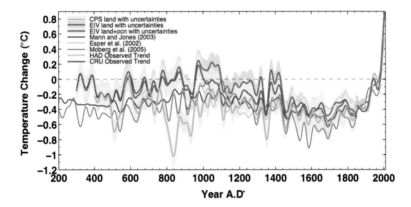

Plate 1 Northern hemisphere temperature change since 200 AD (reconstructed for the period up to the initiation of actual measurement). Source: http://www.copenhagendiagnosis.com. (*See Fig. 1.6a.*)

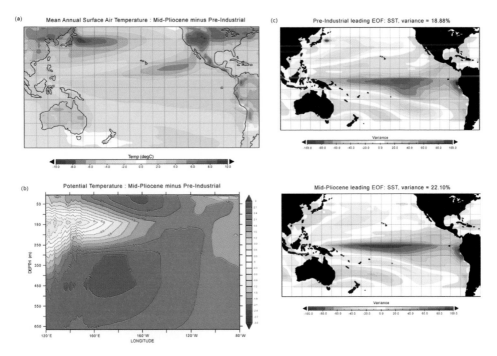

Plate 2 HadCM3 coupled simulations for the mid-Pliocene (MP) and the pre-industrial (PI) period: (a) mean annual surface air temperature (°C) for the MP minus PI over the Pacific, with MP surface temperatures 2–3°C warmer in the EEP; (b) mean annual potential temperature (°C) to 650 m depth along 120°E to 80°W, averaged over 5°N to 5°S, for the MP minus PI, highlighting the warmer sub-surface ocean temperatures and upwelling water in the EEP; and (c) leading empirical orthogonal function (EOF) for December, January, February and March Pacific SSTs for the PI (top) and the MP (bottom), showing a ~15% increase in ENSO variability in the mid-Pliocene simulation from the pre-industrial period. (*See Fig. 2.2.*)

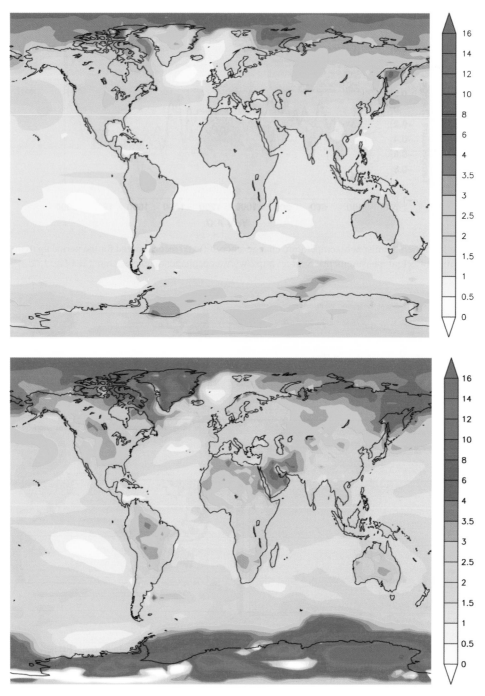

Plate 3 (Top) Temperature response to an increase in CO$_2$ from 280 to 400 p.p.m., as traditionally calculated. (Bottom) Earth System sensitivity to the same CO$_2$ forcing. Source: Lunt *et al.* (2010). Reprinted by permission from Macmillan Publishers Ltd: *Nature Geoscience*, copyright 2010. (*See Fig. 1.11.*)

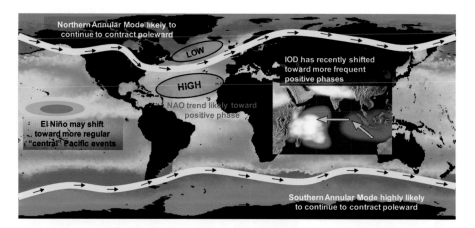

Plate 4 Schematic diagram showing the major modes of climate variability and how they are likely to change in the future. The high-latitude modes have already undergone significant change over the past century. Trends in the tropical modes (ENSO, IOD) have been detected in the more recent climatological record. (*See Fig. 2.3.*)

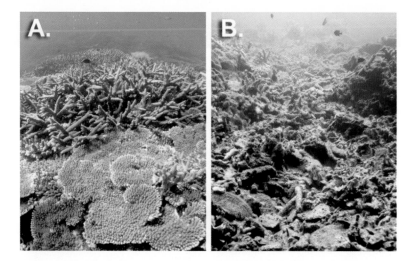

Plate 5 Coral reefs face an uncertain future if their ability to form carbonate skeletons is diminished under high atmospheric carbon dioxide. A. Reef-building coral reefs like this one at Heron Island (on the southern Great Barrier Reef) have relatively healthy populations of reef-building corals and red coralline algae that are able to maintain the three-dimensional framework of the reef. This framework provides habitat for an estimated million species and ultimately food, livelihoods and coastal protection to hundreds of millions of people worldwide. B. Increasing concentrations of atmospheric carbon dioxide, however, will lead to a sharp decline in the rate of calcification, causing a decrease in the ability of reefs to calcify and maintain reefs relative to the rate of physical and biological erosion. Under these circumstances, reefs may crumble and disappear. This example shows a section of a reef near Karimunjawa in the Java Sea, Indonesia, where the reef framework has been weakened by the loss of reef calcifiers. While a different set of drivers (i.e. poor water quality and destructive fishing) lie behind the loss of coral in this particular example, the outcome may be similar to that anticipated under the impact of warming and acidifying seas on corals, coralline algae and reef calcification if current rates of CO_2 emissions continue. Source: modified from Hoegh-Guldberg *et al.* (2007). (*See Fig. 2.4.*)

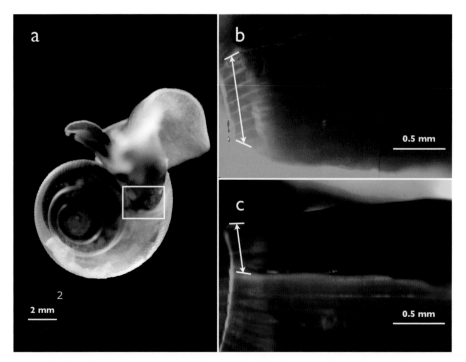

Plate 6 Representative example of live Arctic *Limacina helicina* (a) stained with calcein, and subsequently maintained at pH_T 8.09 (b) and 7.8 (c). Most calcification occurs near the shell opening (white rectangle). The arrow indicates the five days linear extent of the shell. Quantitative estimates obtained with a radiotracer indicate a significant 30% reduction of the calcification rate. Source: Comeau *et al.* (2009) *Biogeosciences* (open access). (*See Fig. 2.5.*)

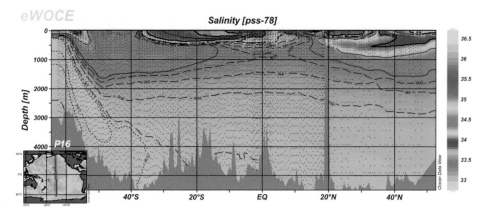

Plate 7 An oceanographic section of salinity from top to bottom from the Pacific Ocean from the Southern Ocean to Alaska. See the insert panel for the precise location of the section. Sources: http://www.ewoce.org/ and Schlitzer (2000). 'PSS' stands for 'practical salinity scale', using the UNESCO 1978 equation between salinity and conductivity. Salinity is dimensionless and reported in parts per thousand. (*See Fig. 2.7.*)

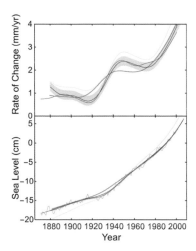

Plate 8 (Top) Observations-based rate of sea-level rise (with tectonic and reservoir effects removed, red) compared to that predicted by Eq. 1 (grey) and a refined version (blue with uncertainty estimate) using observed global mean temperature data. Also shown is the refined estimate using only the first half of the data (green) or the second half of the data (light blue). (Bottom) The integral of the curves in the top panel, i.e. sea level proper. In addition to the smoothed sea level used in the calculations, the annual sea level values (light red) are also shown. The dark blue prediction almost obscures the observed sea level due to the close match. Source: Vermeer and Rahmstorf (2009). (*See Fig. 3.4.*)

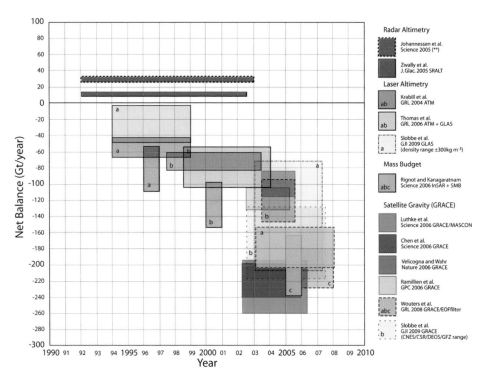

Plate 9 Results from the recent large area total balance measurements, placed into common units and displayed, versus the time intervals of the observations. The heights of the boxes cover the published error bars or ranges in mass change rate over those intervals. Source: AMAP (2009). (*See Fig. 3.7.*)

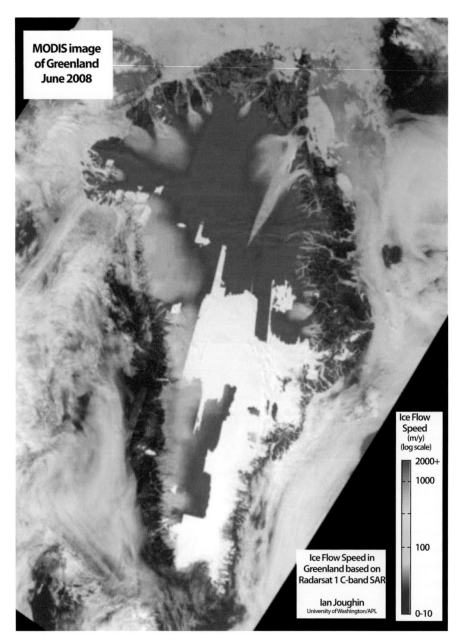

Plate 10 Velocity (m yr⁻¹) mosaic of the Greenland Ice Sheet derived from radar satellite data. The mosaic shows how the slow-flowing inland ice transitions into the fast-moving outlet glaciers, and the large variability of flow along the margin. Source: I. Joughin and M. Fahnestock, unpublished, presented in AMAP (2009). (*See Fig. 3.6.*)

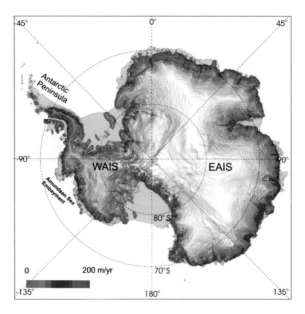

Plate 11 Surface topography (shaded grey) and ice velocities (colour) for Antarctica. Purple to red colours indicate ice streams where flow is generally dominated by sliding at the bed. Source: adapted from Bamber *et al.* (2007). (*See Fig. 3.9.*)

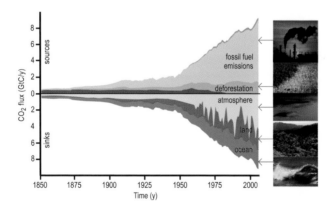

Plate 12 Changes in the global carbon budget from 1850 to 2000. CO_2 emissions to the atmosphere (sources) as the sum of fossil fuel combustion, land-use change, and other emissions (upper); the fate of the emitted CO_2, including the increase in atmospheric CO_2 plus the sinks of CO_2 on land and in the oceans (lower). Flux is in gigatonnes: Gt yr^{-1} carbon. For the deforestation source the light brown colour refers to tropical forests and the dark brown temperate forests. Sources: adapted from Canadell *et al.* (2007), with additional data from C. Le Quéré and the Global Carbon Project. Copyright 2007 National Academy of Sciences, USA. (*See Fig. 4.2.*)

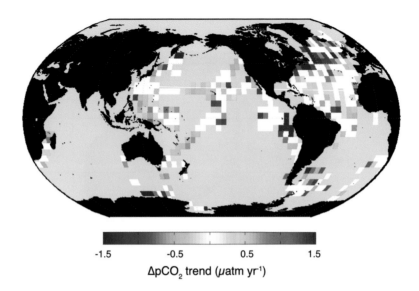

ΔpCO_2 trend (μatm yr^{-1})

Plate 13 Map of long-term trends in the difference of the partial pressure of CO_2 between the atmosphere and ocean surface. A positive value (warm colours) indicates that the oceanic pCO_2 is increasing at a faster rate than that in the atmosphere, i.e. indicating a reduction of the oceanic sink strength. Negative values (cold colours) indicate an increase in the oceanic sink strength. Source: figure from Oberpriller and Gruber, personal communication. (*See Fig. 4.6.*)

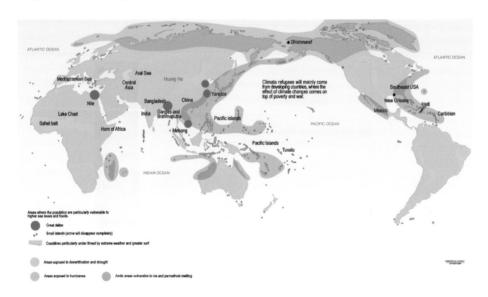

Plate 14 Regions predicted to be particularly vulnerable to different aspects of climate change, such as higher sea levels/floods, extreme weather, desertification and drought, ice/permafrost melting and hurricanes. Source: http://maps.grida.no/go/graphic/fifty-million-climate-refugees-by-2010. Reproduced with permission of UNEP/GRID-Arendal, cartography by Emmanuel Bournay. (*See Fig. 5.1.*)

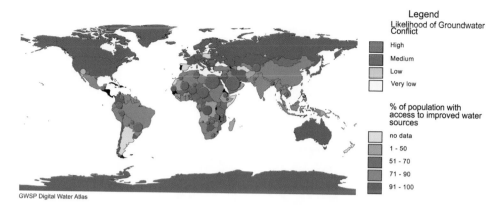

GWSP Digital Water Atlas

Plate 15 Figure showing the percentage of a population in a given country with access to improved water resources and identifying regions where there is a possible likelihood of groundwater conflicts. Source: modified from GWSP Digital Water Atlas (2008). Map 23: Likelihood of Groundwater Conflict 1 (V1.0). Available online at http://atlas.gwsp.org. (*See Fig. 5.8.*)

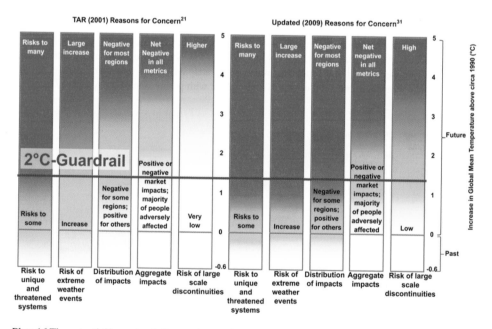

Plate 16 The potential impacts of climate change shown as a function of the rise in global average temperature. Zero on the temperature scale corresponds approximately to 1990 average temperature (i.e. 0.6 °C above pre-industrial levels), and the bottom of the temperature scale to pre-industrial average temperature. The level of risk or severity of potential impacts increases with the intensity of shading. The panel on the left is from Smith *et al.* (2001) as published in the IPCC Third Assessment Report. The panel on the right is an updated version from Smith *et al.* (2009), using the same methodology as the Third Assessment Report, based on expert judgement. The 2 °C guardrail is shown for reference. (*See Fig. 5.9.*)

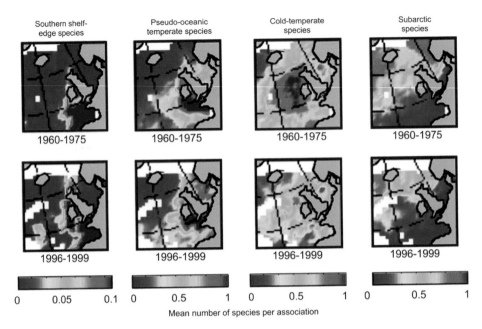

Plate 17 Changes in phytoplankton abundance in the Eastern North Atlantic at the end of the twentieth century based on the Continuous Phytoplankton Recorder database (Beaugrand *et al.*, 2002). Warm colours indicate high abundance and cool colours indicate low abundance. Both cold water-adapted species (cold-temperate and subarctic) and warm-water adapted species (southern shelf-edge and pseudo-oceanic temperate) moved northward. Reproduced with permission of AAAS. (*See Fig. 6.2.*)

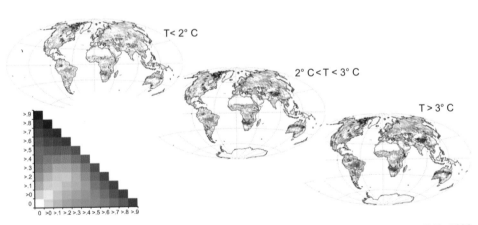

Plate 18 Probability of exceeding critical levels of change for wildfire frequency between 1961–1990 and 2071–2100 for three levels of global warming. Critical change is defined where the change in the mean of 2071–2100 exceeds ±1 σ of the observed (1961–1990) interannual variability (see inset for scale of probability of critical change: red for increase (vertical axis), green for decrease (horizontal axis), mixed colours for both exceeding the critical level by +1 σ as well as by −1 σ). Colours are shown only for grid cells with less than 75% cultivated and managed areas. Source: modified from Scholze *et al.* (2006). Copyright 2006 National Academy of Sciences, USA. (*See Fig. 6.6.*)

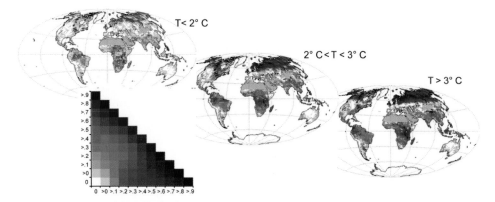

Plate 19 Same as Figure 6.6 but for freshwater runoff (blue for increase, red for decrease, mixed colours for both exceeding the critical level by +1σ as well as by −1 σ). Grey areas denote grid cells with less than 10 mm yr⁻¹ mean runoff for 1961–1990. Source: modified from Scholze *et al.* (2006). Copyright 2006 National Academy of Sciences, USA. (*See Fig. 6.7.*)

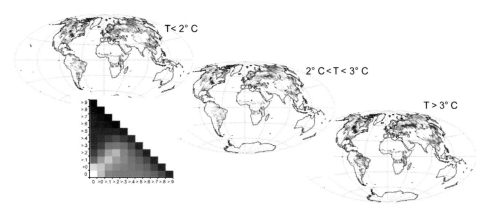

Plate 20 Biome change from forest to non-forest (blue) or vice versa (green). Colours are shown only for grid cells with less than 75% cultivated and managed areas. Source: modified from Scholze *et al.* (2006). Copyright 2006 National Academy of Sciences, USA. (*See Fig. 6.8.*)

Plate 21 Extant examples of reefs from the Great Barrier Reef that are used as analogues for the eco-logical structures anticipated by Hoegh-Guldberg *et al.* (2007) under increasing amounts of atmos-pheric CO_2. (A) If concentrations of CO_2 remain at today's level, many coral dominated reefs will survive although there will be a compelling need to increase their protection from local factors such as deteriorating coastal water quality and overfishing. (B) If CO_2 concentrations continue to rise as expected, reefs will become less dominated by corals and increasingly dominated by other organisms such as seaweeds. (C) If CO_2 levels continue to rise as we burn fossil fuels, coral reefs will disappear and will be replaced by crumbling mounds of eroding coral skeletons. In concert with the progression from left to right is the expectation that much of the enormous and largely unexplored biodiversity of coral reefs will disappear. This will almost certainly have major impacts on the tourist potential of coral reefs as well as their ability to support fisheries, both indigenous and industrial. Source: modi-fied from Hoegh-Guldberg *et al.* (2007). (*See Fig. 6.11.*)

Plate 22 Great human monuments of the past demonstrate a capability to take a long-term perspective and to organise societies to achieve long-term goals. (Left to right) The Great Wall of China (image used under license from Big Stock Photo); Salisbury Cathedral, UK (image copyright FrankfurtDave, 2010. Used under license from Shutterstock.com); the ancient city of Bagan, Myanmar (image used under license from Big Stock Photo). (*See Fig. 10.1.*)

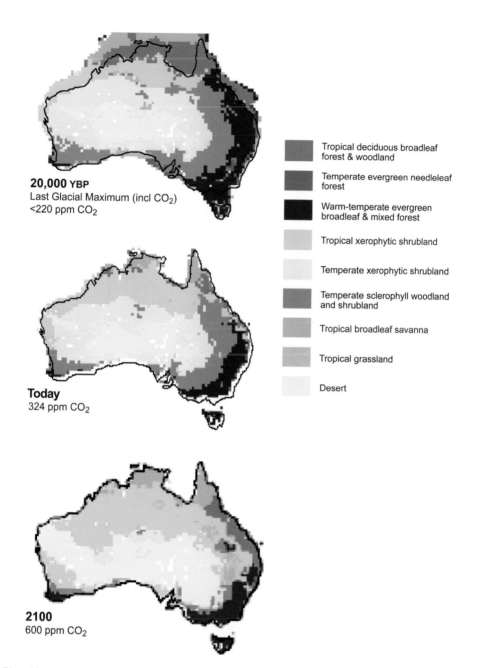

20,000 YBP
Last Glacial Maximum (incl CO_2)
<220 ppm CO_2

Tropical deciduous broadleaf forest & woodland

Temperate evergreen needleleaf forest

Warm-temperate evergreen broadleaf & mixed forest

Tropical xerophytic shrubland

Temperate xerophytic shrubland

Temperate sclerophyll woodland and shrubland

Tropical broadleaf savanna

Tropical grassland

Desert

Today
324 ppm CO_2

2100
600 ppm CO_2

Plate 23 Modelled historical Australian biome distribution. Model data show distribution at the Last Glacial Maximum (approximately 200 p.p.m. CO_2), the 'present' (>324 p.p.m. CO_2) and simulated for 2100 (600 p.p.m. CO_2 scenario) using the dynamic global vegetation model BIOME 4.0. The details of the simulated biome distributions in 2100 are sensitive to the nature of the particular climate and vegetation models used for the projections; shown is a typical example of change. Source: S. Harrison, Bristol University, personal communication, modified from Steffen *et al*. (2009). (*See Fig. 6.14.*)

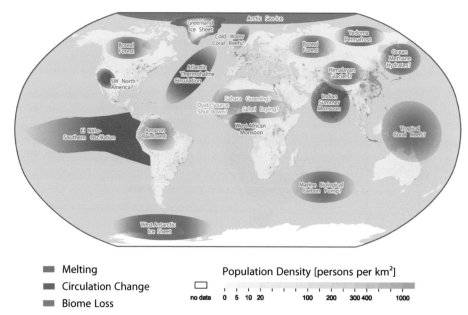

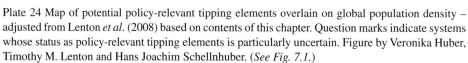

Plate 24 Map of potential policy-relevant tipping elements overlain on global population density – adjusted from Lenton *et al*. (2008) based on contents of this chapter. Question marks indicate systems whose status as policy-relevant tipping elements is particularly uncertain. Figure by Veronika Huber, Timothy M. Lenton and Hans Joachim Schellnhuber. (*See Fig. 7.1.*)

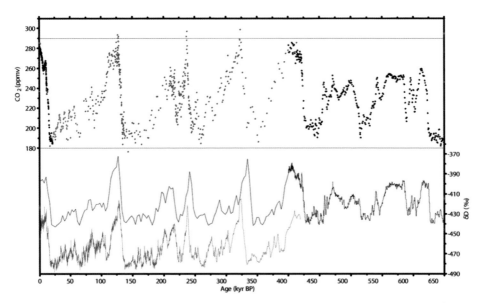

Plate 25 A composite CO_2 record over six and a half ice age cycles, back to 650 000 years b.p. (before present) (Siegenthaler *et al.*, 2005). The record results from the combination of CO_2 data from three Antarctic ice cores: Dome C (black), 0 to 22 k years b.p. and 390 to 650 k years b.p.; Vostok (blue), 0 to 420 k years b.p., and Taylor Dome (light green), 20 to 62 years b.p. Black line indicates δD from Dome C, 0 to 400 k years b.p. and 400 to 650 k years b.p. Blue line indicates δD from Vostok, 0 to 420 k years b.p. δD (deuterium) is a proxy for temperature. Source: references to individual ice core data are found in Siegenthaler *et al.* (2005). (*See Fig. 6.12.*)

Plate 26 Climate change challenges us to think about our relationship to the rest of nature. Do we have the right to eradicate other living species with whom we share the planet? (Left to right): Polar bear (image used under licence from Big Stock Photo); Hooded grebe (image copyright James C. Lowen; reproduced with permission); Mountain pygmy possum (image copyright Glen Johnson; reproduced with permission). (*See Fig. 10.4.*)

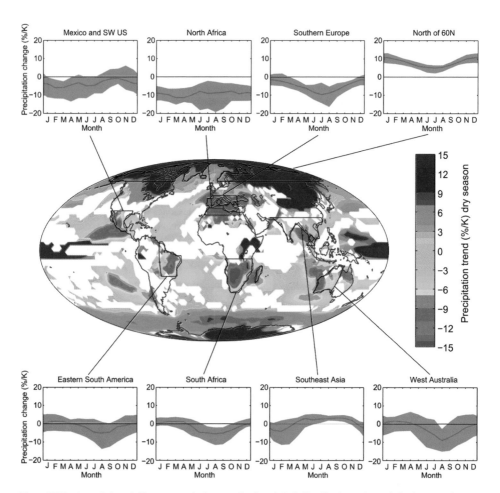

Plate 27 Projected decadally averaged changes in the global distribution of precipitation per degree of warming in the dry period, expressed as a percentage of change from a 1900–50 baseline period. Source: Solomon *et al*. (2009), which includes more details on the methods. (*See Fig. 10.2.*)

A vigorous energy efficiency research program is needed to spur these advances – but as we have seen over the past decades, will pay for itself many times over in energy savings, avoided supply deficits and in greenhouse gas emissions.

11.6 Low-carbon energy supply systems

The last few years have seen a tremendous expansion in interest in renewable energy supply technologies. Technological and cost advances in solar, wind, biofuel, geothermal and ocean energy systems have made some renewable energy supply options competitive with fossil fuel technologies in an increasing number of locations.

Wind energy, in particular, is now often directly cost competitive and, at times, is a least-cost supply option. Prices for delivered wind energy are as low as 3.2 cents per kWh for the 120 MW San Juan Mesa wind farm in New Mexico. Ownership and financial structures are particularly important for wind projects, with privately owned projects averaging 4.95 cents per kWh including the federal production tax credit (PTC), and 6.56 cents per kWh without the PTC. Investor-owned utility projects with corporate financing averaged 3.53 cents per kWh including PTC, and 5.9 cents per kWh without. Projects with public utility ownership and project financing can be as inexpensive as 3.43 cents per kWh including renewable energy production incentives, and 4.89 cents per kWh without (Wiser and Bolinger, 2007). The recent volatility in natural gas prices makes renewables an even better relative deal. Utility-supported projects can be cheaper where the projects can obtain more favourable grid connection costs.

In the US, the performance of renewable energy technologies, the need to respond to climate change and the job creation potential of a new energy sector – including, but not limited to, wind – has encouraged 29 states and the District of Columbia to enact Renewable Energy Portfolio Standards (RPS) (Figure 11.7), which each call for a specific percentage of electricity generated to come from renewable energy. Some of the most aggressive state standards call for over 20% of total electricity generation to come from renewable sources by 2020; in 2007, these technologies generated almost 150 TWh of electricity. States that are likely to meet their RPS goals – including Texas, California and Colorado – have, in turn, increased their call for renewable installations based on the performance, cost-effectiveness and benefits of supply diversity that renewables provide.

An important feedback effect exists in the call for, and installation of, clean energy supply and pollution control technologies. As more solar, wind, biofuel and trace-gas emissions control technologies (e.g. NO_x) have been constructed and deployed, the price has consistently fallen by 10%–20% per doubling of the total number of units ordered (Figure 11.8). This effect, termed the 'learning curve', has held remarkably constant over a wide range of technologies for many years of technology experience (Duke and Kammen, 1999). An outcome of this process of industrial learning is that, as we invest more in the clean energy sector, the products we desire have become increasingly affordable, further increasing their significance in the market. This effect has occurred for technologies that

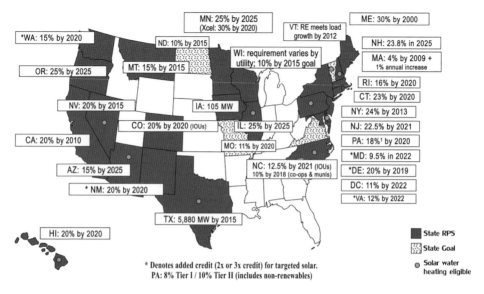

Figure 11.7 Map of states with renewable energy portfolio standards as of September 2007; 29 states and the District of Columbia have enacted or voted to adopt either an RPS or a state goal. These plans represent a diversity of approaches and levels, but each reflects a commitment to clean and secure energy that could be emulated at the federal level. In addition, 13 states have specific measures to increase the amount of solar photovoltaic power in use. These range from specific solar energy targets, to double (Maryland) or up to triple credit (Delaware, Maine and Nevada) for solar. Electricity rates fell in portions of Colorado after voters approved legislation establishing a 20% RPS limited to the investor-owned utilities. The RPS target in Maine is complicated because natural gas was counted as low-carbon/renewable. Source: data from http://apps1.eere.energy.gov/states/maps/renewable_portfolio_states.cfm.

can be mass-produced. Our investment and deployment of energy efficiency, renewable energy and pollution control technologies are important drivers of future innovation.

11.7 The feed-in tariff

While the Renewable Energy Portfolio Standard has been the market-pull policy of choice in the USA, the most common global electricity policy to date has been the feed-in tariff (FIT). By mid 2005, some type of direct capital investment subsidy, rebate or grant was offered to renewable energy in at least 30 countries (REN21, 2009). By far the dominant mechanism, however, has been the FIT. FITs provide fixed-rate, fixed-duration contracts for renewable energy delivery, at rates determined to be sufficient for profitability for the emerging energy technology. As of mid 2010, FITs had been enacted in at least 45 countries and 18 states, provinces or territories (Mendonça, 2007; Rickerson *et al.*, 2007). As shown in Figure 11.9, renewable energy's share of new global electricity generation has risen in line with the increase in FIT and RPS policies.

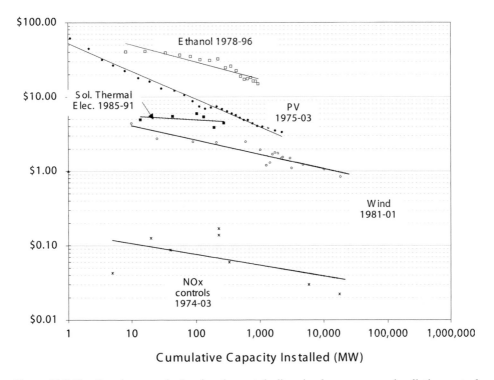

Figure 11.8 The 'learning curve', showing the cost declines in clean energy and pollution control technologies that accompany expanded commercial production. A 10–20% decrease in per-unit cost typically accompanies each doubling of cumulative production. Source: Nemet and Kammen (2007).

FIT policies have been credited with providing the most cost-effective dissemination of solar and wind energy. An example has been the comparison between California and Germany. In both, strong support for solar energy has been expressed and demonstrated. However, despite there being 70% more solar insolation in California, where there is a combination of a RPS and a targeted solar home and business subsidy package to encourage solar energy introduction, the state has only one-fifteenth the installed solar power than does Germany, which relies on a FIT for encouraging the technology (Mendonça, 2007). The primary concerns with respect to FITs are (i) setting the tariff correctly and not overpaying, as has been asserted for both Germany and Spain in the case of solar energy, and (ii) finding the public will to devote ratepayer funds or funds from the public coffers to the FIT payments.

11.8 The job creation dividend from greenhouse gas abatement

A number of analysts have charted an additional benefit of developing the clean energy options of efficiency and renewable energy technologies: job creation. Kammen *et al.* (2004), for example, conducted a study of job growth in the clean energy industry across

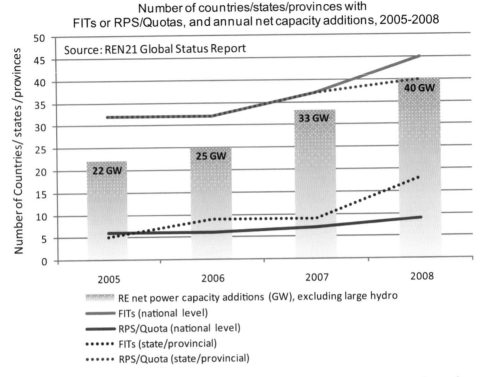

Figure 11.9 Number of countries/states/provinces with feed-in tariffs or RPS/quotas, and annual net electric capacity additions (excluding large hydropower), 2005–08. Sources: REN21 (2009); United Nations Environment Programme and New Energy Finance Limited (2009).

the USA relative to that seen in the fossil-fuel sector. They found that, on average, three to five times as many jobs were created by a similar investment in renewable energy than when the same investment was made in fossil-fuel energy systems. This 'green jobs dividend' is based on empirical studies of company hiring (Wei *et al.*, 2010).

A job creation model for the US power sector from 2009 to 2030 has also been examined (Wei *et al.*, 2010). The model synthesises data from 15 job studies covering renewable energy (RE), energy efficiency (EE), carbon capture and storage (CCS), and nuclear power. Job losses in the coal and natural gas industry were modelled to project net employment impacts. Benefits and drawbacks of the methodology are assessed, and the resulting model is used for job projections under various RPS, EE and low-carbon energy scenarios. It was found that all non-fossil fuel technologies (RE, EE, low carbon) create more jobs per unit energy than coal and natural gas. Several approaches generate over 500,000 full-time equivalent jobs by 2020 and a combination of RE, EE and low carbon approaches can yield three million job-years by 2025.

A key result that emerges from this work can be seen in Figure 11.10. Across a range of scenarios, the renewable energy sector generates more jobs than the fossil fuel-based

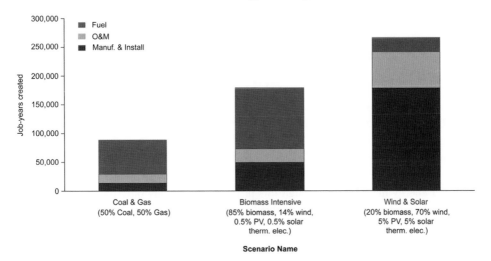

Figure 11.10 Comparison of the estimated employment created by meeting the equivalent of 20% of current US electricity demand via an expansion of fossil or renewables-based electricity generation. These totals use the jobs per megawatt numbers from Kammen *et al.* (2004) and Wei *et al.* (2010). These scenarios are for different fuel mixtures that could comprise state or federal Renewable Energy Portfolio Standards. The use of biofuel assumes that a range of biomass sources are either mixed with coal and combusted as a solid fuel, or as a gasified feedstock that is combined with natural gas. In either case, the net greenhouse gas emissions are reduced over the coal-only or gas-only case because of the amount of biomass that is regrown in subsequent years (based on US Department of Energy work in The Billion Ton Feedstock Supply; Perlack *et al.*, 2005). Source: Wei *et al.* (2010).

energy sector per unit of energy delivered (i.e. per average megawatt). In addition, it was found that supporting renewables within a comprehensive and coordinated energy policy which also supports energy efficiency and sustainable transportation will yield far greater employment benefits than supporting one or two of these sectors separately.

The US Government Accounting Office conducted its own study (US Government Accountability Office, 2004) of the job creation potential of a clean energy economy. In an important assessment of rural employment and income opportunities, they found that:

… a farmer who leases land for a wind project can expect to receive $2,000 to $5,000 per turbine per year in lease payments. In addition, large wind power projects in some of the nation's poorest rural counties have added much needed tax revenues and employment opportunities.

11.9 Combining technology and financial innovation: the next wave of greenhouse gas abatement

As states and cities explore GHG emission reduction opportunities, new and important financial models are also emerging. The City of Berkeley, California – and now several other cities – provide examples that have already attracted international attention.

The financing mechanism is loosely based on existing 'underground utility districts' where the city serves as the financing agent for a neighbourhood when they move utility poles and wires underground. In this case, individual property owners would contract directly with qualified private solar installers and contractors for energy efficiency and solar projects on their building. The city provides the funding for the project from a bond or loan fund that it repays through assessments on participating property owners' tax bills for 20 years. Cities may also be able to aggregate bonds, and state governments can facilitate this program in a number of ways.

No property owner would pay an assessment unless they had work done on their property as part of the program. Those who choose to pay for energy efficiency first, and then solar and energy installations through this program, would pay only for the cost of their project, interest and a small administrative fee.

The Financing District solves many of the financial hurdles facing property owners. First, there is little upfront cost to the property owner. Second, the total cost of the solar system and energy improvements may be less when compared to financing through a traditional equity line or mortgage refinancing because the well-secured bond will provide lower interest rates than are commercially available. Third, the tax assessment is transferable between owners. Therefore, if an individual sells their property before the end of the 20-year repayment period, the next owner takes over the assessment as part of their property tax bill.

This mechanism, announced publicly on 23 October 2007, has attracted statewide attention as other cities – and now the state government and federal government – have been looking to find ways to expand the Financing District model statewide.

Fuller *et al.* (2009) developed a cash flow model to examine the efficacy of this mechanism, which shows the cash flows from energy efficiency improvements that save 5% of electricity use and 25% of gas use from a $4000 investment. An annual 2% real increase in energy prices and 2% inflation rate is assumed, with a Property Assessed Clean Energy (PACE) term of 20 years. The cash flow is positive in every period (Fuller *et al.*, 2009).

While there is a dearth of large-scale studies using measured data on energy savings from multiple efficiency measures in homes, and savings potential varies widely with climate and local energy prices, this modelled level of savings is achievable according to the experience of programs such as Midwest Energy in Kansas and Long Island Green Homes in New York, which have cash flow-positive program requirements, and is supported by the experience of the Weatherization Assistance Program (Fuller *et al.*, 2009).

11.10 A roadmap to low-carbon energy efficiency and clean energy supply options

A useful view of low-carbon energy options has been developed by the Vattenfall McKinsey & Company. Such a roadmap can focus and contextualise considerations of alternative power generation proposals and their related costs. This presentation includes

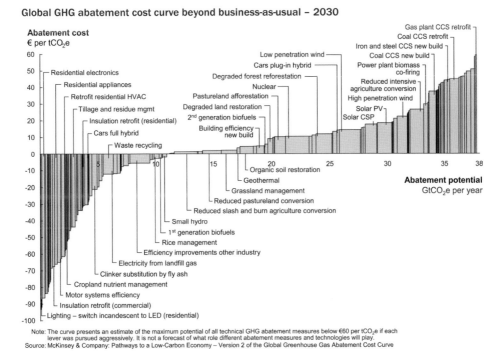

Global GHG abatement cost curve beyond business-as-usual – 2030

Note: The curve presents an estimate of the maximum potential of all technical GHG abatement measures below €60 per tCO₂e if each lever was pursued aggressively. It is not a forecast of what role different abatement measures and technologies will play.
Source: McKinsey & Company: Pathways to a Low-Carbon Economy – Version 2 of the Global Greenhouse Gas Abatement Cost Curve

Figure 11.11 Estimated carbon reduction possibilities in euros per ton of carbon dioxide equivalent (EUR/t CO_2-e). The vertical axis shows the cost (or benefit) of the specific mitigation option, while the horizontal axis shows the estimated mitigation potential in gigatons of CO_2 in the target year (2030). Source: redrawn from Enkvist *et al.* (2007).

a comprehensive view of greenhouse gas (GHG) reduction measures, whereas power is the sector with the largest contribution. The abatement cost curve in Figure 11.11 shows:

- estimated costs (some of which are negative, implying overall *savings* to the economy through the adoption of clean energy options) of potential climate solutions and the amount of carbon they may be able to offset, or avoid, by a specific time (in this case, 2030);
- the potential for a wide range of energy options to play a significant role in CO_2 reduction which, in turn, leads to distinct technology and management strategies that individual states may pursue;
- a clear conclusion that we will need to adopt policies that support and encourage development and deployment of a *diversity* of clean energy options.

The key insight from this analysis is that a portfolio of low-carbon opportunities exists and that state and local governments – if supported by analysis and policies at the federal level – can enact a wide range of cost-effective dissemination programs. While these will require a combination of research and development to become major components of the economy, the tools to begin this process exist today. The abundance of cost-effective energy efficiency options and of low-carbon sources in the USA is a rich resource from which we can draw.

Transportation is possibly the sector of energy use and carbon emissions that presents the greatest challenge with respect to emissions reductions. In the following section, some of the dramatic changes possible in this area are described.

11.11 Decarbonising transportation: integrating science and policy

Transport is currently responsible for 13% of global GHG emissions and it contributes 23% of global CO_2 emissions from fuel combustion (Creutzig and Kammen, 2010). Global transport-related CO_2 emissions are expected to increase by 57% in the period 2005–30 – making this the fastest growing sector globally (Richter *et al.*, 2008). At the same time, there is consensus in science and politics that global GHG emissions must be reduced rapidly and deeply to avoid dangerous anthropogenic interference with the climate system (Chapter 8). It is clear that the transport sector will need to be central to mitigation efforts. One important contribution towards this goal can be to reduce the carbon content of fuels or, more generally, vehicle propellants (Figure 11.12). In this section, the potential of biofuels and electric mobility to decarbonise passenger vehicle transportation is investigated. As with most areas of a sustainable energy economy, large improvements are possible, but they require a 'systems science' approach that works across disciplines, and considers traditional vehicle approaches and stationary power.

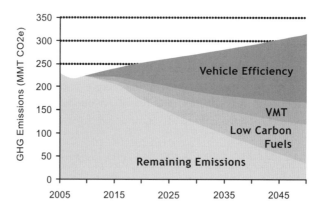

Figure 11.12 A schematic of carbon reduction options in the transportation sector, including vehicle efficiency and mode-switching to greater aggregate efficiency in mass-transit modes, reductions in vehicle miles travelled, and introduction of low-carbon fuels (green biofuels and electricity). Note: VMT – vehicle miles travelled. Source: the NRDC pathways discussed in Figure 11.3 (Natural Resources Defense Council, 2009). Reprinted with permission of NRDC, the Natural Resources Defense Council (http://www.nrdc.org).

Science, technology, policy, economics and cultural awareness must be utilised in concert with each other.

11.12 Innovations in response to challenges: from lead to carbon

During the late 1970s and early 1980s, the highest-profile environmental issue in the vehicle and fuel industries was the establishment of a ban on lead additives in petrol – encapsulated by the catchphrase *get the lead out*. After initial uncertainty and some opposition based on the fear that prices would rise and vehicle performance would suffer, the transition to unleaded fuels proved remarkably easy and effective. The average blood-lead level in the US population dropped by 75%, and the blood-lead levels of up to two million children were reduced every year to below toxic levels between 1970 and 1987 as the use of leaded petrol was reduced. In direct response to the reduction in atmospheric lead, the IQ levels of previously lead-exposed urban children increased (Creutzig and Kammen, 2010).

The US Congress also enacted the Corporate Average Fuel Economy (CAFE) regulations, a sustained effort to raise average vehicle efficiency standards in response to the 1973 Arab oil embargo. This measure increased vehicle mileage standards by more than 25%. Such examples demonstrate that ambitious, yet achievable, targets can be codified, enforced and adjusted as technological, economic and environmental needs change. These targets set a precedent for what is possible. In other words, technological innovation combined with economic and environmental necessity is altering the landscape of vehicle efficiency. Today's innovation is reminiscent of the effort to *get the lead out*, only this time the goal is to *get the carbon out* of transportation fuels. One policy measure that supports ambitious emission reduction targets is the low-carbon fuel standard.

The low-carbon fuel standard is a simple and elegant concept that targets the amount of GHGs produced per unit of energy delivered to the vehicle; i.e. the vehicle's so-called 'carbon intensity'. In January 2007, California Governor Arnold Schwarzenegger signed an Executive Order, which called for a 10% reduction in the carbon intensity of California's transportation fuels by 2020. Eight months later, a coalition at the University of California, Berkeley, responded with a technical analysis of low-carbon fuels that could be used to meet that mandate (Farrell *et al.*, 2006; Arons *et al.*, 2007). The report relies upon life-cycle analysis of different fuel types, taking into consideration the ecological footprint of all activities included in the production, transport, storage and use of the fuel.

If a low-carbon fuel standard were established, fuel providers would track the 'global warming intensity' (GWI) of their products and express it as a standardised unit of measure – the grams of CO_2 equivalent per megajoule (gCO_2-e per MJ) of fuel delivered to the vehicle. This value measures not only direct vehicle emissions but also indirect emissions, such as those induced by land-use changes related to biofuel production. The global warming intensity also provides a common frame of reference to compare propellants as diverse as petrol and electricity. Before discussing the GWI of biofuels and 'electromobility', let us contrast the low-carbon fuel standard with current policies on biofuels.

Unfortunately, the first biofuel policies were developed before the true global warming impacts – including indirect effects of biofuel production and life-cycle use (Searchinger *et al.*, 2008) – were known, with the main examples coming from the USA and EU. In the USA, two current policies promote biofuels: a US$0.51 tax credit per gallon of ethanol used as motor fuel; and a mandate that up to 7.5 billion gallons (5%–6% of total US fuel demand) of 'renewable fuel' be available at US petrol stations by 2012. The EU aims that by 2020 biofuels and electrified transportation will account for 10% of fuels used in the transportation sector.

Government policies to promote biofuels intend to improve environmental quality (e.g. to reduce the impact of global warming), and aim to support agriculture and to reduce petroleum imports. In practice, however, current government biofuel policies tend to function most directly as agricultural support mechanisms, involving measures such as subsidies or mandates for the consumption of biofuels. By contrast, the environmental impacts of biofuels – and, more specifically, the GHG emissions related to fuel production – are often not measured, let alone used to adapt financial incentives or to guide government regulation. Yield maximisation for a number of agricultural staple crops often involves high levels of fossil-fuel inputs (e.g. for fertilisers), further complicating the mix of rationales for biofuel support programs. It is important to apply a fairly broad framework on biofuel policies to avoid repeating past mistakes.

11.13 Sustainability and economic path dependency

The biofuel industry is growing rapidly and can be very profitable when world oil prices are high. Government policies to further subsidise, mandate and otherwise promote biofuels are being implemented, and more are proposed. Given the large investments in research and capital that continue to flow into the biofuels sector, it is time to carefully assess the types and magnitudes of the incentives that are meant to mitigate global warming. By engaging in this analysis, we can reward sustainable biofuel efforts, and avoid the very real possibility that the economy could be further saddled with the legacy costs of short-sighted investments.

Biofuels are often proposed as a solution to environmental problems, especially climate change. However, biofuels can have a positive or negative global warming impact relative to petrol, depending on the precise production pathway (Farrell *et al.*, 2006), as we will discuss in the next section. To distinguish between these two cases – and the myriad of other feedstock-to-fuel pathways as illustrated in Figure 11.13 – clear standards, guidelines and models are needed.

Many new fuels, feedstocks and processing technologies are now emerging and numerous others are under consideration. These are being developed as biofuel technologies *per se*; they are not merely adaptations of pre-existing agricultural production methods. If these developments can be managed to achieve high productivity while minimising negative environmental and social impacts, the next generation of biofuels could avoid the

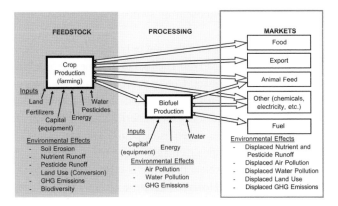

Figure 11.13 Simplified general biofuel pathway with inputs and environmental impacts. Many effects are displaced; i.e. they occur at different locations due to market-mediated forces. Source: adapted from Kammen *et al.* (2007).

disadvantageous properties of a number of current biofuels (e.g. low energy/low density, corrosiveness, poor performance at low temperatures).

A transparent set of data on the standards we wish biofuels to meet, as well as clear and accessible analytic tools to assess different fuels and pathways, are critical to efforts aimed at providing appropriate incentives for the commercialisation of cleaner fuels. This entire analysis, however, needs further elaboration.

11.14 What is the carbon impact of biofuels… and other new fuels?

Biofuels and related GHG emissions are a contentious issue, in both the political and research arenas. A variety of different GHG emission values have been reported, ranging from a 20% increase to a 32% decrease when switching from petrol to ethanol in the USA (Farrell *et al.*, 2006). A model has been developed, EBAMM (The Energy and Resources Group Biofuel Meta-Analysis Model; Farrell *et al.*, 2006; Kammen *et al.*, 2008), to compare and reconcile these different values. A major reason for inconsistencies was the choice of different system boundaries; i.e. the choice of which processes to include in biofuel GHG emission accounting, and which to exclude. Harmonisation of boundaries – for example, excluding emissions induced by human labour but including the displacement of GHG emissions by energy-valuable co-products of ethanol – brings the GWI of the different processes closer together. Any significant remaining uncertainty is due mostly to the unknown and not-well-studied effect of lime application (lime is added to correct the pH of acidic soils; it is applied only once, and it is crucial to account for GHG emissions over the full yield period). According to the updated EBAMM (http://rael.berkeley.edu/ebamm) ethanol produced using a CO_2-intensive refining process (e.g. a lignite-powered ethanol plant) has a marginally better GWI than petrol (i.e. 91 gCO_2-e per MJ instead of 94 gCO_2-e per MJ), while average ethanol production has a GWI of 77 gCO_2-e per MJ.

Biofuel generated by harvesting cellulose from switch grass is projected to have a GWI of only 11 gCO_2-e per MJ.

The EBAMM meta-analysis points out that not only specific processes, but also agricultural practices, largely determine the GWI. The fuel used to power biorefineries is decisive for the absolute climate change impact. Coal-powered biorefineries barely reduce GHG emissions (but shift emissions from petroleum to coal, thus decreasing energy dependency in OECD countries). Natural gas-powered biorefineries are already having a positive net effect; i.e. fewer GHG emissions than when using petrol. The highest potential in terms of GHG emissions is, however, in cellulosic ethanol.

From this discussion, it is already clear that there is substantial need to evaluate each fuel using a detailed life-cycle analysis. However, land-use changes further complicate matters. Recent studies indicate that expanding biofuel production induces large GHG emissions from land-use change for biofuels, in particular when biofuel production competes with other land uses such as the production of food. Indirect effects are difficult to evaluate but highly significant. Commodity substitutability and competition for land transmit land-use change across global markets; for example, when US ethanol production increases the global corn price, making it profitable to clear rainforests for additional corn or crop production in Brazil (Searchinger *et al.*, 2008). These market-mediated land-use change emissions are separated from the biofuel production process by several economic links as well as physical distance. Some studies claim that such indirect land-use changes induce GWI above petrol emissions on a century timespan (Searchinger *et al.*, 2008).

If grassland is converted to crops, both land conversion (e.g. by fire) and land cultivation cause significant emissions. For example, if one acre of land is devoted to bioethanol production, which involves the conversion of 0.6 acres of forest and 0.24 acres of grassland to agricultural land, then 30 metric tonnes of CO_2 are released. One acre produces approximately 400 gallons of ethanol per year, saving one tonne of CO_2 annually. Hence, the GHG payback time is 30 years (O'Hare *et al.*, 2009). Searchinger *et al.* (2008) estimated that GHG payback time is over 150 years in some cases. In particular, expansion of US bioethanol production will cause previously uncultivated land to be utilised for crop production, both in the USA and elsewhere – primarily in Brazil, China and India. Hence, there will be significant loss of pristine grasslands and forests, as well as lost opportunities for carbon sequestration on idle arable land. It is generally recognised that there are significant GHG emissions related to indirect land-use changes. However, the extent of this effect is disputed, as (i) model assumptions cannot easily be verified, and (ii) the system is highly complex; e.g. deforestation is multi-causal (there are also local drivers of deforestation; Chapter 9). The following factors produce major uncertainty:

- carbon emission factors related to agro-ecological zones and new land (i.e. the precise location of biofuel production, and the carbon content of the land prior to conversion to biofuel plants);
- future land-use trends, such as the total global demand on food production, which itself depends on population growth;

• policies and competition for different land-use types (e.g. the existence and effectiveness of rainforest protection measures).

Another issue is the accounting of time. To obtain a GWI, most studies average the total indirect emissions over the total fuel produced during a production period and add these to the direct emissions. This straight-line amortisation has been proposed for California's Low-Carbon Fuel Standard (Arons *et al.*, 2007). Hence, a unit of GHG emissions released today is treated as though it had the same consequences as one released decades in the future. Annual GHG flows are, in general, a poor proxy for economic costs; most climate change costs, as discussed in earlier chapters, are imposed by GHG stocks in the atmosphere (Chapter 4). Furthermore, consideration of long timeframes involves assumptions about technological innovation and land-use changes over that timeframe, including post-cultivation changes in land use. A proper accounting of time, recognising the physics of atmospheric carbon dioxide decay, significantly worsens the GWI of any biofuel that causes land-use change in comparison to fossil fuels (O'Hare *et al.*, 2009). The key point is that a lot of emissions appear due to land-use changes at the beginning of biofuel cultivation, while emission savings occur later. Emissions occur 'up front' and, as a result, cumulative warming – global warming produced by emissions over a fixed analytical horizon, e.g. 50 years – and associated damages in the near-term are more severe than future ones.

Biofuel production has also been criticised for competing with global food supply, and for raising global corn prices as a consequence. For the world's poor a marginal price increase can have devastating effects. The corn required to fill the fuel tank of a sport utility vehicle with bioethanol contains enough calories to feed one person for a year; the sport utility vehicle driver will often pay more for the corn (indirectly as fuel) than people in poor countries can afford. From a narrow market perspective, the starvation of the poor can in fact be an efficient market outcome, making bioethanol policies in the USA and EU even more questionable. To understand the relevance of policies in specific world regions, we should note that, for example, 40% of global corn (maize) production is in the USA, and that the indirect land-use impacts of shifting some of this production from human and animal feed to biofuel production can be significant (for two differing estimates of this indirect land-use effect, see Searchinger *et al.*, 2008 and Hertel *et al.*, 2009).

One way out of this problem is to decouple biofuel cultivation, first from food production by using waste products (second generation) and, in the long run, from land use; for example, by relying on biofuels produced from algae (third generation).

Overall, major uncertainties about the sustainability of current biofuel production persist. Indirect land-use change effects are too diffuse and subject to too many arbitrary assumptions to be useful for rule-making. To ascertain a minimum environmental quality of biofuels, a suggested low-carbon fuel standard can include evolving minimum criteria related to GHG emissions. One could start by placing restrictions on biorefineries, requiring improved agricultural practices such as conservation tillage, and in a few years time allow only biodiesel and biofuels of the second generation. The Roundtable on Sustainable Biofuels (http://cgse.epfl.ch/page65660.html) develops criteria according to which a third party could perform a life-cycle assessment of biofuels and certify the fuels according to established standards.

11.15 Electromobility

Biofuels represent a minor modification in vehicle propulsion. Electromobility is a more radical and rapidly evolving technological change that dates back to the nineteenth century. Electromobility requires not only a different propellant, but also different vehicle technology (an electric motor) and storage system (e.g. a battery). There are two main advantages of electromobility:

- an electric motor has 70%–80% well-to-wheel efficiency and, hence, is far superior to the combustion engine (with 15%–25% well-to-wheel efficiency);
- in principle, it is a straightforward process to get the carbon out of electromobility by increasing the deployment of renewable energies for electricity generation.

A significant challenge for large-scale electromobility is battery technology. Current batteries need to be improved in terms of storage capacity and also in terms of cost. All-electric cars must also be relatively light to reduce overall energy demand. Altogether, the electricity used by a battery-powered electric vehicle in California has a GWI value of 27 gCO_2-e per MJ (Kammen *et al.*, 2008), a considerable improvement on petrol and ethanol. Other comparable technologies, based on the current electricity mix and different storage media – such as compressed air or hydrogen – have, at present, a worse GWI than petrol (Creutzig *et al.*, 2009).

The evaluation of the GWI of electric cars is not a trivial issue. Rather than the GHG emissions of the average power plant, it is the marginal power plant (added when there is additional electricity demand) that must be evaluated for climate change impact. Potentially, car batteries can be used for demand management (for example, cars can be charged by wind energy at night, when there is no other electricity demand). Electromobility is not merely synonymous with electric cars, but also includes smaller vehicles such as electric bikes. For OECD countries, electric bikes are still relatively exotic. However, in China – by 2009 the world's largest market for cars – more electric bikes than conventional cars are sold.

It is important to consider the full spectrum that lies between conventional petrol-operated cars and all-electric cars. For example, average fuel savings in the US can easily be doubled (and fleet emissions halved) by deployment of existing technological advances, weight reductions and a reasonable market penetration of hybrid vehicles (Kammen and Lemoine, 2009). In contrast, plug-in (hybrid) electric vehicles (relying on a battery for short distances and petrol for longer distances) are not expected to contribute substantially to total emission savings until 2030. In the case of urban transportation, even more can be gained (Kammen and Lemoine, 2009).

One of the most important lessons of the rapidly expanding mix of energy efficiency, solar, wind, biofuels and other low-carbon technologies is that the costs of deployment are lower than many forecasts and, at the same time, the benefits are larger than expected. This seeming 'win-win' claim deserves examination, and continued verification, of course.

Over the past decade, the solar and wind energy markets have been growing at rates over 30% per year, and in the last several years growth rates of over 50% per year have taken place in the solar energy sector. This rapid and sustained growth has meant that costs have fallen steadily, and that an increasingly diverse set of innovative technologies and companies have been formed. Government policies in an increasing number of cities, states and nations are finding creative and cost-effective ways to build these markets still further.

At the same time that a diverse set of low-carbon technologies are finding their way to the market, energy efficiency technologies (e.g. 'smart' windows, energy efficient lighting and heating/ventilation systems, weatherisation products, efficient appliances) and practices are all in increasingly widespread deployment. Many of these energy efficiency innovations demonstrate *negative* costs over time, meaning that when the full range of benefits (including improved quality of energy services, improved health and worker productivity) are tabulated, some energy efficiency investments are vehicles for net creation of social benefits over time.

Many more innovations are on the near-term horizon, including those that use innovative municipal financing to remove *entirely* the up-front cost of energy efficiency, and renewable energy investments through loans that are repaid over the duration of the services provided by clean and efficient energy products.

11.16 Conclusion: can the economy be decarbonised?

A growing number of studies are now coming to the conclusion that the US and, indeed many nations, can drastically reduce reliance on fossil fuels with an impact to the economy that is either insignificant or *beneficial*. In turn, this would reduce both GHG emissions and other pollutants. The highly influential *Stern Review* (Stern, 2007) suggested that the necessary changes to the energy infrastructure could be achieved with less than a 1% loss of global world product. Input to the *Stern Review* included model runs of more than 1500 case study scenarios that achieved 25% renewable energy penetration by 2025. The majority indicated that this could be achieved with, and indicated negligible additional cost over, a business-as-usual case. In assessing the range of clean energy and energy-efficient technologies and policy options, the clear message is that we have more than enough tools to achieve a sustainable, low-carbon economy. The critical and missing ingredients are leadership and the will to act.

References

Ahn, J. and Apted, M. (eds.) (2010). *Geological Repository Systems for Safe Disposal of Spent Nuclear Fuels and Radioactive Waste*. Cambridge, UK: Woodhead Publishing.

Arons, S. R., Brandt, A. R., Delucchi, M. A. *et al.* (2007). *A Low-Carbon Fuel Standard for California Part 1: Technical Analysis*. Berkeley, CA: UC Berkeley Transportation Research Sustainability Center.

Bailis, R., Ezzati, M. and Kammen, D. M. (2005). Mortality and greenhouse gas impacts of biomass and petroleum energy futures in Africa. *Science*, **308**, 98–103.

BioLite (2009). Thermoelectrics in biomass stove systems. Presented at the *UN-ASEAN Next Generation Cook Stove Workshop*. 18 November, Asian Institute of Technology, Bangkok.

Congressional Budget Office (2008). *Nuclear Power's Role in Generating Electricity*. Washington, D.C.: Congressional Budget Office, p. 28.

Creutzig, F., Papson, A., Schipper, L. and Kammen, D. M. (2009). Economic and environmental evaluation of compressed-air cars. *Environmental Research Letters*, **4**, 1–9.

Creutzig, F. S. and Kammen, D. M. (2010). Getting the carbon out of transportation fuels. In *Global Sustainability: A Nobel Cause*, eds. H.-J. Schellnhuber, M. Molina, N. Stern, V. Huber and S. Kadner. Cambridge, UK and New York, NY: Cambridge University Press, pp. 307–18.

Duke, R. and Kammen, D. M. (1999). The economics of energy market transformation programs. *The Energy Journal*, **20**, 15–64.

Enkvist, P.-A., Nauclér, T. and Rosander, J. (2007). A cost curve for greenhouse gas reduction. *The Mckinsey Quarterly*, February.

European Climate Foundation (2010). *Roadmap 2050: A Practical Guide to a Prosperous, Low Carbon Europe*. Den Haag, The Netherlands: European Climate Foundation, http://www.roadmap2050.eu/downloads.html.

Farrell, A. E., Plevin, R. J., Turner, B. T. *et al.* (2006). Ethanol can contribute to energy and environmental goals. *Science*, **311**, 506–08.

Fuller, M. C., Portis, S. C. and Kammen, D. M. (2009). Toward a low-carbon economy: Municipal financing for energy efficiency and solar power. *Environment*, **51**, 22–32.

Galperin, A. and Raizes, G. (1997). A pressurized water reactor design for plutonium incineration: Fuel cycle options. *Nuclear Technology*, **117**, 125–132.

Garcia-Frapolli, E., Schilmann, A., Berrueta, V. M. *et al.* (2010). Beyond fuelwood savings: Valuing the economic benefits of introducing improved biomass cookstoves in the Purhépecha region of Mexico. *Ecological Economics*, **69**(12), 2598–2605.

Granade, H. C., Creyts, J., Derkach, A. *et al.* (2009). *Unlocking Energy Efficiency in the U. S. Economy*. New York, NY: McKinsey & Company.

Hertel, T. W., Golub, A. A., Jones, A. D. *et al.* (2009). Effects of US maize ethanol on global land use and greenhouse gas emissions: Estimating market-mediated responses. *BioScience*, **60**, 223–31.

International Atomic Energy Agency (IAEA) (2008). *Nuclear Safety Review*. Vienna: IAEA.

International Energy Agency (2007). *World Energy Outlook 2007*. Paris: International Energy Agency.

Ingersoll, D. T. (2009). Deliberately small reactors and the second nuclear era. *Progress in Nuclear Energy*, **51**, 589–603.

Jacobson, M. Z. and Masters, G. M. (2001). Exploiting wind versus coal. *Science*, **293**, 1438.

Johnson, M., Rufus, E., Ghilardi, A. *et al.* (2009a). Quantification of carbon savings from improved biomass cookstove projects. *Environmental Science & Technology*, **43**, 2456–62.

Johnson, T. M., Alatorre, C., Romo, Z. and Liu, F. (2009b). *Proyecto MEDEC 'México: Estudio Sobre la Disminución de Emisiones de Carbono'*, Colombia: Banco Mundial, Mayol Mediciones.

Kammen, D. M. and Lemoine, D. (2009). Commentary: The transition from ICVs to PHEVs and EVs. In *Betting on Science: Disruptive Technologies in Transport Fuels.* USA: Accenture, pp. 219–20.

Kammen, D. M. and Nemet, G. (2005). Reversing the incredible shrinking energy R&D budget. *Issues in Science & Technology*, **Fall**, 84–88.

Kammen, D. M. and Nemet, G. F. (2007). Energy myth 11 – Energy R&D investment takes decades to reach the market. In *Energy and American Society – Thirteen Myths*, eds. B. K. Sovacool and M. A. Brown. The Netherlands: Springer, pp. 289–310.

Kammen, D. M. and Pacca, S. (2004). Assessing the costs of electricity. *Annual Review of Environment and Resources*, **29**, 301–44.

Kammen, D. M., Kapadia, K. and Fripp, M. (2004). *Putting Renewables to Work: How Many Jobs Can the Clean Energy Industry Generate?* Berkeley, CA: University of California, Renewable and Appropriate Energy Laboratory.

Kammen, D. M., Farrell, A. E, Plevin, R. J. *et al.* (2007). Energy and greenhouse impacts of biofuels: A framework for analysis. *OECD Research Roundtable: Biofuels: Linking Support to Performance*. Berkeley, CA: University of California – Institute of Transportation Studies.

Kammen, D. M., Clabaugh, M., Kerr, A. C. and Portis, S. C. (2008). *Securing a Clean Energy Future: Opportunities for States in Clean Energy Research, Development, & Demonstration.* Washington, D.C.: National Governors Association.

Keystone Center (2007). *Nuclear Power Joint Fact-finding*. Keystone, CO: The Keystone Center.

Marburger, J. H. (2004). *Science for the 21st Century*. Washington, D.C.: U.S. Office of Science and Technology Policy.

Margolis, R. M. and Kammen, D. M. (1999). Underinvestment: The energy technology and R&D policy challenge. *Science*, **285**, 690–92.

Mendonça, M. (ed.) (2007). *Feed-In Tariffs: Accelerating the Deployment of Renewable Energy*. London: Earthscan.

Natural Resources Defense Council (2009). *Issues: Global Warming*, http://www.nrdc.org/globalwarming/Default.asp.

Nemet, G. F. and Kammen, D. M. (2007). U.S. energy research and development: Declining investment, increasing need, and the feasibility of expansion. *Energy Policy*, **35**, 746–55.

Nuclear Energy Institute (2008). *New Nuclear Plant Status*. Washington, D.C.: Nuclear Energy Institute.

O'Hare, M., Plevin, R. J., Martin, J. I. *et al.* (2009). Proper accounting for time increases crop-based biofuels' greenhouse gas deficit versus petroleum. *Environmental Research Letters*, **4**, 024001.

PCAST (1997). *Report to the President on Federal Energy Research and Development for the Challenges of the Twenty-First Century*. Washington, D.C.: President's Committee of Advisors on Science and Technology.

Perlack, R. D., Wright, L. L., Turhollow, A. F., Graham, R. L. and Stokes, B. J. (2005). *Biomass as Feedstock for a Bioenergy and Bioproducts Industry: The Technical Feasibility of a Billion-ton Annual Supply*. US Department of Energy and US Department of Agriculture.

REN21 (2009). *Renewables Global Status Report 2009 Update*. Paris: Renewable Energy Policy Network for the 21st Century.

Richter, B., Goldston, D. L., Crabtree, G. *et al.* (2008). How America can look within to achieve energy security and reduce global warming. *Reviews of Modern Physics*, **80** S1–S109.

Rickerson, W. H., Sawin, J. L. and Grace, R. C. (2007). If the shoe FITs: Using feed-in tariffs to meet U.S. renewable electricity targets. *The Electricity Journal*, **20**, 73–86.

Riddel, M. (2009). Risk perception, ambiguity, and nuclear-waste transport. *Southern Economic Journal*, **75**, 781–97.

Rosenthal, E. (2008). Europe turns back to coal, raising climate fears. *New York Times*, 23 April.

Sailor, W. C., Bodansky, D., Braun, C., Fetter, S. and van der Zwaan, R. (2000). A nuclear solution to climate change? *Science*, **288**, 1177–78.

Searchinger, T., Heimlich, R., Houghton, R. A. *et al.* (2008). Use of U.S. croplands for biofuels increases greenhouse gases through emissions from land-use change. *Science*, **319**, 1238–40.

Smith, K. R. and Haigler, E. (2008). Co-benefits of climate mitigation and health protection in energy systems: Scoping methods. *Annual Review of Public Health*, **29**, 11–25.

Stern, N. H. (2007). *The Economics of Climate Change: The Stern Review*. Cambridge, UK: Cambridge University Press.

Stokes, D. E. (1997). *Pasteur's Quadrant: Basic Science and Technological Innovation*. Washington, D.C.: Brookings Institution Press.

United Nations Environment Programme and New Energy Finance Limited (2009). *Global Trends in Sustainable Energy Investment 2009*. n.p.: United Nations Environment Programme.

US Government Accountability Office (2004). *Renewable Energy: Wind Power's Contribution to Electric Power Generation and Impact on Farms and Rural Communities*. Washington, D.C.: US Government Accountability Office.

Wei, M., Patadia, S. and Kammen, D. M. (2010). Putting renewables and energy efficiency to work: How many jobs can the clean energy industry generate in the US? *Energy Policy*, **38**, 919–31.

Wiser, R. and Bolinger, M. (2007). *Annual Report on U.S. Wind Power Installation, Cost, and Performance Trends: 2006*. Washington, D.C.: US Department of Energy – Energy Efficiency and Renewable Energy.

World Health Organization (WHO) and United Nations Development Programme (UNDP) (2009). *The Energy Access Situation in Developing Countries. A Review Focusing on the Least Developed Countries and Sub-Saharan Africa*. New York, NY: United Nations Development Programme, Environment and Energy Group.

12

Economic approaches and instruments

'Business-as-usual is dead'[1]

12.1 Introduction

Climate change is now understood as a central issue for economic prosperity and development. The Stern Review (Stern *et al.*, 2006) can be seen as the turning point after which climate change has been considered an important economic issue, in addition to being an environmental one. In-depth reviews of the economics of climate change have been conducted for various countries and regions, for example for Australia (Garnaut, 2008) and South-East Asia (Asian Development Bank, 2009), and the development community now sees climate change as one of its main challenges (Chapter 15; World Bank, 2009).

Climate change could affect the fundamentals of economic systems. Meanwhile, curbing greenhouse gas emissions by the extent required to limit the risk of dangerous climate change (e.g. the 2 °C guardrail; Chapter 8) will require comprehensive changes in technologies, energy systems, industrial production practices and locations, and consumption patterns.

Countries will strive to effect such change with the minimum of economic cost; that is, keeping any sacrifices in economic prosperity as small as possible. While it can be argued that excessive consumption is one of the root causes of high greenhouse gas emissions, it is equally clear that few, if any, societies will readily make big cuts to their levels of material well-being. To achieve emissions cuts effectively and efficiently will require sound economic policies, integrated across different sectors of the economy and harmonised with other objectives of economic policy.

Mitigation policies also need to overcome difficult problems of coordination. Climate change is an almost pure externality, so the naked self-interest of individuals or groups is to free-ride on others' efforts to curb emissions. Effective climate policy requires national

---

[1] Professor Lord Nicholas Stern, Plenary Talk, IARU International Scientific Congress on Climate Change, Copenhagen, 10–12 March 2009.

Climate Change: Global Risks, Challenges and Decisions, Katherine Richardson, Will Steffen and Diana Liverman *et al.* Published by Cambridge University Press. © Katherine Richardson, Will Steffen and Diana Liverman 2011.

governments to impose emissions reductions policies, and it requires mechanisms to achieve cooperation between nations.

The international dimension also extends to equity and financing issues. Many developing countries will need external financing to reorient their economies to a low-carbon trajectory, and poor and vulnerable countries will need external funding for adaptation action. To achieve sustainable financial flows, and to make them economically beneficial, again requires sound economic policy design.

A different set of challenges arises from the need to adapt to climate change impacts. Effective and efficient adaptation requires that economic policy settings support individuals and businesses in making forward-looking adaptation decisions. In many instances, specific public investments will be needed to deal with climate impacts.

The importance of economic instruments for responding to climate change is reflected in finance and economics ministries now being centrally involved in climate policymaking in many countries, in particular for mitigation. International economic fora, for example, the G20 group – a forum of finance ministers created originally to manage international financial and economic crises – have also been increasingly prominent in climate change discussions.

12.2 Modelling the economics of climate change

Modelling plays a key role in the analysis of climate change economics and in recommendations for policy responses to climate change. Modelling is commonly used for evaluating the relative merits of different policy approaches to mitigation, costs of mitigation action and the benefits of mitigation in terms of avoided climate change damages.

It can be argued that the usefulness of economic modelling as a guide to policy is greatest for questions around policy design for the short term, and much less so for assessing the economics of possible future climate impacts. For example, some mainstream economic climate models that are used for making inferences about the optimal extent of emissions reductions represent neither uncertainty about future climate change impacts nor the value that people place on nature or other factors (Chapter 10) that are not traded in markets (Ackerman *et al.*, 2009).

The different tiers of economic impacts are schematically represented in Figure 12.1, following Garnaut (2008). Climate change will have impacts that can be readily measured (tier 1) and others that are measurable in principle but not necessarily in practice (tier 2). Society's utility is further diminished through the risk of unexpectedly high climate damages (tier 3). The possibility of catastrophic impacts, identified by Weitzman (2009a) as dominating any other aspects of climate change cost-benefit analysis, is part of this category (Chapter 7). Finally, there are impacts that are not reflected at all in markets, but which people nevertheless care about, for example the existence of species or natural icons (tier 4). Many economic models of climate change only capture tier 1, and none fully cover the range of impacts within each tier.

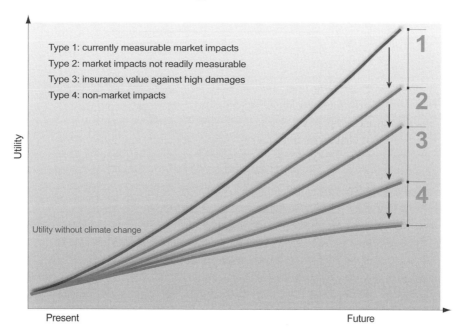

Figure 12.1 Four tiers of climate change impacts. Source: Garnaut (2008).

Furthermore, any long-term economic analysis runs into the problem of how to weight climate impacts decades or centuries in the future relative to the costs of mitigation incurred now. The choice of discount rate in cost-benefit analysis of climate change needs to stem from ethical considerations (Chapter 10), and the role of economics is essentially limited to show the effects of different ethical choices (Box 12.1).

One important issue that almost all models broadly agree on is that, although comprehensive climate change mitigation requires extensive economic restructuring especially in the energy sector (Chapter 11), the macro-economic cost of mitigation is small relative to expected future economic growth. For example, Stern *et al.*'s (2006) median estimate of global mitigation costs for a '500 to 550 p.p.m.' scenario by the year 2050 is around 1% of GDP, and similar numbers are reported from other modelling exercises (IPCC, 2007). Only a few models show global abatement costs above 5% of GDP even for more ambitious mitigation scenarios. A meta-analysis of model cost projections (Barker *et al.*, 2006) shows the vast majority of cost estimates for a 60% emissions reduction from the baseline between 0% and 5% of GDP.

If stretched over a 40-year horizon, a 5% reduction in GDP amounts to a reduction in annual growth of just over 0.1%. This compares to an underlying annual increase in global GDP of around 3% on average over the past several decades, and continued growth expected in future (though slowing with diminished population growth). On this basis, the entire cost of climate change mitigation would mean that doubling of global GDP (adjusted for inflation) would be delayed by between about half a year and two years – so doubling of

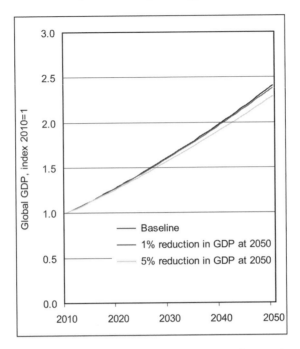

Figure 12.2 Illustration of mitigation costs and underlying economic growth over time. Assumes annual baseline GDP growth of 2.6% in 2010 declining to 1.8% in 2050 (1.5% annual growth in per capita GDP, using United Nations (2008) medium variant population projections. Source: F. Jotzo.

GDP would have to wait until sometime between mid 2040 and late 2042, rather than the beginning of 2040 under a 'no mitigation' scenario (Figure 12.2).

Taking into account the economic benefits of avoided climate change (difficult as they are to quantify) further reduces the net cost. In many analyses, economic growth in the long run is higher with mitigation than without mitigation (Stern *et al.*, 2006; Garnaut, 2008).

However, the assumption underlying these results is that mitigation action is undertaken cost-effectively, using low-cost options in preference to higher-cost ones across the economy, and with minimal distortions or transaction costs. How this could be achieved is discussed in the next section.

Box 12.1

Economic modelling of climate change: its role and limits as a guide to policy

CAMERON HEPBURN AND JISUNG PARK

Economic analysis is now commonly employed in debates on climate change policy
(Nordhaus, 2001; IPCC, 2001, 2007; Stern, 2007). This is because economics provides a useful

analytical framework for navigating the multi-dimensional problem of formulating climate policy. By clarifying the consequences of any particular set of assumptions or policy options, in terms of 'efficiency' and 'equity', economics provides guidance about what to aim for and how best to get there. In other words, it helps us decide upon: (i) an appropriate policy target, and (ii) the choice of policy instruments that will achieve the target most efficiently (Hepburn, 2006). But economic analysis only serves as one input into a political decision-making process.

Guidance to optimum target

Doubling the concentration of GHGs from pre-industrial levels would trap additional heat at the Earth's surface, which is likely to have enormous negative consequences for humans. The crux of the climate policy problem is to determine a pathway for GHG emissions which balances the net benefits of different climates with the benefits from burning cheap carbon-based energy sources. Economists are almost unanimously agreed that this balance requires significantly reducing emissions below business-as-usual levels. But to what level? Cost–benefit analysis can help to clarify the issues. By comparing the costs associated with reducing emissions on the one hand, with the benefits of doing so – in terms of future damages avoided – on the other, cost–benefit analysis can in theory guide us towards a socially 'optimal' level of greenhouse gas emissions.

Actually conducting a cost–benefit analysis for climate change policy is a non-trivial matter, however. Climate change is *global in scope* and the effects *span multiple generations*, so cost–benefit analysis must aggregate the costs and benefits from reducing GHG emissions internationally and inter-temporally. Because climate change is such a long-lived phenomenon, our ethical assumptions regarding future generations are an essential feature of the analysis (Chapter 10; Beckerman and Hepburn, 2007; Stern, 2007; Hulme, 2009) and economics teases out the implications of different ethical assumptions. Recent debates over the discount rate – the economist's proxy for the tradeoff between the welfare of different generations – are illustrative of how important such underlying ethical assumptions are to climate policy.

Similarly, economic analysis tells us that ethical assumptions regarding the welfare of people living in other countries are also critical (Anthoff *et al.*, 2009). Greater concern for people in other countries translates into a greater concern about climate change.

Two other features of climate change also create real intellectual challenges for cost-benefit analysis, namely the treatment of risk and uncertainty (Stern, 2007; Weitzman, 2009a, 2009b), and the sheer magnitude (or 'non-marginality') of the problem (Dietz and Hepburn, 2010).

Do these ethical and intellectual challenges imply that economics is unable to dictate the 'right' target for mitigation? Indeed, they do. While economic models can *clarify* the policy-relevant consequences of any particular set of ethical or political assumptions, they cannot determine which ethical assumptions are 'correct'.[2] Economic analysis helps to ensure consistency, and it can spell out the logical consequences of taking a given political, moral, or ethical position, which may be useful in evaluating their reasonableness. The point is that

---

[2] More often than not, economists take a standard welfarist approach, which involves the assumptions of consequentialism and welfarism. But this is merely a descriptive point; just because economic analysis of climate change tends to use these assumptions does not mean that it is limited to them.

Box 12.1 (*cont.*)

although climate policy cannot be properly conducted without considering a range of ethical perspectives, this consideration is assisted by engaging in economic thinking.

Guidance on how to get there: instrument choice

Given a climate policy target, economics can also provide guidance to policymakers on which 'policy instruments' should be used to achieve it effectively.

The first key insight from economics is that when the costs of reducing emissions vary greatly between different entities (as they do for GHG emissions), economic instruments are likely to be more efficient compared to command-and-control regulation (Baumol and Oates, 1971). Economic instruments include price-based instruments, such as carbon taxes; and quantity-based instruments, such as tradable allowance schemes; and various hybrids of the two (Hepburn, 2006). Both types of instruments put a price on carbon emissions and give firms an incentive to reduce emissions efficiently. Various factors, including economic uncertainty, international dimensions and political economy, may suggest that one instrument is preferable to the other in specific cases. For instance, if marginal costs of reducing emissions increase quickly and damages from climate change are relatively insensitive to emissions over short time periods, a price instrument (tax) is more appropriate over short periods (Weitzman, 1974; Hoel and Karp, 2001; Hepburn, 2006). However, in practice, various other considerations have favoured trading or hybrid systems.

If a trading system is employed, economics provides useful guidance on its design. For instance, economists are largely agreed that efficient results are achieved if the allowances are auctioned, rather than given away for free (Hepburn *et al.*, 2006).

Finally, the sheer scale of investment required to reduce emissions means that asset prices will adjust, competitiveness will change, and even international financial flows and currencies might be affected. Economic modelling can inform policymakers of the potential synergistic characteristics of climate policy. As recent analysis suggests, climate mitigation policies can be implemented to greater effect as part of a package of counter-cyclical macro-economic stimulus spending to positive effect (Bowen and Stern, 2010).

12.3 Economic instruments for mitigation

The economic policymaker's toolbox for curbing greenhouse gas emissions contains a large array of instruments. Some of them are already in operation and need only a change in settings to become effective for mitigation, while others are purpose-built. The instruments can be roughly divided into ones that create a pervasive price incentive to cut emissions, and others that use regulation or fiscal measures for more specific interventions.

A key consideration in instrument choice and design is the cost-effectiveness of mitigation, the aim being to achieve the greatest amount of emissions reductions for a given cost, or to achieve a given reduction for the lowest cost. Distributional effects are also important, and sometimes crucial, in the design and for the political viability of mitigation policies.

12.3.1 Emissions pricing

Putting a price on greenhouse gas emissions is generally considered the backbone of efficient climate policy, building on the fundamental concept of using taxation to internalise economic externalities (Pigou, 1938). The idea is to put a price on greenhouse gas emissions, payable by those producers that emit them. This has the effect of raising the prices of goods that have associated emissions in their production, with this price effect flowing through to all goods and services within the economy. In this way, prices are changed by an amount that reflects the emissions embodied in the entire production chain, creating an incentive to cut emissions throughout the economy. Businesses and consumers shift to lower-emissions processes or products because it saves them money.

The overall response is cost effective because the lowest cost options are chosen automatically by the market. There is no need for a government planner to decide which sectors should reduce emissions, how and by how much.

The two main pricing instruments are an emissions tax and emissions trading (Figure 12.3). Under a tax, the government sets and enforces a charge for each tonne of emissions. The amount of emissions reductions is determined by market forces in response to the price signal. Abatement options that cost less per tonne of emissions savings than the tax rate will

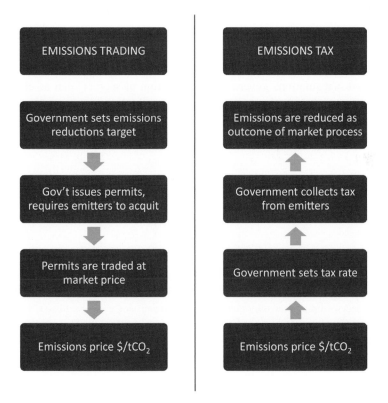

Figure 12.3 Schematic representation of emissions trading and emissions taxes. Source: F. Jotzo.

be implemented because they are cheaper than paying the tax; higher-cost options will not, as it is cheaper to pay the tax than to abate. Emissions taxes have been favoured by many prominent economists (see Nordhaus, 1991).

Under emissions trading ('cap-and-trade'), the government sets a cap on total emissions, issues tradable permits or certificates that add up to the cap, and requires the acquittal of one permit for each unit of emissions. The permits will tend to trade at a price equal to the marginal cost of achieving the targeted emissions level, with that price level determined in markets. The same abatement options will be implemented as for a tax of the same rate.

The two approaches are equivalent in that they both result in a price on emissions that pervades the economy, and achieve a broadly cost-efficient response.[3] But they differ in their effect under uncertainty about the economy's response. A tax gives certainty about the cost of emitting but leaves open what level emissions will be in response. Cap-and-trade, by contrast, provides certainty over emissions levels, but the permit price and hence compliance cost remains uncertain. Debates continue among economists over which approach is preferable, and 'hybrid' instruments with elements of both permit trading and controlling the price have been proposed (see Box 12.1; McKibbin and Wilcoxen, 2002; Pizer, 2002).

12.3.2 Fiscal and regulatory instruments

Governments also have on hand many economic policy levers that more directly affect what technologies get deployed, how businesses operate, and what types of products and services consumers buy. These include taxes and subsidies for specific products and investments; regulations governing the operation of sectors such as energy, transport, industry, agriculture or forestry; technology standards; and support for research and development (Box 12.2).

Box 12.2
Regulatory support for carbon prices in policy portfolios

TERRY BARKER

Carbon-price instruments

The low-cost trajectories towards 2 °C stabilisation explored in the literature involve a strong expectation that carbon prices will have to rise substantially. The carbon prices are assumed to apply in all sectors through the use of carbon taxes and emission trading schemes. The models suggest prices per tonne of CO_2 from US\$10 in 2013, to US\$100 in 2020 to US\$400 in 2050 (all in 2000 prices). In addition, the price signal must be credible, reliable and announced in advance.

---

[3] Interactions with other existing taxes and subsidies can mean that fully efficient carbon pricing would require differentiated carbon pricing for different goods or activities (Babiker *et al.*, 2004).

Such prices appear infeasible to those in many industries and countries, especially where energy prices are low or subsidised. The high prices, coming directly from carbon taxes or indirectly from emission trading schemes, are required in the models to bring about investments in low-GHG-emitting technologies to replace fossil fuel investment and induce technological change. If the carbon prices are low in the near term, this reduces the cost of premature obsolescence, while the expectation that they will be high later encourages research and development, and investment, in long-lived low-emissions capital and reduces the risk of lock-in.

Subsidies for low-GHG-emitting technologies are also justified because they can address a market failure in innovation. Those investing, even allowing for patents, are unable to capture all the benefits that accrue to those able to copy and exploit the innovation, so insufficient innovation takes place in a market system (Jaffe *et al.*, 2005).

Command-and-control instruments

One complement to the market-based carbon prices is the use of traditional regulatory command-and-control measures, which involve agencies (such as Pollution Inspectorates) to fix and force energy and greenhouse gas standards. Climate, air quality and energy-security objectives are all served by technology-forcing policies of the sort pioneered in California over the past 15 years (Chapter 11; Jänicke and Jacob, 2004). The main objection to these has been their potential inefficiency, but they can still be targeted to correct market failures and support investments that are profitable, where social – as opposed to private – costs and discount rates are applied.

Regulation can also accelerate the technological change induced by rising carbon prices because investments are increased and costs fall. However, technology alone is unlikely to produce effective climate change mitigation partly through the 'rebound effect' (Sorrell, 2007). This arises where improvements in energy efficiency reduce the cost of a technology, which then prompts higher use of particular services (e.g. heat or mobility) that energy helps to provide, so that the energy saving from the innovation is offset by increased energy consumption. Although rebound effects will vary widely in size, energy efficiency policies and technological breakthroughs may bring about only a weak reduction, if any, in emissions, especially at a global level. The study by Barker *et al.* (2009) suggests that the energy efficiency policies included in the IPCC's 2007 Fourth Assessment Report may be associated with a 50% rebound; that is, the policies may reduce GHG emissions by half the amount initially expected.

Portfolios of policy instruments

Portfolios of environmental policies allow policymakers to combine the strengths of different instruments and adjust the policies to suit the specific conditions of different countries (Box 12.4; OECD, 2007). In addition, carbon prices for stringent mitigation may be much lower if the carbon-price instrument is combined with other policy instruments (Barker *et al.*, 2008).

The most effective policies appear to be those that combine the carbon price signal with regulation for energy efficiency for key technologies such as low-CO_2 power generation and

Box 12.2 (*cont.*)

vehicles; and subsidies to fund innovation, and research and development, on low-GHG-emitting technologies – for instance, by auctioning emission permits and using a proportion of the revenues to provide additional technological incentives. The appropriate policy portfolios for greenhouse gas mitigation will, of course, be specific to countries depending on their climate, political systems, available renewable and other energy resources, and the energy efficiency of existing building and equipment stock.

Fiscal and regulatory instruments are used widely, with a large and increasing number of policies in place in both developed and developing countries (Figure 12.4). According to the data in International Energy Agency (2009a), there has been particularly strong growth in the number of incentive/subsidy schemes over recent years.

A perfect emissions pricing scheme in a world with perfect markets would not need such 'complementary' policy measures.

However, there are important market failures other than the climate change externality that need to be addressed for climate policy to be efficient. The prime example is externalities in knowledge generation that warrant public investment in research.

There are also many practical limitations of carbon pricing policies. Emissions pricing may not be practicable in sectors with a large number of small-scale emitters, because accurate monitoring of emissions and enforcement taxes or permit systems may not be possible or be too costly; and here may also be political reasons to exclude specific sectors. For example, most existing and proposed emissions pricing schemes exclude agriculture. In the

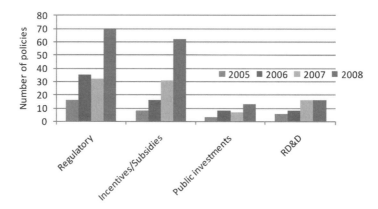

Figure 12.4 Number of national fiscal and regulatory policies in place globally, 2005–2008. Source: F. Jotzo, with data from International Energy Agency (2009a). National-level policies in force in IEA (OECD) member countries and in Brazil, China, the European Union, India, Mexico, Russia and South Africa. Total number of policies in the four categories is lower than the sum of policies under each category, because some policies are listed under multiple categories.

face of partial coverage, regulations and budgetary measures can ensure that low-cost emission reductions outside the covered area contribute to the overall abatement effort.

Even within the sectors covered by the carbon price mechanism, it is likely that obstacles will remain to a least-cost response. They include high transaction costs for small investments, legal and bureaucratic barriers, lack of information or financial credit, and misaligned economic incentives.

Energy efficiency is a classic case of such features, with most studies of economy-wide abatement options and costs identifying large options to save emissions through better energy efficiency, which would also save operators money (Enkvist *et al.*, 2007). In these cases, charging for emissions makes the cost savings larger but will have only limited effect. Complementary ways to encourage energy efficiency include awareness campaigns, minimum standards for buildings and equipment, and budgetary programs such as grants and low-interest loans for investment in energy efficiency.

Supporting deployment of renewable energy is another field where many governments are providing subsidies, or have regulated for higher electricity prices to be paid to producers of renewable electricity, resulting in large investments in renewable electricity supply in some countries (Box 12.3).

These and many other regulatory measures often come at high economic cost relative to the abatement induced by a price signal. The reasons are that regulators do not have the myriad of pieces of information about the cost of each of the individual abatement options that markets can discover; that regulation may easily result in abatement costs per tonne of emissions differing greatly between regulated sectors and so drive up average costs; and that cost effectiveness may not even be a primary objective. The large differences between tariffs paid for different kinds of renewable electricity (Box 12.3) are a case in point.

The upshot for cost-effective climate policy is that fiscal and regulatory instruments need to be used judiciously, complementing and supporting price-based approaches where needed, but not generally as the dominant policy intervention.

Box 12.3
Renewable energy support mechanisms

GREGORY BUCKMAN

About 18% of the world's electricity is currently generated from renewables of which over 85% comes from hydro generation (International Energy Agency, 2009b), but the prospects for hydro's expansion are limited in developed countries in particular. Of the non-hydro renewables generation, about half comes from wind with the balance from biomass, solar and geothermal.

The two most popular ways of stimulating renewable electricity generation are government subsidies (installation grants or loans, and tax subsidies paid to renewable electricity developers) and electricity market support mechanisms. Examples of the former include

Box 12.3 (*cont.*)

low-interest loans provided by the German government for solar photovoltaic installation in the late 1990s (Jacobsson and Lauber, 2006) and the renewable electricity production tax credit provided by the United States government (Wiser and Bolinger, 2008).

Increasingly, electricity market support mechanisms, in the form of either feed-in tariffs or renewable portfolio standards, are becoming the favoured type of support. Under both, the additional cost of cleaner electricity is generally ultimately paid for by electricity consumers in the form of higher prices.

Feed-in tariffs are a mechanism where governments mandate a level of generation subsidy (generally differentiated according to renewable source) that must be paid to renewables generators but where the level of renewables generation is left to the market (Mendonça, 2007). By 2008, the mechanism was used in 43 countries (Renewable Energy Policy Network, 2009). Feed-in tariffs are particularly popular in Western Europe where major users include Denmark, Spain and Germany. By 2020, these countries aim to generate 30%, 20% and 18%, respectively, of their electricity from renewables. In 2007, their shares of renewable electricity were 29%, 20% and 15% (European Commission, 2010). In Germany, onshore wind gets the lowest feed-in tariff subsidy while biomass gets double its rate, offshore wind gets three times and solar photovoltaic gets up to eight times (all rates dependent on capacity and development date) (Federal Ministry for the Environment, Nature Conservation and Nuclear Safety, 2008).

Renewable portfolio standards are a mechanism where governments mandate a level of renewable electricity generation but leave subsidy levels to the market (determined via tradable certificates or contracts between electricity retailers and generators) (Nielsen and Jeppesen, 2003; van der Linden *et al.*, 2005). Renewable portfolio standards are used in North America and in some countries in Europe, Asia and the Pacific. Major applications of it occur in the UK, California and Australia. The UK aims to generate 15% of its electricity from renewables by 2015; in 2007 its share was 5% (European Commission, 2010). California aims to generate 33% by 2020 from a 2007 share of 24% (California Public Utilities Commission, 2009). Australia aims for 20% by 2020 from a 2007 share of 8%.

So far, the development of renewable electricity has had mixed results under either mechanism. In generation terms, some countries have significantly increased their renewable-generated share of electricity but others have not. Spain's share fell from 24% in 1996 to 20% in 2007, while the US share fell from 11% in 1995 to 9% in 2008 (European Commission, 2010; US Energy Information Administration, 2010). Reasons for falling shares include electricity demand rising faster than expansion of renewable generation and falling hydro generation because of rainfall decline.

The cost of renewable electricity subsidies under either mechanism can be controversial. Often renewable electricity targets are blamed but, invariably, the amount of investor risk demanded of renewables developers is a major driver (Held *et al.*, 2006). Both mechanisms can also suffer from flaws such as too many exemptions from the obligation to buy renewable electricity or weak non-compliance penalties.

Greenhouse gas emissions from global electricity generation currently amount to about 10 Gt yr^{-1} CO_2-e, about a third of all non land-use emissions (World Resources Institute, 2009). They will have to shrink to a small fraction of that if strong global mitigation objectives are to be achieved, with a shift to renewables generation a crucial part of the solution in the face of rising energy demand. A tentative start has been made on this but most countries have a long way to go. The stimulation of large-capacity, high-cost/less mature types of renewable electricity – such as offshore wind in Europe; and solar in Africa, North America, South America and Asia – is an issue many countries are only starting to come to terms with. It will significantly influence how large a proportion of their electricity can eventually come from renewables and, therefore, how far they will be able to reduce their greenhouse gas emissions from electricity generation.

12.3.3 Political economy and practice of mitigation policy

In practice, the choice and design of mitigation instruments is not just predicated on their cost effectiveness in reducing greenhouse gas emissions. Other distinct policy objectives almost always have a role, and often political factors are at play.

In the energy sector, an important factor for mitigation policy can be the quest for a more secure energy supply. This tends to work in favour of renewable energy sources, which are not vulnerable to fuel supply disruptions, and partly explains governments' appetites for renewable energy support policies even if they are relatively costly. On the flipside, energy security considerations can also tip the balance in favour of high-carbon options, in particular where coal is seen as more secure than the lower-emissions option of gas. Energy security policy usually uses regulatory approaches.

Striving for 'green jobs' and leadership in emerging technologies can be another reason for 'strategic' government support for low-carbon technology, particularly in the energy sector. As US President Obama said in his 2010 State of the Union Address: '… even if you doubt the evidence [about climate change], providing incentives for energy efficiency and clean energy are the right thing to do for our future – because the nation that leads the clean energy economy will be the nation that leads the global economy.'

The approaches between pricing instruments and regulation have differed between countries, and have also changed within countries over time. For example, the European Union in the early 2000s made a transformation from strident opposition to emissions markets, to implementing the world's largest CO_2 trading scheme (Christiansen and Wettestad, 2003). In another example, Australia in the late 1990s planned the introduction of emissions trading, then in 2002 (coinciding with repudiation of the Kyoto Protocol) renounced all market- and price-based policy approaches in favour of subsidies for technology development (Pezzey *et al.*, 2008) and, under a new government since 2008, has again worked towards a comprehensive emissions trading scheme.

Perhaps the most important aspect of political economy of mitigation relates to the distributional implications of policies within economies. Emissions pricing schemes create large new assets and/or large government revenues, which are available for compensation, redistribution or other uses, and so create large opportunities for rent-seeking. The influence of politically powerful industry groups in the design of greenhouse gas policy has been evident, for example, in Europe (Markussen and Svendsen, 2003) and in Australia (Pezzey *et al.*, 2009) and, at the time of writing, is playing out in the United States, where it was earlier observed in the sulphur dioxide trading program (Joskow and Schmalensee, 1998).

The total value of permits in the EU emissions trading scheme, for example, was around €30 billion in 2009,[4] shared among around 12,000 participating emitters. Under the first two phases of the EU scheme (2005–07 and 2008–12), the majority of the permits are given for free (or 'grandfathered') to existing emitters, even though many emitters can pass through much of their cost increases by charging higher prices for their products (Grubb and Neuhoff, 2006). This led to substantial windfall profits to a range of industries, at the expense of consumers (Sijm *et al.*, 2006). In general, the question of distribution – who pays and who profits from emissions pricing – is separate from the effectiveness and effi-ciency of the policy, but it is possible that free allocations can compromise incentives to reduce emissions, in particular if free allocations are linked to the continued operation of old plants.

Assisting industry through free permits or cash is economically justified insofar as emissions-intensive industries are exposed to international trade, and their competitors in other countries do not face similar carbon penalties, so they cannot pass on their carbon costs to the market. There is a case to assist such producers in order to avoid economically inefficient relocation ('carbon leakage') of industries from one country to another without emissions savings, but to assist them only to the extent that international product prices do not go up as they would if all countries had carbon policies in place (Garnaut, 2008).[5]

Other permits can be auctioned, and revenue used to support the development of clean technologies, climate change adaptation or to cut other taxes that are economically more distorting – potentially yielding a 'double dividend' of simultaneous environmental and economic efficiency gains (Pearce, 1991). However, this tends to be politically difficult to achieve in the face of strong lobbying pressure from incumbent emitters for free permits or monetary compensation. This is borne out in the experience with the emissions trading schemes under development in the United States and Australia, both of which were in the legislative process at the time of writing. In both cases, professional economic advice has been to allocate only a small share of permits or of permit revenue to emitters (Garnaut, 2008; Goulder *et al.*, 2009), but successive policy proposals and legislative drafts have allocated larger shares of total permits and revenue to domestic industries. The European Commission, meanwhile, is attempting to steer toward lower shares of free permits under the third phase of its trading scheme, from an auctioning share of less than 10% of the total from 2008–12.

Retaining revenue for the state may be easier with emissions taxes than with permit trad-ing, because there is not necessarily an expectation of sweeping tax exemptions.[6] However, the practical experience to date with carbon taxes is that they have only been used in some

---

[4] The total amount of EU ETS permits was around 2 billion tonnes of CO_2, and permits traded at a market price of around €15 per tonne (Lewis and Curien, 2010).

[5] The efficient free permit allocation should compensate firms to the extent that the price of their products in international mar-kets would go up if all major competitors faced similar carbon penalties. Under this principle, free permits would be gradually phased out as more countries phased in emissions pricing.

[6] Note, however, that despite the common notion that an emissions tax would apply to each tonne of carbon emitted by covered entities, it would be possible – and indeed might be politically desirable – to exempt some share of emitters' emissions liabilities from the tax by way of tax-free thresholds (Pezzey, 1992). This would affect distribution while leaving the price incentive intact, much as partial free permits under emissions trading.

countries, and typically with restricted coverage – noting that many countries have energy taxes in place, which can have similar effect as carbon taxes depending on differentiation between fuels. The reason that emissions trading is in practice preferred to emissions taxation in most developed countries may, ultimately, have to do with public perceptions. In the words of Nordhaus (2007) '"tax" is almost a four-letter word'.

12.4 Economic approaches for adaptation

In the face of already unavoidable climate change, societies will need to take adaptive actions (Chapter 14). Adaptation to future climate change will be costly but the economic costs of doing nothing would be far greater.

12.4.1 Adaptation actions and their economic cost

In some cases, adaptation will mean to simply change practices in response to changes in conditions. In others, it will require wholesale changes in service systems such as public health; and require the early retirement of existing infrastructure and construction of new facilities, including in housing, transport, water supply, energy and so forth.

The impacts and economic costs of climate change could be large, and their exact nature is uncertain. Recent detailed studies for particular countries, regions and sectors have shown that when taking risk and uncertainty into account, the potential economic damages from climate change, as well as the potential gains from adaptation, are larger than previously thought (Hallegatte, 2009; Isoard and Swart, 2008; Esteban *et al.*, 2009). Furthermore, economic costs and adaptation needs are going to be distributed very unevenly. For example, a large-scale study of potential climate impacts and adaptation for Europe found that south-eastern European and Mediterranean countries are likely to be worse affected in many respects than northern European countries (Ciscar *et al.*, 2009).

Developing countries, which have lesser economic resources, are typically also at risk of greater damages from climate change (Chapter 9), and so could be hit especially hard. Hence, support from rich to poor countries will be important (Hof *et al.*, 2010). Estimates of the future financial needs for climate change adaptation in developing countries alone could be of the same order of magnitude as total current global aid flows (Parry *et al.*, 2009). Table 12.1 shows sectoral abatement cost estimates from a well-known study published by the UN Climate Secretariat (UNFCCC, 2007), and assessment of these estimates by Parry *et al.* (2009). These studies suggest that the largest costs will arise from adapting infrastructure and that there are large upside risks in costs, especially in developing countries.

12.4.2 Roles for policy

Adaptation will occur through both individual as well as policy-driven actions, and may require substantial economic resources. Adaptation will typically consist of local or

Table 12.1. *Estimates of financial needs for adaptation, 2030*

	Estimates by UNFCCC (2007): Additional annual investment needs for adaptation, 2030, US$ billion per year			
	Total	Developed countries	Developing countries	Assessment of UNFCCC estimates by Parry *et al.* (2009)
Agriculture	14	7	7	Reasonable approximation but does not cover all impacts
Water	11	2	9	Substantial costs omitted
Health	5	n.a.	5	Risk of higher costs under alternative development scenarios
Coastal zones	11	7	4	Could be 2–3 times higher
Infrastructure	8–130	6–88	2–41	Likely to be substantially higher, in particular in developing countries
Total	49–171	22–104	27–66	

Note: n.a.= Not available.
Source: F. Jotzo.

regional actions, dealing with observed or anticipated environmental changes in a way that is in the direct self-interest of those individuals and communities affected by the changes. Hence, the problems of externalities and policy coordination to be dealt with are generally at a smaller scale than for mitigation, relating mostly to the questions of how to provide infrastructure and other public goods that are familiar from non-climate contexts.

A polar view that adaptation is an exclusively private good and efficiently provided by the market, however, is often wrong. For example, improvements to critical infrastructure or to agricultural systems need coordinated government policies.

But decision-making on adaptation is hampered by uncertainty about future climate impacts, and the lack of reliable information on the economic tradeoffs in adaptation. Planning for adaptation investments requires data with a high level of detail and reliability on expected future climate impacts to be useful to decision-making. As a result, the role for planning is limited where there is little certainty in climatic and socio-economic projections, and for longer time horizons (Hunt, 2008).

Under these circumstances, the crucial task for economic policymaking for adaptation is to screen existing policy settings for whether they support adaptation if, where and when it might become necessary; or whether they would hinder adaptive responses. Many aspects of the existing policy framework in any country will need to be examined for their likely effect on climate change adaptation. New policies will be needed to support adaptation that would otherwise not come about, and some existing policies will need to be modified or scrapped. The adaptation policy response will need to be 'mainstreamed' to become part of the work of all relevant sectoral portfolios, and at the different levels of government and administration.

12.5 International mitigation mechanisms and climate financing

Effective international mitigation will require overcoming what Stern *et al.* (2006) called 'the greatest market failure the world has seen', securing cooperation in the face of strong incentives for individual countries to free-ride on each others' efforts, and solving the conundrum that costly action needs to be taken in relatively poor countries that cannot afford it and that have not caused the problem in the first place (Chapter 9). These issues are the subject of a large literature in economics, politics, international relations, philosophy and other fields. Here, selected current issues of international economic mechanisms and financing are briefly discussed.

12.5.1 International mitigation mechanisms

The international climate policy architecture, to date, has provided only a limited range of economic and financing mechanisms. The principal of these are the Kyoto Protocol's Clean Development Mechanism (Chapter 9), where developed countries invest in emissions reductions projects in developing countries; and international emissions trading between countries that have emissions targets.

The CDM is channelling several billion dollars per year to mitigation activities in developing countries, but has come under increasing criticism for its environmental effectiveness (Box 9.1; Wara and Victor, 2008). In addition, its reach is limited, to date excluding policy-based abatement and efforts to reduce deforestation. Emissions trading between countries has not yet taken place at a meaningful scale but could occur in future years, when countries aim for compliance with their commitments under the Kyoto Protocol.

New mechanisms are under discussion and negotiation, in particular ones that facilitate private sector finance from developed countries for mitigation in developing countries. These include a mechanism for 'reducing emissions from deforestation and forest degradation' (REDD), which is seen as a priority area for achieving large-scale emissions reductions in the short to medium term in tropical developing countries like Indonesia and Brazil (Chapters 9 and 15, and Boxes 9.3 and 9.4; Parker *et al.*, 2009b).

Proposals have also been made in the negotiations for 'sectoral crediting or trading mechanisms' (Ward, 2008) for developing countries, which would be a first step for developing countries to systematically apply mitigation policies (possibly including emissions pricing) to whole sectors or industries, and be able to sell emissions credits for any reductions below an agreed baseline. A future UN climate agreement is also expected to provide avenues for financing nationally appropriate mitigation actions (NAMAs) in developing countries, with modalities to be agreed.

In the lead-up to the 2009 UNFCCC COP15 in Copenhagen, China and India put forth emissions intensity targets, which link future targeted emissions levels to economic growth and so provide an automatic cushion for unexpected economic boom or bust (Jotzo and Pezzey, 2007). Other developing countries including Indonesia, Brazil and South Africa have committed to specified reductions below a business-as-usual emissions trajectory. The role of commitments by developing countries in the global climate policy architecture – including on what principles countries' commitments might be set, and whether and under what rules emissions reductions will be traded – is a defining question for the ongoing climate negotiations.

This, in turn, relates to the overarching issue of whether there will or should be a single, highly integrated international system of mitigation commitments and extensive trading of emissions allocations between countries; or whether a more disaggregated system of individual but coordinated national approaches may emerge (Box 12.4).

Box 12.4
An integrated global mitigation system versus national systems

WARWICK J. MCKIBBIN

There are two broad positions in the debate about the nature of a global policy regime (or framework) for coordinating climate policy across countries. There are those who believe that the global commitment to climate change mitigation requires a global carbon mitigation system which is highly integrated across economies (Stern *et al.*, 2006; Garnaut, 2008). Others argue that a fully integrated global system is not feasible (and can be dangerous to sustaining effective policy in key countries) and it would be more realistic to focus on creating national systems or national policy responses that are coordinated through the UNFCCC process to build up to a global system (McKibbin and Wilcoxen, 2007). These alternative approaches are comprehensively examined in a series of papers contained in Aldy and Stavins (2007).

The policies advocated by those who support the comprehensive global approach include a global emissions trading system where permits are freely traded between countries, or a common carbon tax to yield a global carbon price. The alternative approach is different national policy frameworks such as carbon taxes, national cap and trade programs (with safety valves), or direct regulation, which are somehow linked loosely around an agreed price for carbon. The core problem is to balance gains from commonality in carbon prices achieved through a global market relative to the losses from instability in one country being transmitted destructively to all countries. In essence, a key problem is how to achieve commonality of effort (through a common price) but with sufficient 'firewalls' between countries to prevent the propagation of destructive shocks through the climate policy framework.

The global financial crisis (GFC) has clearly demonstrated the need to get the design of the climate policy framework correct from the start. The GFC has also shown the importance of unexpected shocks in the global economy (or in large countries). Thus, climate policy must be evaluated not only on its ability to reduce greenhouse gas emissions, but also on its ability to achieve reductions at low cost and in a way that does not cause instability in the global economy that could undermine the political support for the policy regime. It is critical that the weakness in some countries in monitoring, enforcement and regulatory design not undermine the climate policy actions undertaken in other countries. The financial crisis has shown that the world has made some large design mistakes in the global financial system, which need to be avoided in designing a climate policy regime that needs to last many decades.

Experience with existing policy frameworks for carbon emission reduction, such as the European Emissions Trading System (EETS) and the Clean Development Mechanism, suggest that these systems, although able to reduce emissions, may not be able to be scaled up into a global system. Although the Europeans have clearly invested a lot of political capital in the EETS, it is probably better for other countries to improve on this experience and design systems that meet their own national circumstances. This is particularly true in developing countries, where factors such as weak domestic institutions underlie both their development problems as well as indicating that it will be some time before the sort of institutional convergence can be achieved in order to create a unified global system.

Developing effective policy to reduce global emissions to address the issue of climate change is too important to risk waiting for a grand design for the global system. The debate will not be settled in the near future but the experience of the global financial system and the GFC clearly tilts the argument towards the school of thought that linking must be done in a way that balances the need for firewalls between national systems at the same time as sustaining effective mitigation in the major emitting countries. The outcomes of the UNFCCC COP15 meeting in December 2009 clearly show that a grand design is some way off, and a better approach may be to build a global system from developing a variety of alternative national policies to reduce emissions under a cooperative umbrella. In following this approach, one way forward is the 'price collar' approach summarised in McKibbin *et al.* (2009), which focuses on coordinating countries around a common price of carbon created by different policies. In this approach, it is critical to implement policies that credibly raise carbon prices as well as developing ways to measure and compare carbon prices across countries. It is important to understand that comparable effort, as espoused in the Bali Roadmap, is better achieved through commonality of carbon prices across countries than by commonality of emission targets.

12.5.2 International climate finance

Climate change mitigation and adaptation is going to require large amounts of money for additional investments, estimated at hundreds of billions of dollars per year globally after a ramp-up period (World Bank, 2009). Much of this will need to be spent in developing countries, which have much scarcer economic resources to spend on the relatively long-term challenge of climate change, and which under the UNFCCC principle of 'common but differentiated responsibilities and respective capabilities' would not be expected to shoulder all of their costs.

This reality is reflected in many analyses of, and proposals for, international climate finance mechanisms. These include the market-based approaches discussed in the previous section, but extend to the provision of public finances for climate change through aid programs, special grants and concessional loans, on a bilateral basis or through international financial institutions. Importantly, climate finance figured prominently in the COP15 climate negotiations, and developed countries have pledged government funding for climate change adaptation and some mitigation in developing countries (Box 12.5 and Chapter 9).

Box 12.5
Climate finance for developing countries

JONATHAN PICKERING AND PETER J. WOOD

The issue of finance for climate change mitigation and adaptation in developing countries has been prominent in international climate negotiations since the UNFCCC was negotiated in 1992. There are three major areas that need to be addressed (Parker *et al.*, 2009a; Persson *et al.*, 2009).

- Generation: how much finance needs to be raised, who should contribute funds and through what mechanisms?
- Delivery: how should funds be delivered and to whom?
- Governance: what institutional arrangements are required to govern the provision of climate finance?

Generation

Estimates of how much finance is needed vary widely but most agree that financing needs for mitigation and adaptation in developing countries will be at least US$100 billion annually by 2020 (Figure 12.5; UNFCCC, 2007; World Bank, 2009).

The Copenhagen Accord (Chapter 13) includes specific aggregate commitments on finance. Developed countries commit to provide in the short term (2010–12) a figure 'approaching' US$30 billion, and US$100 billion annually by 2020 on the basis of substantial action on mitigation and improved transparency (Copenhagen Accord, para. 8). The Copenhagen Accord states that funding for these commitments will come from a wide range of sources – including bilateral and multilateral, and public and private (para. 8) – but agreement on specific sources is yet to be reached.

Some finance could come from private sources, including carbon markets, with reform and expansion to the currently available mechanisms. A framework for using public finance mechanisms to leverage private investment in mitigation has also been discussed (Ward *et al.*, 2009).

In addition to public finance from existing national budgets, it is widely agreed that 'alternative' or 'innovative' sources are also needed. These could include national auctioning of allowances; a US$2 per tonne global carbon tax, as proposed by Switzerland; international auctioning of Kyoto assigned amount units, as proposed by Norway; a levy on carry-over of surplus assigned amount units between the first and second Kyoto commitment periods; raising funds from international maritime and aviation fuels ('bunker fuels'), which could consist of emissions trading schemes, levies, or a combination of both; a tax on sovereign wealth funds, or a framework for sovereign wealth funds to invest in an international climate fund; and a currency transactions tax (Parker *et al.*, 2009a). In February 2010, the UN Secretary-General announced the formation of a high-level Advisory Group on Climate Change Financing to develop proposals for raising funds (with a focus on new and innovative sources) and report to parties before COP16 in late 2010.

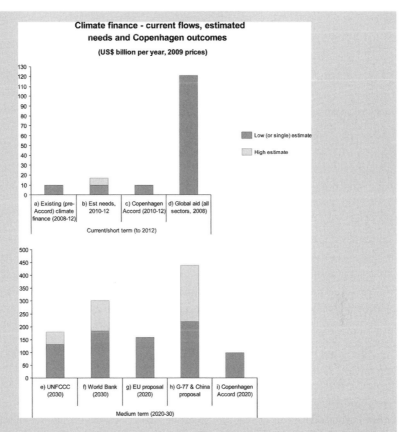

Figure 12.5 Financing for mitigation and adaptation. Sources: Jonathan Pickering and Peter J. Wood, based on World Bank (2009) (columns a, f); Project Catalyst (2009) (b); UNFCCC (2009) (c, i); OECD (2010) (d); UNFCCC (2007) (e); European Commission (2009) (g); and Parker *et al.* (2009b) (h). Commitments in Accord and estimated needs are assumed to be largely additional to pre-Accord funding.

There seems to be broad agreement that, consistent with principles set out in the UNFCCC, countries that are wealthier and have contributed more to causing climate change ought to shoulder more of the financing burden (Dellink *et al.*, 2009; Pendleton and Retallack, 2009). Disagreement remains over whether this implies contributions exclusively from developed countries or also (as envisaged in Mexico's financing proposal) from more advanced developing countries.

Delivery

As mitigation finance produces a global public good, cost-effectiveness of emissions reductions will play a role in determining where and how it is allocated. It is likely that a substantial

Box 12.5 (*cont.*)

proportion of short-term mitigation funding will be directed towards reducing forest-based emissions.

It is widely agreed that vulnerability should be a primary criterion in the allocation of adaptation finance (Copenhagen Accord, para 8). Given the mixed results achieved from development assistance to date, the varying capacities of countries to absorb significant amounts of new funding is also an issue (World Bank, 2009).

Governance

Key questions in establishing institutional arrangements for climate finance include: whether to use new institutions or reform existing ones; whether funding is channelled through a single fund or fragmented into different streams; and the respective roles of recipient and contributing countries in determining funding priorities (Parker *et. al.*, 2009a).

Reflecting proposals launched by Mexico and Norway preceding and during COP15 (including a joint proposal by the two countries and a joint non-paper also co-sponsored by Australia and the UK), the Accord establishes a 'Copenhagen Green Climate Fund' through which a significant portion of new global funding would flow (Copenhagen Accord, para. 10). Fast-start financing between 2010 and 2012 is likely to rely heavily on existing institutions, including those associated with the UNFCCC (such as the CDM, the Global Environment Facility, and the Adaptation Fund), and bilateral and multilateral development finance institutions.

If the UNFCCC agrees on a framework for long-term cooperative action including NAMAs by developing countries, a mechanism such as a registry could be established to facilitate matching finance with NAMAs that require international support. Institutional elements of the Adaptation Fund, such as majority developing country representation and direct access for recipient countries, could provide precedents for a longer-term future architecture.

12.6 Summary and conclusions

Climate change is a highly relevant issue for economic policy and consideration of economics is essential in the response to climate change. This is true not only because of the potential for climate change to disrupt economic growth and prosperity, but also because effective climate change mitigation requires fundamental changes in economic systems. They, in turn, require sound economic analysis and well-designed economic policies.

The key economic policy instrument for mitigation is for governments to put a price on emissions, through either cap-and-trade systems or emissions taxes, with emissions trading the more widely used approach. Emissions pricing facilitates a broad, cost-effective abatement response. Regulatory and fiscal instruments have important complementary roles where there are non-climate externalities or additional policy objectives, for example to foster renewable energy generation.

In practice, a mix of different policies is employed, often chosen and designed in response to political factors. The allocation permits under emissions trading is a case in point: governments often give free permits to large emitters well beyond what is economically justifiable, at the expense of others in society.

Economic models agree that, although comprehensive climate change mitigation requires extensive economic restructuring – especially in the energy sector – the macro-economic cost of mitigation is small relative to expected future economic growth. The vast majority of cost estimates for a 60% emissions reduction from the baseline lie between 0% and 5% of GDP. If stretched over a 40-year horizon, a 5% reduction in GDP amounts to a reduction in annual growth of just over 0.1%, compared to an underlying annual increase in global GDP of around 3%.

Promoting sound climate change adaptation strategies presents a different set of challenges to economic policymakers. Adaptation sometimes will not cost much. In other cases, large investments will be needed, often under uncertainty. Adaptation will often be local or regional in scale, reducing the need for governments to facilitate cooperative action but exacerbating the need for reliable climate information at fine detail. For economic policy to facilitate efficient adaptation action by individuals, communities and businesses, existing policy settings need to be screened for whether they support or hinder adaptive responses.

Finally, a defining aspect of climate change as an economic problem is the international dimension of the problem, in particular how to achieve cooperation among nations on mitigation and how to overcome stark differences in economic capabilities to deal with climate change. On this front, the international community has been making progress. New forms of international mitigation mechanisms are emerging, and climate financing for developing countries is increasingly becoming available. However, the pace of progress in these areas, and the magnitude of financing provided, needs to dramatically increase for the world to effectively tackle climate change.

References

Ackerman, F., DeCanio, S. J., Howarth, R. B. and Sheeran, K. (2009). Limitations of integrated assessment models of climate change. *Climatic Change*, **95**, 297–315.

Aldy, J. E. and Stavins, R. N. (eds.) (2007). *Architectures for Agreement: Addressing Global Climate Change in the Post-Kyoto World*. Cambridge, UK: Cambridge University Press.

Anthoff, D., Hepburn, C. and Tol, R. S. J. (2009). Equity weighting and the marginal damage costs of climate change. *Ecological Economics*, **68**, 836–49.

Asian Development Bank (2009). *The Economics of Climate Change in Southeast Asia: A Regional Review*. n.p.: Asian Development Bank.

Babiker, M. H., Reilly, J. M. and Viguier, L. L. (2004). Is international emissions trading always beneficial? *Energy Journal*, **25**, 33–56.

Barker, T., Qureshi, M. S. and Köhler, J. (2006). *The Costs of Greenhouse Gas Mitigation with Induced Technological Change: A Meta-analysis of Estimates in the Literature*. Cambridge, UK: Cambridge Centre for Climate Change Mitigation Research.

Barker, T., Scrieciu, Ş. S. and Foxon, T. (2008). Achieving the G8 50% target: Modelling induced and accelerated technological change using the macro-econometric model E3MG. *Climate Policy*, Special Issue on 'Modelling long-term scenarios for low-carbon societies', **8**, S30–S45.

Barker, T., Dagoumas, A. and Rubin, J. (2009). The macroeconomic rebound effect and the world economy. *Energy Efficiency*, **2**, 411–27.

Baumol, W. J. and Oates, W. E. (1971). The use of standards and prices for protection of the environment. *Swedish Journal of Economics*, **73**, 42–54.

Beckerman, W. and Hepburn, C. J. (2007). Ethics of the discount rate in the Stern Review on the economics of climate change. *World Economics*, **8**, 187–210.

Bowen, A. and Stern, N. (2010). Environmental policy and the economic downturn. *Oxford Review of Environmental Economics and Policy*, forthcoming.

California Public Utilities Commission (2009). *California's Renewable Energy Programs*. San Francisco: California Public Utilities Commission.

Christiansen, A. C. and Wettestad, J. (2003). The EU as a frontrunner on greenhouse gas emissions trading: how did it happen and will the EU succeed? *Climate Policy*, **3**, 3–18.

Ciscar, J.-C., Iglesias, A., Feyen, L. *et al.* (2009). *Climate Change Impacts in Europe*. Final Report of the PESETA Research Project. Seville: European Commission Joint Research Centre.

Dellink, R., den Elzen, M., Aiking, H. *et al.* (2009). Sharing the burden of financing adaptation to climate change. *Global Environmental Change*, **19**, 411–21.

Dietz, S. and Hepburn, C. (2010). On non-marginal cost-benefit analysis. Mimeo. London School of Economics, Grantham Research Institute on Climate Change and the Environment, Working Paper No. 18, March 2010.

Enkvist, P.-A., Naucler, T. and Rosander, J. (2007). A cost curve for greenhouse gas reduction. *McKinsey Quarterly*, **1**, 35–45.

Esteban, M., Webersick, C. and Shibayama, T. (2009). Estimation of the economic costs of non adapting Japanese port infrastructure to a potential increase in tropical cyclone intensity. *IOP Conference Series: Earth and Environmental Science*, **6**, 322003.

European Commission (2009). *Stepping Up International Climate Finance: A European Blueprint for the Copenhagen Deal*. COM(2009) 475/3. Brussels: Commission of the European Communities.

European Commission (2010). *EU Energy in Figures 2010: Electricity Generation from Renewables. Extended Time Series*. Brussels: European Commission Directorate-General for Energy and Transport.

Federal Ministry for the Environment, Nature Conservation and Nuclear Safety (2008). *Act Revising the Legislation on Renewable Energy Sources in the Electricity Sector and Amending Related Provisions*. Berlin: German Federal Ministry of Economics and Technology.

Garnaut, R. (2008). *The Garnaut Climate Change Review*. Melbourne: Cambridge University Press.

Goulder, L. H., Hafstead, M. A. C. and Dworsky, M. S. (2009). *Impacts of Alternative Emissions Allowance Allocation Methods under a Federal Cap-and-Trade Program*. NBER Working Paper No. 15293, August. Cambridge, MA: National Bureau of Economic Research.

Grubb, M. J. and Neuhoff, K. (2006). Allocation and competitiveness in the EU emissions trading scheme: Policy overview. *Climate Policy*, **6**, 7–30.

Hallegatte, S. (2009). A roadmap to assess the economic cost of climate change with an application to hurricanes in the United States. *IOP Conference Series: Earth and Environmental Science*, **6**, 322001.

Held, A., Haas, R. and Ragwitz, M. (2006). On the success of policy strategies for the promotion of electricity from renewable energy sources in the EU. *Energy and Environment*, **17**, 849–68.

Hepburn, C. (2006). Regulation by prices, quantities, or both: A review of instrument choice. *Oxford Review of Economic Policy*, **22**, 226–47.

Hepburn, C., Grubb, M., Neuhoff, K., Matthes, F. and Tse, M. (2006). Auctioning of EU ETS phase II allowances: How and why? *Climate Policy*, **6**, 137–60.

Hoel, M. and Karp, L. S. (2001). Taxes and quotas for a stock pollutant with multiplicative uncertainty. *Journal of Public Economics*, **82**, 91–114.

Hof, A. F., den Elzen, M. G. J. and van Vuuren, D. P. (2010). Including adaptation costs and climate change damages in evaluating post-2012 burden-sharing regimes. *Mitigation and Adaptation Strategies for Global Change*, **15**, 19–40.

Hulme, M. (2009). *Why We Disagree About Climate Change: Understanding Controversy, Inaction and Opportunity.* Cambridge, UK: Cambridge University Press.

Hunt, A. (2008). Informing adaptation to climate change in the UK: Some sectoral impact costs. *Integrated Assessment*, **8**, 41–71.

International Energy Agency (2009a). *Addressing Climate Change. Policies and Measures Database*. December 2009 version. Paris: International Energy Agency.

International Energy Agency (2009b). *Electricity Information 2009*. Paris: International Energy Agency.

Intergovernmental Panel on Climate Change (IPCC) (2001). *Mitigation. Contribution of Working Group III to the Third Assessment Report of the Intergovernmental Panel on Climate Change*, eds. B. Metz, O. Davidson, R. Swart and J. Pan. Cambridge, UK: Cambridge University Press.

Intergovernmental Panel on Climate Change (IPCC) (2007). *Fourth Assessment Report of the Intergovernmental Panel on Climate Change*. Cambridge, UK and New York, NY: Cambridge University Press.

Isoard, S. and Swart, R. (2008). Adaptation to climate change. In *Impacts of Europe's Changing Climate – 2008 Indicator-based Assessment*. Copenhagen: European Environment Agency, pp. 161–66.

Jacobsson, S. and Lauber, V. (2006). The politics and policy of energy system transformation – explaining the German diffusion of renewable energy technology. *Energy Policy*, **34**, 256–76.

Jaffe, A. B., Newell, R. G. and Stavins, R. N. (2005). A tale of two market failures: technology and environmental policy. *Ecological Economics*, **54**, 164–74.

Jänicke, M. and Jacob, K. (2004). Lead markets for environmental innovations: A new role for the nation state. *Global Environmental Politics*, **4**, 29–46.

Joskow, P. L. and Schmalensee, R. (1998). The political economy of market-based environmental policy: the U. S. acid rain program. *Journal of Law and Economics*, **41**, 37–83.

Jotzo, F. and Pezzey, J. C. V. (2007). Optimal intensity targets for greenhouse emissions trading under uncertainty. *Environmental and Resource Economics*, **38**, 259–84.

Lewis, M. C. and Curien, I. (2010). *The Crying of Lot 2013: Phase-3 Auctions and the EUA Price Outlook*. London: Deutsche Bank.

Markussen, P. and Svendsen, G. T. (2003). Industry lobbying and the political economy of GHG trade in the European Union. *Energy Policy*, **33**, 245–55.

McKibbin, W. J. and Wilcoxen, P. (2007). A credible foundation for long term international cooperation on climate change. In *Architectures for Agreement: Addressing Global Climate Change in the Post-Kyoto World*, eds. J. E. Aldy and R. N. Stavins. Cambridge University Press, pp. 185–208.

McKibbin, W. J. and Wilcoxen, P. J. (2002). *Climate Change Policy after Kyoto: Blueprint for a Realistic Approach*. Washington, D.C.: Brookings Institution Press.

McKibbin, W. J., Morris, A. and Wilcoxen, P. (2009). *A Copenhagen Price Collar: Achieving Comparable Effort through Carbon Price Agreements*. Washington, D.C.: Brookings Institution Press.

Mendonça, M. (2007). *Feed-in Tariffs: Accelerating the Deployment of Renewable Energy*. London: Earthscan.

Nielsen, L. and Jeppesen, T. (2003). Tradable green certificates in selected European countries – overview and assessment. *Energy Policy*, **31**, 3–14.

Nordhaus, W. (2001). Climate change: global warming economics. *Science*, **294**, 1283–84.

Nordhaus, W. D. (1991). *Economic Approaches to Greenhouse Warming*. Cambridge, MA: MIT Press.

Nordhaus, W. D. (2007). To tax or not to tax: Alternative approaches to slowing global warming. *Review of Environmental Economics and Policy*, **1**, 26–44.

OECD (2007). *Instrument Mixes for Environmental Policy*. Paris: OECD.

OECD (2010). *International Development Statistics (IDS) online databases on aid and other resource flows*, Paris: OECD, http://www.oecd.org/dac/stats/idsonline.

Parker, C., Brown, J., Pickering, J. *et al.* (2009a). *The Little Climate Finance Book*. Oxford: Global Canopy Foundation.

Parker, C., Mitchell, A., Trivedi, M., Mardas, N. and Sosis, K. (2009b). *The Little REDD+ Book*. Oxford: Global Canopy Foundation.

Parry, M., Arnell, N., Berry, P. *et al.* (2009). *Assessing the Costs of Adaptation to Climate Change: A Critique of the UNFCCC Estimates*. London: IIED.

Pearce, D. (1991). The role of carbon taxes in adjusting to global warming. *The Economic Journal*, **101**, 938–48.

Pendleton, A. and Retallack, S. (2009). *Fairness in Global Climate Change Finance*. London: Institute for Public Policy Research, http://www.boell.de/downloads/ecology/fairness_global_finance.pdf.

Persson, Å., Klein, R. J. T., Siebert, C. K. *et al.* (2009). *Adaptation Finance Under a Copenhagen Agreed Outcome*. Research Report. Stockholm: Stockholm Environment Institute.

Pezzey, J. (1992). The symmetry between controlling pollution by price and controlling it by quantity. *Canadian Journal of Economics*, **25**, 983–91.

Pezzey, J. C. V., Jotzo, F. and Quiggin, J. (2008). Fiddling while carbon burns: Why climate policy needs pervasive emission pricing as well as technology promotion. *Australian Journal of Agricultural and Resource Economics*, **52**, 97–110.

Pezzey, J. C. V., Mazouz, S. and Jotzo, F. (2009). *The Logic of Collective Action and Australia's Climate Policy*. EERH Research Report No. 24. Canberra: The Australian National University.

Pigou, A. C. (1938). *The Economics of Welfare*. London: Weidenfeld and Nicolsen.

Pizer, W. A. (2002). Combining price and quantity controls to mitigate global climate change. *Journal of Public Economics*, **85**, 409–34.

Project Catalyst (2009). *Potential Uses of 'Fast Start' Funding in 2010–12*. Briefing Paper, December 2009. San Francisco: Project Catalyst, http://www.project-catalyst.info/images/publications/fast_start_funding.pdf.

Renewable Energy Policy Network (2009). *Renewables Global Status Report: 2009 Update*. Paris: Renewable Energy Policy Network.

Sijm, J., Neuhoff, K. and Chen, Y. (2006). CO_2 cost pass-through and windfall profits in the power sector. *Climate Policy*, **6**, 49–72.

Sorrell, S. (2007). *The Rebound Effect: An Assessment of the Evidence for Economy-wide Energy Savings from Improved Energy Efficiency*. London: UK Energy Research Centre.

Stern, N., Peters, S., Bakhshi, V. *et al.* (2006). *Stern Review: The Economics of Climate Change*. London: HM Treasury.

Stern, N. H. (2007). *The Economics of Climate Change: The Stern Review*. Cambridge, UK: Cambridge University Press.

UNFCCC (2007). *Investment and Financial Flows to Address Climate Change*. Bonn: UNFCCC, http://unfccc.int/files/cooperation_and_support/financial_mechanism/application/pdf/background_paper.pdf.

UNFCCC (2009). *Copenhagen Accord*, http://unfccc.int/files/meetings/cop_15/application/pdf/cop15_cph_auv.pdf.

United Nations (2008). *World Population Prospects: The 2008 Revision Population Database*. United Nations Population Division.

US Energy Information Administration (2010). *Official Energy Statistics from the US Government: Electricity*. Washington, D.C.: US Energy Information Administration.

van der Linden, N. H., Uyterlinde, M. A., Vrolijk, C. *et al.* (2005). *Review of International Experience with Renewable Energy Obligation Support Mechanisms*. Amsterdam: Energieonderzoek Centrum Nederland.

Wara, M. and Victor, D. G. (2008). *A Realistic Policy on International Carbon Offsets*. Program on Energy and Sustainable Development Working Paper #74. Stanford: Stanford University.

Ward, M. (ed.) (2008). *The Role of Sectoral No-lose Targets in Scaling Up Finance for Climate Change Mitigation Activities in Developing Countries*. UK: Department for Environment, Food and Rural Affairs.

Ward, J., Fankhauser, S., Hepburn, C., Jackson, H. and Rajan, R. (2009). *Catalysing Low-carbon Growth in Developing Economies: Public Finance Mechanisms to Scale Up Private Sector Investment in Climate Solutions*. Report for UNEP and Partners, http://www.unep.org/PDF/PressReleases/Public_financing_mechanisms_report.pdf.

Weitzman, M. (2009a). On modeling and interpreting the economics of catastrophic climate change. *The Review of Economics and Statistics*, **91**, 1–19.

Weitzman, M. L. (1974). Prices vs quantities. *Review of Economic Studies*, **41**, 477–91.

Weitzman, M. L. (2009b). What is the 'damages function' for global warming and what difference might it make? Mimeo, Harvard University.

Wiser, R. and Bolinger, M. (2008). *Annual Report on US Wind Power Installation, Cost, and Performance Trends: 2007*. Washington, D.C.: US Department of Energy – Energy Efficiency and Renewable Energy.

World Bank (2009). *World Development Report 2010: Development and Climate Change*. Washington, D.C.: The World Bank, http://siteresources.worldbank.org/INTWDR2010/Resources/5287678–1226014527953/WDR10-Full-Text.pdf.

World Resources Institute (2009). *Climate Analysis Indicators Tool*. Washington, D.C.: World Resources Institute.

13

Geopolitics and governance

*'It's also why the world must come together to confront climate change.
There is little scientific dispute that if we do nothing, we will face more
drought, more famine, more mass displacement – all of which will fuel
more conflict for decades. For this reason, it is not merely scientists and
environmental activists who call for swift and forceful action – it's military leaders in my own country and others who understand our common
security hangs in the balance.'*[1]

Politicians have observed regularly in recent years that climate change first emerged as an
'environmental' issue, which later became reconceptualised as an 'economic' issue, and is
now perceived as a 'security' issue. What exactly this means is not always clear – notably
whether it means that climate change can cause violent conflicts or that climate change as
such is the biggest threat humankind faces (Section 13.6; Wæver, 2009a). Nevertheless,
this transformation in the perception of climate change carries with it the unmistakable
message that the importance of climate change has moved up some kind of ladder of politicisation. Climate change is not a technical issue, but a political one – and it belongs to the
category of 'high politics' (Hoffmann, 1966) in that it affects the survival of the state (as
opposed to 'low politics', which deal with more mundane or practical issues).

Nevertheless, this increased focus on conflict in climate change has tended, so far, to
be treated as being separate from the central 'political science' stake in climate change
research, i.e. understanding negotiations and the formation of efficient and appropriate
regimes for dealing with climate change. It has been as if conflict and cooperation operate
on different planes.[2] A major change occurred through the processes related to the COP15
in Copenhagen in December 2009, whereby high politics and security were integrated with
negotiation and regime formation. A key element of this transformation of climate change
into high politics was – as the present chapter will elucidate – the concern about *relative*
(i.e. comparative) economic growth and, thereby, power ascendancy among the leading

[1] Nobel Peace Prize speech by Barack H. Obama, Oslo, 10 December 2009.
[2] A somewhat similar observation about two hitherto unconnected literatures about the international politics of the climate – one
on cooperation and another on security – is made by Giddens (2009).

Climate Change: Global Risks, Challenges and Decisions, Katherine Richardson, Will Steffen and Diana Liverman *et al.* Published
by Cambridge University Press. © Katherine Richardson, Will Steffen and Diana Liverman 2011.

global powers. From being treated as a question of 'climate disagreements', i.e. specific disagreement as to how to deal with the issue, climate politics became too consequential to remain technical. At the level of international affairs, climate policy thereby became caught in a triangle between the actual climate situation, national power and economic growth.

As noted in the previous chapter, climate change has, in recent years, moved from being the domain of environmental (or climate) ministers to increasingly involving ministers of finance, not least of which in domestic politics. However, understanding the events occurring at COP15 is best achieved by interpreting behaviour in relation to general foreign policy lines, e.g. in China–US relations. Such overarching policy integration was symbolised by climate change moving from the hands of environmental or economic ministers to the heads of states or government.

Moving from a 'how to create cooperation' perspective to 'the geopolitics of climate change' is reflected in a key feature of the COP15 course of events: a central element in the strategy for the meeting was to lift it to the level of heads of states and governments. The reasoning behind this included two key ideas: first, this would signal importance and commit countries to more responsible behaviour; and second, that it would be easier to make difficult decisions when the real decision-makers, rather than their delegates, were negotiating. Slightly simplified, the expectation was that the lower levels in the hierarchy would, during the months before and the first week of the COP15 meeting, prepare the major part of an ambitious agreement, and then the heads of states and governments would be able to put the last pieces in place when they came to Copenhagen.

This strategy did not, however, play out as hoped in that it was more difficult for the lower echelons of the hierarchy to agree when they knew that the products of their labours would be scrutinised by their leaders and that their leaders would be held responsible for the text in the final phase of the negotiations. In addition, when the political leaders met in the final phase, they brought with them not only the power of agreement-making, but also the numerous conflicts and concerns already existing between and within the individual countries. This illustrates an underlying flaw in the image many have of the international politics concerning climate. When the climate change response is simply viewed as being a task to execute, it seems obvious that the more political power is assembled, the easier the task becomes. However, in reality, dealing with climate change is not simply a task of execution. It is political in the much deeper and more troublesome sense that it involves conflicts of interest and perspective, not only conflicts over climate change per se, but also the general existing conflicts among parties. The COP15 strategy was justified and successful in the sense that it 'raised the stakes' and made climate inaction more costly politically for leaders. However, the other side of this issue is that the conflicts between nations also are important in domestic politics. As climate change considerations move from being treated in isolation to the centre stage of international politics, different analytical and political approaches need to be developed that are more appropriate for this new situation.

Several of the relevant lines of research currently point towards the importance of gaining a truly political understanding of the interconnected domestic and international politics of climate change (Christoff, 2006; Barnett, 2008; Delmas and Young, 2009; Giddens,

2009; Paterson, 2009; Rajamani, 2009; Aldy and Stavins, 2010). The difficulty is to build a dynamic political process in global decision-making concerning climate change, where this dynamism is anchored in the way these international issues are seen and processed in domestic politics.

Based on this recent realisation of the need to integrate institutional cooperation and security conflicts, this chapter has no strict equivalent in the assessment literature, such as in the IPCC AR4 (but there is a more closely related one in the outline for IPCC AR5). Therefore, it is not enough to summarise an established sub-field as we do in most of the other chapters. This chapter has important links with Chapter 15 on mobilisation and multi-level action, Chapter 12 on economics and Chapter 9 on equity.

A final caveat is that this chapter will not cover all climate relevant regimes, only those directly geared to 'preventing dangerous anthropogenic interference with Earth's climate system' by restraining CO_2 emissions. Geoengineering, as well as adaptation and financing, are side issues brought in when particularly relevant. However, the chapter cannot include general analyses of, for example, all types of financial cooperation which may be relevant for the development of mitigation and adaptation measures.

13.1 Designing optimally structured agreements for curbing greenhouse gas emissions

Existing and variants of possible future agreements designed to reduce greenhouse gas emissions in order to avoid dangerous human interference with the climate system can be assessed according to a number of criteria (e.g. Chapters 9 and 12; Aldy and Stavins, 2007, 2010; Gupta *et al.*, 2007; Kuik *et al.*, 2008; Whalley and Walsh, 2009; Gainza-Carmenates *et al.*, 2010). One of the most comprehensive and systematic analyses (Aldy and Stavins, 2007) uses six criteria for 'evaluating climate policy architectures': (i) environmental outcome, (ii) dynamic efficiency (maximises the aggregate present value of net benefits of taking actions to mitigate climate change impacts), (iii) dynamic cost-effectiveness (identification of the least costly way to achieve a given environmental outcome), (iv) distributional equity, (v) flexibility in the presence of new information, and (vi) participation and compliance. Or in more streamlined form: the policy response must be 'scientifically sound, economically rational, and politically pragmatic' (Aldy and Stavins, 2010).

This literature indicates, amongst other things, that the Kyoto Protocol is cost-effective in the sense that it creates market-oriented institutions (including emission trading). In addition, it scores well on some elements of equity. However, it is weak primarily because four of the five largest emitters are *de facto* not restrained by the protocol: China and India, as they are not parties to the relevant annex; the US due to non-ratification; and Russia as a result of having received such generous targets as not to be constrained. This threatens the environmental outcome of the protocol, its dynamic efficiency, cost-effectiveness, and the incentives for participation and compliance. Analyses of alternative architectures – either other forms of targets and timetables (e.g. carbon taxes), or harmonised domestic actions or coordinated and unilateral policies – show different mixes of strengths and weaknesses.

Much of this has to do with technical and economic aspects of these schemes, which are treated in other chapters of this book. However, such analyses also include political elements dealing with feasibility, and the architecture's ability to generate and sustain participation. Most importantly, the Kyoto model does not provide strong incentives for states to participate (or comply). This is further discussed later in this section.

It has been demonstrated that radically different approaches can yield serious candidates for a climate change agreement structure (Aldy and Stavins, 2007, 2010; Delmas and Young, 2009). All contain different advantages and weaknesses. Notably, the model that the world has, so far, chosen for its agreements, i.e. the Kyoto Protocol, has advantages in terms of speed and cost-effectiveness, if only it could achieve a much higher rate of participation and compliance (and a tightening of targets according to new scientific findings). This 'if' is, however, very large, for reasons the literature identifies well. The failure to achieve high rates of participation and compliance is often presented in the press as being a result of a lack of will. In fact, the current architecture of the political agreements dealing with climate change is simply not good at generating will.

Seen in a political science context, the literature develops parts of an understanding of not only comparative statics in the sense of different possible orders, but also the politics of getting each to work, and the question of whether the architectures have the necessary political dynamics built into their structure in order to succeed. However, a limitation of this literature is that it – possibly as a necessary thought experiment – frames the political response as being a question of what architecture 'we' shall choose. This is unrealistic both because the choice of architecture is, itself, political, i.e. part of what parties struggle over, and because it is a non-option to start anew. At this point it is unrealistic to assume that the political process already established would stop, change track, annul all commitments to Kyoto and start afresh with another format for agreement. For better or worse, it is clear that continuity from the current system has to be built into future decision-making and adoption of response systems – not least of which because momentum, mobilisation and commitment would be lost through scrapping Kyoto. This is taken into account in the various kinds of combination response models discussed later in this chapter, where different regimes coexist. In such combination models, one always needs analyses such as the one presented above of the 'architecture for agreement' in order to assess the strength and weaknesses of the different contributing regimes.

Some of the most systematic analyses in this field have been carried out by Barrett (2003, 2007, 2008a, b). Through a comprehensive review of international environmental agreements, he shows that enforcement is critical and has to be built into the basic architecture of an agreement (i.e. is not easily added later). Thus, an agreement has to enable states to make credible threats, both to deter free-riding and to enforce compliance. These points can be illustrated by comparing the Kyoto Protocol with the Montreal Protocol. Under the latter, which restricts the use of CFC gases demonstrated to damage the ozone layer (Chapter 17), trade restrictions with non-parties function to make the regime self-enforcing and make punishment rational. In contrast, the Kyoto Protocol fails to restructure the climate game.

As shown in Chapter 12, economic cost-benefit calculations are complicated and controversial. In any event, they often demonstrate a rather narrow margin of benefit in adopting measures to curb greenhouse gas emissions (Barrett, 2003). This is in contrast to the Montreal Protocol, where economic benefits of reducing CFC emissions for a key country like the USA far outweighed the economic costs of the action. Therefore, the Kyoto regime becomes vulnerable to free-riding and non-compliance. Benefits are dependent on the action of others and, when the margin of benefit-to-cost is narrow, the failure of other countries to deliver their share can more easily overturn the cost–benefit calculation of a state considering its participation. The compliance mechanism in the Kyoto Protocol is weak (and can only be strengthened by amendments approved by three quarters of the parties and approved amendments are binding only for the countries that ratify the amendment). The compliance mechanism in the Kyoto Protocol consists primarily of suspension from emissions trading, an obligation to explain non-compliance, presentation of a plan to get back on track, and a penalty of having to deliver 1.3 times the missing reduction in the next period. This 'punishment' amounts to a borrowing from future emission cuts at a quite low interest rate, and easily turns into indefinite postponement. In addition, the derelict party is likely to try to negotiate down its target by a similar magnitude for the next period. There is no mechanism for enforcement of the penalty beyond voluntary fulfilment of the tightened obligation in the next period by the party that has violated its obligations. Finally, it is possible for countries to withdraw from the Kyoto Protocol without penalty (Grubb *et al.*, 1999; Victor, 2001; Freestone and Streck, 2005).

The major difference between the Montreal and Kyoto protocols is that, if one were to use trade restrictions in relation to climate change policies, first it would require some very complex and unwieldy calculations to decide the carbon content of products and, second, it would create tensions with the international trade regime, the World Trade Organization (WTO). Third, and most importantly, however, carrying out the punishment would often be detrimental to those carrying it out and thus not rational, because the penalising state would run the risk of trade wars due to the ambiguous basis for deciding on taxes. Finally, a usage of taxes against non-signatories would be controversial and, if constrained to participants, would de-motivate participation.

Enforcement of international agreements is generally weak if it takes the form of central rulings convincing a state in violation to come back in line (and maybe even pay a penalty as under Kyoto). It is equally vulnerable if enforcement demands some kind of costly, collective enforcement by all other states (as in collective security systems), and much more efficient if the punishment is produced by other states in a way that aligns with their self-interest as best known from the WTO, where a WTO decision against a country permits the complaining country to rebalance through its own tariffs. Such threats are credible and make states comply out of self-interest. Kyoto does not have this built-in mechanism.

Barrett's own solution for climate change is a technology-focused one (Chapter 10) including the imposition of technology standards (Barrett, 2003, 2007, 2008a, 2010). A natural companion is international collaborative research and development, but the decisive element – and that which brings out the key question of dynamics – is that of mandating technological standards for key emission-reducing processes. Among the advantages of this strategy is that it can be launched by a smaller group, but would have its own pull mechanisms making it attractive for others to join. The standard can create a tipping effect. Tipping can be assisted further by trade restrictions much more easily (and compatible with the WTO) than CO_2-based border taxation. Enforcement is less of a problem, because incentives shift in favour of following the standards.

This strategy has its own problems, however. First of all, it is not clear that it is able to work quickly enough in relation to actual climate effects occurring. It also rests on assumptions about technologies yet to emerge, and raises demands on rich countries for financing research and development, transferring technologies, and assisting in adaptation. In addition, the process of choosing standards will be difficult to design and will become highly politicised. In any case, the task of the present chapter is not to assess the different competing models and architectures; the main point taken from Barrett, and in line with most of the contributions to the comprehensive volumes on 'architectures for agreement' by the Harvard Project on International Climate Agreements (Aldy and Stavins 2007, 2010), is that it is crucial the international regime has a self-propelling dynamic of attraction, which stimulates membership and compliance.

Barrett's findings correspond to one of the most immediate conclusions from more abstract research on rational choice and game theory, one of the most dynamic and influential branches of the discipline in the last decades. Reduction in CO_2 emissions is a global public good; that is, non-rivalrous and non-excludable, and available for consumption by all – and, therefore, prone to problems of free-riding. Large-scale market failure is to be expected and can be remedied according to conventional wisdom only by explicitly collective action (Chapter 12; Grasso, 2004; Brennan, 2009; Grundig, 2009; de Mesquita, 2009).

From the theoretical arsenal of game theory, two more unusual and idiosyncratic ideas relevant to governance issues in relation to climate change are, in particular, worth mentioning.

- Nobel Laureate Schelling (1978) developed the concept of a 'k-group' – a core of parties small enough that each member's contribution to the public good can be made conditional on the contribution of others. Such a core can produce the necessary leadership to produce a wider agreement. Notably, this kind of agreement does not come about through universal, multilateral procedures. A small group of actors has a sufficient interest in creating a public good that they go ahead and absorb the costs in order for the good to be created, even if they know that free-riders cannot and will not be excluded from enjoying the benefits of the resulting public good.
- Nobel Laureate Ostrom (1990, 2009) demonstrated how common pool resources – such as forests, fisheries and grazing lands – are dealt with through institutional arrangements that avoid the expected 'tragedy of the commons' outcome. Humans can – under the right circumstances – develop trust in the action of others, and deal with an issue in terms of not 'me against you' but 'all of us against ourselves'. Typically, this is done in local

settings with attention to specifics of the case, delineating clearly the parties and with monitoring, sanctions and conflict resolution anchored in the community. Ostrom is wary of top-down arrangements that the population does not identify with, and emphasises the importance of building stronger commitments by cooperation at multiple levels linked together through information networks and monitoring at all levels. This polycentric approach underpins the case for both multiple fora at the international level and for multilevel systems (Chapters 14 and 16). Finally, this analysis shows the importance of framing or social construction. An agreement has to have a self-sustaining dynamic not only by way of incentives as emphasised by game theory and rational choice perspectives, but also in its cognitive dimension in terms of the understanding and attitude it generates. That is, if a set-up stimulates 'us-them' competitive perspectives, this makes it self-undermining, in contrast to an arrangement that through its decision stimulates social constructions of collective management of a joint problem.

Regime theory has been one of the predominant approaches of the past 30 years to the study of international cooperation among states, covering both organisations in the narrow sense and cooperation in the broader sense of conventions and informal coordination. Globalisation has increasingly presented problems that the nation-states are not in a position to solve individually, and this has presented a rising demand for governance. The most common response has been the negotiation of intergovernmental agreements, whereby governments get a role in the supply of governance, but 'now as players in multilateral agreements rather than purveyors of governance on their own' (Young, 2009). As noted by Young, one of the leading proponents of regime theory, the area of climate change is one where the traditional type of regime formation centred on international agreements seems unable to deliver enough, quickly enough (Young, 1998, 2009). Regime theorists have increasingly moved into either game theory-inspired analysis of the possibility of transforming the incentive structures (cf. previous discussion), or given attention to broader forms of governance – coordinated systems involving actors in the public sector, the private sector and civil society (Box 13.3 and Chapter 15; Biermann *et al.*, 2009; Delmas and Young, 2009; Okereke *et al.*, 2009).

A recent outgrowth of regime theory is the concept of 'regime complex', which proves eminently applicable to the case of climate change. Keohane and Victor (2010) argued that there is no integrated, comprehensive regime governing mitigation efforts on climate change. However, there is a regime complex, a loosely coupled set of regimes. This is seen as inhabiting a middle position between comprehensive international regulatory institutions focused on a single legal instrument and, on the other hand, highly fragmented arrangements. The climate change challenge encompasses a number of different problems (adaptation funds for developing countries, emission reductions, research and development, etc.), which create functional reasons for different regimes. In addition, a multiplicity of regimes is produced by strategic and organisational mechanisms. Owing to diversity in interests and structural variation, it would be unlikely that one global regime would form in response to human-induced climate change.

Presently, the regime complex constituting the societal response to climate change is a combination of the cluster around the UNFCCC, comprising near-universal membership with a number of other initiatives (i.e. bilateral, regional or organisational) that are not tied together in any hierarchy or integrated structure. Such a regime complex 'if it meets specified criteria, has advantages over any politically feasible comprehensive regime, particularly with respect to adaptability and flexibility' (Keohane and Victor, 2010). These standards relate to coherence (compatible and mutually reinforcing regimes), effectiveness (compliance with appropriate rules), determinancy (clear normative content), sustainability (coherent equilibrium point), accountability (relevant audiences can hold actors to standards) and epistemic quality (consistency between regime rules and scientific knowledge). Some of the specific clubs and institutions will be discussed in Section 13.4, and the benefit from the regime complex analysis is that one should not look at these different elements as competitors for *the* regime, but – at least for the foreseeable future – see them as elements that have to play together in a loosely coupled system (Biermann *et al.*, 2009).

A related case was made by Haas, who argued that multilateral diplomacy has been successful in creating international regimes in other areas of global environmental politics but that this general model is not likely to work for climate change. 'Climate change is the limiting case… for the multilateral diplomacy approach. Climate change is economically and politically more difficult than other issues yet addressed, so it is not surprising that the diplomatic efforts to date have been disappointing' (Haas, 2008). The core problem is, in a sense, well known but usually wished away: the short- to medium-term victims are mostly in the global South (Chapter 9), which has little political clout internationally, and those asked to act in the short to medium term are in the global North (Haas, 2008; Schelling, 2009).

The constellation is even more vicious than that – as shown in Section 13.3. However, the basic feature is that (in contrast to the case of ozone and the Montreal Protocol) the distribution of costs and benefits is not beneficial for agreement. In addition, there is time inconsistency, which makes it tempting to devote fewer resources at any given time than required to meet the long-term target, e.g. due to doubt about the credibility of the long-term policy (Hovi *et al.*, 2009). This obstacle would operate even for a unitary actor ('world government') but would, obviously, be aggravated by interaction with the problems produced by domestic politics and the challenge of achieving international agreement (Hovi *et al.*, 2009). Meaningful action on climate change has been averted by elevating 'cooperation' to the status of end, rather than it being a means of achieving an end (Haas, 2008), and weak institutionalisation has been deliberately chosen. The modest Secretariat and budget of the UNFCCC, for example, is a deliberate choice that has been made.

As observed by the IPCC in its Fourth Assessment Report: 'there are no authoritative assessments of the UNFCCC or its Kyoto Protocol that assert that these agreements have succeeded – or will succeed without changes – in fully solving the climate problem' (Gupta *et al.*, 2007).

The analyses presented previously in this chapter are useful for thinking systematically about possible architectures for international governance in the response to the climate change challenge. However, their main limitation is that they are mostly abstract, about 'parties'. Thus, they are not as specific as one could be if modelling the actual relevant main parties and their kinds of conflicts. We turn to this in Section 13.3. First, however, a more principled addition to the picture has to be made.

13.2 The 'relative gains' side of climate negotiations: competing for growth and wealth

One major consideration that has been underplayed in much of the existing research, but which is likely to gain more prominence as more focus is given to conflict and geopolitics in relation to climate change, is an aspect that could even be formulated in the language of rational choice theory – that of relative gains (Grieco, 1990, building on Waltz, 1979). Theorists of international relations, typically of the classical Realpolitik school, have pointed out that decisions about when to enter agreements are calculated fundamentally differently in international relations than in domestic settings because of the eventual self-help situation for states. Because the security and survival of states are their ultimate concerns and they, in the final instance, have to rely on themselves for this – rather than on a system of law, courts and police as in domestic affairs – states are concerned about their relative power vis-à-vis other states. Another state might, in the future, become a threat and a situation could ultimately be decided by power – not necessarily military, but power nonetheless. The state would not want to be weakened relative to that other state.

Therefore, this theory suggests that a state will assess a possible cooperation not in a cost-benefit manner (absolute gains), but in terms of 'who gains most' (relative gains). It is not enough that a cooperative arrangement would leave the individual state being better off compared to a situation of non-cooperation. If the arrangement benefits another major or relevant state more, the gain would be forfeited. Therefore, the typical game theoretical analysis of collective goods problems only points to one kind of weakness, i.e. the risk of others free-riding and, thereby, the cooperating state pays for something that does not work out, thus incurring more costs and fewer benefits. As, for instance, in the famous model of the 'prisoner's dilemma', it is then not rational for the state in question to cooperate. However, as if this is not bad enough for cooperation, the Realpolitik relative gains analysis adds the distinct mechanism that cooperative arrangements are so difficult to achieve in international affairs because states assess them not in terms of cost-benefit for themselves (with outcome for others as neutral or potentially valued with a steep discount rate), but viewing gains for others as a negative value.

Grundig (2006) applied this principle to climate change cooperation (compared to ozone depletion and international trade). The model established explains well the differences found between these areas, and it underlines a simple but often overlooked factor: that relative gains only matter if the stakes of issue are large enough to translate into power effects and, thus, be security relevant: 'cooperation is less likely *ceteris paribus* if potential gains are big enough to be security relevant, thus inducing relative gains concern, and if at the same time the good provided does not permit defectors to be excluded from consumption, than in cases where either exclusion is possible or where possible relative gains are small and thus do not induce relative gains concern' (Grundig, 2006). Thus, the regime dedicated to avoiding ozone depletion benefited – in addition to the other factors concerned with dangers falling in the rich countries and narrow technical solutions – from the fact that the possible relative advantages for some states in relation to others could in no way be seen as influencing general levels of growth and, thereby, future power and security. In contrast, it is possible to see climate change mitigation as carrying costs that influence overall growth. If sufficiently unevenly distributed, this could affect the future balance of power in international affairs.

Events at COP15 emphasised the importance of this type of consideration. The negotiations that came (with an unfortunate risk of becoming a self-confirming hypothesis) to be seen as a USA–China disagreement, clearly had a strong element of concern for who would get the better deal. The important question became not whether one's own negotiated arrangement was a fair or economically viable combination of costs and benefits but, rather, how it looked compared to that of the other. Actually, this was quite explicit already in the Byrd-Hagel Resolution in the US Senate (105th Congress, 1st Session, S.Res.98) that *de facto* killed US ratification of Kyoto: 'the Senate strongly believes that the proposals under negotiation, because of the disparity of treatment between Annex I Parties and Developing Countries and the level of required emission reductions, could result in serious harm to the United States economy, including significant job loss, trade disadvantages, increased energy and consumer costs, or any combination thereof'; and 'the United States should not be a signatory to any protocol … which would – (A) mandate new commitments to limit or reduce greenhouse gas emissions for the Annex I Parties, unless the protocol or other agreement also mandates new specific scheduled commitments to limit or reduce greenhouse gas emissions for Developing Country Parties within the same compliance period, or (B) would result in serious harm to the economy of the United States'.

Similarly, the press in both China and India around COP15 depicted the attempt to tie their countries to emission targets as a sophisticated weapon by the developed countries to prevent China and India from achieving growth and wealth.

When the COP15 process ran aground on a Chinese–American rock, this was not, as expounded by numerous pundits, because of an emerging rivalry between these two as the new superpowers in a new bipolar system. (The international system is not bipolar, but rather defined power-wise by one superpower, the USA, and a number of global great powers – China, Russia, Japan, the EU and possibly India, Brazil and South Africa – so the difficulty of getting agreement has a lot to do with a delay in the West of coming to terms with a more diffuse power structure, the rise of Asia and a post-Western system, but not a USA–China bipolarity; Buzan and Wæver, 2003.) When the two countries ended up in this situation and became so obsessed with each other, it was only indirectly a reflection of the global power structure, in the sense that these countries are indispensable. If negotiations had proceeded further, it would have been revealed that other countries were indispensable too. More importantly, their respective domestic political agendas framed the issue in a way that made agreement very difficult.

For the USA, the economic crisis was the dominant issue. A large part of the American public sees American economic woes as, ultimately, being caused by unfair competition from China. In this context, an American leader had to show maximum attention to relative effects on China and the USA and, actually, could benefit politically by 'standing up to' and confronting China. The USA, therefore, arrived in Copenhagen with an excessive emphasis on monitoring and verification. Thus, the role of China as the COP15 villain was more or less assured. Similarly, the dominant agenda in China is economic growth, including the threats to it from the global crisis stemming from the USA. China has no general geopolitical desire to become seen as a global superpower or peer competitor to the USA; quite the contrary. However, it certainly played well domestically to have the Chinese prime minister fend off these threatening demands on China.

In this type of political situation, dealing with human-induced climate change gets tied into a zero-sum competition for growth advantages, motivated partly by concerns about power. Arriving at cooperative solutions is extremely difficult as long as the issue is framed in this way.

It is also in this perspective that COP15 increased worry about a risk of climate tariffs. Eco-punitive tariffs were brought up by many commentators; but less, as discussed in some of the 'architecture' literature, as a structural element to make the regime operate more efficiently, and more as being a way to make the most CO_2 intensive imports more expensive; read: China (*Spiegel Online*, 24 December 2009).

The consequences of the quite abstract relative gains theory have been simulated in game theoretical models (e.g. Snidal, 1991a, 1991b), calculating expected behaviour with various mixes of absolute and relative gains motivation and varying different conditions. For some purposes, however, it is more helpful to see this as a question of two possible overarching approaches that compete. Often, they are not consolidated nor an average result calculated. Instead, the utilitarian, rational, maximising, economistic attitude of optimising the outcome of negotiations can be pushed aside whenever an issue is 'securitised', and redefined in terms of threats and a potentially hostile 'other'. Only by including this dimension does it become possible to account for the consistent 'irrationality' of international history – the numerous unhelpful wars, rivalries and missed cooperative opportunities. Much of international affairs is negatively driven, motivated not by what one tries to achieve, but by what is feared (Wæver, 2009b). Recurrent over-optimism is produced by the expectation that actors try to optimise when, in reality, they often define specific developments as completely unacceptable and simply to be prevented; these then overrule the normal calculations. The question then becomes what developments are given this status as absolute necessities – economic growth, territorial security or the prevention of dangerous climate change?

To get to cooperation, it is necessary to not only secure a positive calculation of costs and benefits associated with cooperation but also to manage how the logic of fear for other states competes with fear of climate change. Security logic is likely to intervene first as a hindrance for climate cooperation because of mutual fears among states but, potentially, security logic could also become a driver for action if climate change is seen as an overriding threat. Thus, one can imagine possible security logic *for* climate policy (Section 13.9). Such security actions against large-scale, joint threats are usually difficult to construct, and within the theory of securitisation have been analysed in terms of macro securitisation, when states align around such threat constructions (Buzan and Wæver, 2009). To be considered in this context, the threat has to be able to integrate and articulate various local and specific agendas and, therefore, needs quite high flexibility – the way that the Cold War and Global War on Terror worked in their days. Along the lines of Ostrom (2009) and Haas (2008), it is critical to manage the dynamics of framing. It is not enough to say 'therefore we should tell this story not that story'; it is necessary also to analyse through realistic political analyses the dynamics of arrangements, negotiations and domestic politics that can sustain particular political constructions of the issue.

13.3 Modelling super games and multiple axes

Often, the negotiations over climate agreements are depicted in terms of one dimension: developed vs developing countries (or USA vs China) (Box 13.1). This explains quite well the basic structure of the 'Kyoto deal' (Chichilnisky and Sheeran, 2009), especially why the global carbon market combined the demands for efficiency of the developed world with the equity concerns of the developing world. However, to understand the current situation, we need to include complex multidimensional factors.

A particularly useful approach is described in a map constructed by Höhne *et al.* (2005) (Figure 13.1). Of course, more complex maps with more dimensions could easily be drawn, but this one actually captures with a limited number of parties the four main axes of disagreement.

The COP15 process illustrated well the interactions depicted in this map. During the first week of COP15, much attention was taken by the least developed countries. In the second week, most action took place along the axis between China (advanced developing countries) and the USA. Had China taken a weaker line, however, then India would most likely have stepped forward. If this hurdle had been passed, another round would have started – the one that took much prominence in early reflections on COP15 and in the post-Kyoto refinements: exactly how much should be done by the USA, EU, Japan, Australia and Russia? The negotiations never got this far, as some of the other hurdles were not passed. The resulting Copenhagen format (Box 13.2 and Table 13.1) was based on national pledges and, thus, avoided this conflict among developed countries.

Other lines of conflict that did not become fully exposed this time around but which, eventually, will have to be dealt with run within the developing countries. China, India, Brazil and (to some extent) South Africa have interests quite different from the least developed countries in many respects and, yet, the latter are often represented by the former. Many of the least developed countries will be particularly hard hit by climate change, but are not at all in line with respect to their own emission targets and, thus, have little

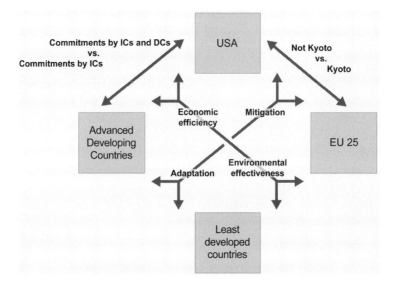

Figure 13.1 Simplified conflict areas between selected countries/groups. Source: redrawn from Höhne *et al.* (2005).

stake in the USA vs China negotiations, or might even be said to be closer to the USA than China in terms of their interests. Even more problematic is the unholy alliance of the Organization of the Petroleum Exporting Countries (OPEC) and G-77 (Barnett, 2008). The internal dynamics of G-77 have not been well studied but OPEC has, naturally, some instruments of significant influence. The agenda of the OPEC countries, however, differs radically from most of the least developed countries, both regarding the favoured price of oil and also, notably, in relation to decisive action on climate change.

Just as generals are accused of preparing for the previous war, there is a risk of the international community now preparing for future negotiations with an excessive focus on the primary axes that constrained progress at the COP15 meeting. Equally problematic is thinking only in abstract terms about parties and cooperation (Section 13.1). It is necessary to carry out a strategic analysis of the main axes and games that overlap in climate negotiations.

Box 13.1

The entanglement of climate change in North–South relations: stumbling blocks and opportunities for negotiation

ARIEL MACASPAC PENETRANTE

The North–South divide was reinvigorated as a political axis in the climate change negotiations when it became clear that the participation of the so-called 'South' on one hand became a precondition for the participation of the 'North' (Penetrante, forthcoming). For instance, President George Bush Jr. announced during his term that the United States will not return to the negotiation table unless developing countries had also formally accepted the same responsibilities as the countries in the North in formulating mitigation targets. On the other hand, developing countries – particularly those with emerging economies such as China, India and Brazil – have discovered resources not only to determine the agenda of the talks, but also more importantly to set the framework for talks and even delay the process through issue-specific negative power. Developing countries have come to form coalitions such as the 'G-77 and China' that could match and actually challenge the resources and knowledge of the developed world.

The climate change negotiations seem to have resuscitated the North–South divide in the international system. The resuscitation of the divide in the climate change negotiations has two faces.

- The divide had come to be regarded as somewhat outdated, deceased, discredited and misleading (Harris, 1986; van Evera, 1990; Sebenius, 1991; Hoogvelt, 1997; Hardt and Negri, 2000). It represents the concept about the previously regarded geographical division that is represented by the socioeconomic gap that exists between the developed (North) and developing (South) countries (Zartman, 1987; Hayes and Smith, 1993; White, 1993). A clear membership to a side seems to be unclear and, therefore, misleading. India, Brazil, China, South Korea, Singapore and Malaysia, among others, can easily be classified as developed, and therefore should belong to the North. For instance, Brazil, the Republic of Korea and China are members of the G-20. However, in climate change negotiations, these states, particularly China and Brazil, identify themselves, and behave, as developing countries. For example, in 1997, South Korea would rather have left the OECD than to take on binding obligations. The self-identification of several formally classified as developed countries is more inclined to the self-image of the G77. The divide has transformed itself as an image.
- The North–South divide originally represents the development gap between countries as described by the Human Development Index and exemplified by Rostow's (2000) model of development. The divide, then, has entered the sphere of the climate change negotiations particularly because of the horizontal inter-linkages between climate change and other negotiated issues – such as security, trade and economics. The divide is like an elephant. Once it enters the room, it is something that cannot be ignored. It determines coalitions, hinders any effective comprehensive leadership, and dictates issues that are supposed to be merely means of achieving the goal of emission abatement and now becoming issues on which states are negotiating. The divide has moved the attention of negotiating parties to 'side-effects' by enlarging the bargaining table. While the North–South divide might be obsolete in the domain of trade, it has gained relevance in the climate change context.

The divide is a state of mind and has become a reality in the current climate talks. The latest COP15 in Copenhagen did not produce the expected outcome as formulated in Bali to agree on a post-Kyoto mechanism on mandatory emission cuts primarily because of the huge distrust between the developed and developing countries. The abovementioned horizontal inter-linkages of issues imply the securitisation and politicisation of the context that led to the increased complexity of climate change talks. Securitisation and politicisation are conducive to

Box 13.1 (*cont.*)

distrust. Therefore, the basis of any decision and policymaking of states in the climate change negotiations will be determined primarily by a 'political' rather than a 'scientific' or 'technical' formula (Penetrante, forthcoming). This unfolding dynamic is a very important aspect of the current climate change negotiations that should be addressed, understood and accommodated.

The application of the North–South divide in the climate change context requires revisiting the divide as a concept. Although the divide comes from the development sphere, the divide addresses a specific focus – the fair procedure of burden-sharing (Penetrante, forthcoming) that is manifested by narratives. The boundary between the North and the South has been determined by positions about who should shoulder the costs of confronting climate change. On the one hand, the North pursues a 'forward-looking' notion of justice and fairness. Present generations should not be punished for 'crimes' they did not commit and 'crimes' that were not crimes when the actions in question were carried out by earlier generations (Caney, 2009). It would be unfair if present and future generations had to carry the costs of coping with climate change. Polluters must pay – both developed and developing countries. In contrast, the South follows a 'backward-looking' notion of justice and fairness, according to which the North 'owes' to the South the environmental space it has contaminated (Muller, 2009); therefore, there should be a scheme of compensation for the developing countries. The self-identification of countries such as China, Mexico, South Korea and Brazil as part of the South – despite their fulfilment of the criteria as developed countries – indicates the identity-building process that is derived from narratives about the past. The difference in the interpretation of what is fair has maintained incompatibilities of behaviour leading to consolidation of the contestation line.

Carefully conceptualising the North–South divide can lead to the proper understanding of the divide as an indicator and not the cause of the delay in reaching a sustainable outcome through negotiations. Reframing the divide as the solution and not as the problem enables parties to address the root-problems that promote sensitivity for the needs of the other side through the exchange of perspectives. Reframing the divide depends on how the divide is understood. One school of thought about the divide, as understood in the current literature, refers to power relationships with asymmetrical power (based on resources) that can be measured formally through development criteria such as per capita income. Another school of thought refers to the divide that is about ideological/cognitive differences between countries. This school of thought has, for instance, reduced the study of the 'third world' to merely an ideology (Norbu, 1992; Latham, 2000). It involves unfair structures in the international system that favours a specific group of countries.

However, the divide can also be understood in an operational manner, and is used in the current text. The divide in the climate change context can be considered from the negotiation perspective according to which through the divide, positions can be seen and this leads to the reduction of uncertainties in the climate change negotiations. Countries are players in the bargaining game where declarations and positions are made. From these positions, coalitions are established, maintained and changed. From this point of view, the divide is therefore not a problem that should be eliminated. The divide is not the conflict subject but, rather, an indicator of the dilemmas that have plagued the international system in the context of climate change. Reframing the divide as the point of departure of the analysis of the system enables the identification of opportunities for achieving a more sustainable, focused and yet comprehensive outcome.

The divide serves as a point of reference in coalition-building. Coalitions do not only serve as a 'negotiation vehicle' (Penetrante, forthcoming) enabling the expansion of resources to pursue interests through cooperation among like-minded actors, they also serve as a channel to moderate extreme views or positions through the various deliberation rounds within coalitions before engaging talks in the international stage. The divide points to the necessity of more facilitators and less leaders in the bargaining process. It would mean a chair not drafting a text, but rather a chair that mediates between groups. It would be very interesting to look more towards threshold states, such as Mexico and South Korea, which might be able to bridge the gaps between the two sides. The constructive usage of the North–South framing enables a future oriented conflict management because the divide points to the issues that must be accommodated to allow future cooperation. It sets the norm. As Zartman (2003) explained, 'discussing the future actually means reconciling two rights, not re-addressing ancient wrong.'

For many, the North–South divide represents a stumbling block in reaching an agreement to confront climate change and moving away from this paradigm seems to be the only viable way (La Viña, 1997), because the paradigm often leads to deadlocks or unsatisfactory compromises. However, the divide, itself, is not the underlying cause of such deadlocks but should be understood as an implication of structural faults and changes in the international system. For instance, there is a paradoxical situation in the climate change context in which 'weakness' (predicated on resource-based power) is translated as 'strength' (based on negative power; that is the capability to delay the process). This changing context of power is based on the perception of one's status in the bargaining game. Another example of changing perceptions is the leaked Danish text that was harshly criticised by most of the developing countries. Similar situations happened in past negotiations – such as the original text on trade and services during the GATT negotiations, which however, was not as harshly criticised as the Danish text. The bargaining, however, has changed with the increasing 'assertiveness' of developing countries.

If the North–South divide is to be understood as a bargaining game, then the divide is a chance to hear positions. Once developing countries perceive that their positions have been adequately heard, then it will pave the way to the next negotiation stage – bargaining on content. The divide is a social reality and, to some extent, a social imperative. It is not a 'disease' that should be cured. It hurts, but it also shows where actions are needed. Developed countries should not feel threatened by the 'unprecedented' assertiveness of developing countries, because this just shows how important the issue for them is, that they cannot afford just to be silent as usually happened in the past. Listening can be the more pragmatic and more effective approach.

13.4 International fora and formats: UNFCCC, the Gs and all the others

As discussed earlier, many different institutions are involved in managing climate change. Notably, a number of 'clubs' are relevant in addition to the more formalised, broader institutions.

- The Asia-Pacific Partnership on Clean Development and Climate was created by the USA with some countries on the Asian Rim. It had no financial resources for delivering on its alleged key element technology and soon became seen mostly as a Kyoto-hostile initiative centred on the USA and Australia.

- The Major Emitters Forum was then created in 2007 with 16 members. It was expanded to 17 and is now named the Major Economies Forum on Energy and Climate. The circle of membership is about right for the purpose of joining a critical mass of main emitters.
- The G8, G8+5 (adding the five most pivotal developing countries) and G8 plus India and China also represent a grouping that makes sense in terms of size. Some attempts to take climate-related action have been made here but, so far, these attempts have been in the form of rather vague, collective promises, and no real attempt has been made to use this setting to negotiate commitments with respect to each other. This is likely due to the fact that the general image, until now, has been that this had to be done in the UNFCCC context and that fora such as G8 should mainly help build the drive for this. With the general change of expectations after COP15, it is possible that the kind of action in G8 or some version of G8+5 could change to become a more autonomous complementary forum for actual negotiations and commitments.

Analysts have also pointed to a potential role for existing, middle-sized institutions – those that are larger than the clubs but smaller than UN/UNFCCC, such as The International Energy Agency and OECD with the addition of India and China (Box 13.4).

These new and seemingly loose institutions are important because there is a need for some forum where the main emitters can commit towards each other in a relatively small and mutually binding way. In addition to multilateralism, where a global arrangement with legitimacy and legalisation can be created, there is also a need for 'minilateralism' (Naím, 2009): the smallest number of countries necessary to achieve the aim. This is in line with k-group logic as well as Ostrom's (1990, 2009) analysis where parties actually make mutual commitments with direct reciprocity.

It is a characteristic of the climate change challenge that the three largest CO_2 emitters are responsible for 45% of global emissions, the largest 10 emitters almost 65%, and the largest 20 emitters constitute 76% of global emissions (Table 13.1). If the EU is counted as one country – which it is in terms of negotiations – the largest 15 actors contribute more than 80% of global emissions. If this is combined with the pattern of countries bound by Kyoto targets, the misfit is obvious. Whereas the group that *de facto* negotiated the COP15 deal – although vilified as self-appointed and illegitimate – corresponded quite well to something close to a k-group: China, the USA, India, Brazil and South Africa (although the EU, and possibly Japan and Russia, usefully could be brought in). More important than finding optimal membership is to get some significant 'climate coalition of the willing' (Christoff, 2006; von Marschall, 2009).

COP15 can also be used as an illustration of this core group formation because it clearly showed the need for, and the difficulty of, a complementary system of universal membership and a k-group. The importance of UNFCCC (or, in principle, some other UN General Assembly-anchored institution) derives not only from concern for legitimacy, but also the

Table 13.1. *Countries' negotiations over climate agreements at the Climate Change Summit, Copenhagen*

Rank	Country	Annual carbon emissions – gigagrams of carbon (estimation)[a]	Percentage of global total[b]	Cumulative summation of the N largest countries as percentage of global total	Cumulative summation of the N largest negotiation parties as percentage of global total	Annex I countries' reduction target under Kyoto: percentage change from 1997 to average of 2008–12[c]	Party to drafting of the accord in Copenhagen	Reductions promised under Copenhagen Accord[d]
1	China	1 922 687 Gg-C	22.17	22.17	22.17	–	X	Emissions reduction in 2020: per unit of GDP of 40 to 45% below 2005 levels
2	USA	1 547 460 Gg-C	17.85	40.02	40.02 from 2 largest	(– 7)	X	Reductions in the range of 17% below 2005 levels by 2020, 42% below 2005 levels by 2030, and 83% below 2005 levels by 2050
–	EU[e]	1 107 421 Gg-C	12.77	–	52.79 from 3 largest	–	–	Emissions reduction in 2020: 20% below 1990 and will increase to 30% if other countries commit to ambitious efforts (The European Union)
3	India	479 039 Gg-C	5.52	45.55	58.92 from 4 largest	–	X	Emissions reduction in 2020: emissions per unit of GDP of 20 to 25% below 2005 levels
4	Russian Federation	435 126 Gg-C	5.02	50.56	63.34 from 5 largest	0		Emissions reduction in 2020: 15–25%
5	Japan	357 534 Gg-C	4.12	54.69	67.46 from 6 largest	– 6		Emissions reduction in 2020: 25%
6	Germany	210 480 Gg-C	2.43	57.11	–	– 8		Emissions reduction in 2020: 20/30% (see under EU)
7	Canada	153 659 Gg-C	1.77	58.89	69.23 from 7 largest	– 6		Emissions reduction: 17%
8	United Kingdom	148 818 Gg-C	1.72	60.60	–	– 8		Emissions reduction in 2020: 20/30% (see under EU)

Table 13.1 (*cont.*)

Rank	Country	Annual carbon emissions – gigagrams of carbon (estimation)[a]	Percentage of global total[b]	Cumulative summation of the N largest countries as percentage of global total	Cumulative summation of the N largest negotiation parties as percentage of global total	Annex I countries' reduction target under Kyoto: percentage change from 1997 to average of 2008–12[c]	Party to drafting of the accord in Copenhagen	Reductions promised under Copenhagen Accord[d]
9	South Korea	142 230 Gg-C	1.64	62.24	70.87 from 8 largest	–		Emissions reduction in 2020: 30% below projected levels, which equates to a target of approximately 4% below 2005 levels
10	Iran	133 961 Gg-C	1.54	63.79	72.42 from 9 largest	–		Emissions reduction in 2020: 20/30% (see under EU)
11	Italy	125 015 Gg-C	1.44	65.23	–	– 8		Emissions reduction in 2020: 20/30% (see under EU)
12	Mexico	124 450 Gg-C	1.44	66.67	73.85 from 10 largest	–		Emissions reduction in 2020: reduce its greenhouse gas emissions up to 30% compared to business-as-usual levels by 2020. And reduce carbon dioxide emissions by 51 million tons by 2012.
13	South Africa	120 520 Gg-C	1.39	68.06	75.24 from 11 largest	–	X	Emissions reduction in 2020: reduce emissions growth 34% below business-as-usual levels and 42% by 2025
14	Saudi Arabia	119 374 Gg-C	1.38	69.43	76.62 from 12 largest	–		
15	Brazil	110 833 Gg-C	1.28	70.71	77.90 from 13 largest	–	X	Expected emission reduction: 36.1–38.9% by 2020
16	France	103 845 Gg-C	1.20	71.91	–	– 8		Emission reduction in 2020: 20/30% (see under EU)

Table 13.1 (*cont.*)

Rank	Country	Annual carbon emissions – gigagrams of carbon (estimation)[a]	Percentage of global total[b]	Cumulative summation of the N largest countries as percentage of global total	Cumulative summation of the N largest negotiation parties as percentage of global total	Amex I countries' reduction target under Kyoto: percentage change from 1997 to average of 2008–12[c]	Party to drafting of the accord in Copenhagen	Reductions promised under Copenhagen Accord[d]
17	Indonesia	99 648 Gg-C	1.15	73.06	79.05 from 14 largest	—		Emissions reduction in 2020: by 26% by 2020 from business-as-usual levels
18	Australia	96 168 Gg-C	1.11	74.17	80.15 from 15 largest	+ 8		Emissions reduction in 2020: 5 up to 15 or 25%
19	Spain	94 468 Gg-C	1.09	75.26	—	– 8		Emissions reduction in 2020: 20/30% (see under EU)
20	Poland	90 072 Gg-C	1.04	76.29	—	– 6		Emissions reduction in 2020: 20/30% (see under EU)
21	Ukraine	84 448 Gg-C	0.97	77.27	81.13 from 16 largest	0		
22	Turkey	80 207 Gg-C	0.93	78.19	82.05 from 17 largest	—		
23	Thailand	76 817 Gg-C	0.89	79.08	82.94 from 18 largest	—		
24	Taiwan	75 066 Gg-C	0.87	79.95	83.81 from 19 largest	—		
25	Kazakhstan	59 016 Gg-C	0.68	80.63	84.49 from 20 largest	—		Emissions reduction in 2020: 15%
26	Argentina	53 822 Gg-C	0.62	81.25	85.11 from 21 largest	—		
27	Venezuela	52 529 Gg-C	0.61	81.85	85.71 from 22 largest	—		

Table 13.1 (*cont.*)

Rank	Country	Annual carbon emissions – gigagrams of carbon (estimation)[a]	Percentage of global total[b]	Cumulative summation of the N largest countries as percentage of global total	Cumulative summation of the N largest negotiation parties as percentage of global total	Annex I countries' reduction target under Kyoto: percentage change from 1997 to average of 2008–12[c]	Party to drafting of the accord in Copenhagen	Reductions promised under Copenhagen Accord[d]
28	Egypt	52 336 Gg-C	0.60	82.46	86.32 from 22 largest	–		
29	Malaysia	50 515 Gg-C	0.58	83.04	86.90 from 23 largest	–		
30	Netherlands	46 202 Gg-C	0.53	83.57	–	– 8		Emission reduction in 2020: 20/30% (see under EU)

Notes:

Percentage of emission by Annex 1 countries: 40.29%.

Percentage of emission by countries behind the negotiation of the Copenhagen Accord: 48.21%.

Percentage of emission by countries which have promised targets for the Copenhagen Accord: 74.59%.

[a] See: http://cdiac.ornl.gov/ftp/trends/emissions/Preliminary_CO$_2$_Emissions_2007_2008.xls (Carbon Dioxide Information Analysis Center (CDIAC)).

[b] The CDIAC scheme presents two different sum categories for each year: 'sum of above' and 'Total world'. The two sums differ – the percentage calculations are based on the latter.

[c] See: http://unfccc.int/resource/docs/convkp/kpeng.pdf

[d] See:
http://www.nrdc.org/international/copenhagenaccords/
http://unfccc.int/home/items/5264.php
http://unfccc.int/home/items/5265.php

[e] Estimates for Cyprus, Estonia, Latvia and Malta are not currently available, which is why they are considered unchanged compared to 2006 data.

ability to coordinate various schemes – particularly for mitigation and adaptation in the least developed countries, and financial transfers for this purpose.

Box 13.2
COP15 and the Copenhagen Accord

KATRINE KROGH ANDERSEN

In December 2007, at the 13th Conference of the Parties to the UN Climate Change Convention (COP13) in Bali, the way had been paved for finalisation of the negotiations of a post-2012 global climate regime in the Bali Action Plan. Negotiations were to continue in two tracks, one Ad hoc Working Group (AWG) under the Kyoto Protocol on 'Further Commitments for Annex 1 Parties under the Kyoto Protocol' (the AWG-KP) and the other under an Ad Hoc Working Group on Long-term Cooperate Action (the AWG-LCA). Negotiations under the AWG-LCA were to conclude by COP15 in 2009. Very importantly, the Bali Action Plan for the AWG-LCA defined five building blocks to be addressed in the negotiations: a shared vision for long-term cooperative action, mitigation, adaptation, technology and financial support.

At the 15th Conference of the Parties to the UN Climate Change Convention (COP15/CMP5)[3] parties 'took note' of the Copenhagen Accord[4] negotiated within a smaller informal group of heads of states called by the COP president. In the Accord, important decisions were made with respect to a maximum long-term human-induced global temperature increase, financial support for developing countries, mitigation, and other issues. However, being a short political agreement, it barely goes into detail on how these decisions are to be operationalised and on the institutional set-up. A deadline of 31 January 2010 was agreed upon for parties to submit their mitigation actions and targets for 2020. As the Copenhagen Accord in the final plenary was not agreed upon by all parties to the UNFCCC, the same deadline applied for parties to associate themselves with the Accord. By the deadline, 55 countries representing 78% of global emissions from energy use had submitted information under the Accord.[5] By mid February about half of the 194 parties to the convention supported the Accord.

In the Copenhagen Accord, climate change is recognised as one of the greatest challenges of our time, and necessary actions to be taken on climate change are highlighted. It is certainly not the comprehensive global legally binding agreement that many had hoped could be achieved at COP15. Nevertheless, major issues in the climate change negotiations were addressed.

- Parties agreed that deep cuts in global emissions are required in pursuit of holding the increase in global temperature below 2 °C, consistent with science and on the basis of equity.
- '…developed countries shall provide adequate, predictable and sustainable financial resources, technology and capacity-building to support the implementation of adaptation action in developing countries'

---

[3] The 15th Conference of the Parties to the UN Climate Change Convention/Fifth Conference of the Parties serving as the meeting of the Parties to the Kyoto Protocol; further on referred to as COP15.

[4] http://unfccc.int/files/meetings/cop_15/application/pdf/cop15_cph_auv.pdf

[5] http://unfccc.int/files/press/news_room/press_releases_and_advisories/application/pdf/pr_accord_100201.pdf

Box 13.2 (*cont.*)

- Annex I[6] countries to the Climate Change Convention committed to quantified economy-wide emission targets for 2020, as subsequently submitted. The delivery of reductions and financing by developed countries will be measured, reported and verified.
- Non-Annex I[7] countries to the Convention will implement mitigation actions, as subsequently submitted. Mitigation actions are to be reported through national communications every two years; and provisions for measurement, reporting and verification of actions are given for national mitigation actions, and actions seeking international support respectively.
- A technology mechanism and a mechanism including REDD+ (Chapter 9) would be established to enable the mobilisation of financial resources from developed countries.
- Developed countries collectively committed to provide new and additional resources, approaching US$30 billion for the period 2010–12 with balanced allocation between adaptation and mitigation. Furthermore, they committed to a goal of mobilising jointly US$100 billion a year by 2020.
- An assessment of the implementation of the Accord is to be completed by 2015, including consideration of strengthening the long-term goal.

The agreement that the global temperature increase should be held below 2 °C is a great achievement, although many parties had worked hard for a limit of 1.5 °C. In the months leading up to COP15, many parties had announced their mitigation efforts for 2020, several of them with a more ambitious level contingent on the achievement of a global agreement. However, the level of ambition was not raised in the course of negotiations, and the collective ambition level for mitigation by 2020 in the submissions to the Copenhagen Accord is very similar to what had been announced ahead of COP15 and lower than various assessments of pathways towards reaching the temperature goal (e.g. IPCC AR4; Rogelj *et al.*, 2010). To achieve the goal of containing the human-caused global temperature increase to a maximum of 2 °C it will, thus, be crucial that global emissions by 2020 be lower than implied by the current pledges, and that relevant systems enabling and ensuring the needed actions are in place. Moreover, the Accord does not contain reduction targets for 2050. While these may be agreed upon at a later stage, the temperature guardrail cannot be respected through mitigation actions by 2020 alone. These must be followed up by the necessary deep cuts in global emissions over the remaining century.

The Accord, for the first time, includes all major emitters, and acknowledges that mitigation efforts are needed by both developed and developing nations – albeit under different circumstances – and with further specifications on measurement, reporting and verification. Although many outstanding issues remain to be clarified, developed countries in the Accord commit to substantial financial support, both up to and beyond 2012.

The Copenhagen Accord is a political agreement and it did not become a decision by either the COP or the CMP. Through the work of the AWGs leading up to COP15/CMP5, very detailed negotiation texts had been elaborated on all issues, which were passed on to the COP and CMP in Copenhagen. However, on many of the issues, parties were still very far apart at the beginning of the high-level segment and, in the end, the Copenhagen Accord

---

[6] Group of developed countries.
[7] Group of developing countries.

was negotiated in an informal group of heads of states. Progress was made on central issues under the Copenhagen Accord, but its status, as well as its operationalisation, need further elaboration. At the same time, substantial issues remain to be resolved under the UNFCCC in the coming years, such as: decisions on a legally binding agreement and a second commitment period under the Kyoto Protocol, the challenge of matching mitigation efforts with the overall long-term temperature guardrail, an adaptation framework, institutional arrangements on financing, the establishment of a REDD mechanism, details of a technology mechanism – as well as many other issues covered by the negotiation texts under the two AWGs, where agreement must be reached before COP and CMP decisions can eventually be made. The Copenhagen Accord did not solve technical and institutional issues but it did supply numbers on a long-term temperature target, mitigation efforts and financing. The work of both AWGs will continue and time will show whether the progress captured by the Accord can, and will, eventually be reflected in, and benefit, the continued negotiations on a post-2012 agreement under the UNFCCC.

Predictably, in the UNFCCC meetings after COP15, controversy kept arising over the status (or lack hereof) of the Copenhagen Accord (CA) vis-à-vis UNFCCC and Kyoto (Bodansky, 2010). Some observers go as far as: 'The United States seems to be the only country that still sees the Copenhagen Accord as having a life of its own. Almost all the rest, including countries that have "associated" themselves with the accord have insisted that the UNFCCC remains the only agreed decision-making forum' (Huq, 2010). On the other hand, it would also be a step backward if the 'package' of the CA begins to come apart, because it does define an inclusivity regarding both contributors and issues that has to be maintained. These controversies illustrate well how the evolution of climate change governance through something akin to the 'regime complex' model is both advantageous – because no single format would stand a chance – but also far from constituting an ultimate resolution because tensions continue among the components of the complex.

Schelling (1997, 1998, 2009) developed a proposal for climate agreement that takes as a surprising parallel the Marshall Plan and the Office of Economic Employment and Development (OEED, later the OECD). The first step of Schelling's argument is that it is central to create 'a new institutional structure to coordinate assistance from advanced industrialized countries to developing countries with the objective of transforming the way that people in the developing world produce and utilize energy' (Schelling, 2009). The second step is to set up an institution to channel funds. Too much control to donors, as well as total discretion to recipients, would ruin efficiency. It is here that the Marshall Plan comes in as a model. In the second year of the Plan, the USA appropriated a lump sum for the Europeans to divide amongst themselves. This was done through questionnaires to countries soliciting plans and an extended nego-tiation until agreement was reached – more easily achieved because it was, after all, about receiving funds. A notable feature here is that developed and developing coun-tries (epitomised by the USA and China) will not be communicating directly, which Schelling noted – already before COP15 – they are not too good at. Thus, through this separation of the phases and the injection of an intermediary institution, there will mostly be negotiations amongst the developed countries about sharing payments, and amongst (the major) developing countries about allocating funds. This proposed plan has a number of obvious weaknesses, including the problem that the USA in particular will not be inclined to make transfers to China for reasons spelled out in the previous section. However, the useful insight from the proposal is that it can be constructive to create several linked institutions where each creates mutually binding commitments within a circle able to cooperate, while the relations with least trust are handled in more distant and technical forms.

13.5 Top-down vs bottom-up, and the domestic–international interface (two-level games)

Levi (2009) has written a number of influential analyses of the politics of climate change arguing the impracticality of a comprehensive treaty and suggesting that much can be achieved at least on a mid-range timeframe by building a global effort from the bottom up 'through ambitious national policies and creative international cooperation focused on specific opportunities to cut emissions'. The central idea is that for each major country, there are specific and asymmetric opportunities to be seized. For example, it makes sense to focus on deforestation in Indonesia and Brazil, tapping into China's concern about safe energy supplies and assist in deep emissions-intensity cuts; while India, on the other hand, has different needs and opportunities, and so on.

The economic advantages and disadvantages of top-down vs bottom-up approaches are discussed in Box 12.4. However, a specific political dimension is important: the bottom-up approach has advantages with regard to ensuring processes in domestic politics that support cooperation. This dimension has also been studied within the literature on so-called two-level games (Putnam, 1988; Evans *et al.*, 1993); that is, how international negotiators are always simultaneously limited by what can be agreed with the other parties and by what can be ratified at home. The win-set for a given negotiation is constrained by (i) national preferences as shaped by domestic struggles between different interest groups, (ii) procedures for ratification of agreements, and (iii) the ability of negotiators to achieve the best possible agreement vis-à-vis other countries (Kroll and Shogren, 2008; Mortensen, 2009).

Often, the two-level theory is used for only its first insight but a more important feature is revealed as one comes further into it. The analysis uses a truncated version of the domestic–international interface, only looking at the double limitation logic. However, Putnam (1988), who invented the theory, was actually fascinated by the potential for actions that dynamically linked the two arenas to transform the game. Through 'reverberation', statesmen from one country can influence expectations in another. Processes like this have only become more important with the increasing global linkage of national and regional public spheres, as well as media around the world.

The most famous kind of domestic and international linkage by statesmen is the so-called Schelling Conjecture (Schelling, 1960), where a state representative intentionally constricts his own domestic space because this strengthens his negotiation position internationally. Counterparts have to meet this country further towards its aims because compromise has been ruled out in the most credible manner by creating *de facto* internal impossibilities. However, for our purposes, it is more interesting to look at the possibility of 'reverse Schelling Conjectures' (Neuhoff, 2009; Grodsky, 2010), where domestic audiences are convinced or constrained by moves on the international arena. 'By setting the international agenda, joining international regimes, or linking issues in international negotiations, statesmen have the power to shape the way in which issues are decided domestically' (Moravcsik, 1993).

Statesmen might also consciously assist each other in strengthening their domestic prestige through international activities. Domestic groups might, similarly, form transnational alliances to influence inter-state processes. The point here is that this is not a separate process of 'global civil society', but must also be assessed in terms of the way broader transnational politics influence state-to-state relations and the respective domestic situations for different statesmen.

The challenge is to think of climate politics as an interconnected, global political system. Although it is crucial to respect that international politics contain a very strong state-to-state element (talking about the 'end of the nation-state' will not help us much here), these state-to-state processes are embedded in domestic politics. Today, domestic arenas are increasingly interconnected and domestic actions echo through other domestic arenas as well, and inter-state relations become embedded in this global political process.

Grundig *et al.* (2001) and Ward *et al.* (2001) presented a revealing model of international negotiations that takes into account how side-payments to players opposed to progress takes place in the context of competition between several leaders. In line with two-level theory, lead nations make side-payments to both their own domestic veto-players and to those of other nations. The model shows, among other things, that it is hard to get progress beyond what the most conservative of the competing leaders want.

In the more specific literature on climate agreement architectures, the model that has most strongly integrated the domestic dimension is 'pledge and review'. Politically associated with former US president George W. Bush (and, therefore, suspected to aim at avoiding serious action), the label is not widely used today, despite the fact that the Copenhagen Accord comes close to this format. Academically, Schelling, Pizer and others have advocated this kind of bottom-up process (Schelling, 1997, 2002; Pizer, 2007; Bodansky and Diringer, 2007; Lewis and Diringer, 2007). Each country determines the kind of commitments it wants to take on. These commitments are negotiated among the parties with the aim to get a – not easily formalised – balance among the states, taking into account differing national circumstances. The proposals are exposed to reciprocal scrutiny and cross-examination; and the stronger the institutions for compiling and analysing data on the adequacy and comparability of mitigation efforts, the more efficient becomes the key compliance mechanism of this system: shaming. Therefore, Aldy (2008) suggested a Bretton Woods Institutions approach to address climate change, i.e. one that presents analyses of data on national efforts and their effectiveness. Still, such procedures are highly unlikely to provide sufficient mitigation to achieve something close to the 2 °C guardrail identified by the Copenhagen Accord.

A partly related argument is put forward by Victor (2010) in an analysis that builds on an integrated Kyoto-like arrangement for developed countries, but addresses the question on how to get developing countries to participate. He points out that this is unlikely to happen through abstract harmonisation of carbon pricing and a global system of emission trading. Instead, Victor points to international economic regimes like GATT/WTO, IMF and OECD where terms of accession are negotiated specifically for each country. They bid policies and programs that make sense in their specific situation. Those who are already members of the 'club' make it sufficiently attractive to join so that a meaningful deal is reached for each country, even if it varies between countries. Thus, a much more complicated total picture is created than usually envisioned.

Much can be said for thinking beyond the levels of national and international, not least to include various forms of local communities (Box 13.3; Delmas and Young, 2009; Lester and Neuhoff, 2009; Ostrom, 1990, 2009). This does not imply that local communities can 'take over' from states, but rather – in line with the way domestic-international relations have here been presented – in terms of how to produce synergies and political processes that become self-sustaining through the interaction of politics at various levels.

Box 13.3
Adaptive governance

AMANDA H. LYNCH AND RONALD D. BRUNNER

Scientists and other academics concerned about climate change have been occupied with large-scale initiatives to mitigate climate change for some time. Over 20 years ago, independent scientists issued the Toronto Conference Statement calling for industrialised nations to reduce their emissions of carbon dioxide by 20% from 1988 levels by 2005. In December 2009, the Conference of the Parties to the UNFCCC – meeting for the 15th time – came together with the parties to the Kyoto Protocol – meeting for the 5th time – in Copenhagen. They tried once again to negotiate mandatory international targets and timetables for reductions of greenhouse gas emissions – reductions sufficient to stabilise concentrations of greenhouse gases in the atmosphere at a level that prevents dangerous human interference in the climate system. Meanwhile, the separate parties are focused primarily on cap-and-trade systems to control emissions. The European Union is implementing its system; the USA and Australia, among others, are attempting to enact systems of their own.

Working directly with understanding climate change and its potential consequences, it becomes clear that large-scale initiatives need to succeed quickly in reducing emissions that have continued to increase at 3.4% per year (excluding the period of the global financial crisis) (Box 4.1; Le Quéré *et al.*, 2009). So far, there is a substantial risk that these initiatives may continue to be stymied in political process unable to generate sufficient political will and commitment. Indeed, during the past two decades, the inability to make real progress on this issue at the national and international levels has been apparent in: the substitution of scientific research and assessments for actions that deploy existing knowledge and technology; the official proclamation of long-term goals that defer the costs of realising them to future officeholders and their constituents; the negotiation of emissions-reduction targets and timetables lacking sanctions severe enough to be enforceable; limited capabilities to measure compliance and otherwise enforce those regulations that do have significant sanctions; and in the neglect of appraisals to terminate policies that have not worked and to improve those that have. In short, progress in mitigating dangerous concentrations of greenhouse gases in the atmosphere has been disappointing, especially given the magnitude of the task ahead. Meanwhile, concerns are growing that the world is running out of time; irreversible changes in climate are imminent, if not already occurring.

In this context, it would be prudent to open climate change science and policy to local communities that have made significant progress in mitigating greenhouse gas emissions or in adapting to already observed climate changes, with or without outside assistance. For example, local efforts have reduced losses from big storms in Barrow, Alaska; drought in Melbourne, Australia; and oceanic changes affecting commercial fish populations in west Greenland (Hamilton *et al.*, 2000; Brunner and Lynch, 2010). Ten major North American cities and counties are in a position to share what works through the Urban Leaders Adaptation Initiative, a network focused primarily on adaptation. Successful local efforts on mitigation include the village of Ashton Hayes in the UK moving toward carbon neutrality; the Danish island of Samsø, which achieved carbon neutrality in five years; and Portland, Oregon, which has reduced its emissions nearly to 1990 levels. For sharing what works, one major network that focused primarily on mitigation is the ICLEI–Local Governments for Sustainability which, in 2006, included 546 cities in 27 countries. It also aspires to influence international environmental policy from the bottom up.

An opportunity to explore progress by local communities can be provided through an effective analogy provided by adaptations in the face of large climate variability such as El Niño (Lynch and Brunner, 2010). In the context of the intense 1997–98 El Niño, the Pacific ENSO Applications Center (PEAC) successfully field tested a model for research in support of action on adaptation. Responding to the recommendations of a 1992 forecast applications workshop held in Honolulu, PEAC was instituted in the early 1990s to test the feasibility of integrating climate variability research, forecasts, and application services 'end-to-end' on an operational basis. The project initially explored how large-scale coupled ocean–atmosphere models could produce seasonal forecasts that could be turned into useful information products for Pacific Islanders. However, the spatial resolution of large-scale models did not meet the needs of the people that PEAC was intended to serve. PEAC supplemented the global-scale model forecast with empirically based statistical models that would forecast rainfall on specific islands using historical rainfall data. PEAC also conducted workshops, focus group meetings and personal briefings; issued a quarterly newsletter; and supported the creation of local task forces and public education campaigns. PEAC was the catalyst for pulling together participants and resources scattered throughout a distributed decision-making structure. By creating this robust network, effective policy responses – differing according to community circumstance and interests – could be enacted. Further, PEAC sought external resources from federal officials to help the affiliated Pacific Islands cope.

Box 13.3 (*cont.*)

Such cases exemplify an approach to reducing net losses from climate change called 'adaptive governance', where 'adaptive' means responsive to differences and changes on the ground (Brunner *et al.*, 2002; Brunner and Lynch, 2010). The term 'adaptive governance' is used for overlapping concepts in Dietz *et al.* (2003) and Folke *et al.* (2005). A grounded interpretation of the term depends on the particular context in which it is used but, not surprisingly, recognition of the pattern in the past decade followed innovations in practice.

This approach depends on factoring climate change – once conceived as 'a globally irreducible problem' – into thousands of local problems, each of which is more tractable scientifically and politically than the global problem, and somewhat different. Local differences encourage policy innovation. Local scale facilitates action on policies to mitigate or adapt to climate change while accommodating other community interests. Indeed, other community interests alone are sometimes enough for action to mitigate or adapt to climate change; these are 'no regrets' policies that make sense regardless of climate change. What works on the ground tends to be made available for voluntary adaptation by local communities through networks; failed policies tend to be set aside unless modified. Networks also tend to clarify shared needs for external resources from central authorities interested in, and able to support, what has worked on the ground. In short, adaptive governance is an opportunity for field-testing in series and in parallel thousands of policies for adapting to those climate changes we cannot avoid, and for mitigating those we can.

Opening climate science and policy to adaptive governance would also advance our interest in democracy. People on the ground – in the communities where they live and work – need not be reduced to mere stakeholders in decisions made in international summits, or in Washington D.C., Copenhagen or other capital cities. They can be participants in local decisions that bear directly on their multiple interests, contributing knowledge of local facts and values, and taking responsibility for direct consequences they cannot avoid without moving away. Climate scientists can help by working with local communities in planning and evaluating local decisions, and by finding and correcting malfunctions that inhibit scaling up and scaling out what works on the ground through networks. For those of us concerned about climate change, there is more to be gained at the margin by encouraging adaptive governance than by relying *exclusively* on large-scale initiatives.

13.6 Security and climate change: the conflict dimension

One aspect of the movement of climate change towards high politics status is an increasing interest in the link between climate change and security. Both politically and academically, climate change as a security issue comes in two main versions. One looks at the link between climate change and security in the classical, narrow, military sense, i.e. whether climate change can become the cause of violent conflict. The second school of literature looks at the possibility that climate change can be addressed politically as a security issue in itself. The former dominated the first phase of research on climate security. As reported in Chapter 5, much work has gone into investigating the statistical question about climate change as a cause of conflict (Box 5.5; Suhrke, 1997; Homer-Dixon, 1999; Barnett, 2001, 2003; Gleditsch, 2001a, 2001b; Gleditsch *et al.*, 2007; Gleditsch and Nordås, 2007, 2009; Raleigh and Urdal, 2007; Reuveny, 2007; Schubert *et al.*, 2007; Salehyan, 2008; Scheffran, 2008; Burke *et al.*, 2009). The most common chains leading from climate change, as such, to conflict are migration and (especially in research on Africa) food shortages. The resulting conflicts should rarely be seen as 'climate wars', but rather in terms of climate change acting as a 'force multiplier' that aggravates other causes where factors like ethnic tension or water shortages already created a risk of conflict (CNA, 2007; Blair, 2009).

The policy implications of this 'causes of climate change' type of security literature are, broadly speaking, within the category of adaptation. Looking at the possibility of climate change entailing military conflicts naturally points to various types of preparation for this eventuality, not least of which is military preparation both by the direct parties and by third parties preparing for intervention. Much of this research does not take the form of public, published, academic research but takes place within intelligence agencies and militaries where climate change is, generally, allocated much higher security significance than most outsiders suspect (Campbell, 2008; Dyer, 2008; Dabelko, 2009). Various non-military preparations are also on the agenda such as preparing for climate refugees, including the possibility of establishing an international convention for climate refugees. Much of this policy area eventually points towards the general issue of financial assistance to adaptation in developing countries. Helping developing countries deal with the consequences of climate change – adapting crops, water management, etc. – is the most important way to break the causal chains from climate change to violent conflict. In addition, increasing knowledge about the likely forms and places of climate-stimulated conflicts can be helpful in concentrating conflict prevention efforts. Also, this (probable) general force for increased level of conflict is a general argument for strengthening international capacities for conflict resolution and conflict management. However, given institutional inertia and the magnitude of resources available to traditional security agencies, it is likely that a climate–security link (of this first causal relationship type) will generally tend to strengthen unilateral, national responses (Barnett, 2009).

The second main type of climate security – climate change as security threat *per se* – is addressed in Sections 13.7 and 13.9 and, generally, can be seen as a special form of mitigation (Wæver, 2009a).

13.7 Energy security meets climate security

One specific link that significantly shapes how different states approach climate change as a potential security issue is how the issue is articulated with energy security. Energy security, on the one hand, serves to intensify the urgency and security rationale of climate policy. On the other hand, close links to energy security also often tie the issue into state-to-state rivalries and non cooperative approaches, especially among the major powers.

This link is made consistently, especially in the USA. As argued by Hayes and Knox-Hayes (2010), the rationale seems to be that climate action is not solidly established politically and, therefore, 'pro-climate' policymakers in the USA resort more often to securitisation than, for example, in the EU. When trying to justify climate action on a security rationale, the dependence on foreign oil more easily qualifies as security-relevant (a tendency that is reinforced by the US preference for a relatively narrow concept of security compared to Europe and several other parts of the world). The links between climate considerations and energy security are substantial. Clean energy measures will, typically, serve the dual purpose of reducing CO_2 emissions and reducing dependence on foreign oil. Furthermore, the global oil economy can be seen as an underlying reason for political and economic structures in the Middle East that cause security problems of all kinds – in addition to the increasing power of countries like Russia and Venezuela on the basis of oil, and the pull that energy enacts on China leading, ultimately, to its investments and involvement in regions like Africa. Consequently, 'energy and climate security' has become a standard rubric in the USA. Similarly for China, the problem of future energy supplies is clearly a high priority in the leadership. Thus, it seems likely that energy conservation programs will get much emphasis in the future for reasons other than climate change.

Concerns for future oil and gas supplies drive various elements of great power policy from the USA re-emphasising naval power to Chinese and Russian policies in Central Asia. There is particularly strong attention being paid to a potential future rivalry between the US and China, possibly supplemented by India (Giddens, 2009; Paskal, 2010). At the same time, Russia is strengthened whenever energy prices rise.

Among other climate–energy linkages is the possibility that the opening of new areas in the far north, with retreating ice, will enable the exploitation of energy resources in areas still contested between Russia, Canada, the USA, Greenland/Denmark and others. Another complex linkage is constituted by Darfur, a tragic conflict sometimes seen as partly stirred by climate change, where international intervention was constrained by Chinese support for the regime in Khartoum – which could well be linked to Chinese energy supplies from Sudan.

To simplify, the interaction between energy security and climate security works *for* international climate policy when addressed at an early stage and in an anticipatory fashion, where both speak for decarbonation of economies. However, the interaction easily weakens cooperation when the states actually get into rivalry and conflict, because then trust decreases and relative gains consideration gets the upper hand, making climate agreements even more difficult. Behind this dynamic is the too-often taken-for-granted condition that energy is an issue area with minimal international institutionalisation (Box 13.4). With energy policy so solidly anchored in sovereign, national policy, its growing precariousness will most naturally (that is, if not counteracted by conscious political intervention) lead to increased rivalry. In that case, the close connection between energy and climate as policy areas will shift from the currently often envisioned energiser of cooperation to a source of dissent among states.

Box 13.4
Global energy governance

SYLVIA I. KARLSSON-VINKHUYZEN

The way that humanity produces and consumes energy accounts for more than half of greenhouse gas emissions. All ambitious mitigation strategies must, therefore, drastically transform the energy sector. The global climate regime based on UNFCCC, the Kyoto Protocol and what agreements will come hereafter could, thus, indirectly lead to changed national energy policies. The treaties themselves are, however, silent on how energy policies should change. Nor do they include any measures for international collaboration in the energy sector. They can, therefore, not be considered part of *global energy governance*, which encompasses explicit multilateral efforts, such as the creation of norms and institutions dedicated to influence the production and consumption patterns of energy. A broader definition would include regulation and collaboration purely in the private sector but this is not discussed here.

There is not much global energy governance. There are some multilateral efforts to collaborate on energy through, for example, the G8 and the International Energy Agency but all have two limitations. First, they are only addressing certain limited dimensions of the energy issue, thus severely limiting their effectiveness to enable the transformation to energy infrastructure for a carbon-constrained world. Second, each of these is located outside the UN and has limited membership. They are, thus, not representing the whole world – but only, for example, major producers or consumers of energy, or only the richest countries – and are therefore facing severe concerns of legitimacy from those who are outside them and who, in many cases, are living in energy poverty and/or are most vulnerable to the impacts of climate change.

Throughout the history of the UN system, there has been a virtual normative and institutional vacuum on energy (Karlsson-Vinkhuyzen, 2010). With the exception of the International Atomic Energy Agency, there is no UN organisation dedicated to energy issues and, only since 2004, has there been a permanent but very weak coordination mechanism (UN-Energy) for the many UN organisations that carry out projects in the energy sector, primarily in developing countries. There is also no overarching normative framework for these organisations regarding, for example, what type of energy sources they should promote. Since 1993, energy is part of the mandate of the Commission on Sustainable Development, which discussed this topic in 2001, 2006 and 2007. Energy was also high on the agenda of the World Summit on Sustainable Development in 2002 but, in all cases, the outcome in the form of normative text was meagre.

The reasons that there is so little global energy governance lie in the historic links between energy, economic development and national security, which have made governments adamant to keep their power over energy policy and very reluctant to create international norms or institutions that address energy. Energy has been part of high politics where national, rather than collective, security has been the primary concern; and access to cheap energy – primarily fossil fuels – has been almost synonymous with military security, industrialisation and economic development (Karlsson-Vinkhuyzen, 2010). Furthermore, the location of energy sources has been dispersed, and the physical infrastructure and markets for energy have, until recently, been primarily national.

Box 13.4 (*cont.*)

The lack of global energy governance is increasingly perceived – by some countries, non-state actors and scholars – as an obstacle for the international community to support the transition to a low-carbon economy and ensuring affordable modern energy for the 1.6 billion people currently lacking it. Nonetheless, efforts to strengthen energy on the UN agenda at the Commission on Sustainable Development in 2007 met so much resistance that the negotiations broke down and ended without a decision. Instead, countries seeking cooperation on renewable energy took the unusual step of creating a new international institution outside the UN – the International Renewable Energy Agency – in early 2009.

The term 'global energy governance' has appeared only recently in the academic literature with a focus on key actors (Florini and Sovacool, 2009; Lesage *et al.*, 2009; Van de Graaf and Lesage, 2009), but limited attention has been given to explore the normative and theoretical rationales for strengthening global energy governance. For example, the energy system has been absent in the analysis of global public goods. Yet it can be argued that the *sustainability* of the global energy system, rather than the energy sources themselves, is a global public good (Karlsson-Vinkhuyzen *et al.*, 2009). Within this perspective, measures at all levels to support, for example, renewable energy and energy efficiency, would be contributing to the production of a global public good.

There is a clear need for more research on which global norms and institutions outside the energy field influence national energy governance, what elements of global energy governance would be most effective as well as legitimate, and which actors and institutions should be responsible for their production. If a role for global energy institutions is identified, either old institutions with other mandates have to be radically transformed or new ones have to be created. Possible valuable functions of such institutions could be to:

• develop research and development capacity for technology and policy on renewable energy and energy efficiency in diverse geographical and socio-political contexts;
• act as a clearinghouse with capacity-building functions;
• evaluate and coordinate the multilateral institutions' actions on energy.

The climate regime needs support from global norms and institutions that address energy 'heads on'. A strengthening of global energy governance would benefit from more analytical and political attention.

13.8 The geopolitics of geoengineering

Geoengineering, the possibility of deliberate modification of the climate – possibly defined more narrowly as modifying it by means other than by changing the atmospheric concentration of greenhouse gases (Barrett, 2008a, b) – is discussed in Sections 7.4 and 17.4.2. In the present chapter, we deal with the particular aspects regarding geopolitics and governance in relation to geoengineering.

There are heated discussions about the advisability of giving the issue legitimacy and investing resources into researching geoengineering. Proponents range from those few preferring this over emission control, to the more common position that it would be better to deal with the root problem (greenhouse gases) than treating the symptom (global warming) – but that it would be irresponsible not to prepare emergency instruments in the likely case that emission reductions fail to materialise in the necessary quantity (Crutzen, 2006). Opponents stress both the direct risks associated with geoengineering and, especially, the indirect risk that it can undermine human resolve to deal with the root problem (Cicerone, 2006). Irrespective of how this debate unfolds, it is worth noting how the incentive structures speak for the likelihood of geoengineering actually being employed (Schelling, 1996; Barrett, 2008a, b). As argued above, the pattern of incentives is very inhospitable towards international cooperation on emission reductions because of both the costs and the need to overcome free-rider problems; whereas the incentive structure is almost the reverse with respect to geoengineering. Comparatively, the most often discussed models are incomparably much cheaper than emission reduction and it could be done unilaterally by most major countries or by small coalitions. Therefore, the problem with respect to geoengineering is, in essence, the opposite to that for emission reductions: 'not how to get countries to do it, [but] the fundamental question of who should decide whether and how geoengineering should be attempted – a problem of governance' (Barrett, 2008b). This situation speaks in favour of bringing the issue out in the open and initiating a discussion of a proper governance structure rather than keeping it stifled.

The fact that geoengineering can be carried out by one or a few states, potentially at relatively low cost, makes it a real risk that this will be done in ways that other states disagree with – because of inherent opposition to climate meddling, unevenly distributed side-effects or even due to preference for warming (which might arguably be rational for countries like Canada and Russia). Most concerning is probably the issue of side-effects of geoengineering schemes that do not directly address the concentration of greenhouse gases in the atmosphere, but only influence solar radiation and, therefore, do not match the greenhouse effect symmetrically. In that case, some effects will go unchecked (i.e. acidification of oceans, Chapter 2) and even the temperature correction may overperform and underperform in different regions of the world. Even more concerning, the measures might impact other natural phenomena such as monsoon patterns. Thus, one could imagine, for example, China 'solving' its climate change problems in a way that had disastrous effects on India (Dyer, 2008; Kintisch, 2009; Lloyd, 2009). If this imagined Chinese geoengineering was performed unilaterally with no international governance framework of regulation justifying or dimensioning it, India might well feel justified in taking its own unilateral steps. Should we prepare for 'geoengineering wars'? Geoengineering is a security issue in the dual sense that, first, it is perceived by most actors as a relevant emergency measure to be adopted only if the situation is clearly getting out of control – climate change proving abrupt or catastrophic and the established policy measures proving impotent. Second, geoengineering can very likely become an object of major conflicts among states exactly because it is too easy to perform with too little dependence on other states – the mirror image of the problem regarding emission control.

13.9 The UN Security Council as panic bottom – or reformation of the UNFCCC?

On 17 April 2007, the UN Security Council had climate change on the agenda at the initiative of the British presidency. This was hugely controversial and a number of countries criticised the idea of defining this subject as a matter for the powerful Security Council, primarily because it was seen as belonging with all states in the UNFCCC and the General Assembly of the United Nations. While Britain and other UN Security Council members were keen to emphasise that the session was in no way intended to lead to any action in the auspices of the UN Security Council, the session did inevitably point to an intriguing option in climate change policy.

Legally speaking, nothing prevents the UN Security Council from denoting climate change a threat to international peace and security. In which case, they could prescribe and impose all necessary means including emission targets and impose these (if need be with military means). The UN Security Council is endowed with quite far-reaching competences under the Charter of the United Nations, and the main criterion is that the members agree to label something a threat to international peace and security – the UN Security Council is free to widen the concept of security as it pleases, as it has done in the cases of humanitarian emergencies and HIV/AIDS (Penny, 2005).

A solution to climate change through UN Security Council action is not very likely – primarily because the same lines of conflict among states that prevent agreement at a COP meeting or at G20 also run down through the UN Security Council. However, it is worth keeping this possibility in the picture for three reasons: (i) emergency solutions can become relevant – in an emergency situation, (ii) the very possibility enacts a pressure on other institutions, and (iii) the card can be played with symbolic effects. Note also that many observers find it realistic and viable to see the leading powers come together in MEF, G8, G20 or some such fora and, if agreement is reached here, it actually also means that the UN Security Council would be beyond its veto situation. The five permanent members of the UN Security Council are all among the top 16 emitters in G20, and almost all in MEF and G8. If they took their agreement through the UN Security Council arena instead of other fora, it would have the disadvantage of arousing resentment from other states. However, it would have the advantage of formally bestowing legality and competences. In addition, it would, in principle, enable decisions that are binding on others.

To invoke this mechanism, we would have to imagine a situation with clear and unmistakable evidence that a powerful tipping point has become unavoidable and is about to take effect (Chapter 7; Lenton *et al.*, 2007). In an emergency situation, discussions would turn to emergency procedures. In practice, the emergency action would probably have to also include (managed) geoengineering and other stop-gap measures, such as policies aimed at ozone and black carbon. However, it is very likely that if a package were imposed through this – the most extreme and centralised procedure – it would also include reduction of carbon emissions.

This would probably only happen in a situation marked by a relatively clear image of consistent failure by the established political institutions (UNFCCC in particular) as well as strong public opinion support for climate change analyses pointing to accelerating climate change, and then topped off by some kind of 'climate 9–11' (Wæver, 2009a) – some dramatic and television-friendly event, say, related to the West Antarctic Ice Sheet. While a route to achieve environmentally efficient governance (the criteria outlined in Section 13.1), this format is unattractive for a number of other reasons – notably the low political legitimacy and participation, and also the very heavy-handed forms of overruling of various other concerns that typically follow whenever measures are taken for security reasons (Buzan *et al.*, 1998).

Therefore, it would in many respects be politically advantageous if emergency options like this would never materialise but, instead, possibly enact a lateral pressure on other institutions and generate institutional reform. The UNFCCC system is relatively weak in terms of budget, bureaucracy, discretionary powers to its Executive Secretary (compared to, say, the Secretary General of the UN or NATO), rhythm of meetings and, most importantly, rules for decision-making (unanimity in almost all major areas). Nothing of this is given by nature, but is decided by member states. There is a misfit between the seriousness, urgency and primacy of the climate challenge and the weakness of the institutions addressing it. Whether a perception of serious threat can lead to construction of stronger institutions before a strict emergency situation is a difficult question to address, and it is only partly understood by security theory. Generally, security action has always been easiest on behalf of limited collectivities like states, nations, religions and alliances – not individuals or humanity. However, under specific conditions, it has proven possible to mobilise action in defence of system-level referents like 'human rights' or 'the liberal international economic order' (Buzan *et al.*, 1998). Therefore, it is an important but, so far, under-researched issue whether climate change can become a new 'macro-securitisation'; that is, an overarching security issue spanning many countries and actors, like the Cold War and the Global War on Terror, where other security concerns get redefined and realigned into a larger constellation (Buzan and Wæver, 2009). As shown in the analysis in Section 13.5, this hinges on an integrated understanding of domestic and international politics; in this case, the politics around threat and danger.

The two forms of 'securitisation' of climate change point to a serious dilemma, where part of the resolution might be regional. The 'causal-military' genre of climate security analysis is prone to policy implications in the format of preparing for wars to come (or at best, accept the link from climate change to intermediate variables and try to break the final link to conflict). This strand, therefore, holds a risk of militarisation and escalation with a national policy angle. The other genre – that of climate change as a 'security issue *per se*' – brings dangers of excessive concentrations of power and authority, due to a dangerous logic of necessity and exceptionalism (Liverman, 2009; Wæver, 2009a; Cohen, 2010). A regional focus, in contrast, might be able to include relevant linkages between, for instance, water shortage, state-to-state tension, migration and various other economic, political and security factors, while pointing to specific avenues for improvement (Brown and Crawford, 2009a, 2009b). The advantage of the regional perspective is that most conflict constellations in practice are regional and not global (Buzan and Wæver, 2003) and, therefore, integration of specific issues at the regional level are connected as direct relations, whereas connections in global analysis become aggregate and statistical. There is, therefore, a certain hope of keeping a focus on concrete connections – neither one super cause, nor everything being connected to everything. When executed most optimally, a regional analysis of climate–security linkages in a regional setting makes it possible to think climate, conflict and development together in specific processes.

13.10 Conclusion

As climate change has moved from a delineated, specific, 'technical' issue to a politicised issue linked to overarching national concerns about growth and power, it has become necessary to integrate two previously disconnected fields of research: one on cooperation, regime formation and design of institutional architectures; and one on geopolitics and security.

This chapter has described how existing research has pointed to weaknesses in relation to both rationalist incentives and cognitive framing with the existing format for cooperation. The addition of an increasingly competitive and confrontational great power link makes international agreements even more difficult. One tendency in much research is to move away from a singular, integrated regime towards a conception in terms of a 'regime complex' or multi-layered governance, where several institutions are directed towards a mutually supportive constellation. Another important tendency is to point to the need for analyses that integrate domestic and international levels in a dynamic format, where changes in foreign policy can be designed to foster future domestic support in key countries.

Climate security is an issue of increasing interest and covers diverse considerations, including the causal connection between climate change and violent conflict, as well as the possibility that climate change becomes defined as the biggest threat per se and thus a major security issue to be addressed with appropriately vigorous instruments. This climate–security linkage contains various risks, where the 'bottom-up' causal connection threatens to illicit unilateral and, possibly, military responses, and the 'top-down' climate threat format points to excessively centralised and authoritarian solutions like UN Security Council action. It remains to be seen whether an in-between option exists, where the threat from climate change leads to a strengthening of international instruments; for instance, through an institutional upgrading of the UNFCCC.

References

Aldy, J. E. (2008). *Designing a Bretton Woods Institution to Address Climate Change.* Cambridge, MA: Harvard Project on International Climate Agreements.
Aldy, J. E. and Stavins, R. N. (2007). *Architectures for Agreement: Addressing Global Climate Change in the Post-Kyoto World.* Cambridge, UK: Cambridge University Press.

Aldy, J. E. and Stavins, R. N. (2010). *Post-Kyoto International Climate Policy: Implementing Architectures for Agreement*. Cambridge, UK: Cambridge University Press.

Barnett, J. (2001). *Security and Climate Change*. Tyndall Centre Working Paper No. 7, www.tyndall.ac.uk/publications/working_papers/wp7.pdf.

Barnett, J. (2003). Security and climate change. *Global Environmental Change*, **13**, 7–17.

Barnett, J. (2008). The worst of friends: OPEC and G-77 in the climate regime. *Global Environmental Politics*, **8**, 1–8.

Barnett, J. (2009). The prize of peace (is eternal vigilance): A cautionary editorial essay on climate geopolitics. *Climatic Change*, **96**, 1–6.

Barrett, S. (2003). *Environment and Statecraft: The Strategy of Environmental Treaty-Making*. Oxford, UK: Oxford University Press.

Barrett, S. (2007). A multitrack climate treaty system. In *Architectures for Agreement: Addressing Global Climate Change in the Post-Kyoto World*, eds. J. Aldy and R. Stavins. Cambridge, UK: Cambridge University Press, pp. 237–59.

Barrett, S. (2008a). *Climate Change Negotiations Reconsidered*. Progressive Governance Papers. London: Policy Network.

Barrett, S. (2008b). The incredible economics of geoengineering. *Environmental and Resource Economics*, **39**, 45–54.

Barrett, S. (2010). A portfolio system of climate treaties. In *Post-Kyoto International Climate Policy: Implementing Architectures for Agreement*, eds. J. E. Aldy and R. N. Stavins. Cambridge, UK: Cambridge University Press, pp. 240–72.

Biermann, F., Pattberg, P., van Asselt, H. and Zelli, F. (2009). The fragmentation of global governance architectures: A framework for analysis. *Global Environmental Politics*, **9**, 14–40.

Blair, D. C. (2009). *Annual Threat Assessment of the Intelligence Community for the Senate Select Committee on Intelligence*, http://www.dni.gov/testimonies/20090212_testimony.pdf.

Bodansky, D. (2010). *The Copenhagen climate change conference: A post-mortem*. *American Journal of International Law*, **104**(2), 230–40.

Bodansky, D. and Diringer, E. (2007). *Towards an Integrated Multi-track Climate Framework*. Arlington, VA: Pew Center on Global Climate Change.

Brennan, G. (2009). Climate change: A rational choice politics view. *Australian Journal of Agricultural and Resource Economics*, **53**, 309–26.

Brown, O. and Crawford, A. (2009a). *Climate Change and Security in Africa: A Study for the Nordic-African Foreign Ministers Meeting*. Winnipeg, ON: International Institute for Sustainable Development.

Brown, O. and Crawford, A. (2009b). *Rising Temperatures, Rising Tensions: Climate Change and the Risk of Violent Conflict in the Middle East*. Winnipeg, ON: International Institute for Sustainable Development.

Brunner, R. D., Colburn, C. H., Cromley, C. M., Klein, R. A. and Olson, E. A. (2002). *Finding Common Ground: Governance and Natural Resources in the American West*. New Haven, CT: Yale University Press.

Brunner, R. D. and Lynch, A. H. (2010). *Adaptive Governance and Climate Change*. Boston, MA: American Meteorological Society.

Burke, M. B., Miguel, E., Satyanath, S., Dykema, J. A. and Lobell, D. B. (2009). Warming increases the risk of civil war in Africa. *Proceedings of the National Academy of Sciences (USA)*, **106**, 20670–74.

Buzan, B. and Wæver, O. (2003). *Regions and Powers: The Structure of International Security*. Cambridge, UK: Cambridge University Press.

Buzan, B. and Wæver, O. (2009). Macrosecuritisation and security constellations: Reconsidering scale in securitisation theory. *Review of International Studies*, **35**, 253–76.

Buzan, B., Wæver, O. and de Wilde, J. (1998). *Security. A New Framework for Analysis.* Boulder, CO: Lynne Rienner.

Campbell, K. M. (ed.) (2008). *Climatic Cataclysm: The Foreign Policy and National Security Implications of Climate Change.* Washington, D.C.: Brookings Institution Press.

Caney, S. (2009). Human rights, responsibilities, and climate change. In *Global Basic Rights*, eds. C. R. Beitz and R. E. Goodin. Oxford, UK: Oxford University Press, pp. 227–47.

Chichilnisky, G. and Sheeran, K. A. (2009). *Saving Kyoto. An Insider's Guide to the Kyoto Protocol: How it Works, Why it Matters and What it Means for the Future.* London: New Holland Publishers.

Christoff, P. (2006). Post-Kyoto? Post-Bush? Towards an effective 'climate coalition of the willing'. *International Affairs*, **82**, 831–60.

Cicerone, R. J. (2006). Geoengineering: Encouraging research and overseeing implementation. *Climatic Change*, **77**, 221–26.

CNA (2007). *National Security and the Threat of Climate Change.* CNA Corporation.

Cohen, M. J. (2010). Is the UK preparing for 'war'? Military metaphors, personal carbon allowances, and consumption rationing in historical perspective. *Climatic Change*, doi: 10.1007/s10584–009–9785-x.

Crutzen, P. J. (2006). Albedo enhancement by stratospheric sulfur injections: A contribution to resolve a policy dilemma? *Climatic Change*, **77**, 211–20.

Dabelko, G. D. (2009). Planning for climate change: The security community's precautionary principle. *Climatic Change*, **96**, 13–21.

Delmas, M. A. and Young, O. R. (eds.) (2009). *Governance for the Environment: New Perspectives.* Cambridge, UK: Cambridge University Press.

Dietz, T., Ostrom, E. and Stern, P. (2003). The struggle to govern the commons. *Science*, **302**, 1907–12.

Dyer, G. (2008). *Climate Wars.* Carlton North, Australia: Scribe.

Evans, P. B., Jacobson, H. K. and Putnam, R. D. (eds.) (1993). *Double-edged Diplomacy: International Bargaining and Domestic Politics.* Berkeley, CA: University of California Press.

Florini, A. and Sovacool, B. K. (2009). Who governs energy? The challenges facing global energy governance. *Energy Policy*, **37**, 5239–48.

Folke, C., Hahn, T., Olsson, P. and Norberg, J. (2005). Adaptive governance of social-ecological systems. *Annual Review of Environment and Resources*, **30**, 441–73.

Freestone, D. and Streck, C. (2005). *Legal Aspects of Implementing the Kyoto Protocol Mechanisms: Making Kyoto Work.* New York, NY: Oxford University Press.

Gainza-Carmenates, R., Altamirano-Cabrera, J. C., Thalmann, P. and Drouet, L. (2010). Trade-offs and performances of a range of alternative global climate architectures for post-2012. *Environmental Science and Policy*, **13**, 63–71.

Giddens, A. (2009). *Politics of Climate Change.* Cambridge, UK: Polity Press.

Gleditsch, N. P. (2001a). Environmental change, security, and conflict. In *Turbulent Peace: The Challenges of Managing International Conflict*, eds. C. A. Crocker, F. O. Hampson and P. Aall. Washington, D.C.: United States Institute of Peace Press, pp. 53–68.

Gleditsch, N. P. (2001b). Resource and environmental conflict: The state-of-the-art. In *Responding to Environmental Conflicts: Implications for Theory and Practice*, eds. E. Petzold-Bradley, A. Carius and A. Vincze. Dordrecht: Kluwer, pp. 53–66.

Gleditsch, N. P. and Nordås, R. (2007). Climate change and conflict. *Political Geography*, **26**, 627–38.

Gleditsch, N. P. and Nordås, R. (2009). IPCC and the climate-conflict nexus conference. *IOP Conference Series: Earth and Environmental Science*, **6**, 562007.

Gleditsch, N. P., Nordås, R. and Salehyan, I. (2007). *Climate Change and Conflict: The Migration Link*. Coping with Crisis Working Paper Series. New York, NY: International Peace Academy.

Grasso, M. (2004). *Climate Change: The Global Public Good*. Working Papers 75. Milan: University of Milano-Bicocca.

Grieco, J. M. (1990). *Cooperation Among Nations: Europe, America, and Non-tariff Barriers to Trade*. New York, NY: Cornell University Press.

Grodsky, B. K. (2010). Exploring the Schelling Conjecture in reverse: 'International constraints' and cooperation with the International Criminal Tribunal for the former Yugoslavia. *European Journal of International Relations*, **16**, (Published on line before print, April 20, 2010).

Grubb, M., Vrolijk, C. and Brack, D. (1999). *The Kyoto Protocol: A Guide and Assessment*. London: Royal Institute of International Affairs.

Grundig, F. (2006). Patterns of international cooperation and the explanatory power of relative gains: An analysis of cooperation on global climate change, ozone depletion, and international trade. *International Studies Quarterly*, **50**, 781–801.

Grundig, F. (2009). Political strategy and climate policy: A rational choice perspective. *Environmental Politics*, **18**, 747–64.

Grundig, F., Ward, H. and Zorick, E. R. (2001). Modeling global climate negotiations. In *International Relations and Global Climate Change*, eds. U. Luterbacher and D. F. Sprinz. Cambridge, MA: MIT Press.

Gupta, S., Tirpak, D. A., Burger, N. *et al.* (2007). Policies, instruments and co-operative arrangements. In *Climate Change 2007: Mitigation. Contribution of Working Group III to the Fourth Assessment Report of the Intergovernmental Panel on Climate Change*, eds. B. Metz, O. Davidson, P. Bosch, R. Dave and L. Meyer. Cambridge, UK and New York, NY: Cambridge University Press, pp. 745–807.

Haas, P. (2008). Climate change governance after Bali. *Global Environmental Politics*, **8**, 1–7.

Hamilton, L., Lyster, P. and Otterstad, O. (2000). Social change, ecology and climate in 20th century Greenland. *Climatic Change*, **47**, 193–211.

Hardt, M. and Negri, A. (2000). *Empire*. Cambridge, MA: Harvard University Press.

Harris, N. (1986). *The End of the Third World. Newly Industrializing Countries and the Decline of an Ideology*. Harmondsworth, UK: Penguin Books.

Hayes, J. and Knox-Hayes, J. (2010). Democracy, security, and carbon markets: Comparative analysis of US and European climate policy. Paper presented at the Annual Meeting of the International Studies Association, New Orleans, February 2010.

Hayes, P. and Smith, K. R. (eds.) (1993). *The Global Greenhouse Regime. Who Pays? Science, Economics and North–South Politics in the Climate Change Convention*. London: Earthscan.

Hoffmann, S. (1966). Obstinate or obsolete? The fate of the nation-state and the case of western Europe. *Dædalus*, **95**, 862–915.

Höhne, N., Phylipsen, D., Ullrich, S. and Blok, K. (2005). Options for the second commitment period of the Kyoto Protocol. Utrecht, The Netherlands: Ecofys.

Homer-Dixon, T. F. (1999). *Environment, Scarcity and Violence.* Princeton, NJ: Princeton University Press.

Hoogvelt, A. M. M. (1997). *Globalization and the Postcolonial World: The New Political Economy of Development.* Baltimore, MD: John Hopkins University Press.

Hovi, J., Sprinz, D. F. and Underdal, A. (2009). Implementing long-term climate policy: Time inconsistency, domestic politics, international anarchy. *Global Environmental Politics*, **9**, 20–39.

Huq, S. (2010). Bonn climate talks: picking up the pieces after Copenhagen. *Guardian*, 12 April.

Karlsson-Vinkhuyzen, S. I. (2010). The United Nations and global energy governance: past challenges, future choices. *Global Change, Peace & Security*, **22** (2), 175–195.

Karlsson-Vinkhuyzen, S. I., Jollands, N. and Staudt, L. (2009). Global energy policy: Transforming governance for the transition to a sustainable energy future. *2009 International Energy Workshop*, 17–19 June, Fondazioni Giorgi Cini, Venice, Italy.

Keohane, R. O. and Victor, D. G. (2010). *The Regime Complex for Climate Change.* Discussion Paper 10–33. Cambridge, MA: The Harvard Project on International Climate Agreements.

Kintisch, E. (2009). The politics of climate hacking: What happens if one country decides to start geoengineering on its own? *Slate*, **29** April, http://www.slate.com/id/2217230/.

Kroll, S. and Shogren, J. F. (2008). Domestic politics and climate change: International public goods in two-level games. *Cambridge Review of International Affairs*, **21**, 563–83.

Kuik, J.A., Aerts, J., Berkhout, F. *et al.* (2008). Post-2012 climate policy dilemmas: A review of proposals. *Climate Policy*, **8**, 317–36.

Latham, M. E. (ed.) (2000). *Modernization as Ideology. American Social Science and 'Nation-building' in the Kennedy Era.* Chapel Hill, NC: University of North Carolina Press.

La Viña, A. G. M. (1997). *Climate Change and Developing Countries: Negotiating a Global Regime.* Quezon City, Philippines: Institute of International Legal Studies, University of the Philippines Law Center.

Lenton, T. M., Held, H., Kriegler, E. *et al.* (2007). Tipping elements in the Earth's climate system. *Proceedings of the National Academy of Sciences (USA)*, **105**, 1786–93.

Le Quéré, C., Raupach, M. R., Canadell, J. G. *et al.* (2009). Trends in the sources and sinks of carbon dioxide. *Nature Geoscience*, **2**, 831–36.

Lesage, D., Van de Graaf, T. and Westphal, K. (2009). The G8's role in global energy governance since the 2005 Gleneagles summit. *Global Governance*, **15**, 259–77.

Lester, S. and Neuhoff, K. (2009). Policy targets: Lessons for effective implementation of climate actions. *Climate Policy*, **9**, 464–80.

Levi, M. A. (2009). Copenhagen's inconvenient truth: How to salvage the climate conference. *Foreign Affairs*, **88**, 92–104.

Lewis, J. and Diringer, E. (2007). *Policy-Based Commitments in a Post-2012 Climate Framework.* Arlington, VA: Pew Center on Global Climate Change.

Liverman, D. (2009). The geopolitics of climate change: Avoiding determinism, fostering sustainable development. *Climatic Change*, **96**, 7–11.

Lloyd, R. (2009). Geoengineering wars: Another scientist teases out a surprising effect of global deforestation. *Scientific American Observations*, 19 October, http://www.scientificamerican.com/blog/post.cfm?id=geoengineering-wars-another-scienti-2009-10-19.

Lynch, A. H. and Brunner, R. D. (2010). Learning from climate variability: Adaptive governance and the Pacific ENSO Applications Center. *Weather, Climate, and Society*.

de Mesquita, B. B. (2009). Recipe for failure: why Copenhagen will be a bust, and other prophecies from the foreign-policy world's leading predictioneer. *Foreign Policy*, **175**, 7. **2** (4). Doi:10.1175/2010 WCAS1049.1.

Moravcsik, A. (1993). Introduction: Integrating international and domestic theories of international bargaining. In *Double-Edged Diplomacy: International Bargaining and Domestic Politics*, eds. P. B. Evans, H. K. Jacobson and R. D. Putnam. Berkeley, CA: University of California Press, pp. 3–41.

Mortensen, J. L. (2009). Det dobbelte klimadiplomati (Double climate diplomacy). In *Klimapolitik: Dansk, Europæisk, Globalt*, eds. P. Nedergaard and P. Fristrup. Copenhagen: DJØF, pp. 109–33.

Muller, B. (2009). Interview on June 26 at the South Centre, Geneva.

Naím, M. (2009). Minilateralism: the magic number to get real international action. *Foreign Policy*, **173**, 2.

Neuhoff, K. (2009). Understanding the roles and interactions of international cooperation on domestic climate policies. *Climate Policy*, **9**, 435–49.

Norbu, D. (1992). *Culture and the Politics of Third World Nationalism*. New York, NY: Routledge.

Okereke, C., Bulkeley, H. and Schroeder, H. (2009). Conceptualizing climate governance beyond the international regime. *Global Environmental Politics*, **9**, 58–78.

Ostrom, E. (1990). *Governing the Commons: The Evolution of Institutions for Collective Action*. Cambridge, UK: Cambridge University Press.

Ostrom, E. (2009). *A Polycentric Approach for Coping with Climate Change*. World Bank Policy Research Working Paper No. 5095. Washington, D.C.: World Bank.

Paskal, C. (2010). *Global Warring: How Environmental, Economic, and Political Crises Will Redraw the World Map*. Houndmills: Palgrave Macmillan.

Paterson, M. (2009). Post-hegemonic climate politics? *British Journal of Politics and International Relations*, **11**, 140–58.

Penetrante, A. M. (forthcoming). Common but differentiated responsibilities: The North–South divide in climate change negotiations. In *Climate Change Negotiations: A Guide to Resolving Disputes and Facilitating Multilateral Cooperation*, ed. G. Sjöstedt. London: Earthscan.

Penny, C. K. (2005). Greening the Security Council: Climate change as an emerging 'threat to international peace and security'. Paper prepared for *Human Security and Climate Change. An International Workshop*, 21–23 June, Asker, Norway.

Pizer, W. A. (2007). *Pizer Proposal: Practical Global Climate Policy*. Policy Brief, Harvard Project on International Climate Agreements, Belfer Center for Science and International Affairs, Harvard Kennedy School, 5 September 2007.

Putnam, R. D. (1988). Diplomacy and domestic politics: The logic of two-level games. *International Organization*, **42**, 427–60.

Rajamani, L. (2009). Addressing the 'post-Kyoto' stress disorder: Reflections on the emerging legal architecture of the climate regime. *International and Comparative Law Quarterly*, **58**, 803–34.

Raleigh, C. and Urdal, H. (2007). Climate change, environmental degradation and armed conflict. *Political Geography*, **26**, 674–94.

Reuveny, R. (2007). Climate change-induced migration and violent conflict. *Political Geography*, **26**, 656–73.

Rogelj, J., Nabel, J., Chen, C. *et al.* (2010). Copenhagen Accord pledges are paltry. *Nature*, **464**, 1126–28.

Rostow, W. (2000). The five stages of economic growth: A summary. In *Development: Critical Concepts in the Social Sciences. Volume 1: Doctrines of Development*, ed. S. Corbridge. London: Routledge, pp. 105–16.

Salehyan, I. (2008). From climate change to conflict? No consensus yet. *Journal of Peace Research*, **45**, 315–26.

Scheffran, J. (2008). Climate change and security. *Bulletin of the Atomic Scientists*, **64**, 19–25, 59–60.

Schelling, T. C. (1960). *The Strategy of Conflict*. Cambridge, MA: Harvard University Press.

Schelling, T. (1978). *Micromotives and Macrobehavior*. New York, NY: W.W. Norton.

Schelling, T. C. (1996). The economic diplomacy of geoengineering. *Climate Change*, **33**, 303–07.

Schelling, T. C. (1997). The cost of combating global warming: Facing the tradeoffs. *Foreign Affairs*, **76**, 8–14.

Schelling, T. C. (1998). *Costs and Benefits of Greenhouse Gas Reduction*. Washington, D.C.: American Enterprise Institute Press.

Schelling, T. C. (2002). What makes greenhouse sense? Time to rethink the Kyoto Protocol. *Foreign Affairs*, **81**, 2–9.

Schelling, T. (2009). International coordination to address the climate challenge. *Innovations: Technology, Governance, Globalization*, **4**, 13–21.

Schubert, R., Schellnhuber, H.-J. and Buchmann, N. (2007). *Climate Change as a Security Risk*. London: Earthscan.

Sebenius, J. K. (1991). Negotiating a regime to control global warming. In *Greenhouse Warming: Negotiating a Global Regime*, ed. J. T. Matthews. Washington, D.C.: World Resources Institute.

Snidal, D. (1991a). International cooperation among relative gains maximizers. *International Studies Quarterly*, **35**, 387–402.

Snidal, D. (1991b). Relative gains and the patterns of international cooperation. *American Political Science Review*, **85**, 701–26.

Spiegel Online (2009). What if global temperatures rose by 4 degrees Celsius? *Spiegel Online*, 24 December, http://www.spiegel.de/international/world/0,1518,662887,00.html.

Suhrke, A. (1997). Environmental degradation, migration, and the potential for violent conflict. In *Conflict and the Environment*, ed. N. P. Gleditsch. Dordrecht: Kluwer Academic, pp. 255–72.

Van de Graaf, T. and Lesage, D. (2009). The International Energy Agency after 35 years: Reform needs and institutional adaptability. *The Review of International Organizations*, **4**, 293–317.

Van Evera, S. (1990). Why Europe matters, why the third world doesn't: American grand strategy after the cold war. *Journal of Strategic Studies*, **13**, 1–51.

Victor, D. G. (2001). *The Collapse of the Kyoto Protocol and the Struggle to Slow Global Warming*. Princeton, NJ: Princeton University Press.

Victor, D. G. (2010). Climate accession deals: New strategies for taming growth of greenhouse gases in developing countries. In *Post-Kyoto International Climate Policy: Implementing Architectures for Agreement*, eds. J. E. Aldy and R. N. Stavins. Cambridge, UK: Cambridge University Press.

von Marschall, C. (2009). Verhandelten in Kopenhagen zu viele Länder? *Zeit Online*, 20 December 2009, http://www.zeit.de/wirtschaft/2009–12/kopenhagen-klimagipfel-scheitern.

Ward, H. F., Grundig, F. and Zorick, E. R. (2001). Marching at the pace of the slowest: A model of international climate-change negotiations. *Political Studies*, **49**, 438–61.

Wæver, O. (2009a). All dressed up and nowhere to go? Securitization of climate change. *ISA's 50th Annual Convention 'Exploring the Past, Anticipating the Future'.* 15 February, New York, NY.

Wæver, O. (2009b). What exactly makes a continuous existential threat existential – and how is it discontinued? In *Existential Threats and Civil–Security Relations*, eds. O. Barak and G. Sheffer. Lanham, MD: Lexington Books, pp. 19–35.

Waltz, K. N. (1979). *Theory of International Politics*. New York, NY: McGraw-Hill.

Whalley, J. and Walsh, S. (2009). Bringing the Copenhagen global climate change negotiations to conclusion. *CESifo Economic Studies*, **55**, 255–85.

White, R. (1993). *North, South, and the Environmental Crisis*. Toronto: University of Toronto Press.

Young, O. R. (1998). *Creating Regimes: Arctic Accords and International Governance*. New York, NY: Cornell University Press.

Young, O. R. (2009). Governance for sustainable development in a world of rising interdependencies. In *Governance for the Environment*, eds. M. A. Delmas and O. R. Young. Cambridge, UK: Cambridge University Press.

Zartman, I. W. (ed.) (1987). *Positive Sum. Improving North–South Negotiations*. New Brunswick, NJ: Transaction.

Zartman, I. W. (2003). Negotiating the rapids: The dynamics of regime formation. In *Getting It Done: Post-Agreement Negotiation and International Regimes*, eds. B. I. Spector and I. W. Zartman. Washington, D.C.: United States Institute of Peace, pp. 13–50.

14

Adapting to the unavoidable

'The people who are bearing the brunt of the effects of climate change are those who can least afford to do so and who have done least to cause the problem. Adaptation is both a practical need and a moral imperative.'[1]

Climate change is a reality. Previous chapters in this volume have presented an up-to-date summary of the science of climate change, a description of some of the impacts that are already occurring, and an analysis of the risks that lie ahead as the climate continues to change. It is clear that even the most vigorous approaches to mitigation will not prevent a further escalation of the risks of climate change. Further impacts are unavoidable.

Adaptation to climate change is thus now considered essential in order to reduce the impacts of climate change that are already happening and to increase the resilience to future impacts, some of which cannot be reversed (Parry *et al.*, 1998). Some countries have already started adaptation programs (Kabat *et al.*, 2005; Swart *et al.*, 2009). The IPCC started to address adaptation more comprehensively in a single chapter of its Third Assessment Report (McCarthy *et al.*, 2001), while in its Fourth Assessment Report, adaptation attracted significant attention throughout the whole document (Parry *et al.*, 2007). The shift from a focus primarily on mitigation towards embracing adaptation too can also be observed in the scientific literature (Figure 14.1), where adaptation and mitigation now get equal attention.

Adaptation refers to adjustments in ecological, social or economic systems in response to actual or expected climatic change and their impacts. It refers to changes in processes, practices and structures to moderate potential damages or to benefit from opportunities associated with climate change (Smit *et al.*, 2001). A number of different types of adaptation approaches have been identified, some of which are closely related to coping mechanisms that individuals and societies have developed to deal with other stressors (Table 14.1).

---

[1] UN Secretary General, Ban Ki-moon. Speech at Government House, Ulaanbaatar, Mongolia, 27 July 2009.

Climate Change: Global Risks, Challenges and Decisions, Katherine Richardson, Will Steffen and Diana Liverman *et al.* Published by Cambridge University Press. © Katherine Richardson, Will Steffen and Diana Liverman 2011.

Table 14.1. *The various types of adaptation that can be distinguished*

Anticipatory adaptation	Adaptation that takes place before impacts of climate change are observed; also referred to as proactive adaptation.
Autonomous adaptation	Adaptation that does not constitute a conscious response to climatic stimuli but is triggered by ecological changes in natural systems and by market or welfare changes in human systems; also referred to as spontaneous adaptation.
Planned adaptation	Adaptation that is the result of a deliberate policy decision, based on an awareness that conditions have changed or are about to change and that action is required to return to, maintain, or achieve a desired state.
Private adaptation	Adaptation that is initiated and implemented by individuals, households or private companies. Private adaptation is usually in the actor's rational self-interest.
Public adaptation	Adaptation that is initiated and implemented by governments at all levels. Public adaptation is usually directed at collective needs.
Reactive adaptation	Adaptation that takes place after impacts of climate change have been observed.

Source: after Smit *et al.* (2001).

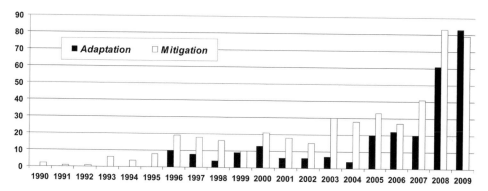

Figure 14.1 Number of publications over the last two decades with 'Adaptation' or 'Mitigation' and 'Climate Change' in the title. Source: ISI Web of Knowledge database, Thomson Reuters, accessed December 2009.

The aim of this chapter is to present and discuss relevant aspects of adaptation. Because adaptation is a complex, multidimensional activity, several aspects of adaptation have been dealt with in other chapters. These treatments will not be repeated here.

Chapters 5 and 6 outlined the impacts of climate change on human societies and natural ecosystems. The focus in these chapters was on estimation of impacts to help define what might be considered dangerous climate change. However, the same knowledge is often useful for adaptation, although the uncertainties around projected impacts and the often generic nature of impact assessments make this information much less useful for adaptation in particular places with particular socio-economic contexts.

Several equity issues associated with the climate change challenge are discussed in Chapter 9. Adaptation plays a prominent role in this chapter in terms of: (i) the greater impacts on developing countries which, therefore, place the largest adaptation 'demand' on those least able to carry out effective adaptation – the world's poorest people and societies, along with the natural ecosystem, generally have the least adaptive capacity; (ii) adaptation in some critical sectors – food, water, health and security – are examined from an equity perspective; and (iii) funding an adaptation safety net, the need for which was strongly endorsed at the COP15 in Copenhagen, is explored in some detail in this chapter.

Although the cost of mitigation is an intensively studied area that generates considerable debate amongst economists and politicians, the costs associated with adaptation or unavoidable impacts are much less studied. Chapter 12 includes a section on the estimation of the economic costs of adaptation.

Chapter 15 raises the crucial issue of integrating adaptation, mitigation and sustainable development. Such integration is an imperative for all countries, but is especially important for developing countries. The chapter deals with several topics that directly relate to adaptation to climate change: (i) poverty alleviation and building resilience; (ii) adaptive capacity; (iii) integrating climate adaptation into a multiple drivers framework that affects vulnerability; and (iv) trade-offs between adaptation and other aspects of sustainable development. Adaptation to climate change is put in the even broader perspective of the human–environment relationship (Rockström *et al.*, 2009) in Chapter 17.

Because of the treatment of these aspects of adaptation in other chapters, this chapter focuses sharply on a few, more generic issues associated with adaptation. First, we describe a few fundamental principles associated with adaptation, followed by a consideration of tools and information that can support adaptation. The core of the chapter is organised around sectors for which adaptation is rapidly becoming critical, ranging from urban areas to biodiversity and natural ecosystems. The chapter concludes with discussions of adaptation challenges in the least-developed countries and of the ways in which we can learn from case studies of on-the-ground adaptation.

14.1 Principles of adaptation

Although it is now becoming common to consider impacts, adaptation and vulnerability together (the so-called 'IAV community'), there are actually significant differences between an impacts-oriented approach to coping with climate change and an adaptation–vulnerability approach. The former normally begins with a climate scenario that then drives a number of potential biophysical impacts which, in turn, prompt responses from individuals and societies. The cascading uncertainties associated with this approach, however – especially those associated with downscaling global climate model information to short time and small spatial scales – render this approach of limited value to on-the-ground adaptation. Starting with vulnerability, on the other hand, emphasises the socio-economic context in which adaptation must occur. In particular, it emphasises the nature of the institutional, cultural, equity, economic, social and governance contexts that help to define vulnerability, as well as the range of external factors that affect people's livelihoods and well-being. In this approach, climate change is considered as an additional external factor, often interacting with existing stressors rather than impacting on people in isolation.

Building resilience is, in general, a robust approach to adaptation, and is closely related to the vulnerability approach because the aim of building resilience is to reduce vulnerability. This approach is not dependent on detailed climate change projections, and can handle the daunting uncertainties associated with local-scale projections. Building resilience, in essence, prepares societies to deal with a wide range of possible climatic futures. In some cases, though, where the direction of climate change is becoming clear based on observations as well as model projections (e.g. an increase in the probability of severe heatwaves as global average temperature rises), adaptation approaches can be usefully guided by this type of information.

Another approach to coping with climate change, based on a different way of using climate information, is sensitivity analysis linked with risk assessment. In this approach, a range of possible climate futures is used to explore the potential impacts on a social–ecological system. A risk assessment is then used to judge at which level climate change becomes difficult to adapt to, and the steps needed to manage the risks at various levels of climate change. An advantage of a sensitivity study is that it can identify some thresholds or tipping points in the dynamics of the social–ecological system, as it is affected by increasing levels of climate change; such valuable information cannot be obtained by using a single climate scenario.

A variant of the sensitivity approach is to use a set of stylised climate scenarios in a 'risk-spreading' approach to adaptation on the long time scales (Figure 14.2). This approach was used in an assessment of the vulnerability of Australia's biodiversity to climate change (Steffen *et al.*, 2009) to inform strategies for adapting long-lived 'ecological infrastructure', such as trees, to a fundamentally uncertain climatic future. For the case of replanting trees following a disturbance, for example, if the 'Recovery' scenario is assumed, then trees adapted to the existing location and present climate would be replanted, whereas under the 'Runaway' scenario, trees suited for the projected future climate would be planted. To spread risks, if the area of replanting is large enough, sections would be planted with trees suited to each of the three stylised climatic futures.

Learning from doing is an important component of most approaches to adaptation. In many cases, experience gained from coping with past climate-related events and natural climate variability provides valuable lessons for dealing with the climatic changes of the future. Learning from the past has shown that successful adaptation depends not only on governments, but also on the active and sustained engagement of stakeholders – including national, regional, multilateral and international organisations; the public and private sectors (private sector initiative); civil society; and other relevant stakeholders.

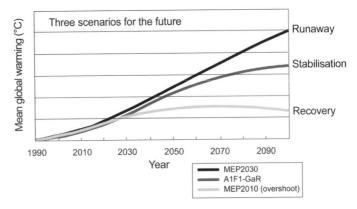

Figure 14.2 Three scenarios of future climate: recovery, stabilisation and runaway. Drawn using the IPCC (2007) suite of scenarios as driven by different assumptions about how humanity responds to the challenge of mitigating CO_2 emissions over the coming century. The shorthand names range from the most optimistic 'recovery', though the more realistic 'stabilisation', to the pessimistic 'runaway' (along which the world is currently tracking). These three scenarios inform risk management against an uncertain future when assessing management options for long-lived 'biological infrastructure'. Source: Steffen *et al.* (2009), by Mark Stafford Smith, based on an original by Roger Jones.

Community-based adaptation, for example, can benefit greatly from knowledge of local coping strategies developed anywhere in the world. The UNFCCC Secretariat, therefore, is developing a 'coping strategies database' to transfer insights on long-standing coping strategies and knowledge from communities that already have adapted to specific hazards or climatic conditions to communities that may be just starting to experience such conditions as a result of climate change. Unfortunately, much of this indigenous coping knowledge is not well documented in the scientific literature or elsewhere, and so is in danger of being lost or has already been lost. A more formalised way of learning by doing is based on the concept of active adaptive management (Box 14.1).

Box 14.1

Active adaptive management

WILL STEFFEN

A formalised way of learning by doing is based on the concept of 'active adaptive management'. The related term 'adaptive management' is commonly used to refer to a *post facto* analysis of a policy intervention or management practice to determine how well the approach worked. The implication is that such an evaluation will lead to improved policy formulation or more effective policy tools. Adaptive management has indeed been useful in many cases and will continue to be an important tool in building successful climate adaptation approaches.

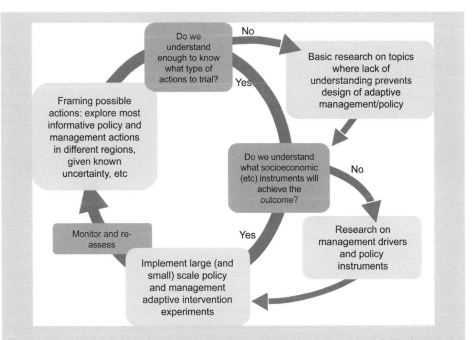

Figure 14.3 A visual representation of adaptive management, an iterative approach built around explicit, experimentally based development of plausible management options. Courtesy of M. Stafford Smith, CSIRO Sustainable Ecosystems, Canberra, Australia.

However, active adaptive management is more than this. This approach goes beyond simply being aware of management and policy outcomes in a trial-and-error fashion to formally include research and monitoring as part of the management/policy cycle (Figure 14.3; Holling, 1978; Bunnell *et al.*, 2003; Brown *et al.*, 2005; Haynes *et al.*, 2006; Steffen, 2009). In a true active adaptive management approach, management procedures are developed as part of the research design itself and outcomes from these procedures are closely monitored against target measures to improve management in the next iterative cycle. The improved management approaches are again monitored and subsequently readjusted, forming a continuous feedback loop. The active adaptive management process can trigger further research in a number of directions; for example, on the fundamental dynamics of the social–ecological system in question, or on the nature of appropriate economic or policy instruments.

This more formalised approach to adaptive management is particularly appropriate for adaptation to climate change, where uncertainties abound, and the design of policy and management will rarely 'get it right' *a priori* (Kitching and Lindenmayer, 2009). In fact, the complexity of adaptation to climate change – the interactions with other stressors, the high potential for 'surprises' in terms of climatic changes at local and regional scales, and the cross-sectoral linkages – almost ensures that a 'research first, policy follows' approach will fail and a more interactive, experimental framework for policy development is required.

Active adaptive management is very challenging, perhaps even confronting to the research, policy and management communities. Many researchers are often pushed out of their comfort

Box 14.1 (*cont.*)

zone, as the research is usually of a highly interdisciplinary nature, and involves stakeholders in the design and evaluation of the research. Engagement in active adaptive management is perhaps even more daunting for the policy and management communities. Undertaking experimental approaches to policy formulation and management practices, with no guarantee of success, can be exceptionally threatening – particularly in risk-adverse employment settings where perceived failure is not tolerated. Ultimately, however, in such a complex arena as adaptation to climate change, initial failure in implementing effective strategies is only really failure if we do not learn from the experience.

Finally, there are undoubtedly limits to adaptive capacity. Where those limits might lie provides crucial information to inform the debate on the level of mitigation required. Basically, if greenhouse gas emissions are mitigated more effectively and rapidly, less adaptation is needed and vice versa; however, this relationship breaks down when various limits to adaptation are reached. For example, many different vulnerability assessments indicate that beyond a global mean temperature increase of about 2 °C, adaptive capacity rapidly declines for many systems that depend on natural processes (Leemans and Eickhout, 2004; Thuiller *et al.*, 2005; Smith *et al.*, 2009). For example, coral reefs will largely disappear in a +2 °C world due to the combination of rising sea-surface temperatures and increasing ocean acidity, no matter how well-managed they are (Hoegh-Guldberg *et al.*, 2007). To stay within adequate adaptation ranges for coral reefs and many other natural systems, mitigation thus needs to be accelerated towards levels that comply with the maximum temperature increase of 2 °C as adopted in the Copenhagen Accord by the UNFCCC. There are also many social components to adaptation limits, many of which are closely related to equity issues. These are discussed in more detail in Chapters 9 and 15.

14.2 Tools and information to support adaptation

Scientific information is essential to support adaptation, but the nature how that information is generated and communicated is crucial for its usefulness. The linear model of linking science with policy – scientists producing scientific knowledge (e.g. regionally downscaled predictions of global climate models) that is then transmitted to the policy world – is not appropriate for adaptation, and is being replaced by iterative processes in which dialogue between the science and the policy and management communities is embedded in the local context (Moll, 2009). The latter approach includes features such as defining the problem together with practitioners and users, respect for different forms of knowledge (not just scientific), ongoing dialogue between knowledge producers and politicians and other decision-makers, and a reflexive process of 'reasoning together' that bridges cultural styles (Beck and Georg, 2009). In short, information for adaptation needs to move from knowledge transfer to co-production of knowledge and integrating knowledge with action (Hinkel *et al.*, 2009).

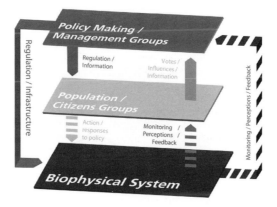

Figure 14.4 Typical interactions in multi-level governance systems, in which citizen groups can play a key role in mediating between policymaking that operates at regional or national scales, and the on-ground management of biophysical systems, which often occurs at the local scale. Such profession-ally organised, multi-level processes can help to reduce scale mismatches and policy incoherence, and to support integrated social and regulatory changes. Source: Daniell *et al.* (2009).

Participatory approaches are an important tool in successfully pursuing the co-production of knowledge concept, often because of the multi-scale nature of the problem from local to global levels (Figure 14.4; Daniell *et al.*, 2009). Different processes and actors are involved at different levels, requiring participatory approaches to define common priority issues and values, stimulate dialogue, explore management options, and coordinate decisions of mutually interacting actors on resource use. Specific techniques employed in this approach include decision-aiding process theory, participatory modelling, scenario planning, risk assessment, role playing and implementation planning. Results of case studies show that professionally organised multi-level processes can reduce scale mismatches and policy incoherence (Daniell *et al.*, 2009).

Constructing scenarios for a range of possible futures – especially when using a participatory approach – is one of the most powerful ways to develop effective adaptation strategies, and to bridge the gap between knowledge and action (Bohunovsky and Omann, 2009). Scenarios facilitate a holistic view of the problem(s) and the context in which it is (they are) embedded and trigger a discussion of 'what would happen if …'. A range of scenarios can aid decision-making, especially when decision-makers are involved in constructing them. In short, scenarios and other visioning processes stimulate, empower and support decision-making in the face of complexity and long timescales.

Scenario development can be placed in the broader context of Integrated Sustainability Assessment (ISA), for example, as introduced in the Methods and Tools for Integrated Sustainability Analysis (MATISSE) project (http://www.matisse-project.net). In this approach, ISA is conceptualised in four stages: scoping, envisioning (including scenario development), experimenting and evaluating/learning (Figure 14.5). A strongly participatory approach, ISA is aimed at initiating and sustaining a transition towards sustainability.

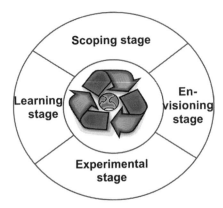

Figure 14.5 The Integrated Sustainability Assessment Cycle. Source: Bohunovsky and Omann (2009), as developed by Weavers and Rotmans (2005).

14.3 Adaptation in key sectors

The adaptation challenge is strongly context-specific, and varies from location to location, even within the same country or province. It intersects with equity issues (Chapter 9), and most effectively in a broader context of development and sustainability (Chapter 15). Nevertheless, there are a number of generic issues associated with adaptation, often considered on a sector-by-sector basis, which are important across both developed and developing countries. This section explores critical features of the adaptation challenge facing a small number of key sectors.

14.3.1 Human health

Although the health research community strongly supports mitigation as an urgent, first-order task, there is recognition that some impacts are now unavoidable, so adaptation is also needed to reduce current and future risks (Ebi, 2009). An approach that can be implemented immediately is to maintain and enhance current public health interventions to reduce the burdens of the major climate-sensitive health outcomes; these include diarrhoeal diseases, malnutrition and malaria. Relevant activities include better water and sanitation infrastructure, flood control, mosquito control, and nutritional supplementation programs. Early warning systems for extreme weather events – such as heatwaves, storms and cyclones – are required, as for many other sectors. More innovative approaches, such as genetic modification to control mosquito populations, may also be necessary. It is crucial to coordinate adaptation activities with other sectors to ensure that strategies implemented in those sectors do not become maladaptive for human health.

14.3.2 Agricultural systems

Agricultural production is likely to be enhanced over the next few decades at least in the mid latitudes and high latitudes, but is likely to be diminished in the tropics and sub-tropics, where food security is already an issue in many areas (IPCC, 2007). The situation is particularly acute for Africa (Box 14.2), where the risk of hunger will increase, although the projected increase in the number of poor people in Africa is a strong contributing factor (Parry, 2009).

In general, there are many generically applicable approaches to adapting agricultural systems to projected changes in climate, but these required sufficient adaptive capacity. Examples include improved information networks to improve pest and pathogen control, alter the timing of cropping activities such as sowing, and improve irrigation and water management (Crimp *et al.*, 2009). Increasingly sophisticated models that link the biophysical drivers of crop growth that predict yield with socioeconomic modules that simulate social and institutional contexts for agriculture are highlighting the potential for adaptive practices in the socioeconomic realm. Such context-specific studies have been carried out for China (Xiong *et al.*, 2009) and for Mozambique and Uganda (Friis-Hansen *et al.*, 2009). In the longer term, genetic modification of important crop species may be required to track a rapidly changing climate.

Box 14.2
Adaptations to enhance agriculture in Sub-Saharan Africa (SSA)

POLLY ERICKSEN

A wide variety of prospective options for adapting African agriculture to the effects of climate change exist, from the effective use of climate information to paying smallholders for ecosystem goods and services to increasing household income. Most African farmers are accustomed to managing climatic risk; typical coping strategies to reduce negative impacts on production include experimenting with alternative planting dates and new crops such as drought-tolerant varieties, replanting, and cultivation at higher elevation in the case of severe rainfall events and flooding. A good deal has been written about the need to recognise and support local coping and adaptive strategies rather than introducing external ideas (Mortimore, 2005). Pastoralists also have traditional strategies for managing climate risk, including movement of feed resources and/or livestock over what may be large distances. In other instances they have experimented with different mixes of breeds (Blench and Marriage, 1999). Increasing yields or productivity is, of course, not the only adaptation option. Expansion of cropped land can also be used to maintain production. However, climate change may heavily modify land suitability. In relatively large areas of sub-Saharan Africa that are currently classified as mixed, rainfed, arid–semiarid systems, cropping may become increasingly risky and marginal – perhaps leading to increased dependence on livestock keeping (Jones and Thornton, 2009). The challenge for a climatically altered future is to improve access to the crop and livestock varieties, and management techniques that will be needed to adapt farming systems to new climatic, land and water constraints.

Adopting any new options requires farmers to take on new risks and explore new markets. They also require access to credit and technical support. The ability of many smallholder farmers in sub-Saharan Africa to obtain access to such support mechanisms is already low

Box 14.2 (*cont.*)

(Ruben and Pender, 2004; Scoones *et al.*, 2005). The need to adapt to climate change may provide a stimulus to increase such support, but it can be achieved only with considerable institutional and political commitments (Eakin and Lemos, 2006; Padgham, 2009). Although many studies in developing country contexts emphasise the importance of supporting local-level, grassroots adaptation, lessons from decades of agricultural development interventions have shown the need for higher-level institutional and policy support as well (Adger *et al.*, 2003; Dorward *et al.*, 2003; Ellis and Freeman, 2004). Currently, many farmers rely on a host of off-farm diversification strategies to support their own agricultural activities and ensure a household income (Barrett *et al.*, 2001). If a changing climate increases the risks associated with agriculture, it is not clear how farmers will adjust the balance of on- and off-farm activities, as they are already struggling to maintain this balance. As climate change affects not only what can be grown but where – and as land suitability shifts – existing successful technological packages and systems may not be that useful. At the moment, climate change seems likely to diminish the viability of agricultural options. If climate continues to change so that no historical analogues exist, then pressures on farming systems, livelihoods, agricultural technology and supporting mechanisms may become intense, as smallholders become increasingly alienated from their realm of experience.

Adaptation occurs in the context of uncertainty, and if that uncertainty is too great then it will be difficult to assess appropriate adaptation options. Some adaptation options are likely to be robust, even given the uncertainties that exist concerning future patterns of climate change. Thus, support for livelihood diversification is consistently stressed as key to climate change adaptation (Padgham, 2009). Payments for carbon sequestration are another new and promising option that may both bolster smallholder livelihoods while enhancing ecosystem services (Bulte *et al.*, 2008). But the lessons of the recent past teach us that the difficulty of implementing many of these options in Sub-Saharan Africa should not be underestimated; massive investment and increases in agricultural productivity will be necessary if economic development is to succeed in Africa in the coming decades (Collier and Dercon, 2009).

14.3.3 *Land systems, including forests*

Agriculture is one example of a land system, and the challenge to adapt to climate change is relevant for a wide range of land systems. In general, adaptation will be a complex process, involving context-specific information on regional and local climate, other environmental conditions, ecosystem services and a thorough understanding of how land-use decisions are made. Nevertheless, some general approaches to adaptation are relevant for a large number of land systems. For example, insights into the drivers of past land-use changes provide important clues as to how land use has responded to environmental change, and how effective these responses have been. Second, an ecosystems approach, or a social–ecological systems approach, is likely to be more effective than focusing on a single, dominant land-use or ecosystem service. Third, there is no substitute for high-resolution studies to capture the local characteristics and dynamics, even if the climate information can only be provided on much larger scales (Moore *et al.*, 2009). Finally, as land use is changed as an adaptive response to climate change, care must be taken not to increase the carbon emissions from landscapes (Ojima and Corell, 2009).

Forests are a particularly important type of land system. However, as for many other land systems, there is no single technical solution to the adaptation challenge, and flexibility will thus be required to account for local biophysical and institutional characteristics as well as take advantage of local knowledge (especially for managing forests during periods of high climate variability). A successful application of sustainable forest management principles at both national and international levels will increase the resilience of forests towards a changing climate, and may comprise an adequate overall adaptation strategy. In essence, sustainable forest management reduces the exposure and sensitivity of forests to climate change, and enhances its resilience. Management can already be modified to deal with some generic impacts of climate change (at least in mid latitude and high latitude forests), such as an increase in the maturation rates of trees and an increase in disturbance from pests and wildfires. Compared to many other ecosystems, though, the adaptive capacity of forests will be reached relatively quickly, given the slow growth rate of trees compared to the rapidity of climate change being experienced now and that projected for the rest of this century at least.

To date, most adaptation responses in the forest sector to climate change have been reactive rather than anticipatory. In addition, adaptation measures are often conditioned by the original management objectives for the forest as well as the legacy of past management practices. One of the most challenging aspects of adaptation to climate change is the need to deal more effectively with disturbances such as wildfire; such adaptation requires considerable investment in infrastructure (e.g. communication, fire detection, transport) as well as training and perhaps modification of anticipatory management practices such as controlled burning. In general, more robust adaptation strategies will emerge from adaptive co-management, in which stakeholder collaboration is a central part of the strategy.

14.3.4 Water resources

The impacts of climate change on water resources are now apparent in many parts of the world, and will intensify over the coming decades at least. Adaptation is now an urgent imperative in many places (Alcamo, 2009), and is often given high priority in adaptation as access to sufficient water of high quality is absolutely central to human well-being.

As for many other sectors, adaptation with respect to water resources is multi-dimensional, and must be based on an appropriate balance between options focusing on infrastructure – such as water storage and transfer – and on governance-based approaches such as demand management and a range of other local measures. Underlying governance of water resources are fundamental issues such as institutional architecture and performance, values and behaviour – all of which affect the consumption, conservation, valuation and distribution of water. Thus, effective adaptation probably depends less on the projected impacts of climate change on water resources and more on the knowledge, communication and understanding of local socio-economic processes (Pfaff *et al.*, 2009).

Democracy			
		High	**Low**
Knowledge	**High**	Higher levels of adaptive capacity	Technocratic Insulation
	Low	Potential for ill informed decisions	Maladaptation

Figure 14.6 Relationship between technical knowledge use and democratic decision-making with adaptive capacity to climate variability and change. Source: Lemos *et al.* (2009).

More than just specific information on climate change itself – for example, knowing that the risk of a drying climate and prolonged droughts is increasing may be sufficient – good governance is the key to successful adaptation, building on integrated approaches where the need for open and transparent sharing of information between all stakeholder groups is recognised and observed. Integration needs to work in three dimensions – horizontally between catchments or river basins in which mutual learning can occur (Goulden and Conway, 2009), horizontally between different sectors (e.g. agriculture, domestic water supply, energy, tourism, industry) (Huntjens *et al.*, 2009) and vertically between national, regional and local governments and civil society. Information sharing is critical for this integration to be effective (Figure 14.6; Feldmann *et al.*, 2009; Lemos *et al.*, 2009). Although democracy is associated with high levels of adaptive capacity, this is not always true and central authority may be able to respond more rapidly to natural disasters, for example.

14.3.5 Urban areas

Urban areas are concentrated sources of anthropogenic greenhouse gases and are thus central to mitigation efforts. However, with more than half of the human population now resident in urban areas, they are coming to play a much more prominent role in the adaptation agenda. It is thus no exaggeration to claim that the challenge of climate change may be won or lost in the cities.

Adaptation to climate change – particularly to ameliorate the impacts of those extreme events whose probability increases with climate change – is particularly acute in urban areas because extreme events can create sudden supply shortages, or trigger major catastrophes that quickly lead to serious bottlenecks or emergencies for a large number of people. However, urban areas perhaps have an advantage in implementing adaptation activities, as most of them have planning authorities that, in principle, can influence the design of the built environment (urban form, infrastructure, buildings, land use, parks, etc.) in ways that adapt them to future climates. Such planning is especially important for infrastructure, which often have life spans of 75 years or longer. Thus, mainstreaming climate adaptation considerations into current urban development agendas has to be a central strategy for dealing with climate change (Chapter 15).

14.3.6 Coastal zone

Sea-level rise is one of the most important components of climate change to impact on the coastal zone. Although our understanding of global average sea-level rise is improving rapidly (Chapter 3), there is much variability at the regional and local levels due to other factors like sediment compaction, tectonics, and the gravitational and isostatic effects of ice sheets. Thus, to help plan for adaptation in a particular place along a coast, global average sea-level rise is of little or no use; much more location-specific information must be generated.

A few countries have begun to undertake detailed analyses of sea-level rise for their particular situations. The Netherlands, for example, under the Dutch Delta Commissions has coupled sea-level rise information with storm surge estimates and Rhine River discharge information to develop scenarios for the risk of inundation (Vellinga *et al.*, 2008a). High-end scenarios yield local sea-level rises ranging from 0.05 m to 1.25 m by 2100 (Katsman *et al.*, 2008). When a projected increase in runoff from the River Rhine catchments is included in the analysis, a current 1-in-1250-year flooding event occurs much more frequently (Vellinga *et al.*, 2008b). A similar analysis for the UK gave a range of absolute sea-level rise around the UK of 0.12–0.76 m for the period 1990–2095, and relative rises of 0.16–0.63 m for London and 0.07–0.54 m for Edinburgh, for the A1B emission scenario (Lowe *et al.*, 2009; Mokrech *et al.*, 2009).

The Venice lagoon is an interesting case where adaptation to high water levels, a component of which is due to rising sea level in general, is already occurring because of the extreme vulnerability of the historical towns and cities in the lagoon. Most of the adaptation that has been carried out so far consists of 'hardening' the defences of the lagoon. More than €15 billion has been allocated already for adaptation, including studies, plans, projects and structures. Much of the effort is directed towards coastal defence infrastructure, such as the reconstruction of eroded beaches and the reinforcement of ancient sea walls and breakwaters. In addition, focus is on the construction of a system of mobile barriers to temporarily isolate the lagoon from the sea, when necessary. Adaptation also includes the construction of raised public walkways in Venice city as well as the restoration of lagoon embankments. Overall, these adaptation activities are now about 50% complete (Munaretto, 2009).

A wide range of climate change-related impacts to fish populations is already being observed, including changes in the distribution of species, changes in the timing of life history events (phenology) and physiological stress in some species (Pörtner and Farrell, 2008). Adaptation of the fishing industry to such climate-related changes is still in its infancy, but one principle is already clear – adaptation must be carried out in the context of the combined and interacting effects of climate change and overfishing (Perry and Barange, 2009).

Two examples illustrate the widely differing challenges facing coastal dwellers in different regions of the world as the climate shifts. In the Arctic, rapid changes to sea ice and coastal erosion are posing significant risks to Inuit communities as they attempt to maintain their culturally important hunting activities. Adaptation strategies include promotion of hunting, fishing and survival skills for young Inuit under changing ice conditions; subsidised insurance schemes for lost or damaged equipment; and modification of hunting quotas in collective decision-making by communities, scientists and wildlife managers (Ford *et al.*, 2009). The mangrove systems of the tropical and sub-tropical coastal zones face a squeeze caused by sea-level rise from the seaward side and by land-use change (i.e. conversion of mangroves to prawn farms) from the landward side (Giri *et al.*, 2008; Walters *et al.*, 2008). Adaptation responses to this squeeze are based more on 'soft' ecosystem approaches rather than 'hard' engineering approaches; the former include the restoration of drained and degraded coastal peatlands through the closure of drainage channels and through reforestation with indigenous tree species (Hale *et al.*, 2009).

14.3.7 Insurance industry

The insurance sector is finding it increasingly difficult to cope with the losses from climate-related disasters and is very concerned about potential global warming impacts, yet insurance against climate-related events represents one of the most important adaptation tools. Experience in developed countries, such as the UK, offers many lessons about the use of insurance – particularly public–private partnerships – to cope with extreme events. Such experience is now being translated into developing country contexts, often in the form of index-based micro-insurance for weather hazards. Insurance also acts indirectly to promote other forms of adaptation. By pricing risk, insurance provides incentives for risk reduction through other measures. Such risk-reducing activities are critical to the use of insurance as an adaptation tool; failure to reduce risk – for example, through appropriate measures to protect property or move assets along the coastal zone – could render large areas uninsurable. Finally, insurance-based approaches, especially the provision of micro-insurance, may play an important role in the implementation of the adaptation safety net (Chapter 9), whose funding was agreed in the Copenhagen Accord (Box 13.2).

14.3.8 Biodiversity and natural ecosystems

Chapter 6 emphasised the limited adaptive capacity of natural ecosystems to rapid climate change and, thus, the need for prompt and effective mitigation efforts to avoid an acceleration in the rate of extinctions and other aspects of biodiversity loss through the twenty-first century and beyond. But the situation is not hopeless. There are many policy and management approaches that can be implemented now to help natural ecosystems adapt to climate change; without such actions, the impacts of climate change on biodiversity are likely to be much more severe.

As foreshadowed in Chapter 6, with a rapidly changing abiotic environment, it is useful to go back to fundamental ecological principles to guide adaptation activities (Steffen *et al.*, 2009). This is particularly important in that estimates of species interactions based on projections of changing distributions may be in serious error, prompting the use of alternative information for management guidance. For example, recognising functional rather than species-specific responses to climate change can be important (Syphard and Franklin, 2009, in relation to fire), as well as analysing changes in landscape-scale functioning. An example of the latter is a classification of sites across sub-regions that face differing impacts, helping managers to devise different conservation goals (e.g. promote resilience or facilitate transformation) for specific contexts (Hole *et al.*, 2009).

Expansion of protected areas, identification of refugia, and linkages across other types of land uses remain central pillars of biodiversity conservation under a changing climate. Examples include the linking of key habitat areas in central Africa to support the migration of bird species (Hole *et al.*, 2009) and the identification and protection of biorefugia for tree species in Italy (Attorre *et al.*, 2009). However, Steffen *et al.* (2009) emphasised that increasing connectivity is not a panacea, but 'appropriate connectivity' must be considered – what are we trying to connect, at what scale and for what purposes?

There are several important management tools that can be deployed to help build resilience in natural ecosystems. These include: regional analysis of the dynamics of reserve networks under climate change (Hole *et al.*, 2009); improved regional conservation planning through integration of socio-economic and climate trajectories (Steffen *et al.*, 2009); the use of 'robust rules of thumb' based on ecological principles and local knowledge to guide local management (Stampfli and Zeiter, 2009); and the abandonment of out-of-date management tools, such as static species lists, use of political boundaries for declaring species endangerment, and conservation driven by single species declarations.

Finally, the use of landscapes for mitigation – via REDD+ forest carbon activities (Box 9.2) and the storage of carbon in soils – offers many opportunities to enhance biodiversity conservation and to increase the delivery of other ecosystem services. Poorly designed mitigation activities, on the other hand, could have deleterious impacts on biodiversity. A systematic analysis of risks and opportunities should be mandatory for all land-based mitigation approaches (Berry *et al.*, 2009). Adaptation in other sectors may also impact on biodiversity. For example, beach nourishment aimed at preventing climate change-related beach erosion can be very harmful to coastal biodiversity if not carefully planned and implemented (Berry *et al.*, 2009).

14.4 Adaptation to climate change in the least-developed countries

The preceding section has outlined the various adaptation approaches that are now being undertaken or are being planned in several sectors, regardless of where this adaptation is taking place. As noted at the outset of this chapter, the very important equity aspects of the adaptation challenge, and the need to mainstream adaptation into sustainable development, have been dealt with in detail elsewhere in this book. Nevertheless, it is useful to briefly review here the challenges, experiences and ways forward for adaptation in the least-developed countries of the world.

Not surprisingly, most of the least-developed countries have identified the current inadequate access to funding for adaptation as a significant constraint. Beyond that, however, is a strong consensus on (i) focusing adaptation strategy development and decision-making strongly at the local level, (ii) communicating learning at the local level to other localities and transmitting it upward to the national level, (iii) scaling up successful projects at the micro-level to bolder programs at the national level, (iv) learning from coping with the present-day climatic extremes and modes of variability, and (v) emphasising the reduction of social vulnerability.

In summary, adaptation in the world's wealthiest countries is backed up by significant resources and much inherent human and institutional capacity. In the least developed countries, adaptation is an urgent necessity to maintain livelihoods in the face of shifts in climatic variability and extremes – but with inadequate financial resources, severe gaps in the capacity of social and institutional systems, and pressure to deal with more immediate stresses and challenges (Box 14.3).

Box 14.3
Challenges in creating an international regime for adaptation

DIANA LIVERMAN

The urgency of negotiating an international agreement to limit anthropogenic climate change by reducing emissions means that, until recently, mitigation has dominated the international climate regime and that concerns about adaptation were less evident. Although the 1992 UNFCCC includes a clause promising that the developed countries will assist developing countries in meeting the costs of adaptation to the adverse effects of climate change the following 15 years saw little attention to adaptation issues, with only pilot projects and little in the way of funds delivered for actual adaptation projects (Liverman and Billet, 2010).

Adaptation rose up the agenda as the risks of dangerous climate change became clearer, yet emissions continued to grow, and as some of the most vulnerable countries – such as the less-developed low-lying island states, Bangladesh, and parts of Africa – became more aware of the risks and of their power as negotiating blocks in the UN system. Humanitarian and environmental NGOs also started to see the need to adapt both human and ecological systems to the onset and future risks of climate change. Also, the private sector – especially insurance – identified the critical need for adaptation.

The 1997 COP15 climate negotiations in Bali placed adaptation as one of the four pillars of the Bali Action Plan to establish a strong climate agreement by 2012. An effort began to design an international adaptation regime that would help the developing world cope with the impacts of climate change. Pilot adaptation funds from the World Bank, UNDP and some bilateral assistance funds had already helped several countries prepare National Adaptation Programmes of Action (NAPAs) and undertake capacity building, but funds were too limited to implement serious reductions in vulnerability. Low-lying islands such as Kiribati, Tuvalu and the Maldives became even more aware of the onset of sea-level rise; the threats to their land, water and culture; and the high costs of responding to these risks.

Underlying the international discussions about adaptation are fundamental questions about how much money is needed for adaptation and who should provide it, as well as debates over who should manage the funds and how they should be allocated. Estimates of the amount of money needed for adaptation are as high as $150 billion a year (Oxfam, 2009; Parry *et al.*, 2009) especially if the unmet needs for adaptation to current climate variability are included and if the costs of climate proofing current development assistance are accounted for. These costs are based on estimates of damages and the costs of adapting infrastructure such as coastal defences, water resources and agriculture to climate change – and may also include some estimates for adapting natural ecosystems (Parry *et al.*, 2009).

Finding sources for this magnitude of response is extremely challenging, especially given possible tradeoffs against the costs of mitigation and other priorities. Alternative proposals include funds generated through auctioning of emissions allowances, a levy on international air and ship transport, increases in official development assistance, carbon taxes, and a fee on international financial transactions (UNDP, 2007; Müller, 2008; Parry *et al.*, 2009; World Bank, 2009). There is considerable concern that any funds be truly additional, and not diverted from current development priorities or relabelling of other promised funds.

In terms of governance some see the World Bank as the obvious manager of the funds, whereas others would like to see a separate fund administered through the UNFCCC. In terms of eligibility for the funds there is a difficult debate about whether adaptation finance should only go to the most vulnerable and small island countries, or more broadly to the developing world; and whether countries must demonstrate their needs by showing evidence for climate change impacts, gain eligibility by also preparing mitigation plans, and document the effective use of the funds. The allocation of adaptation funding is also confounded by scientific uncertainties about the exact regional nature of climate change – for example, whether rainfall will increase or decrease – and the lack of good techniques for assessing vulnerability across large regions (Klein, 2009).

14.5 Learning from case studies

A synthesis of a large number of case studies on climate adaptation from the freshwater sector, many of them in developing countries, ties together many of the principles and approaches described earlier in this chapter and provides a robust set of general guidelines for adaptation to climate change (Leary *et al.* 2007; Moench and Stapleton, 2007; Pittock, 2009). The choice of freshwater-oriented cases studies was based on the facts that (i) many of the most important impacts of climate change on human systems occur through water, and (ii) adaptation to changes in water resources will require responses that address both the climatic and non-climatic drivers of vulnerability.

The synthesis identified five key lessons.

(1) Despite considerable uncertainties in the projection of climate change impacts, adaptation needs to begin now, often based on risk management approaches.
(2) Adaptation and development are usually compatible, and so adaptation should be mainstreamed into sustainable development (Chapter 15).
(3) Knowledge sharing, strengthening of institutions, and human and social capacity-building more generally are vitally important for successful adaptation.
(4) Adaptation is strengthened by community ownership and subsidiarity; consistent and, if possible, substantial funding; concurrent, linked action across sectors and at different geopolitical scales; and long-term, iterative programs.
(5) The role of national governments in adaptation includes allocating funds for adaptation, supporting sub-national institutions to act and removing barriers to funding them, and communicating the risks and opportunities for adaptation.

14.6 Summary and conclusions

Until recently, adaptation has been overshadowed by mitigation as a response to climate change. However, with many impacts of climate change already visible and with predictions for rapid increases in impacts, adaptation is now widely recognised as essential, urgent, growing rapidly in importance, and very complex in implementation – intersecting strongly with dimensions of development, equity, economic and social policy and climate mitigation. Several principles to guide effective adaptation have evolved through experience: (i) adaptation is not necessarily driven by, or dependent on, knowledge of impacts or specific climate projections; (ii) it is almost always context- and location-specific; (iii) building resilience is an effective general approach; (iv) risk-spreading is a robust strategy in the face of daunting uncertainties about the future; and (v) learning-by-doing, particularly active adaptive management, are central features of many effective adaptation activities. An unsuccessful attempt at adaptation is not really a failure unless we do not learn from it.

Most experience in adaptation to date has been organised around sectors, ranging from human health through land systems and freshwater resources to urban areas and the coastal zone. Many important lessons have been learned. First, the most effective action is often focused strongly at the local scale, with higher scales important in a support role, often in terms of financial resources, information sharing and particular areas of expertise (e.g. scientific, legal, financial). Pursuing adaptation within sectors in a silo fashion is almost always a mistake; considerable opportunities for synergies arise from cross-sector integration, and unwanted collateral damage can occur when adaptation in one sector leads to maladaptive outcomes in another. Finally, if there is one mandatory, generic feature of all successful adaptation approaches – regardless of sector or developing or developed status – it is building, nurturing and maintaining sufficient capacity: individual, social, institutional and economic/financial.

References

Adger, W. N., Huq, S., Brown, K., Conway, D. and Hulme, M. (2003). Adaptation to climate change in the developing world. *Progress in Development Studies*, **3**, 179–95.

Alcamo, J. (2009). Climate change and the changing frequency of floods and droughts: scenario analysis of risk and adaptation in Europe. *IOP Conference Series: Earth and Environmental Science*, **6**, 292016.

Attorre, F., Vitale, M., Tomasetto, F. *et al.* (2009). Effect of climate change on tree species distribution to support the elaboration of adaptive management strategies in natural protected areas. *IOP Conference Series: Earth and Environmental Science*, **6**, 312012.

Barrett, C. B., Reardon, T. and Webb, P. (2001). Nonfarm income diversification and household livelihood strategies in rural Africa: Concepts, dynamics, and policy implications. *Food Policy*, **26**, 315–31.

Beck, S. and Georg, C. (2009). One size fits all? Can the IPCC serve as blueprint for scientific advice on adaptation to climate change? *IOP Conference Series: Earth and Environmental Science*, **6**, 392010.

Berry, P., Paterson, J., Cabeza, M. *et al.* (2009). Climate change adaptation and mitigation: Synergisms, antagonisms and trade-offs for biodiversity. *IOP Conference Series: Earth and Environmental Science*, **6**, 312002.

Blench, R. and Marriage, Z. (1999). *Drought and Livestock in Semi-arid Africa and Southwest Asia*. Working Paper 117. London: Overseas Development Institute.

Bohunovsky, L. and Omann, I. (2009). Participatory scenario development for integrated sustainability assessment. *IOP Conference Series: Earth and Environmental Science*, **6**, 392005.

Brown, K., Mackensen, J., Rosendo, S. *et al.* (2005). Integrated responses. In *Ecosystems and Human Well-being. Volume 3. Policy Responses. Findings of the Responses Working Group*, eds. K. Chopra, R. Leemans, P. Kumar and H. Simons. Washington, D.C.: Island Press, pp. 425–64.

Bulte, E. H., Lipper, L., Stringer, R. and Zilberman, D. (2008). Payments for ecosystem services and poverty reduction: Concepts, issues and empirical perspectives. *Environment and Development Economics*, **13**, 245–54.

Bunnell, F., Dunsworth, G., Huggard, D. and Kremsater, L. (2003). *Learning to Sustain Biological Diversity on Weyerhauser's Coastal Tenure*. Vancouver: Weyerhauser Company.

Collier, P. and Dercon, S. (2009). African agriculture in 50 years: Smallholders in a rapidly changing world? Background paper for the *Expert Meeting on How to Feed the World in 2050*, 24–26 June, Food and Agriculture Organization of the United Nations, Rome, http://www.fao.org/docrep/012/ak542e/ak542e00.htm.

Crimp, S., Howden, M., Laing, A. *et al.* (2009). Managing future agricultural production in a variable and changing climate. *IOP Conference Series: Earth and Environmental Science*, **6**, 372006.

Daniell, K. A., Manez Costa, M.A., Ferrand, N. *et al.* (2009). Aiding multi-level decision-making processes for climate change mitigation and adaptation. *IOP Conference Series: Earth and Environmental Science*, **6**, 392006.

Dorward, A., Poole, N., Morrison, J., Kydd, J. and Urey, I. (2003). Markets, institutions and technology: missing links in livelihoods analysis. *Development Policy Review*, **21**, 319–32.

Eakin, H. and Lemos, M. C. (2006). Adaptation and the state: Latin America and the challenge of capacity-building under globalization. *Global Environmental Change*, **16**, 7–18.

Ebi, K. (2009). Health adaptation choices at local and regional scales. *IOP Conference Series: Earth and Environmental Science*, **6**, 142019.

Ellis, F. and Freeman, H. A. (2004). Rural livelihoods and poverty reduction strategies in four African countries. *Journal of Development Studies*, **40**, 1–30.

Feldman, D. and Ingram, H. (2009). The use and application of knowledge networks for water systems' management of climate change. *IOP Conference Series: Earth and Environmental Science*, **6**, 292010.

Ford, J., Gough, B., Laidler, G. *et al.* (2009). Sea ice, climate change, and community vulnerability in northern Foxe Basin, Canada. *Climate Research*, **37**, 138–54.

Friis-Hansen, E., Stage, O. and Aben, C. (2009). Importance of local institutions and downwards accountability of local government for mitigating adverse effects of global climate change for poor farmers in South-Eastern Africa. *IOP Conference Series: Earth and Environmental Science*, **6**, 372002.

Giri, C., Zhu, Z., Tieszen, L.L. *et al.* (2008). Mangrove forest distributions and dynamics (1975–2005) of the tsunami-affected region of Asia. *Journal of Biogeography*, **35**, 519–28.

Goulden, M. and Conway, D. (2009). Cooperation and adaptation to climate change in transboundary river basins in Africa: Evidence from the Nile Basin. *IOP Conference Series: Earth and Environmental Science*, **6**, 292005.

Hale, L. Z., Meliane, I., Davidson, S. *et al.* (2009). Ecosystem-based adaptation in marine and coastal ecosystems. *Renewable Resources Journal*, **25**, 21–8.

Haynes, R. W., Bormann, B. T., Lee, D. C. and M. J. R. (2006). *Northwest Forest Plan – The First 10 Years (1994–2003): Synthesis of Monitoring and Research Results.* Pages General Technical Report PNW-GTR-651. Portland, OR: USDA Agriculture, Forest Service, Pacific Northwest Research Station.

Hinkel, J., Hofmann, M. E., Lonsdale, K. *et al.* (2009). Knowledge for adaptation: The ADAM digital adaptation compendium. *IOP Conference Series: Earth and Environmental Science*, **6**, 392004.

Hoegh-Guldberg, O., Mumby, P. J., Hooten, A. J. *et al.* (2007). Coral reefs under rapid climate change and ocean acidification. *Science*, **318**, 1737–42.

Hole, D., Turner, W., Brooks, T. *et al.* (2009). Towards an adaptive management framework for climate change across species, sites and land/seascapes. *IOP Conference Series: Earth and Environmental Science*, **6**, 312006.

Holling, C. S. (ed.) (1978). *Adaptive Environmental Assessment and Management.* Toronto: John Wiley and Sons.

Huntjens, P., Pahl-Wostl, C., Jun, X., Camkin, J. and Kranz, N. (2009). Comparison of the processes for developing climate change adaptation strategies for dealing with floods and droughts in the Netherlands, South Africa, China and Australia. *IOP Conference Series: Earth and Environmental Science*, **6**, 292013.

Intergovernmental Panel on Climate Change (IPCC) (2007). Summary for policymakers. In *Climate Change 2007: Impacts, Adaptation and Vulnerability. Contribution of Working Group II to the Fourth Assessment Report of the Intergovernmental Panel on Climate Change*, eds. M. L. Parry, O. F. Canziani, J. P. Palutikof, P. J. van der Linden and C. E. Hanson. Cambridge, UK and New York, NY: Cambridge University Press, pp. 7–22.

Jones, P. G. and Thornton, P. K. (2009). Croppers to livestock keepers: Livelihood transitions to 2050 in Africa due to climate change. *Environmental Science & Policy*, **12**, 427–37.

Kabat, P., van Vierssen, W., Veraart, J., Vellinga, P. and Aerts, J. (2005). Climate proofing the Netherlands. *Nature*, **438**, 283–84.

Katsman, C.A., Church, J., Kopp, R. *et al.* (2008). High-end projection for local sea level rise along the Dutch coast in 2100 and 2200. In *Exploring High-end Climate Change Scenarios for Flood Protection of the Netherlands: An International Scientific Assessment*, eds. P. Vellinga, C. Katsman, A. Sterl and J. J. Beersma. The Netherlands: KNMI/Alterra.

Kitching, R. L. and Lindenmayer, D. B. (2009). Adaptive management, monitoring and climate change. In *Australia's Biodiversity and Climate Change*, eds. W. Steffen, A. Burbidge, L. Hughes, R. Kitching, D. Lindenmayer, W. Musgrave, M. Stafford Smith and P. Werner (2009). Melbourne: CSIRO Publishing, p. 153.

Klein, R. J. T. (2009). Identifying countries that are particularly vulnerable to the adverse effects of climate change: an academic or a political challenge? *Carbon & Climate Law Review*, **3**, 284–91.

Leary, N., Adejowan, J., Barros, V. *et al.* (2007). A stitch in time: lessons for climate adaptation from the AIACC Project. AIACC Working Paper 48. Washington.

Leemans, R. and Eickhout, B. (2004). Another reason for concern: regional and global impacts on ecosystems for different levels of climate change. *Global Environmental Change*, **14**, 219–28.

Lemos, M. C., Bell, A., Engle, N., Formiga-Johnsson, R. and Nelson, D. (2009). Technical knowledge and democratic decision-making: Building adaptive capacity of freshwater systems in Brazil. *IOP Conference Series: Earth and Environmental Science*, **6**, 292015.

Liverman, D. and Billett, S. (2010). Copenhagen and the Governance of Adaptation. *Environment*, May/June. **52** (3): 28–36.

Lowe, J., Tinker, J., Howard, T. *et al.* (2009). Informing adaptation: A new set of marine climate change scenarios. *IOP Conference Series: Earth and Environmental Science*, **6**, 352002.

McCarthy, J. J., Canziani, O. F., Leary, N. A., Dokken, D. J. and White, K. S. (eds.) (2001). *Climate Change 2001. Impacts, Adaptation, and Vulnerability*. Cambridge, UK: Cambridge University Press.

Moench, M. and Stapleton, S. (2007). Water, Climate, Risk and Adaptation Working Paper 2007, Delft, Cooperative Program on Water and Climate.

Mokrech, M., Hanson, S., Nicholls, R. J. *et al.* (2009). The Tyndall coastal simulator. *Journal of Coastal Conservation*, doi: 10.1007/s11852–009–0083–6.

Moll, P. (2009). From knowledge to action? Either via science as usual or 'just do it'. *IOP Conference Series: Earth and Environmental Science*, **6**, 392011.

Moore, N., Alagarswamy, G., Pijanowski, B. *et al.* (2009). Food production risks associated with land use change in East Africa. *IOP Conference Series: Earth and Environmental Science*, **6**, 342003.

Mortimore, M. (2005). Dryland development: Success stories from West Africa. *Environment*, **47**, 8–21.

Müller, B. (2008). *International Adaptation Finance: The Need for an Innovative and Strategic Approach*. Oxford, UK: Oxford Institute for Energy Studies.

Munaretto, S. (2009). Adaptation to climate change in coastal zones: An analysis of the intervention to safeguard Venice and its lagoon. *IOP Conference Series: Earth and Environmental Science*, **6**, 352012.

Ojima, D. and Corell, R. (2009). Adaptation to land use and global change thresholds. *IOP Conference Series: Earth and Environmental Science*, **6**, 342012.

Oxfam (2009). *Suffering the Science: Climate Change, People, and Poverty*. Oxfam Briefing Paper No. 103. Oxford: Oxfam International.

Padgham, J. (2009). *Agricultural Development Under a Changing Climate: Opportunities and Challenges for Adaptation.* Washington, D.C.: World Bank, Agriculture and Rural Development and Environment Departments.

Parry, M. (2009). Adapting future agricultural production to climate change: A survey of the challenge to increase food security and reduce risk of hunger. *IOP Conference Series: Earth and Environmental Science*, **6**, 372001.

Parry, M., Arnell, N., Hulme, M., Nicholls, R. and Livermore, M. (1998). Adapting to the inevitable. *Nature*, **395**, 741.

Parry, M., Canziani, O., Palutikof, J., van der Linden, P. and Hanson, C. (eds.) (2007). *Climate Change 2007. Impacts, Adaptation and Vulnerability. Contribution of Working Group II to the Fourth Assessment Report of the Intergovernmental Panel on Climate Change.* Cambridge, UK and New York, NY: Cambridge University Press.

Parry, M., Arnell, N., Berry, P. *et al.* (2009). *Assessing the Costs of Adaptation to Climate Change: A Review of the UNFCCC and Other Recent Estimates.* London: IIED and Grantham Institute for Climate Change.

Perry, I. and Barange, M. (2009). Policy options for adapting marine ecosystems to climate change. *IOP Conference Series: Earth and Environmental Science*, **6**, 352003.

Pfaff, A., Velez, M. A., Broad, K., Hercio, J. and Taddei, R. (2009). Water bargaining & climate adaptation: Asymmetric access to information affects equity. *IOP Conference Series: Earth and Environmental Science*, **6**, 292014.

Pittock, J. (2009). Lessons for climate change adaptation from better management of rivers. *Climate and Development*, **1**, 194–211.

Pörtner, H. O. and Farrell, A. P. (2008). Physiology and climate change. *Science*, **322**, 690–92.

Rockström, J., Steffen, W., Noone, K. *et al.* (2009). A safe operating space for humanity. *Nature*, **461**, 472–75.

Ruben, R. and Pender, J. (2004). Rural diversity and heterogeneity in less-favoured areas: the quest for policy targeting. *Food Policy*, **29**, 303–20.

Scoones, I., Devereux, S. and Haddad, L. (2005). New directions for African agriculture. *IDS Bulletin*, **36**, 1–12.

Smit, B., Pilifosova, O. V., Burton, I. *et al.* (2001). Adaptation to climate change in the context of sustainable development and equity. In *Climate Change 2001. Impacts, Adaptation, and Vulnerability*, eds. J. J. McCarthy, O. F. Canziani, N. A. Leary, D. J. Dokken and K. S. White. Cambridge, UK: Cambridge University Press.

Smith, J. B., Schneider, S. H., Oppenheimer, M. *et al.* (2009). Assessing dangerous climate change through an update of the Intergovernmental Panel on Climate Change (IPCC) 'reasons for concern'. *Proceedings of the National Academy of Sciences (USA)*, **106**, 4133–37.

Stampfli, A. and Zeiter, M. (2009). Prediction and mitigation of undesirable change in semi-natural grassland due to extreme droughts and hay management. *IOP Conference Series: Earth and Environmental Science*, **6**, 312027.

Steffen, W. (2009). Interdisciplinary research for managing ecosystem services. *Proceedings of the National Academy of Sciences (USA)*, **106**, 1301–02.

Steffen, W., Burbidge, A., Hughes, L. *et al.* (eds.) (2009). *Australia's Biodiversity and Climate Change.* Melbourne: CSIRO Publishing.

Swart, R., Biesbroek, R., Binnerup, S. *et al.* (2009). *Europe Adapts to Climate Change. Comparing National Adaptation Strategies.* Helsinki: Partnership for European Environmental Research.

Syphard, A. D. and Franklin, J. (2009). Expanding the evaluation of species distribution models to guide their assessment of climate impacts. *IOP Conference Series: Earth and Environmental Science*, **6**, 312028.

Thuiller, W., Lavorel, S., Araújo, M. B., Sykes, M. T. and Prentice, I. C. (2005). Climate change threats to plant diversity in Europe. *Proceedings of the National Academy of Sciences (USA)*, **102**, 8245–50.

UNDP (2007). Fighting climate change: Human solidarity in an unequal world. In *Human Development Report 2007/2008*. New York, NY: United Nations Development Programme.

Vellinga, P., Katsman, C., Sterl, A. and Beersma, J. (eds.) (2008a). *Exploring High-end Climate Change Scenarios for Flood Protection of the Netherlands. International Scientific Assessment*. The Netherlands: KNMI/Alterra.

Vellinga, P., Katsman, C., Sterl, A. *et al.* (2008b) *Exploring High-end Climate Change Scenarios for Flood Protection of the Netherlands: International Scientific Assessment*. Scientific report WR 2009–05. Wageningen: KNMI.

Walters, B. B., Rönnbäck, P., Kovacs, J. M. *et al.* (2008). Ethnobiology, socio-economics and management of mangrove forests: a review. *Aquatic Botany*, **89**, 220–36.

Weavers, P. M. and Rotmans, J. (2005). Integrated sustainability assessment: what it is, why do it and how. *Project Assembly on Methods and Tools for Integrated Sustainability Assessment*, 8–11 November, Barcelona, Spain, available as Matisse Project Working Paper 1, http://www.matisse-project.net/projectcomm/uploads/tx_article/Working_Pager_1_03.pdf.

World Bank (2009). *The Costs to Developing Countries of Adapting to Climate Change: New Methods and Estimates. The Global Report of the Economics of Adaptation to Climate Change Study*. Washington, D.C.: World Bank Environment Department.

Xiong, W., Conway, D., Jinhe, J. *et al.* (2009). Future cereal production in China: The interaction of climate change, water availability and socio-economic scenarios. *IOP Conference Series: Earth and Environmental Science*, **6**, 372004.

Part V

Meeting the challenge

If the societal transformation required to meet the climate change challenge is to be achieved, then a number of significant constraints must be overcome and critical opportunities seized…

15

Integrating adaptation, mitigation and sustainable development

'Forget about making poverty history. Climate change will make poverty permanent'[1]

Climate change and sustainable development are urgent problems that are closely related and, therefore, can be most effectively addressed in an integrated fashion. The challenge of sustainable development must confront the unmet development goals of many poorer nations, and this creates additional tensions in north–south negotiations about climate change (Box 13.1). Climate change is one indicator that contemporary society is currently on an unsustainable pathway, and so climate change can be viewed as one component of the broader challenge of achieving global sustainability. In fact, climate change impacts are perhaps most appropriately framed not as isolated phenomena but, rather, as potent threat multipliers that will exacerbate the multiple problems we already face – such as poverty, hunger, illness, water and energy scarcities, environmental degradation, and conflict. Conversely, unsustainable development patterns will not only increase greenhouse gas emissions that worsen climate change, but also increase vulnerability to the impacts of climate change.

The need to address climate risks, climate response and development has been recognised by major international institutions as well as a growing number of NGOs and national development agencies. The international development community has started to confront the challenge of integrating adaptation, mitigation and development through several major reports on climate and development, and a debate about 'mainstreaming' climate into development policies. For example, the 2010 World Development Report (World Bank, 2010) has a major focus on development and climate change, and the UNDP explicitly called for an integrated approach to development in the 2007–08 Human Development Report (Watkins, 2007). Several authors have argued that climate change is a major threat to the achievement of the Millennium Development Goals,[2] especially those related to

---

[1] Nazmul Chowdhury, Practical Action, Bangladesh (2009). http://www.christianaid.org.uk/images/annual_review.pdf.

[2] A set of 8 goals and 18 targets agreed upon by all UN Member countries and a number of international organizations; the first of these goals being to eradicate extreme hunger and poverty (Millennium Ecosystem Assessment, 2005).

Climate Change: Global Risks, Challenges and Decisions, Katherine Richardson, Will Steffen and Diana Liverman *et al.* Published by Cambridge University Press. © Katherine Richardson, Will Steffen and Diana Liverman 2011.

reducing hunger, reducing disease, protecting natural resources and increasing access to safe water (Davidson *et al.*, 2003; Halsnæs and Verhagen, 2007). The OECD and EU are also starting to mainstream climate concerns into their development assistance programs (Few *et al.*, 2006; Baudoin and Zaccai, 2009; Munasinghe, 2003).

The problems faced in dealing with climate change in the broader context of sustainable development stem, in part, from approaching the challenge from the perspective of 'disciplinary tunnel vision' (Høyer and Næss, 2008). That is, the problem is often narrowed down to an emphasis on a single, major approach (e.g. an appropriate economic instrument, a 'magic bullet' technology or replanting forests). Instead, the climate change problem should be understood in terms of biophysical and socioeconomic processes – as well as types of normative frameworks and contexts operating at different spatial scales and changing over time. More specifically, the underlying structural causes of climate change need to be understood as they drive a range of non-sustainable practices, and responses to climate change should be integrated into the broader sustainable development agenda (Munasinghe, 2009).

Adaptation and mitigation are the two main responses to climate change. Rather than considering adaptation and mitigation as elective supplements to traditional development activities, it is now widely recognised that bringing adaptation and mitigation measures into the core of sustainable development strategies (sometimes called mainstreaming) is most likely to give the most effective outcome (Munasinghe, 2001; Klein *et al.*, 2005; Eriksen *et al.*, 2007; IPCC, 2007a; Kok *et al.*, 2008).

Adaptation refers to adjustments in human and natural systems as a response to changes in climate and their impacts. Such adjustments reduce vulnerability, moderate damage and enhance benefits through actions such as building higher sea walls or developing salt- and drought-resistant crops.

Mitigation typically includes activities that reduce greenhouse gas emissions, such as introducing low-emissions energy technologies, reducing energy use, halting deforestation and absorbing atmospheric CO_2 by growing biomass. Although adaptation and mitigation can be highly integrative activities, they are often considered in isolation from one another.

The overarching challenge, then, is to make development – whether in industrialised or developing countries – more sustainable by undertaking adaptation and mitigation measures and integrating them with each other, and by mainstreaming them into sustainable development strategies (Munasinghe, 2001; IPCC, 2007a). Meeting such a challenge will require more tightly coordinated activities and measures, but these will need to be backed up by more highly integrated knowledge systems and a much more effective, interactive coupling between knowledge and policy.

15.1 Frameworks for integration

The well-known 'triple bottom line' or sustainable development triangle (Munasinghe, 1992; Foran *et al.*, 2005) – integrating social, environmental and economic concerns – provides

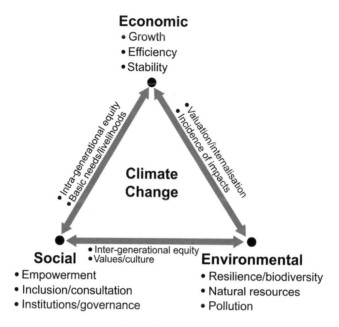

Figure 15.1 Climate change interacts with all three dimensions of the sustainable development triangle, which require balanced treatment. Source: Munasinghe (1992).

an obvious framework for addressing climate change from a broader sustainability perspective (Figure 15.1). There have been several related frameworks – such as the 'sustainomics' framework (Munasinghe, 1997; 2009), originally proposed at the 1992 Rio Earth Summit – which have been developed to help mainstream issues like climate change into a sustainable development framework.

Climate change is linked to all three dimensions of the triple bottom line and sustainable development triangle approaches. First, economic growth drives emissions that cause climate change, while climate change impacts will undermine future economic development prospects. Second, the impacts of climate change have severe social implications, worsening poverty and equity in many regions of the world. Third, climate change will exacerbate ongoing ecological damage, while environmental harm (like deforestation) will reinforce climate change. Approaches that holistically integrate natural and socio-economic systems often provide the best foundation for effective actions. Win-win options that satisfy all criteria can best integrate adaptation, mitigation and development. In other cases, judicious trade-offs would be required to resolve potential conflicts, for example, between mitigation and economic growth. Such tradeoffs are often political, and create winners and losers.

Making development more sustainable empowers people to take immediate action, which is practical because many unsustainable ongoing activities are easy to recognise and eliminate (Munasinghe, 2001). Techniques for identifying unsustainable development involve a wide variety of methods such as life cycle analysis, which reveals greenhouse gases, pollutants and resource use in various products and activities, and can suggest options for reducing them (Hendrickson *et al.*, 2005; Searchinger *et al.*, 2008).

One approach to integration is the Action Impact Matrix, which has been used to integrate climate change policies into sustainable development strategy (Box 15.1). This tool identifies and prioritises the two-way interaction: how (a) the main national development policies and goals affect (b) the key adaptation and mitigation options; and vice versa. It determines the practical and priority macro-strategies in economic, environmental and social spheres that mainstream climate change adaptation and mitigation policies. A variety of other tools can supplement the analysis, including cost-benefit analysis, multi-criteria analysis, environmental assessment, social assessment, poverty analysis, green national income accounts, environmental-macroeconomic modelling, sustainable resource pricing and environmental valuation.

Box 15.1
Action Impact Matrix methodology

MOHAN MUNASINGHE

Adaptation and mitigation measures ultimately must be implemented by nations, and will receive attention from decision-makers only if they are successfully integrated into national sustainable development strategies, which usually focus on more immediate issues like growth, poverty and food security. As shown in Figure 15.2, climate change is seen as a minor subset of the environment, which is itself only one element of sustainable development that is considered a special aspect of conventional development. The Action Impact Matrix (AIM) is a strategic analytical tool to link climate change directly with national policy goals (Munasinghe and Swart, 2005; 2007; Munasinghe, 2009).

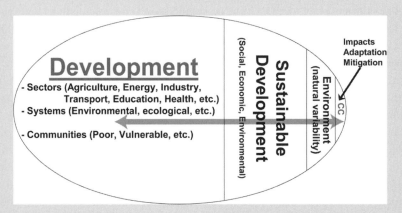

Figure 15.2 Mainstreaming climate change into sustainable development strategy. Source: modified from Munasinghe (2010).

The *adaptation AIM* identifies and prioritises interactions among three key elements, at the country-specific level:

(a)　national development policies and goals,
(b)　key vulnerability areas (VA) – economic sectors, ecological systems, etc.,
(c)　climate change adaptation.

Similarly, the *mitigation AIM* links the three key elements:

(a)　national development policies and goals,
(b)　key mitigation options,
(c)　climate change adaptation.

First, the two-way linkages between elements (a) and (b) are explored, in the context of natural climate variability. Then, we impose the additional impacts of climate change – element (c) – on the interactions between elements (a) and (b).

We illustrate the method, focusing here on adaptation; mitigation is treated in exactly the same way. Two adaptation matrices are derived: (i) DEV – Development effects on VA; and (ii) VED – VA effects on Development. The rows are development goals/policies and the columns are vulnerable areas. Thus, the matrix looks at linkages flowing both ways (i.e. impacts of columns on rows and vice versa). Figure 15.3 shows one sample VED–AIM, with the arrows indicating the impact direction.

Assessment 3 = High 2 = Medium 1 = Low + = good - = bad		Vulnerable areas (VA)									
		(1)	(2)	(3)	(4)	(5)	(6)	(7)	(8)	(9)	(10)
		Agriculture	Hydro power	Deforestation	Biodiversity (flora & fauna)	Wetlands, coastal ecosystems	Water resources	Impoverished communities	Human health	Infrastructure	Industries & tourism
(S0)	Status (No CC impacts)*	-1	0	-2	-1	-1	-2	-1	0	2	2
(S1)	Status (+CC impacts)**	-2	-1	-2	-2	-2	-3	-2	-1	-1	-1
Development goals/policies (+CC impacts)											
(A)	Growth	-2	-1	-1	-1	-1	-2	-2	-1	-1	-1
(B)	Poverty alleviation	-1	0	-1	-1	-1	-2	-2	-2	-1	-1
(C)	Food security	-3	0	-1	-1	-1	-3	-1	-1	0	0
(D)	Employment	-1	0	-1	0	-1	-2	-1	-2	-1	-2
(E)	Trade & globalisation	-2	-1	0	0	0	-1	-1	0	-2	-1
(F)	Budget deficit reduction	-1	-1	0	0	0	0	0	-2	0	-1
(G)	Privatisation	0	1	1	0	0	1	0	0	-1	-1
(H)	Other										

Effects that are highly damaging
Effects that cause moderate damage

Figure 15.3 Sri Lanka sample AIM–Vulnerable Areas Effects on Development (VED). Source: modified from Munasinghe (2009).

Box 15.1 (*cont.*)

The top two lines show the status of VA, both without and with climate change. The cells in the matrix indicate the severity of future impacts and whether they are beneficial or harmful. The analysis suggests that only a few shaded cells should be prioritised – the great majority of cells showing zero or low impacts may be ignored for simplicity's sake. Each key cell is hyperlinked to a lengthier description giving details justifying the value assigned, including citations of research articles.

The matrix identifies broad relationships, provides qualitative magnitudes of key interactions, helps to prioritise the most important links and identifies potential policy remedies. More detailed post-AIM studies are usually commissioned to investigate the priority issues and make specific policy recommendations.

For example, Figure 15.3 indicates that the food security–agriculture–water nexus will be critical. After this AIM exercise, a more comprehensive six-month study of the problem was carried out by experts, and detailed adaptation policies for agriculture were determined.

The AIM methodology relies on a fully participative stakeholder exercise to generate the AIM itself. Between 10 and 50 experts are drawn from government, academia, civil society and the private sector, who represent various disciplines and sectors relevant to both sustainable development and climate change. In the initial exercise, they usually interact intensively over a period of about two days to build a preliminary AIM. This participative process is as important as the product (i.e. the AIM), as important synergies and cooperative team-building activities emerge. The collaboration helps participants to better understand opposing viewpoints, resolves conflicts and ultimately facilitates implementation of agreed policy remedies. On subsequent occasions, the updating or fine-tuning of the initial AIM can be done within a few hours by the same group, as they are already conversant with the methodology.

For maximum effectiveness, the AIM meeting needs careful preparation in terms of trained instructors to conduct the exercise; document, screen and preselect a balanced group of participants; and gather relevant background data.

The AIM approach determines priority strategies, policies and projects in the economic, environmental and social spheres that facilitate implementation of measures to manage and restore vulnerable areas. After completing a national-level AIM exercise, it is possible to apply the process at a sub-national or community level to fine-tune the analysis.

Another approach is the World Bank's Strategic Framework for Development and Climate Change, which calls for an integration of climate considerations in development strategies, and promotes the mainstreaming of climate change policies in business plans and lending programs (World Bank, 2010). The Framework is based on six action areas, each addressing both adaptation and mitigation: (1) support climate actions in country-led development processes; (2) mobilise additional concessional and innovative finance; (3) facilitate the development of market-based financing mechanisms; (4) leverage private-sector resources; (5) support accelerated development and deployment of new technologies; and (6) step up policy research, knowledge and capacity-building. In Ethiopia the framework combines efforts to reduce vulnerability to extreme events with community-level sustainable land and watershed initiatives, and efforts to increase agricultural production in the face of climate change (World Bank, 2010).

Similarly, existing poverty reduction strategy programs provide an opportunity to link poverty reduction to mitigation and adaptation, but so far incorporate climate change in an uneven fashion – often focusing on responses to short-term climate variability rather than adaptation to longer-term climate change or mitigation (Kishore, 2007; Griebenow and Kishore, 2008).

Another integrating framework is proposed by the AdMit program (IIED, 2010), which would enable communities that are vulnerable to climate change to adapt to its impacts with the support of voluntary investments by developed country parties who would like to offset and take care of the impacts of their emissions.

Rarieya and Fortun (2010) have proposed the 'agrocomplexity' framework to integrate social, economic, ecological, political, cultural, institutional, biophysical, climatic and technological factors underlying sustainable development in Kenya. They use this framework to show how climate forecasts might be made useful through partnerships with local institutions and attention to local knowledge, and contribute to an overall increase in food security.

These are just a few examples of frameworks that can be used to include climate considerations into sustainable development.

15.2 Integrating adaptation and sustainable development

Adaptation policy and strategies are an important component of climate and development (Chapter 14). The poorest countries often have very limited capacity to adapt to climate change due to low income levels, inadequate skills and technological resources. Ethical perspectives suggest that it is unfair that these countries will suffer the worst impacts, as they have contributed least to the existing GHG concentrations (Adger *et al.*, 2005). Strategies for adapting to climate change include preventing losses (e.g. barriers against sea-level rise), reducing losses (e.g. changing the crop mix), spreading or sharing losses (e.g. government disaster relief), changing land use (e.g. relocating away from steep slopes) or restoring a site (e.g. historical monument prone to flood damage). Adaptive capacity, vulnerability and resilience are three key concepts in the links between climate and sustainable development.

Adaptive capacity is the ability of a system to adjust to climate change as well as other stresses (Brooks *et al.*, 2005; Smit and Wandel, 2006). Strengthening adaptive capacity is a key policy option for integrated adaptation and development, especially for vulnerable and disadvantaged groups. Adaptive capacity, itself, will depend on the availability and distribution of economic, natural, social and human resources; institutional structure and decision-making processes; information, public awareness and perceptions; the menu of technology and policy options; and the ability to spread risk through insurance and other collective responses. These variables are linked to location-specific patterns of socioeconomic and social development. Adaptation should address both biophysical factors and social vulnerability. Building the capacity of human systems, institutional strengthening and participatory approaches are all important tools that can be used to enhance adaptation strategies. One important link between development and adaptation is that of responding to natural disasters where disaster risk management has focused on reducing vulnerability.

Adaptation and sustainable development should include a focus on maintaining the quality and quantity of asset stocks such as soil and water savings, or human resources (Munasinghe and Swart, 2005). Maintaining and enhancing economic, social and natural capital is more effective than mere accumulations of economic outputs in decreasing vulnerability to climate change and reducing irreversible harm (Pelling and High, 2005; Reid and Vogel, 2006; Tschakert, 2007; Agrawal, 2008). In the context of climate change, vulnerability is the extent to which human and natural systems are susceptible to, or unable to cope with, the adverse effects of climate change. It is a function of the character, magnitude and rate of climate variation, as well as the sensitivity and adaptive capacity of the system concerned. The most vulnerable ecological and socio-economic systems are those with the greatest sensitivity, the greatest exposure and the least ability to adapt to climate change. Ecosystems already under stress are particularly vulnerable to climate change. Social and economic systems tend to be more vulnerable in developing countries with weaker institutions and economies (e.g. high population density; and low-lying coastal, flood-prone and arid areas).

A synthesis of 24 climate change assessments in the developing world provides important insights into the dynamics of vulnerability and adaptation, demonstrating the strong influence of non-climatic drivers on vulnerability to extreme events, the important role of development in either building or reducing adaptive capacity, and the importance of understanding barriers to adaptation (Leary *et al.*, 2007, 2008). Obstacles common across the projects included competing priorities in resource-scarce settings, entrenched poverty, lack of knowledge and information, lack of financial resources, weak institutions, degraded natural resources, inadequate infrastructure, insufficient financial resources, distorted incentives, and poor governance.

System resilience, vigour, organisation and ability to adapt will depend dynamically on the asset endowment, and magnitude and rate of change of a shock (Munasinghe and Swart, 2005). Resilience is the degree of change a system can undergo without changing state. Emphasis needs to be placed on building resilient systems to withstand shocks and surprises to manage current climate risks, by increasing the capacity of vulnerable communities to predict risks and to anticipate and plan – such as enhancing agricultural productivity, optimising land use, managing forests and conserving water (Vogel, 2006; Nelson *et al.*, 2007).

Learning from adaptation projects that are successful and work at the local level will only be effective if that learning is transposed at the national level, and allows policymakers to link adaptation with development priorities and challenges. Scaling-up successful projects from the micro level, and incorporating lessons learned into bigger programs at the national level, have considerable potential.

For example, local communities have developed strategies that can serve as a basis to facilitate, enable and support local adaptation. Adaptation experience and learning can be rooted in local knowledge, which serves as a foundation (Nyong *et al.*, 2007). Adaptation is more successful when local leadership is empowered to make decisions in ways that would support the integration of climate adaptation within broader sustainable development strategies. Collective learning of key stakeholders that supports demand-driven adaptation is necessary for successfully implementing adaptation projects.

Emphasis needs to be placed on capacity building and knowledge sharing/collective learning as well as encouraging local leadership. Capacity building and institutional strengthening is, in itself, a key adaptation response – particularly in regions where institutions are poorly resourced, and scientific capacity for understanding and communicating climate change risks is low (Eakin and Lemos, 2006). Sharing of knowledge between scientific, policy and practitioner communities is essential to sustain adaptation. This will require processes that enhance knowledge generation, interpretation, sharing and application. Such efforts should be targeted at a broad range of sectoral interests, such as agriculture, forestry, fisheries, water resource management, meteorology and climatology, energy, public health, disaster management, urban planning, and rural development. Given the great variation between regions and even within communities, identification of specific interventions depends heavily on *shared learning processes* supported by a combination of technical, economic, social and institutional analysis to enable the integration of local and global information that will identify effective response strategies.

Effective communications tools are crucial to popularise complex adaptation messages in a language accessible to policymakers. Societies respond to multiple stressors, not just those generated by climate risks; thus, effective adaptation will require integrating climate change vulnerability and risks within a multiple-drivers framework.

The IPCC highlights the difference between spontaneous (autonomous) and planned adaptation (IPCC, 2007b). Planning can minimise the costs of negative impacts and maximise the benefits of positive impacts. Adaptation efforts can be combined with mitigation, as controlling emissions is vital to minimise future impacts. Better adaptation requires advances in technology, management, law, finance, public education, training and research, and institutional changes, and must focus on issues of infrastructure.

Incorporating climate change concerns into sustainable development plans can help ensure that new investments in infrastructure reflect likely future climate conditions. For example, the design of coastal cities, reservoirs and new agricultural zones can take climate change into account. While uncertainty complicates the crafting of adaptation policies, many such policies will make development more sustainable (e.g. by improving natural resource management or social conditions).

There is a need to focus on Millennium Development Goals, and include mainstream policymakers in adaptation planning. When policymakers understand the linkages between adaptation and development, they will be most receptive to identifying policy windows that will enable effective integration (Box 15.1). Identifying barriers and costs to implementing climate change response strategies is important for successful mainstreaming.

Adaptation at scale in the developing world will require considerable funding, with estimates ranging up to $150 billion a year (Parry *et al.*, 2009) (see also Chapter 9 and Box 14.3). These funds should be in addition to current development funding. Effective mainstreaming will require mobilising funds at international and national levels to implement a series of National Adaptation Programmes of Action (NAPA) projects prepared under the auspices of the IPCC. There are many difficulties that have arisen with the current Adaptation Funds, United Nations Capital Development Fund and Special Climate Change

Fund including few countries paying in to the fund and inadequate criteria for distribution (Müller, 2008). There is a clear need for an increase in funds and improved access for developing countries vulnerable to climate change impacts.

15.3 Integrating mitigation and sustainable development

Integrating mitigation into development pathways can be achieved by making development more sustainable through pathways, such as those in Figure 15.4, which promote decarbonisation (C–E) and leapfrogging (B–D–E) (Munasinghe, 2010). These pathways are not fully reflected in the IPCC reference scenarios, which show a wide variety of alternative development pathways over the next century – each yielding a very different pattern of GHG emissions (IPCC, 2007a). Lower emission scenarios require less carbon-intensive energy resource development than in the past (World Bank, 2010). Emission reduction technologies are developing rapidly and improved methods of land use (especially forests) offer significant potential for carbon sequestration. Ultimately, mitigation options will depend on the sustainability of the development path; differences in the distribution of natural, technological and financial resources; and mitigation costs across nations and generations.

The effectiveness of future mitigation could be improved by strengthening *mitigative capacity* (defined as the ability to reduce GHG emissions or enhance sinks), which depends on making development more sustainable by building social, political and economic structures and conditions (Yohe, 2001). More research and analytical capacity is needed to build mitigative capacity, especially in developing countries. Social learning and innovation, and changes in institutional structure, will strengthen mitigative capacity.

Although future low-emission paths will vary by country, appropriate and positive socio-economic changes – combined with known mitigation technology and policy options – could help to achieve atmospheric CO_2 stabilisation levels around 400–450 p.p.m. by 2100, at modest costs (IPCC, 2007a). Policy options that yield no-regrets outcomes will help to

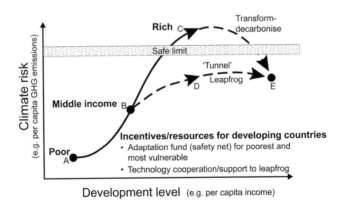

Figure 15.4 Integrating climate response and development aspirations. Source: modified from Munasinghe (2010).

reduce GHG emissions at no or negative social cost. However, there are many technical, social, behavioural, cultural, political, economic and institutional barriers to implementing mitigation options within development policy.

Policies to minimise risk by reducing GHG emissions will also come with a price tag and vary widely due to uncertainty. Although immediate action may seem more expensive, delays could lead to greater risks and therefore greater long-term costs (Stern, 2007; van Vliet *et al.*, 2009). An early effort towards controlling emissions would increase the long-term sustainability and flexibility of human responses to work towards stabilising atmospheric GHG concentrations.

Many cost-effective technologies and policies are available in all sectors and regions, which not only mitigate GHG emissions, but also address other development objectives by increasing resource efficiency and decreasing environmental pressure (World Bank, 2010). For example, programs to promote energy efficiency can reduce emissions and help reduce poor people's energy costs. Governments need to promote these solutions actively, by addressing institutional and other barriers (Box 15.2), and using economic incentives to influence investors and consumers. For example, deposit funds could encourage people to trade in their cars and appliances for more energy-efficient models; manufacturers could be rewarded for selling climate-friendly goods, or penalised if they do not. Prices could incorporate climate change concerns by changing taxes or subsidies. For example, a tax on oil, coal or gas would discourage fossil fuel use and help reduce CO_2 emissions. Tradable emission permits could also offer a cost-efficient, market-driven approach for controlling emissions (see Chapter 12 for more detail on economic instruments). New guidelines on measures like carbon footprints could be linked with corporate social responsibility to facilitate the successful integration of climate mitigation strategies into a more holistic framework for corporate sustainability.

Box 15.2
Building the solar energy market in Kenya through product quality

DANIEL M. KAMMEN AND ARNE JACOBSON

Kenya is home to one of the largest and most dynamic per capita solar photovoltaic markets among developing countries. Cumulative sales since the mid 1980s are estimated to be in excess of 300 000 systems, and annual sales growth has regularly topped 15% since 2000 (Acker and Kammen, 1996; Jacobson and Kammen, 2007). Household systems account for an estimated 75% of solar equipment sales in Kenya. This unsubsidised market arose to meet demand for reliable power in rural areas through relatively low-cost and dependable solar home systems. Solar is the largest source of new electrical connections in rural Kenya and, starting in about 2000, has also been spreading to neighbouring countries (Jacobson and Kammen, 2007).

Despite this commercial success, product quality threatened to derail the market in the 1990s, when reports began to emerge about problems with low-quality amorphous silicon

Box 15.2 (*cont.*)

Figure 15.5 'Solar panel scandal' advertisement in the *Daily Nation* newspaper, 2 May 2004.
Source: modified with permission from Sollatek.

(a-Si) modules, which were indistinguishable from high-quality modules (Duke *et al.*, 2000; Hankins, 2000; Duke *et al.*, 2002). It was not clear initially if this performance gap was related to inherent properties of the solar technology (Staebler and Wronski, 1977) or to issues in the manufacturing and/or field performance (Duke *et al.*, 2000; Hankins, 2000; Duke *et al.*, 2002; Fairman *et al.*, 2003). Advertisements in local newspapers sparked a heated debate about quality, consumer rights and the ethics of negative advertising.

In 1999, a set of private studies on the performance of the solar modules for sale in Kenya indicated clearly which brands were performing well, and which were not (Jacobson *et al.*, 2000). This information – made public in local newspapers and in regional trade journals by the academic research team from the University of California, Berkeley (Jacobson *et al.*, 2001) – had a major impact on the industry, inducing manufacturers to improve product quality (Jacobson and Kammen, 2007).

Several years after the 1999 study, a new line of low-performing a-Si modules began to enter the market in significant quantities (Figure 15.5). The approach to weeding out these panels was a close repeat of the earlier episode (Jacobsen and Kammen, 2007). Re-emergence of quality problems in the Kenyan market confirmed that the issue could not be solved decisively by one-time testing efforts or by focusing on the improvement of individual low-performing brands. Rather, institutional solutions that persistently require high performance for all brands are needed to ensure quality.

As a result of these events, the Kenya Bureau of Standards (KBS, 2003a, 2003b) collaborated with the Kenya Renewable Energy Association to draft performance standards for a range of solar products, including a-Si modules. The government drafted and adopted new standards, drawing heavily from codes established by the International Electrotechnical Commission.[3]

However, because the Kenya Bureau of Standards lacked access to the necessary equipment and technical capacity to carry out all specified tests, continued involvement of local solar groups and international academic teams was critical to the dissemination, and at times enforcement, of the Kenyan national solar standards. Thus, while the move to adopt national performance standards represented a positive step towards an institutionalised approach to quality assurance, the adoption of unenforced standards requires continued vigilance and partnerships among research and testing groups, the solar industry, and the government.

This Kenyan solar story highlights a number of key issues. First, an 'enabling environment' for a clean energy technology can evolve during or even after the market begins to expand. Second, there is often a need for continued assessment and analysis to build what initially can be fragile renewable energy markets. In addition, science and engineering inputs can be critical at many stages of the evolution of a renewable energy system and market. At present the Kenyan solar market has, with some ups and downs, continued to expand; as of 2007 over 35 000 new systems were sold annually in Kenya (Jacobson and Kammen, 2007).

---

[3] The relevant IEC standard for crystalline silicon PV modules is IEC 61215, while the standard for amorphous modules is IEC 61646. The corresponding Kenyan standards are KS 1674 and KS 1675. For more information about the International Electrotechnical Commission see http://www.iecee.org/pv/html/pvcntris.htm.

Energy policy is a key to enhancing the cost effectiveness of decreasing emissions. Incentives for investing in cost-effective and energy-efficient technologies are essential – for example, to promote improved building design, new chemicals for refrigeration and insulation, and more efficient refrigerators and cooling/heating systems. Technology innovation, energy efficiency and more renewable energy can stabilise GHG concentrations, while also providing other co-benefits. Governments need to remove barriers that slow the spread of low-emission technologies (World Bank, 2010).

The transport sector is the most rapidly growing source of GHG emissions in many developing countries. Heavy reliance on fossil fuels makes controlling GHG emissions particularly difficult. Climate-friendly transport policies, especially public transport, can make development more sustainable while minimising the local costs of air pollution, traffic congestion and road accidents (IPCC, 2007a). New technologies can increase the efficiency of automobiles and reduce the emissions per kilometre travelled. Switching to less fuel-intensive cars will reduce carbon emissions as well as local and regional air pollution – especially particulate matter and ozone precursors in urban areas. Some policies to reduce emissions from transport include (i) use of renewable energy technologies, (ii) better maintenance and operating practices, (iii) policies to reduce traffic congestion, (iv) encouraging urban planners to encourage low-emissions transport (e.g. trams and trains, bicycles, walking), and (v) imposing user fees.

Land use is another source of greenhouse gas emissions where policies can foster sustainability and reduce emissions. Deforestation and agriculture account for increases in local and regional air pollution as well as GHG emissions. Forests need to be protected and better managed to simultaneously support mitigation, adaptation and sustainable development. For example, integrating REDD+ (Reducing Emissions from Deforestation and Forest Degradation) activities within forest management more generally may yield better results than treating it as a separate, climate-driven mitigation activity (Box 15.3). Deforestation may be controlled by decreasing pressures of agriculture on forestry, slowing down population growth, involving local people in sustainable forest management, harvesting commercial timber sustainably, and reducing migration into forest areas. Sustainable forest management can generate renewable biomass as a substitute for fossil fuels. Agriculture can be a source and a sink for greenhouse gas emissions (McCarl and Schneider, 2001). Improving management to increase agricultural productivity will enable soils to absorb more carbon, while leading to higher organic matter content and enhanced productivity. Methane emissions from livestock could be reduced with new feed mixtures that improve the efficiency of digestion, and also increase the overall efficiency of livestock management. Methane emissions from wet rice cultivation can be reduced significantly through changes in irrigation practices and fertiliser use, while improving or safeguarding current productivity, and reducing environmental pressures. Nitrous oxides from agriculture can be minimised with new fertilisers and practices, some of which not only reduce nitrous oxide emissions, but also reduce nitrogen losses (in the form of polluting nitrates, nitrogen oxides and ammonia), and enhance fertiliser efficiency (IPCC, 2007a).

Box 15.3
Post-Copenhagen strategies for the implementation of REDD+

PETER KANOWSKI, CONSTANCE MCDERMOTT
AND BENJAMIN CASHORE

The 2009 Copenhagen Climate Change Summit illustrated vividly the profound difficulties of reaching a post-Kyoto comprehensive global climate agreement. The conjunction of the scale and urgency of emissions reductions required to limit global warming (Angelsen, 2009; Prince's Rainforest Project, 2009), and the uncertainty of a timetable for progress in international climate negotiations, together suggest that those committed to addressing climate change should look – at least in the short term – to strategies that already enjoy, or are nearing, global consensus.

Such is the case for initiatives to limit greenhouse gas emissions associated with forest loss and degradation, which contribute 15% of global anthropogenic CO_2 emissions (van der Werf *et al.*, 2009). 'Reducing emissions from deforestation and forest degradation in developing countries' (REDD) (Angelsen, 2008), and extending this goal to include enhancing forest-based carbon stocks through activities such as forest restoration and sustainable forest management (REDD+) (Angelsen, 2009; Campbell, 2009), were amongst the few points of general agreement at Copenhagen (UNFCCC, 2009a, 2009b).

Pursuing REDD+ in ways that promote, rather than detract from, other important environmental and social objectives is now widely recognised as an international priority for a number of reasons. The first is that rates of deforestation and forest degradation have remained persistently high, despite international efforts since the 1992 Earth Summit to curtail them. The second is that it is now widely recognised, even by countries that do not agree about broader climate change mitigation strategies, that reducing forest-based emissions can 'buy time' for action in other sectors (Stern, 2006; Tavoni *et al.*, 2007; Prince's Rainforest Project, 2009).

One of the principal constraints to reaching agreement on measures to address forest loss and degradation has been the reluctance of many countries to forego sovereignty over their forests (Humphreys, 2006). Consequently, the focus of international efforts for forest conservation and sustainable forest management has progressively shifted to that of enabling 'good forest governance' at national and sub-national levels (Contreras-Hermosilla *et al.*, 2008; World Bank, 2008; Cashore, 2009).

A number of strands of recent research suggest how REDD+ implementation might be advanced in the absence of a new global climate change agreement. The first draws from scholarship in environmental and forest governance, and emphasises aspects of the 'new collaborative environmental governance' (Gunningham, 2009a): the 'localisation of regulation' (Gunningham, 2009a) through the better integration of governance functions at different levels, from international to local; greater collaboration between public, private and civil society actors; and use of a variety of policy instruments, from traditional regulation to market-based mechanisms (Agrawal *et al.*, 2008; Cashore, 2009; Gunningham, 2009b). Some of these approaches have already been applied through preliminary REDD+ projects (Angelsen, 2009).

Box 15.3 (*cont.*)

A second strand of learning emerges from comparative international environmental policy analysis, such as that we conducted of forest practice policies globally[4] (McDermott *et al.*, 2009, 2010), which are relevant to the implementation of REDD+. The most important result of this analysis was that forest-rich developing countries such as Brazil and Indonesia, which account for the majority of forest-based emissions and are therefore the focus of REDD+, already have environmentally oriented forest management requirements that are comparable to those in industrialised countries. In contrast, the least-demanding forest practice requirements globally are those for private forests in some industrialised countries. There was also little overall difference between developing and industrialised countries in some other policy settings directly relevant to REDD+, such as the proportion of formally protected areas.

These results suggest that many forest-rich developing countries have already made forest management-related policy commitments as demanding as could be expected from any global agreement, and that there is little to be gained from pursuing further global agreement about what those commitments should be. Rather, international efforts intended to promote REDD+ initiatives should focus on enabling and streamlining implementation of existing national and sub-national forest policy commitments, as well as on addressing conflicts and seeking synergies between forest-related policies and those in other sectors such as agriculture (Campbell, 2009).

This conclusion is consistent with those emerging for environmental and forest governance more generally (Agrawal *et al.*, 2008; Gunningham, 2009a, 2009b), and with related initiatives already underway in some arenas – for example, in supporting community forestry (Agrawal and Angelsen, 2009) and in addressing illegal logging (Tacconi, 2007). It suggests that the goals of REDD+, and the implementation of the REDD+ mechanisms envisaged by the Copenhagen Accord (UNFCCC, 2009b), might best be realised by strengthening national and sub-national capacity to implement existing forest policy settings in ways that are consistent with the established principles of good forest and environmental governance.

Thus, it is possible to decrease emissions while generating many other environmental and socio-economic benefits. 'No regrets' emissions reduction strategies are most effective in minimising the costs of climate change policies. Their benefits often exceed costs, even excluding the benefits of avoided climate change – for example, removing market imperfections like fossil fuel subsidies, or generating double dividends by using tax revenues to reduce other distortionary taxes. Low cost adaptations, such as improving water use efficiency, can provide easy benefits. Public participation (i.e. individuals, communities, businesses) is important for effective policies. Education and training are also vital (e.g. conserving energy, switching to renewables). Equity aspects of policy, such as cost efficiency and fairness, also need to be considered because the costs and benefits often fall on different groups (Chapter 9).

---

[4] The forest policy settings assessed were allowable harvest levels, biodiversity protection, clear-cutting and harvesting rules, reforestation requirements, riparian buffer specifications, and roading rules. The 20 countries sampled represent 70% of global forest area and 61% of global forest products trade, and include those with the highest deforestation rates.

15.4 Implementation of integration approaches

While much knowledge exists to address climate change, effective action to address this threat is piecemeal and sparse. It is most important to analyse and choose win-win options that yield joint climate adaptation and mitigation benefits while making development more sustainable.

Agriculture and land use offer good potential for integrating adaptation and mitigation. Adaptation is crucial to feed a continuously growing world population, because climate change will reduce the soil quality and water availability in many regions, and also increase variability of temperature and rainfall. Meanwhile, as world food demand grows, the already large contribution of agriculture to global greenhouse gas emissions will increase, unless more sustainable and climate-friendly farming methods are adopted.

15.4.1 Trade-offs

The interdependence of sustainable development, climate change and human well-being suggests that it is more effective to address multiple challenges in an integrated fashion, rather than individually. While win-win outcomes that yield gains for both climate and development are the best, in other cases, difficult trade-offs need to be made between conflicting objectives.

There is a clear need for integrative analytical frameworks that can engage decision-makers in analysing trade-offs. Research results can be translated into effective policies. Learning should take place at all levels as well as take advantage of existing processes like the NAPAs, to further deepen the understanding of policymakers on the adaptation–development nexus. New approaches are being developed that will help track and predict climate impacts, including global analyses of hot spots for water scarcity and vulnerability of crop systems. Such analyses will help decision-makers identify the most vulnerable people and places, and could potentially assist in setting priorities for investment in adaptation or vulnerability reduction strategies.

An example from bioenergy illustrates the opportunities and trade-offs involved (Box 15.4). Biomass from agricultural residues or dedicated crops can be an important biomass feedstock, but its contribution to mitigation depends on demand for bioenergy from transport and energy supply, on water availability, and on requirements of land for food and fibre production (FAO, 2008). Widespread use of agricultural land for biomass production may compete with other land uses and have broader environmental impacts. Furthermore, substituting biofuels for gasoline may reduce GHG emissions – but the analyses often failed to count additional emissions that occur as farmers worldwide respond to higher prices, and convert forest and grassland to new cropland, and replace the cropland diverted to biofuels (Scharlemann and Laurance, 2008; see also discussion in Chapter 11).

Global agricultural model simulations have shown that corn-based ethanol in the USA, instead of producing 20% emissions savings would nearly double GHG emissions over 30 years (Searchinger *et al.*, 2008). Biofuel from switchgrass, if grown on US corn lands, increase emissions by 50% (Searchinger *et al.*, 2008). This highlights the need to integrate agriculture and biofuel production with the increased use of agricultural wastes for

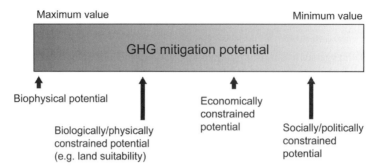

Figure 15.6 Impacts of different constraints on reducing GHG mitigation potential from its theoretical biophysical maximum to the lower achievable potential. Source: Smith *et al.* (2007).

bioenergy, or substantially increase productivity of bioenergy crops to avoid excessive land-use changes (Box 15.4).

15.4.2 Policy instruments and barriers to implementation

A crucial issue with regard to integration is the applicability and effectiveness of different policy tools available – and cultural, economic, technical, institutional and physical constraints to implementing them.

Increases in mitigative capacity could effectively limit GHG emissions, while maximising the developmental co-benefits of mitigative actions. Analyses of mitigation options have clearly shown that there is considerable potential for reducing emissions at very low costs or even a net profit. Such mitigation options are not being adopted because of implementation barriers and constraints.

Barriers are likely to be regionally and often locally specific, depending on biophysical, social and economic conditions (Figure 15.6). Three examples involving land use, described as follows, give an insight into barriers and constraints.

15.5 Sectoral approaches to integrating adaptation, mitigation and development

15.5.1 Agriculture

As noted earlier, agriculture provides many opportunities for combining adaptation and mitigation with sustainable development (Figure 15.7). Farmers can sequester carbon in the soil while increasing yields and food security, or can benefit from protecting natural ecosystems that enhance biodiversity and provide water essential for crops.

Synergies between soil carbon sequestration and sustainable agriculture can significantly reduce climate change vulnerability and reduce emissions, while increasing crop production and providing employment. Agricultural practices could make significant low-cost contributions to increasing soil carbon sinks, reducing emissions and increasing biomass

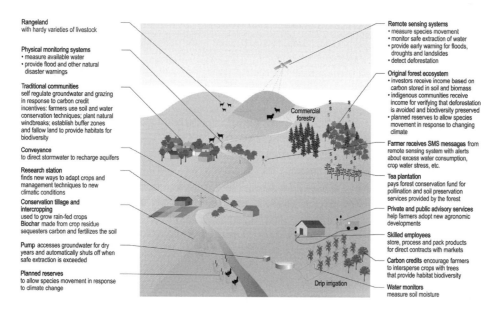

Figure 15.7 An ideal climate-smart agricultural landscape where farmers are maximising yields and natural systems are integrated into the farming system. It includes the use of techniques for both adaptation and mitigation, and the use of cheap but advanced technologies and local knowledge. Source: World Bank (2010).

for energy use. A large proportion of the economic mitigation potential (at US$100 per tonne CO_2-e and excluding bioenergy) arises from soil carbon sequestration, which has strong synergies with sustainable agriculture and vulnerability reduction (Niles *et al.*, 2002). Permanent reductions in methane and nitrous oxide emissions are also possible, and enhancing soil carbon storage may help to adapt both to more intense rainfall events (less erosion) and to prolonged droughts (more soil moisture retention).

Soil carbon sequestration illustrates some of the barriers to effective integration of adaptation, mitigation and development – especially the challenges of demonstrating that soil mitigation potential is a permanent and additional way of reducing emissions, and because of potential trade-offs if soil management for carbon reduces yields (Smith *et al.*, 2007, 2008). Integrated approaches may also have high up-front costs, and may need to be supported by information, insurance, incentives, education and secure property rights.

Box 15.4

Bioenergy: opportunities and trade-offs

JØRGEN E. OLESEN AND PETE SMITH

Bioenergy is derived from biomass from agricultural energy crops; forestry and wood-based industries; farm, municipal and industrial organic waste; and aquatic (micro and macro algae)

Box 15.4 (*cont.*)

and marine sources (e.g. seaweed). Biomass can be used in the generation of electricity, heat and biofuels. Using bioenergy can be beneficial to achieve environmental objectives, reduce CO_2 emissions compared to fossil fuels and support rural development efforts – but there are also some risks and negative impacts linked to extensive use.

Current production processes for liquid biofuels follow the first-generation conversion technologies relying on sugar, starch or vegetable oil components of crops. These are extensively used in Brazil (sugar cane for bioethanol), the US (cereals – mainly maize – for bioethanol) and the EU (oilseeds, mainly rapeseed for biodiesel). Feedstocks utilised for these first-generation technologies are primarily food and feed plants, and there is thus a direct competition with land required for food production. These technologies generate both fuel and various by-products that are used for livestock feed or in industry.

A substantial expansion of biofuel production will require an expansion of the range of feedstocks and introduction of advanced (so-called second-generation) conversion technologies. Second-generation biofuels will be based on lignocellulosic biomass comprising cellulose, hemicelluloses and lignin. Such feedstock sources exist in agriculture and forestry residues (e.g. straw, maize stover, wood) and dedicated energy crops (e.g. miscanthus, switchgrass, willow). Such perennial energy crops can potentially be grown on land that is not suitable for intensive agriculture, thus reducing competition with food production. Environmental impacts as well as input intensity are also lower in these crops compared with first-generation feedstocks.

Life cycle assessments have shown a large variation among different types of biofuels in their substitution of CO_2 emissions compared with use of petrol or diesel. Using a so-called well-to-wheels assessment, the emissions associated with conventional gasoline or diesel are about 164 and 152 g CO_2 km^{-1}, respectively (JRC, 2007). For first-generation biofuels, emissions have been calculated as 58–110 g CO_2 km^{-1} for sugar beet in Europe and 22 g CO_2 km^{-1} for sugar cane in Brazil. Second-generation biofuel technologies have emissions of 33–43 g CO_2 km^{-1} from bioethanol straw and wood, and emissions from synthetic diesel and the advanced diesel fuel dimethyl ether (DME) from farmed wood of 14–15 g CO_2 km^{-1}. These estimates do not include emissions associated with land-use change.

Current biofuel targets are projected to result in a share of biofuels in transport fuel of 12% in developed countries and of 8% in developing countries by 2030 (Fischer *et al.*, 2009). Although current indications are that biofuel targets could be reached by 2020, caution is needed because first-generation biofuels, through competition with food crops, are not sustainable in the long term (Table 15.1). Arable land resources are limited and expansion into forest, grassland and woodland areas will result in significant carbon emissions, negating the primary justification for carbon savings with biofuels (Searchinger *et al.*, 2008).

Carbon losses due to land-use change occur at the time of land conversion, but greenhouse gas savings from biofuels substituting fossil oil accumulate only gradually over time (Fargione *et al.*, 2008). As a consequence, net greenhouse gas savings resulting from rapid expansion of first-generation biofuels will only be reached after several decades.

Rapid development of first-generation biofuels can have drastic negative spillover effects through effects on food security, biodiversity, social development, and soil and water resources.

Table 15.1. *Sources of greenhouse gas emissions (grams CO_2-e per MJ of energy in fuel) for gasoline compared with bioethanol based on maize grain (first-generation) or switchgrass (second-generation)*

Source	Gasoline	Maize	Switchgrass
Making feedstock	4	24	10
Refining fuel	15	40	9
Vehicle operation	72	71	71
Feedstock carbon uptake	0	−62	−62
Land-use change	0	104	111
Total	**91**	**177**	**139**
% change relative to gasoline		+94	+52

Source: modified from Searchinger *et al.* (2008). Reproduced with permission of AAAS.

Increased demand for biofuels will tend to raise world market prices for biofuels which, in turn, will increase food prices as less land is available for food production. Higher food prices increase poverty and cause more people to become affected by hunger.

A different outlook on bioenergy is provided when the use of biomass for heat and power, biomaterials and second-generation biofuels is considered, taking into account different potential biomass resources as residues and organic wastes, and perennial crops cultivated on arable, pasture and marginal and degraded lands. The ecological, social and economic impacts of further deployment of bioenergy are also fully conditional, and there may be many benefits from developing sustainable biomass production.

Based on the current state-of-the-art analyses that take key sustainability criteria into account, the upper bound of the biomass resource potential halfway through this century can amount to over 400 EJ per year (Dornburg *et al.*, 2008). The global primary energy demand is projected at about 600–1040 EJ per year in 2050. Thus biomass has the potential to meet a substantial share of the world's energy demand. The larger part of the potential biomass resource base is interlinked with improvements in agricultural management, investment in infrastructure, good governance of land use and introduction of strong sustainability frameworks.

However, if the conditions for sustainable expansion of biomass production are not met, the biomass resource base may be largely constrained to a share of the biomass residues and organic wastes, some cultivation of bioenergy crops on marginal and degraded lands, and some regions where biomass is evidently a cheaper energy supply option compared to the main reference options. Biomass supplies may then remain limited to an estimated 100 EJ in 2050.

Optimal use and performance of biomass production is regionally specific. Policies therefore need to take regionally specific conditions into account, and need to incorporate the agricultural and livestock sector as part of good governance of land-use and rural development, interlinked with developing bioenergy.

15.5.2 Forests

Deforestation is a significant source of carbon emissions and thus protecting forests is a key to any global mitigation strategy. However, in the developing world, there is often pressure to cut forests to make way for agriculture and other economic opportunities. Forests are also important to adaptation because they provide key supporting, provisioning, regulating and cultural ecosystem services to support human well-being and buffer climate variability. Climate change threatens the supply of such services – especially the carbon-regulating functions where the loss of forests could accelerate climate change. In subtropical forests, climate change may decrease productivity, but short-rotation plantations provide opportunities. However, such opportunities will increase the likelihood of conflict between short-rotation forests for carbon storage and biodiversity conservation, because such plantation forests potentially threaten biodiversity. Long-term adaptation and mitigation actions can address such conflicts, and large emissions reductions from avoided deforestation will strengthen both adaptive and mitigative capacities of forests (Box 15.5).

Box 15.5

Integrated solutions to the future of the world's tropical rainforests: linking climate adaptation and mitigation with sustainable development

CHRISTIAN PILEGAARD HANSEN, JENS FRIIS LUND
AND NIELS ELERS KOCH

Deforestation and forest degradation account for some 18% of global carbon dioxide emissions (IPCC, 2007c). Reducing these emissions is seen as a fast and cost-effective strategy that, in addition to emission reductions, holds potential for co-benefits such as poverty alleviation, biodiversity conservation and rural enfranchisement (Stern, 2007; Angelsen, 2008; Palmer and Engel, 2009). To be effective, policies and measures for reducing emissions from deforestation and forest degradation must target the underlying causes of these processes. Research has documented that deforestation and forest degradation is the result of multiple underlying causes (Figure 15.8) that combine in context-specific ways (Geist and Lambin, 2002; Kanninen *et al.*, 2007). Hence, solutions will be context-specific.

   Much deforestation is driven by national macro-economic frameworks that subsidise agricultural inputs, provide tax credits or subsidies for land clearing or logging, or transport subsidies – all in pursuit of national growth. Hence, stopping deforestation will require the elimination (or reduction) of such policies and the creation of positive incentives for forest protection – including the development of markets for hitherto non-valued forest products and services, such as biodiversity and carbon dioxide retention. Much attention has lately centred on transfer payment regimes that compensate agents for foregone revenues from deforestation and forest degradation (Wunder, 2007). Yet such schemes essentially require that actors have legal and enforceable rights to the forest; factors that are wanting in many areas. Additional challenges are how to clearly identify the would-be actors of deforestation and forest degradation; and the design of appropriate and cost-effective

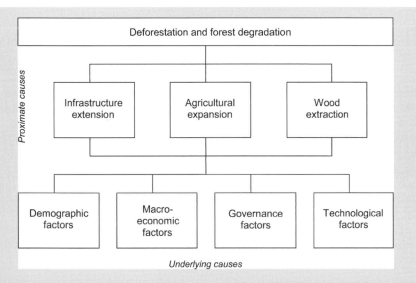

Figure 15.8 Four main clusters of underlying causes result in deforestation and forest degradation through human actions in the form of infrastructure extension, agricultural expansion and wood extraction. Integrated solutions will have to address underlying causes vis-à-vis development needs. Source: modified from Geist and Lambin (2002).

monitoring, reporting and validation practices. Another alternative is certification which, over the past two decades, has spread to more than 300 million hectares. Furthermore, while North consumers' awareness is likely to grow, the majority of timber from tropical forests is traded in domestic and international markets with restricted consumer awareness and willingness to pay.

Governance factors may also lead to deforestation and forest degradation. Central governments administer 74% of the world's forests (Sunderlin *et al.*, 2008), and although currently giving way to communities and private sector, clearly their role cannot be overlooked. In some cases, governments face insurmountable constraints in enforcing rules governing forests. Forests are usually found in the periphery, where the presence of the state is weak and the costs of enforcement are high. Furthermore, in many cases, forest rules are neither designed nor implemented with an aim to secure sustainable forest management, but rather to serve political and personal interests of elite groups – resulting in excessive deforestation and forest degradation (Ascher, 1999; Colchester *et al.*, 2006). Clearly, solutions to this demise will not come fast and easily. Increased public participation in forest management (e.g. community-based approaches) may present one potential solution, although the evidence from practical experiences is scarce and mixed (Lund *et al.*, 2009; Roe *et al.*, 2009).

In the past, demographic factors – such as population density, growth and distribution – have been given much emphasis. Yet, research has shown that their effect is relatively weak (Angelsen and Kaimowitz, 1999). In some contexts, migration to the agricultural frontier and the growing need for agricultural land to meet increasing demands for consumer goods arising from population growth may lead to more deforestation, while not in others (Leach and

Box 15.5 (*cont.*)

Fairhead, 2000). Population control policies, therefore, do not appear to be a general solution to
the future of the world's tropical forests. Finally, the effect of technological factors, especially
changes in agricultural technologies leading to changes in the demand for agricultural land, is
ambiguous and dependent on the particular practices and context (Angelsen and Kaimowitz,
2001). Generally, solutions in favour of the forest are to be sought among those technologies
that result in agricultural intensification.

In sum, solutions to safeguard the world's remaining tropical rainforests will require an
integrated approach, as the drivers of deforestation and forest degradation are multiple and
many factors are outside the forest sector. Only in this way can the potential for mitigation
of climate change be realised, as well as the potential for adaptation. Yet, implementing such
integrated solutions is a tremendous challenge that will require the willingness of governments
in developing countries to implement reforms and governments in developed countries to
support such processes, including by financial means.

Forest adaptation and mitigation interact with important elements of sustainable development, including the livelihoods and socio-economic well-being of forest-dependent
people. Forests are also a safety-net for farmers, providing alternative food to some and
income. Thus, climate change will impact significantly on the already vulnerable, forest-
dependent poor.

The best strategy is to take advantage of mitigation opportunities while reducing vulnerabilities. An example from Africa is the planting of drought-resistant fruit trees, which
reduces erosion, provides an additional fuel source, and enhances food security and nutrition (Lykke *et al.*, 2009). In India and Bangladesh, non-timber forest products are more
vulnerable to climate changes than timber products (Basu, 2009). Here, the adaptive strategies include vulnerability assessment, capacity building and in-depth analysis of social-
ecological systems. Thus, adaptation must go beyond technical solutions and consider
social-economic and ecological systems.

Sustainable forest management practices may work well in reducing vulnerability, but
the current failure to implement sustainable forest management limits the benefits. There
is no universally applicable individual measure for adapting forests to climate change, and
forest managers will need sufficient flexibility to choose locally appropriate adaptation
measures. Several European studies (see, for example, Lindner and Kolström, 2009) report
that the most important adaptation tools include:

- active management of suitable tree species and provenances,
- planning and managing forests in the context of overall land use,
- clarifying the use of exotic species in relation to the goals of nature conservation.

Genetic management to support continued adaptation and selection of suitable genotypes are also important tools, for both local adaptation and transplanting tree species to
new regions.

15.5.3 Biodiversity conservation

Addressing climate change impacts on biodiversity is a challenge, mainly because of perverse incentives and outcomes of mitigation activities that influence carbon sequestration in natural ecosystems such as biofuels. Mitigation and adaptation to climate change together may either provide opportunities for biodiversity conservation, or greatly increase risks to biodiversity (Kühn *et al.*, 2008; Paterson *et al.*, 2008), depending on the approach and its application (Box 15.6). Therefore, approaches that simultaneously store more carbon and protect biodiversity need to be harnessed. For example, beach nourishment aimed at preventing climate change-related beach erosion (thereby reducing coastal flood risk) can be deleterious to coastal biodiversity unless carefully planned, whereas many wetland restoration measures are positive overall (e.g. contributing to flood protection, carbon storage and ecosystem protection).

Box 15.6

Integrating biodiversity conservation into other mitigation and adaptation activities: how to build synergistic outcomes that benefit both

PAM BERRY

There has been increasing recognition that co-benefits can be achieved between climate change mitigation (e.g. through REDD (O'Connor, 2008) or adaptation and biodiversity (Secretariat of the Convention on Biological Diversity (SCBD), 2009; World Bank, 2009), and also between mitigation and adaptation. It is only recently that the possibility of bringing all three together to achieve even greater synergies has been promoted (Forner, 2005; Paterson *et al.*, 2008). This is due partly to the difficulties consequent upon the differences between mitigation and adaptation in terms of their spatial and temporal scales of operation, actors involved, and distribution of costs and benefits (Tol, 2005).

  To build synergistic outcomes, there is a need for formal mechanisms for the recognition of interactions between mitigation and adaptation objectives and biodiversity, for example through strategic environmental assessments, environmental impact assessments or risk assessments (SCBD, 2009). These should involve the identification of complementary and antagonistic actions, and an explicit – ideally quantified – assessment of the extent to which each objective might be achieved. The former could take the form of a table (see SCBD, 2009, Annexes III and IV), which identifies the climate change impacts in different sectors that require a response, and lists mitigation and adaptation options (both technical and more ecosystem-based), and their effects on biodiversity – focusing on how positive aspects can be enhanced and negative ones minimised. Maximisation of positive outcomes that result in synergies can depend on where and how measures are implemented and managed (Berry, 2009). For example, production of bioenergy crops on degraded land has the potential to increase biodiversity, while the opposite would occur if they replace semi-natural habitats (Firbank, 2008). The disadvantage of such an approach is that it is difficult to identify the cross-sectoral interactions of proposed measures, yet these can be important to the outcome. Thus, it is

Box 15.6 (*cont.*)

necessary once synergistic outcomes have been identified to check them against mitigation opportunities and adaptation options in other sectors in order to ensure that antagonisms have not been created elsewhere, which might negate the synergies.

Where there is conflict between objectives, then trade-offs will need to occur. Valuation is one way of dealing with these, but direct and quantitative comparisons of adaptation and mitigation strategies are rare, and cost–benefit analysis to examine trade-offs is not thought to be practicable given the differences in the two strategies as outlined previously (Tol, 2007). An alternative is to explore the trade-offs between the costs of emission reduction and the costs of adaptation plus the residual damages, or explore their interaction in a risk avoidance framework (Kane and Shogren, 2000). Even more problematic is the incorporation of biodiversity into cost-benefit analyses. First, there is the identification of the contribution of biodiversity for valuation, as it can be part of the adaptation and mitigation strategies (e.g. climate regulation) but also can provide other ecosystem services such as food production or soil erosion regulation. Second, valuation of these services, not to mention biodiversity's intrinsic current and future value, is fraught with difficulties (Skourtos *et al.*, 2010). Another approach is multi-criteria analysis, which would require the identification of alternatives for achieving the various objectives and the establishment of criteria against which these can be measured. Both, ultimately, need prioritisation of the objectives but, as pointed out earlier, this will raise a question of 'by whom', as usually different actors are involved in mitigation and adaptation activities.

Increasing recognition of the potential policy synergies between climate change mitigation and adaptation, and also with biodiversity and sustainable development (Robinson *et al.*, 2006; Barker *et al.*, 2007), is vital if synergies are to be realised. Nevertheless there are a number of issues regarding not just the capacity for their implementation, but also the processes, mechanisms, institutions and constituencies involved; as without realistic pathways for implementation, analysis and advocacy they are unlikely to lead to demonstrable results (Wilbanks and Sathaye, 2007).

As highlighted earlier, there are methodological difficulties in valuing outcomes of adaptation and mitigation activities, and the burden of mitigation and capacity for adaptation are differentially distributed according to actor, activity and location (Dowlatabadi, 2007). The capacity for action and implementation in many areas is limited, and may be constrained by socio-economic, technical and physical factors. For many low-income countries, where people may be more dependent on natural ecosystems and the capacity for action is less, the consideration of potential synergies could be particularly important (Barker *et al.*, 2007). The Clean Development Mechanism, Adaptation Fund and Special Climate Change Fund could provide opportunities for financing mitigation and/or adaptation measures that are compatible with biodiversity and sustainable development, and which may redress some of these inequalities.

Economic and non-economic incentives can be used to influence human behaviour and decision-making to ensure that desired synergistic outcomes are achieved and adverse outcomes avoided. The SCBD (2009) suggested that financial incentives could include payment for ecosystem services, and additional resources that could provide alternative livelihoods. Non-financial incentives could include the use of laws, regulations, property rights and education. They summarised instruments and incentives in the context of ecosystem-based adaptation (SCBD, 2009, Table 4.2), but this approach could be adapted to assess those appropriate to achieving the desired synergistic outcomes, while recognising the importance of context and scale dependency. The removal of economic barriers to implementation (e.g. perverse subsidies), or the use of environmental taxes, are possible neoclassical economic ways of furthering the desired objectives.

There are a number of alternative paradigms for achieving greater synergy between mitigation, adaptation and biodiversity, and also sustainable development. These focus around building institutions, policies, processes and infrastructure that concurrently enhance economic growth, technological change, and human and social capital (Goklany, 2007; Halsnæs and Verhagen, 2007), to which should be added building natural capital, which is the foundation of human well-being. Nevertheless, the realisation of the possibility of achieving multiple objectives through the promotion of synergistic activities could lead to more effective and efficient use of resources, and greater conservation of biodiversity.

Biofuel studies also illustrate the importance of collateral impacts on biodiversity. Biofuels may mitigate climate change, but there are large negative impacts from increased land use on highly diverse ecosystems (FAO, 2008) (Figure 15.9). Positive outcomes are possible if deployment is limited to degraded lands or developed on currently intensively cultivated lands. Such approaches, therefore, need to be urgently integrated in mitigation and adaptation planning. Furthermore, the land-use impacts of biofuels demand in developed countries may have biodiversity impacts in developing countries.

The Millennium Ecosystem Assessment also showed the interdependence of biodiversity and climate with sustainability issues, economic development, ecosystem services and human well-being (Millennium Ecosystem Assessment, 2005). Millennium Ecosystem Assessment future scenarios indicate continuing degradation of ecosystem services unless effective policy options based on interdependence of systems are adopted to achieve more sustainable pathways.

15.6 Broader perspectives

Munasinghe *et al.* (2009) illustrated one way to integrate mitigation into sustainable development strategy by mobilising the power of sustainable consumers and producers. The consumption of 1.2 billion richer humans accounts for some 75% of total greenhouse emissions. Instead of merely viewing these consumers as part of the problem, they should be persuaded to contribute to the solution by choosing consumption choices with lower emissions. The private sector can offer and promote more sustainable products to the consumer.

Making consumption patterns more sustainable will reduce emissions significantly if development promotes obvious measures like the use of energy-saving light bulbs, eating less meat, planting trees or using fuel-efficient cars or public transport. Such steps towards sustainable consumption can save money, and are also often faster and more achievable than

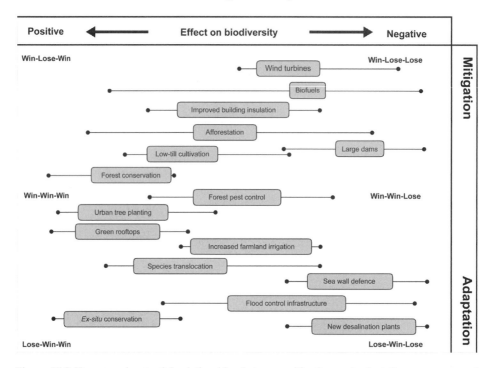

Figure 15.9 Known and potential relationships between mitigation and adaptation measures, and their impacts on biodiversity. The position of the boxes on the biodiversity axis was determined from a literature review of the biodiversity impacts of various mitigation and adaptation schemes and represents the typical outcome. The whiskers show the potential range of impacts. For example, not all afforestation projects are the same: monoculture plantations with high water demands could have detrimental effects, and encouraging natural regeneration around existing species-rich woodland could be beneficial. The win–lose trilogies are ordered: mitigation, adaptation, biodiversity (e.g. lose–win–lose means a loss for mitigation, a win for adaptation, and a loss for biodiversity). Source: modified from Paterson *et al*. (2008).

many big technology solutions and top-down government policies. Many existing 'best' practice examples can be replicated widely, while innovative businesses are already developing the future 'next' practice products and services.

Structuring knowledge will increase the adaptive and transformative capacity of social-ecological systems. A 'network of networks' approach can link structured knowledge and long-range planning for sustainable development and climate change adaptation and mitigation – by transforming fragmented disciplinary knowledge in universities and research institutions into linked networks of knowledge that facilitate innovation. Knowledge innovation also requires communications between academia and the public to promote societal reforms involving all stakeholders (Komiyama and Takeuchi, 2006; Kajikawa *et al*., 2007).

Knowledge networks for sustainable development can also include indigenous people's knowledge that lies outside conventional boundaries of 'modern' analysis. Indigenous peoples have been recognised as powerful knowledge holders who can make significant contributions by reminding us of the lessons of the past, improving strategies for the future and building the political will necessary to implement them (Galloway *et al.*, 2009).

Applying the concept of resilience of social-ecological systems (their ability to bounce back from disturbance), and promoting stewardship and transformation of ecological and social systems, provides opportunities for policy interventions. A resilience perspective on the climate change challenge, which stresses adaptability and transformability, will move us towards more sustainable paths (Folke *et al.*, 2002).

15.7 Summary and conclusions

Mainstreaming mitigation and adaptation into sustainable development is an important task. Developing country researchers and policy-makers have reiterated that development must remain the first priority, while adaptation and mitigation need to be pursued in line with development policy (Winkler *et al.*, 2002; Munasinghe and Swart, 2005). They suggest that coordinated actions across countries and sectors could reduce both adaptation and mitigation costs, and limit concerns about competitiveness, conflicts over international trade regulations and carbon leakage. They also argue that early actions on adaptation and mitigation measures, technology development, and better scientific knowledge about climate change are crucial.

All human beings are stakeholders when it comes to sustainable development and climate change, and they can and must strive to make development more sustainable – economically, socially and environmentally. By acting together now, to mainstream climate adaptation and mitigation policies into sustainable development strategy, humanity will make the planet a better and safer place for future generations.

References

Acker, R. and Kammen, D. M. (1996). The quiet (energy) revolution: Analysing the dissemination of photovoltaic power systems in Kenya. *Energy Policy*, **24**, 81–111.

Adger, N. W., Paavola, J., Huq, S. and Mace, M. J. (2005). *Fairness in Adapting to Climate Change*. Cambridge, MA: MIT Press.

Agrawal, A. (2008). The role of local institutions in adaptation to climate change. Paper prepared for the *Social Dimensions of Climate Change*, Social Development Department. Washington, D.C.: World Bank.

Agrawal, A. and Angelsen, A. (2009). Using community forest management to achieve REDD+ goals. In *Realising REDD+: National Strategy and Policy Options*, eds. A. Angelsen, M. Brockhaus, M. Kanninen, E. Sills, W. D. Sunderlin and S. Wertz-Kanounnikoff. Bogor: Centre for International Forestry Research, pp. 201–11.

Agrawal, A., Chhatre, A. and Hardin, R. (2008). Changing governance of the world's forests. *Science*, **320**, 1460–62.

Angelsen, A. (ed.) (2008). *Moving Ahead with REDD: Issues, Options and Implications*. Bogor: Center for International Forestry Research.

Angelsen, A. (ed.) (2009). *Realising REDD+: National Strategy and Policy Options*. Bogor: Centre for International Forestry Research.

Angelsen, A. and Kaimowitz, D. (1999). Rethinking the causes of deforestation: Lessons from economic models. *The World Bank Research Observer*, **14**, 73–98.

Angelsen, A. and Kaimowitz, D. (eds.) (2001). *Agricultural Technologies and Tropical Deforestation*. Wallingford, UK: CABI Publishing.

Ascher, W. (1999). *Why Governments Waste Natural Resources: Policy Failures in Developing Countries*. Baltimore, MD: The John Hopkins University Press.

Barker, T., Bashmakov, I., Alharthi, A. *et al.* (2007). Mitigation from a cross-sectoral perspective. In *Climate Change 2007: Mitigation. Contribution of Working Group III to the Fourth Assessment Report of the Intergovernmental Panel on Climate Change*, eds. B. Metz, O. R. Davidson, P. R. Bosch, R. Dave and L. A. Meyer. Cambridge, UK and New York, NY: Cambridge University Press, pp. 619–90.

Basu, J. P. (2009). Adaptation, non-timber forest products and rural livelihood: An empirical study in West Bengal, India. *IOP Conference Series: Earth and Environmental Science*, **6**, 382011.

Baudoin, M.-A. and Zaccai, E. (2009). How does the EU integrate climate change issues in its development aid policies? Short term reaction versus 'capacity approaches'. In *IHDP Open Meeting 2009. 7th International Science Conference on the Human Dimensions of Global Environmental Change*. 26–30 April, Bonn.

Berry, P. (ed.) (2009). *Biodiversity in the Balance: Mitigation and Adaptation Conflicts and Synergies*. Sofia, Bulgaria: Pensoft.

Brooks, N., Adger, W. N. and Kelly, P. M. (2005). The determinants of vulnerability and adaptive capacity at the national level and the implications for adaptation. *Global Environmental Change*, **15**, 151–63.

Campbell, B. M. (2009). Beyond Copenhagen: REDD+, agriculture, adaptation strategies and poverty. *Global Environmental Change*, **19**, 397–99.

Cashore, B. (2009). *Key Components of Good Forest Governance*. Jakarta: ASEAN Forests Clearing House Mechanism, ASEAN Secretariat, http://www.aseanforest-chm.org.

Colchester, M., Boscolo, M., Contreras-Hermosilla, A. *et al.* (2006). *Justice in the Forest: Rural Livelihoods and Forest Law Enforcement*. Bogor: Center for International Forestry Research.

Contreras-Hermosilla, A., Gregersen, H. M. and White, A. (2008). *Forest Governance in Countries with Federal Systems of Government*. Bogor: Centre for International Forestry Research.

Davidson, O., Halsnæs, K., Huq, S. *et al.* (2003). The development and climate nexus: The case of sub-Saharan Africa. *Climate Policy*, **3**, S97–S113.

Dornburg, V., Faaij, A., Verweij, P. *et al.* (2008). *Biomass Assessment: Assessment of Global Biomass Potentials and their Links to Food, Water, Biodiversity, Energy Demand and Economy*. Report No. WAB 500102012. Bilthoven, The Netherlands: Netherlands Environmental Assessment Agency, http://www.bioenergytrade.org/downloads/wab-biomassmainreportbiomassassessment.pdf.

Dowlatabadi, H. (2007). On integration of policies for climate and global change. *Mitigation and Adaptation Strategies for Global Change*, **12**, 651–63.

Duke, R., Graham, S., Hankins, M. *et al.* (2000). *Field Performance Evaluation of Amorphous Silicon (a-Si) Photovoltaic Systems in Kenya: Methods and Measurements in Support of a Sustainable Commercial Solar Energy Industry*. Washington, D.C.: World Bank Energy Sector Management Assistance Program (ESMAP).

Duke, R. D., Jacobson, A. and Kammen, D. M. (2002). Product quality in the Kenyan solar home systems market. *Energy Policy*, **30**, 477–99.

Eakin, H. and Lemos, M. C. (2006). Adaptation and the state: Latin America and the challenge of capacity-building under globalization. *Global Environmental Change*, **16**, 7–18.

Eriksen, S. E. H., Klein, R. J. T., Ulsrud, K., Næss, L. O. and O'Brien, K. (2007). *Climate Change Adaptation and Poverty Reduction: Key Interactions and Critical Measures*. Report prepared for the Norwegian Agency for Development Cooperation (Norad). Oslo: University of Oslo.

Fairman, D., Bukobza, D., Kabalo, S. *et al.* (2003). Amorphous, mono- and poly-crystalline silicon PV modules: A comparative study of their relative efficiencies under various outdoor conditions. *Proceedings of 3rd World Conference on Photovoltaic Energy Conversion*, 11–18 May 2003, Osaka, Japan.

FAO (2008). *The State of Food and Agriculture 2008: Biofuels, Prospects, Risks and Opportunities*. Rome: Food and Agriculture Organization of the United Nations.

Fargione, J., Hill, J., Tilman, D., Polasky, S. and Hawthorne, P. (2008). Land clearing and the biofuel carbon debt. *Science*, **319**, 1235–38.

Few, R., Osbahr, H., Bouwer, L. M., Viner, D. and Sperling, F. (2006). *Linking Climate Change Adaptation and Disaster Risk Management for Sustainable Poverty Reduction: Synthesis Report*. Brussels: European Union.

Firbank, L. G. (2008). Assessing the ecological impacts of bioenergy projects. *BioEnergy Research*, **1**, 12–19.

Fischer, G., Hizsnyik, E., Prieler, S., Shah, M. and van Velthuizen, H. (2009). *Biofuels and Food Security: Implications of an Accelerated Biofuels Production*. Vienna: The OPEC Fund for International Development (OFID).

Folke, C., Carpenter, S., Elmqvist, T. *et al.* (2002). Resilience and sustainable development: Building adaptive capacity in a world of transformations. *AMBIO: A Journal of the Human Environment*, **31**, 437–40.

Foran, B., Lenzen, M., Dey, C. and Bilek, M. (2005). Integrating sustainable chain management with triple bottom line accounting. *Ecological Economics*, **52**, 143–57.

Forner, C. (2005). Forests, adaptation and mitigation under the UNFCCC. In *Tropical Forests and Adaptation to Climate Change: In Search of Synergies*, eds. C. Robledo, M. Kanninen and L. Pedroni. Bogor: Center for International Forestry Research.

Galloway McLean, K., Ramos-Castillo, A., Gross, T. *et al.* (2009). *Report of the Indigenous Peoples' Global Summit on Climate Change*. 20–24 April, Anchorage, Alaska. Darwin, Australia: United Nations University – Traditional Knowledge Initiative.

Geist, H. J. and Lambin, E. F. (2002). Proximate causes and underlying driving forces of tropical deforestation. *BioScience*, **52**, 143–50.

Goklany, I. M. (2007). Integrated strategies to reduce vulnerability and advance adaptation, mitigation, and sustainable development. *Mitigation and Adaptation Strategies for Global Change*, **12**, 755–86.

Griebenow, G. and Kishore, S. (2008). Mainstreaming environment and climate change in the implementation of poverty reduction strategies. *Environmental Economics Series*, **119**. Washington, D.C.: World Bank.

Gunningham, N. (2009a). The new collaborative environmental governance: the localization of regulation. *Journal of Law and Society*, **36**,145–66.

Gunningham, N. (2009b). Environment law, regulation and governance: shifting architectures. *Journal of Environmental Law*, **21**, 179–212.

Halsnæs, K. and Verhagen, J. (2007). Development based climate change adaptation and mitigation – conceptual issues and lessons learned in studies in developing countries. *Mitigation and Adaptation Strategies for Global Change*, **12**, 665–84.

Hankins, M. (2000). A case study on private provision of photovoltaic systems in Kenya. In *Energy Services for the World's Poor*. Washington, D.C.: World Bank Energy Sector Management Assistance Program (ESMAP), pp. 92–99.

Hendrickson, C. T., Lave, L. B. and Matthews, H. S. (2005). *Environmental Life-cycle Assessment of Goods and Services: An Input-Output Approach*. Washington, D.C.: RFF Press.

Høyer, K. G. and Næss, P. (2008). Interdisciplinarity, ecology and scientific theory: The case of sustainable urban development. *Journal of Critical Realism*, **7**, 179.

Humphreys, D. (2006). *Logjam: Deforestation and the Crisis of Global Governance*. London: Earthscan.

International Institute for Environment and Development (IIED) (2010). AdMit, http://www.iied.org/climate-change/key-issues/economics-and-equity-adaptation/admit.

Intergovernmental Panel on Climate Change (IPCC) (2007a). *Climate Change 2007: Mitigation. Contribution of Working Group III to the Fourth Assessment Report of the Intergovernmental Panel on Climate Change*, eds. B. Metz, O. R. Davidson, P. R. Bosch, R. Dave and L. A. Meyer. Cambridge, UK and New York, NY: Cambridge University Press.

Intergovernmental Panel on Climate Change (IPCC) (2007b). *Climate Change 2007: Impacts, Adaptation and Vulnerability. Contribution of Working Group II to the Fourth Assessment Report of the Intergovernmental Panel on Climate Change*, eds. M. Parry, O. Canziani and J. Palutikof. Cambridge, UK and New York, NY: Cambridge University Press.

Intergovernmental Panel on Climate Change (IPCC) (2007c). *Climate Change 2007: Synthesis Report. Contribution of Working Groups I, II and III to the Fourth Assessment Report of the Intergovernmental Panel on Climate Change*, eds. Core Writing Team, R. K. Pachauri and A. Reisinger. Geneva, Switzerland: IPCC.

Jacobson, A. and Kammen, D. M. (2007). Engineering, institutions, and the public interest: Evaluating product quality in the Kenyan solar photovoltaics industry. *Energy Policy*, **35**, 2960–2968.

Jacobson, A., Duke, R., Kammen, D. M. and Hankins, M. (2000). Field performance measurements of amorphous silicon photovoltaic modules in Kenya. *Conference Proceedings of The American Solar Energy Society (ASES)*, Madison, WI, 16–21 June.

Jacobson, A., Duke, R. and Kammen, D. M. (2001). Update on the performance of 14 watt Phoenix gold solar panels. *Solarnet*, **3**.

JRC (2007). *Well to Wheels Analysis of Future Automotive Fuels and Powertrains in the European Context*. n.p.: European Commission, Joint Research Centre.

Kajikawa, Y., Ohno, J., Takeda, Y., Matsushima, K. and Komiyama, H. (2007). Creating an academic landscape of sustainability science: an analysis of the citation network. *Sustainability Science*, **2**, 221–31.

Kane, S. and Shogren, J. F. (2000). Linking adaptation and mitigation in climate change policy. *Climatic Change*, **45**, 75–102.

Kanninen, M., Murdiyarso, D., Seymour, F. *et al.* (2007). *Do Trees Grow on Money?: The Implications of Deforestation Research for Policies to Promote REDD*. Bogor: Center for International Forestry Research.

Kenya Bureau of Standards (KBS) (2003a). *Kenya Standard: Crystalline Silicon Terrestrial Photovoltaic (PV) Modules – Design Qualification and Type Approval. KS1674–2003*. Nairobi: KBS.

Kenya Bureau of Standards (KBS) (2003b). *Kenya Standard: Thin-film Terrestrial Photovoltaic (PV) Modules – Design Qualification and Type Approval. KS1675–2003.* Nairobi: KBS.

Kishore, S. (2007). *Mainstreaming Environment in the Implementation of PRSPs in Sub-Saharan Africa.* Environment Department Paper No. 112. Washington, D.C.: World Bank.

Klein, R. J. T., Schipper, E. L. F. and Dessai, S. (2005). Integrating mitigation and adaptation into climate and development policy: Three research questions. *Environmental Science & Policy*, **8**, 579–88.

Kok, M., Metz, B., Verhagen, J. and van Rooijen, S. (2008). Integrating development and climate policies: national and international benefits. *Climate Policy*, **8**, 103–18.

Komiyama, H. and Takeuchi, K. (2006). Sustainability science: Building a new discipline. *Sustainability Science*, **1**, 1–6.

Kühn, I., Sykes, M. T., Berry, P. M. *et al.* (2008). MACIS: Minimisation of and adaptation to climate change impacts on biodiversity. *GAIA – Ecological Perspectives for Science and Society*, **17**, 393–95.

Leach, M. and Fairhead, J. (2000). Challenging neo-Malthusian deforestation analyses in West Africa's dynamic forest landscapes. *Population and Development Review*, **26**, 17–43.

Leary, N., Conde, C., Kulkarni, J., Nyong, A. and Pulhin, J. (2007). *Climate Change and Vulnerability*. London: Earthscan.

Leary, N., Adejuwon, J., Barros, V. *et al.* (2008). *Climate Change and Adaptation*. London: Earthscan.

Lindner, M. and Kolström, M. (2009). A review of measures to adapt to climate change in forests of EU countries. *IOP Conference Series: Earth and Environmental Science*, **6**, 382007.

Lund, J. F., Balooni, K. and Casse, T. (2009). Change we can believe in? Reviewing studies on the conservation impact of popular participation in forest management. *Conservation & Society*, **7**, 71–82.

Lykke, A. M., Barfod, A. S., Tinggaard Svendsen, G., Greve, M. and Svenning, J.-C. (2009). Climate change mitigation by carbon stock: The case of semi-arid West Africa. *IOP Conference Series: Earth and Environmental Science*, **8**, 012004.

McCarl, B. A. and Schneider, U. A. (2001). Greenhouse gas mitigation in U.S. agriculture and forestry. *Science*, **294**, 2481–82.

McDermott, C. L., Cashore, B. and Kanowski, P. (2009). Setting the bar: An international comparison of public and private forest policy specifications and implications for explaining policy trends. *Journal of Integrative Environmental Sciences*, **6**, 217–37.

McDermott, C., Cashore, B. and Kanowski, P. (2010). *Global Environmental Forest Policies: An International Comparison*. London: Earthscan.

Millennium Ecosystem Assessment (2005). *Ecosystems and Human Well-being. General Synthesis*. Washington, D.C.: Island Press.

Müller, B. (2008). *International Adaptation Finance: The Need for an Innovative and Strategic Approach*. Oxford: Oxford Institute for Energy Studies.

Munasinghe, M. (1992). *Environmental Economics and Sustainable Development*. Washington, D.C.: World Bank.

Munasinghe, M. (2001). Sustainable development and climate change: applying the sustainomics transdisciplinary meta-framework. *International Journal of Global Environmental Issues*, **1** (1), 13–55.

Munasinghe, M. (2003). *Analysing the Nexus of Sustainable Development and Climate Change: an Overview*, Environment Directorate, Organisation for Economic Cooperation and Development, Paris, France.

Munasinghe, M. (2009). *Sustainable Development in Practice: Sustainomics Framework and Applications*. Cambridge, UK: Cambridge University Press, http://www.cup.es/catalogue/catalogue.asp?isbn=9780521719728.http://www.earthportal.org/?p=95.

Munasinghe, M. (2010). *Making Development More Sustainable: Sustainomics Framework and Applications*, 2nd edn. Colombo: MIND Press, Munasinghe Institute for Development.

Munasinghe, M. and Swart, R. (2005). *Primer on Climate Change and Sustainable Development: Facts, Policy Analysis and Applications*. Cambridge, UK: Cambridge University Press.

Munasinghe, M., Dasgupta, P., Southerton, D., Bows, A. and McMeekin, A. (2009). *Consumers, Business and Climate Change*. Manchester: Sustainable Consumption Institute, University of Manchester, http://www.sci.manchester.ac.uk/publications/related/consumers.

Nelson, D. R., Adger, W. N. and Brown, K. (2007). Adaptation to environmental change: Contributions of a resilience framework. *Annual Review of Environment and Resources*, **32**, 395–419.

Niles, J. O., Brown, S., Pretty, J., Ball, A. S. and Fay, J. (2002). Potential carbon mitigation and income in developing countries from changes in use and management of agricultural and forest lands. *Philosophical Transactions of the Royal Society A – Mathematical, Physical & Engineering Sciences*, **360**, 1621–39.

Nyong, A., Adesina, F. and Osman Elasha, B. (2007). The value of indigenous knowledge in climate change mitigation and adaptation strategies in the African Sahel. *Mitigation and Adaptation Strategies for Global Change*, **12**, 787–97.

O'Connor, D. (2008). Governing the global commons: linking carbon sequestration and biodiversity conservation in tropical forests. *Global Environmental Change*, **18**, 368–74.

Palmer, C. and Engel, S. (eds.) (2009). *Avoided Deforestation. Prospects for Mitigating Climate Change*. London: Routledge.

Parry, M., Arnell, N., Berry, P. *et al.* (2009). *Assessing the Costs of Adaptation to Climate Change: A Review of the UNFCCC and Other Recent Estimates*. London: IIED and Grantham Institute for Climate Change.

Paterson, J. S., Araújo, M. B., Berry, P. M., Piper, J. M. and Rounsevell, M. D. A. (2008). Mitigation, adaptation and the threat to biodiversity. *Conservation Biology*, **22**, 1352–55.

Pelling, M. and High, C. (2005). Understanding adaptation: What can social capital offer assessments of adaptive capacity? *Global Environmental Change A*, **15**, 308–19.

Prince's Rainforest Project (2009). *An Emergency Package for Tropical Rainforests*, http://www.rainforestsos.org.

Rarieya, M. and Fortun, K. (2010). Food security and seasonal climate information: Kenyan challenges. *Sustainability Science*, **5**, 99–114.

Reid, P. and Vogel, C. (2006). Living and responding to multiple stressors in South Africa: Glimpses from KwaZulu-Natal. *Global Environmental Change*, **16**, 195–206.

Robinson, J., Bradley, M., Busby, P. *et al.* (2006). Climate change and sustainable development: Realizing the opportunity. *AMBIO: A Journal of the Human Environment*, **35**, 2–8.

Roe, D., Nelson, F. and Sandbrook, C. (eds.) (2009). *Community Management of Natural Resources in Africa: Impacts, Experiences and Future Directions*. London: IIED.

Scharlemann, J. P. W. and Laurance, W. F. (2008). How green are biofuels? *Science*, **319**, 43–44.

Searchinger, T., Heimlich, R., Houghton, R. A. *et al.* (2008). Use of U.S. croplands for biofuels increases greenhouse gases through emissions from land-use change. *Science*, **319**, 1238–40.

Secretariat of the Convention on Biological Diversity (SCBD) (2009). *Connecting Biodiversity and Climate Change Mitigation and Adaptation: Report of the Second Ad Hoc Technical Expert Group on Biodiversity and Climate Change.* Technical Series No. 41. Montreal: SCBD.

Skourtos, M., Kontogianni, A. and Harrison, P. A. (2010). Reviewing the dynamics of economic values and preferences for ecosystem goods and services. *Biodiversity and Conservation*, **19** (10), 2855–72.

Smit, B. and Wandel, J. (2006). Adaptation, adaptive capacity and vulnerability. *Global Environmental Change*, **16**, 282–92.

Smith, P., Martino, D., Cai, Z. *et al.* (2007). Policy and technological constraints to implementation of greenhouse gas mitigation options in agriculture. *Agriculture, Ecosystems and Environment*, **118**, 6–28.

Smith, P., Martino, D., Cai, Z. *et al.* (2008). Greenhouse gas mitigation in agriculture. *Philosophical Transactions of the Royal Society B: Biological Sciences*, **363**, 789–813.

Staebler, D. L. and Wronski, C. R. (1977). Reversible conductivity changes in discharge-produced amorphous Si. *Applied Physics Letters*, **31**, 292–94.

Stern, N. (2006). The challenge of stabilisation. In *Stern Review: The Economics of Climate Change*, eds. Stern, N., Peters, S., Bakhshi, V., Bowen, A., Cameron, C., Catovsky, S., Crane, D., Cruickshank, S., Dietz, S., Edmonson, N., Garbett, S.-L., Hamid, L., Hoffman, G., Ingram, D., Jones, B., Patmore, N., Radcliffe, H., Sathiyarajah, R., Stock, M., Taylor, C., Vernon, T., Wanjie, H. and Zenghelis, D. London: HM Treasury, pp. 193–210.

Stern, N. H. (2007). *The Economics of Climate Change: The Stern Review.* Cambridge, UK: Cambridge University Press.

Sunderlin, W. D., Hatcher, J. and Liddle, M. (2008). *From Exclusion to Ownership? Challenges and Opportunities in Advancing Forest Tenure Reform.* Washington, D.C.: Rights and Resources Initiative.

Tacconi, L. (2007). Illegal logging and the future of the forest. In *Illegal Logging: Law Enforcement, Livelihoods and the Timber Trade,* ed. L. Tacconi. London: Earthscan.

Tavoni, M., Sohngen, B. and Bosetti, V. (2007). Forestry and the carbon market response to stabilize climate. *Energy Policy*, **35**, 5346–53.

Tol, R. S. J. (2005). Adaptation and mitigation: trade-offs in substance and methods. *Environmental Science & Policy*, **8**, 572–78.

Tol, R. S. J. (2007). The double trade-off between adaptation and mitigation for sea level rise: an application of FUND. *Mitigation and Adaptation Strategies for Global Change*, **12**, 741–53.

Tschakert, P. (2007). Views from the vulnerable: understanding climatic and other stressors in the Sahel. *Global Environmental Change*, **17**, 381–96.

UNFCCC (2009a). *Methodological Guidance for Activities Relating to Reducing Emissions from Deforestation and Forest Degradation and the Role of Conservation, Sustainable Management of Forests and Enhancement of Forest Carbon Stocks in Developing Countries,* http://unfccc.int/meetings/cop_15/items/5257.php.

UNFCCC (2009b). *Copenhagen Accord*, http://unfccc.int/meetings/cop_15/items/5257. php.

van der Werf, G. R., Morton, D. C., DeFries, R. S. *et al.* (2009). CO_2 emissions from forest loss. *Nature Geoscience*, **2**, 737–38.

van Vliet, J., den Elzen, M. G. J. and van Vuuren, D. P. (2009). Meeting radiative forcing targets under delayed participation. *Energy Economics*, **31**, S152–S162.

Vogel, C. (2006). Foreword: Resilience, vulnerability and adaptation: A cross-cutting theme of the International Human Dimensions Programme on Global Environmental Change. *Global Environmental Change*, **16**, 235–36.

Watkins, K. (2007). *Human Development Report 2007/2008. Fighting Climate Change: Human Solidarity in a Divided World*. Palgrave Macmillan.

Wilbanks, T. J. and Sathaye, J. (2007). Integrating mitigation and adaptation as responses to climate change: A synthesis. *Mitigation and Adaptation Strategies for Global Change*, **12**, 957–62.

Winkler, H., Spalding-Fecher, R., Mwakasonda, S. and Davidson, O. (2002). Sustainable development policies and measures: starting from development to tackle climate change. In *Building on the Kyoto Protocol: Options for Protecting the Climate*, eds. K. Baumert, O. Blanchard, S. Llosa and J. F. Perkaus. Washington, D.C.: World Resources Institute, pp. 61–87.

World Bank (2008). *Forests Sourcebook*. Washington, D.C.: World Bank.

World Bank (2009). *Convenient Solutions to an Inconvenient Truth: Ecosystem-based Approaches to Climate Change.* Washington, D. C.: World Bank.

World Bank (2010). *World Development Report 2010: Development and Climate Change.* Washington, D.C.: World Bank.

Wunder, S. (2007). The efficiency of payments for environmental services in tropical conservation. *Conservation Biology*, **21**, 48–58.

Yohe, G. W. (2001). Mitigative capacity – the mirror image of adaptive capacity on the emissions side. *Climatic Change*, **49**, 247–62.

16

Mobilising the population

'Never doubt that a small group of thoughtful, committed citizens can change the world; indeed, it's the only thing that ever has.'[1]

An effective response to the risks of climate change can build on the ideas and actions of many different individuals – both as citizens and consumers, and as leaders of business and other organisations. Individual citizens can play an important role in the response to climate change, especially when they make decisions to reduce their greenhouse gas emissions or adapt to climate change encouraged by institutional structures and access to credible, understandable and relevant information. There is considerable evidence that individual behavioural change can contribute to reductions in emissions, especially from households and transportation and when supported by government policies, incentives and private sector activities.

Individuals are also engaged with climate change as decision-makers, members of non-governmental organisations (NGOs) and voters in ways that influence the actions of governments and corporations. Public support is critical in the success of national and regional government actions, and public perceptions can impede the acceptance of mitigation technologies. Behavioural and attitudinal changes are also important in terms of political and corporate leadership where, for example, business leaders and city mayors have made significant commitments to emission reductions that go far beyond national political obligations or simple cost–benefit analysis. In terms of adaptation, millions of farmers and herders have adjusted their practices to past climate shifts and are already making decisions in response to the onset of warming and other shifts associated with climate change.

Although many individual decisions are driven by economic factors, the role of culture, ethics, information and perceptions in mobilising people to respond (or not) to climate change should not be underestimated. As policymakers design larger-scale institutional and policy responses to climate change – such as carbon markets (Chapter 12), low-carbon technologies (Chapter 11), and international agreements (Chapter 13) – they must take account of possible public and private sector responses to these policies and technologies.

---

[1] Margaret Mead (1901–78), American cultural anthropologist.

Climate Change: Global Risks, Challenges and Decisions, Katherine Richardson, Will Steffen and Diana Liverman *et al.* Published by Cambridge University Press. © Katherine Richardson, Will Steffen and Diana Liverman 2011.

These responses may include resistance to new energy sources and regulations, concerns about social justice and risks, voluntary efforts that go beyond those anticipated by the market, and broad changes in cultural norms. International policy needs to support – and be sure not to constrain – the agency of individuals to respond to climate change, and to recognise the importance of providing relevant information to citizens so that they can make informed decisions about supporting policies and changing their own behaviour.

This chapter focuses on the range of human responses to climate change, especially the role of 'non-state actors' – individuals, NGOs and corporations – in mitigation and adaption and in influencing the response of governments. Discussion of climate policy often focuses on the actions of nation states and the goal of this chapter is to explore the ways in which 'people matter' in the response to climate change.

16.1 What do we know about attitudes to climate change?

Insights into how people understand and respond to climate change are offered by a series of academic surveys and public opinion polls of attitudes to climate change. These surveys include questions about awareness and concern about climate change and policy preferences and, although many studies are country-specific, several provide international comparisons. Leiserowitz (2007) reviewed several global surveys conducted between 1998 and 2006 and concluded that the majority of people surveyed have heard about global warming and are increasingly seeing it as a serious problem. They are especially concerned about health risks, water shortages, species loss and extreme weather, and most see climate change as caused by human activity. Significant majorities want their governments to respond to climate change, even if the response is costly. Leiserowitz noted, however, that many people confuse global warming with the ozone depletion problem, and that people in the Middle East and the United States are less concerned than those in other parts of the world, especially Europe.

Figure 16.1 shows results from a recent survey of international attitudes to climate change, which shows half or more of the sampled population in 17 countries seeing global warming as a 'very serious' problem (Pew Global Attitudes Project, 2009). The survey also asked whether people were willing to pay more to address climate change and found most countries divided – in the USA, Indonesia, Egypt and Mexico, 55% or more of those surveyed were unwilling to pay more. It is important to note that the survey was conducted in the midst of global recession. A similar study of 23 countries conducted between June and October 2009 found that concern about climate change was at its highest since polling began in 1998, with 64% of those surveyed saying climate change was a 'very serious problem' (Globescan, 2009). However, concern had fallen in the USA, with now only 45% of Americans seeing climate change as being very serious. Focusing on the December 2009 COP15 climate negotiations in Copenhagen, the survey found 44% of those surveyed wanting their governments to lead in setting targets to address climate change and only 6% wanting their governments to oppose an international agreement (with slightly larger

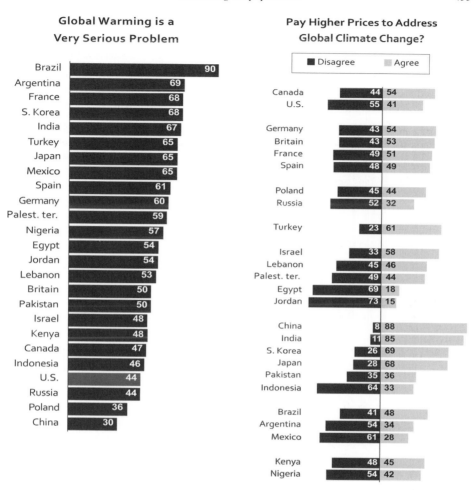

Figure 16.1 Attitudes to climate change in early 2009 based on public opinion surveys in 25 countries totalling 26 397 interviews. Source: Pew Global Attitudes Project (2009).

proportions with a negative view of the Copenhagen negotiations being found in the USA and China).

Other surveys examine the deeper roots of attitudes to climate change, the specific actions people say they have taken, and the views of corporate leaders. An in-depth survey of actions in response to climate change in the USA found that 51% of respondents considered energy efficiency in buying products and 45% had weatherised their homes (Leiserowitz, 2007). Figure 16.2 summarises the responses of consumers in eight countries about their willingness to act on climate change, with many people across all countries willing to use energy-efficient appliances or recycle; and lower proportions willing to use public transport, reduce flying and meat consumption, or purchase offsets (Bonini *et al.*,

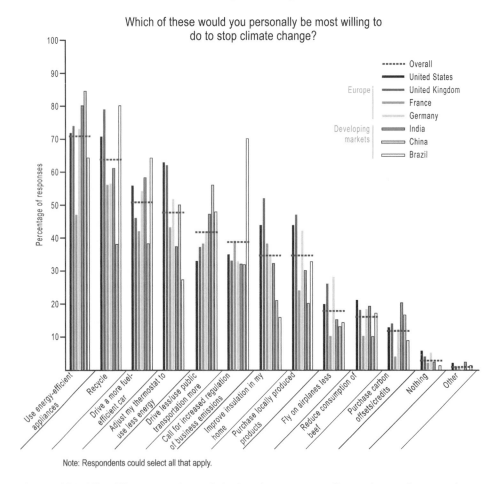

Figure 16.2 Public willingness to change behaviour in response to climate change. Source: redrawn from Bonini *et al.* (2008).

2008). A global survey of corporate opinions about climate change with more than 2000 respondents found that executives consider climate change as both a risk and opportunity, with more than half mentioning that it is important to overall corporate strategy, and significant for brand reputation and throughout supply chains (McKinsey, 2007).

Social scientists interpret such surveys cautiously for reasons that include a likely bias to urban and literate sample populations in developing countries, a tendency for people to report socially acceptable environmental attitudes, the influence of recent events and the media on attitudes (Box 16.1), the lack of in-depth explanation of reported attitudes and, most importantly, the significant gaps that have been identified between reported attitudes and actual behaviour (Stern, 2000; Dietz *et al.*, 2005; Leiserowitz, 2007). This so-called value-action gap is difficult to overcome. The assumption that providing more information (an information-deficit model) will encourage appropriate behaviour has been criticised by scholars who see culture, norms and leadership as being more important in environmentally significant behavioural and organisational change.

Box 16.1
The role of the media

MAXWELL BOYKOFF

Media representations – from news to entertainment – are critical links between the everyday realities of how people experience climate change and the ways in which these are discussed at a distance between science, policy and public actors (Boykoff, 2009). Many studies, surveys and polls have found that the public frequently learn about science (and more specifically climate change) from the mass media. Mass media range from entertainment to news media, and spanning television, films, books, flyers, newspapers, magazines, radio and the internet (websites, blogs, Youtube, Facebook, etc.). In the past decade, there has been a significant expansion from consumption of 'new social media' – such as the internet, and mobile phone communications – where a great deal of climate communications now happen on blogs and tweets. This movement has signalled substantive changes in how people access and interact with information, who has access and who is seen as an 'authority' as well as an 'expert'. Together, these media are constituted by a diverse and dynamic set of institutions, processes and practices that, together, serve as 'mediating' forces between communities such as science, policy and public citizens. Members of the communications industry and profession – publishers, editors, journalists and others – produce, interpret and communicate images, information and imaginaries for varied forms of consumption (Carvalho, 2007).

Media coverage of climate change first emerged in the late eighteenth century, and was linked to discussions of weather events: an example is news coverage linking climate and weather in the winter of 1788–89, when the Thames River in London froze. Coverage expanded in North America and Europe through the nineteenth century with a focus on the links between climate and agriculture. These themes of climate, weather and agriculture continued to play out in the twentieth century, such as during the 1930s in relation to the Dust Bowl in North America. Moving through the twentieth century, connections were increasingly made between climate change and scientific studies. As international and domestic climate policy began to take shape in the mid 1980s, sporadic media coverage of climate change science and policy gave way to a steady flow of climate news. Climate scientists were widely quoted and called upon in the media as 'authorised' speakers on behalf of the climate, and business and environmental groups also started to speak out on the issue. In the process of understanding changes in the climate, many entities, organisations, interests and individuals battled to shape awareness, engagement and possible action.

Media coverage of climate change reached a peak in 2006 and into 2007. Figure 16.3 shows the ebbs and flows of news articles on climate change or global warming from January 2004 through January 2010 in 50 newspapers across 20 countries. Abundant coverage of 'climate change' or 'global warming' can be attributed to a number of key and concatenate events.

Box 16.1 (*cont.*)

Among them, mid-2006 marked the global release of the Al Gore film 'An Inconvenient Truth'. Moreover, the much anticipated, discussed and criticised *Stern Review* was released on 30 October 2006. Intense media coverage of the *Stern Review* then fed into media attention on the UNFCCC COP12 meeting in Nairobi, Kenya, which began approximately a week later. Following on in 2007 were highly fluctuating oil and gasoline prices, as well as the releases of the highly influential IPCC Fourth Assessment Report. Figure 16.3 shows that there has been a discernible levelling off or decrease in the amount of coverage in later 2007, through 2008 and into 2009. The decline could be attributed to a number of intersecting influences including: (i) media attention on the global economic recession may have displaced climate change reporting; (ii) issues formerly discussed explicitly as 'climate change' or 'global warming' are now treated as 'energy' issues, 'sustainability' considerations, and other associated themes such as 'carbon trading'; and (iii) upon the 2007 release of various Working Group reports for the IPCC Fourth Assessment Report, fewer fundamental issues were deemed as 'controversial' as in previous assessments (although this apparently changed in late 2009 with the controversial hacking of emails about climate science from the University of East Anglia (UEA) Climate Research Unit (CRU)). However, another peak was no doubt reached during the Copenhagen COP15 in December 2009. The highly anticipated United Nations climate talks in Copenhagen, Denmark (COP15), along with news about the released UEA CRU emails, played key parts in this dramatic rise to close out 2009.

Figure 16.3 also shows the low number of stories on climate change or global warming in the regions of South America and Africa. This can be attributed partly to the low number of sources sampled in these regions. This points to an 'information gap' in reporting on these issues, and relates to capacity issues and support for reporters in these regions and countries (developing and poorer regions/countries). Those most at risk from the impacts of climate change typically have had access to the least information about it through mass media. Some organisations and networks are seeking to overcome these challenges, including the Climate Change Media Partnership, Internews, PANOS Institute and the Earth Journalism Network.

At the same time, capacity in climate change reporting is declining in developed countries, as all over Europe and America staff is being cut and budgets for getting out of the office slashed. From 1989 to 2006, the number of newspapers featuring weekly science sections shrunk by nearly two-thirds; while since 2001, nearly one newspaper journalist in five in the USA has been laid off. In December 2008, CNN cut its entire science, technology and environment news staff.

In the past decade, questions raised by the mass media have largely moved away from 'Is the climate changing?' and 'Do humans play a role in climate change?' to more textured considerations of governance and economics. For instance, many articles have addressed questions regarding how to effectively govern the mitigation of GHG emissions, and how to construct and maintain initiatives to help vulnerable communities adapt to already unfolding climate impacts.

As mass media serve a vital role in communication processes, media portrayals of climate change shape ongoing perceptions and considerations for action. Representations construct and negotiate meaning, and shape how people make sense of the world. Choices

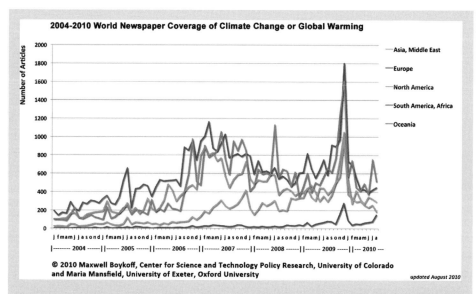

Figure 16.3 Newspaper coverage of climate change or global warming in 50 newspapers across 20 countries and six continents over a five-year period (January 2004–January 2010). The newspapers used in this figure (appearing alphabetically by newspaper) are *The Age* (Australia), *The Australian* (Australia), *Business Day* (South Africa), *Clarín* (Argentina), *The Courier-Mail* (Australia), the *Daily Express* (and *Sunday Express*) (United Kingdom), *Daily Mail* (*Mail on Sunday*) (United Kingdom), the *Daily News* (United States), the *Daily Telegraph* (Australia), *Dominion Post* (New Zealand), *Fiji Times* (Fiji), the *Financial Mail* (South Africa), *Globe and Mail* (Canada), the *Guardian* (and *Observer*) (United Kingdom), *The Herald* (United Kingdom), the *Hindu* (India), *Hindustan Times* (India), the *Independent* (and *Sunday Independent*) (United Kingdom), *Indian Express* (India), the *Irish Times* (Ireland), *Japan Times* (Japan), the *Jerusalem Post* (Israel), the *Jerusalem Report* (Israel), the *Korea Herald* (South Korea), the *Korea Times* (South Korea), the *Los Angeles Times* (United States), the *Mirror* (*Sunday Mirror*) (United Kingdom), the *Moscow News* (Russia), the *Nation* (Pakistan), the *Nation* (Thailand), *National Post* (Canada), the *New Straits Times* (Malaysia), the *New York Times* (United States), *New Zealand Herald* (New Zealand), the *Prague Post* (Czech Republic), *The Press* (New Zealand), *The Scotsman* (and *Scotland on Sunday*) (United Kingdom), the *South China Morning Post* (China), the *South Wales Evening Post* (United Kingdom), *The Straits Times* (Singapore), *The Sun* (and *News of the World*) (United Kingdom), *Sydney Morning Herald* (Australia), the *Telegraph* (and *Sunday Telegraph*) (United Kingdom), the *Times* (and *Sunday Times*) (*United Kingdom*), *The Times of India* (India), the *Toronto Star* (Canada), *USA Today* (United States), the *Wall Street Journal* (United States), the *Washington Post* (United States) and *Yomiuri Shimbun* (Japan). Source: http://sciencepolicy.colorado.edu/media_coverage/.

by journalists regarding how they represent climate science and policy through the media hinge on interpretation, perspectives and available information – as well as contextual social, political, economic and environmental factors. While media interventions seek to enhance understanding of complex and dynamic human–environment interactions, vague and

Box 16.1 (*cont.*)

de-contextualised reporting instead can enhance bewilderment. Better reporting has critical implications for understanding, meaning and potential public engagement, and possible support for policy action.

  Ultimately, a more informed public space and better supported links between science, policy and media are in our collective self-interest.

Encouraging pro-environmental behaviour through shifting norms may be better achieved through social and community networks than through top-down government communication campaigns. Analyses of barriers to behaviour examine both external/contextual and internal/psychological factors. For example, Jackson identified three main sets of barriers – external conditions such as infrastructure and institutions, social context, and lock in to habits and preferences (Jackson, 2006). Examples of the significance of infrastructure include the presence or absence of public transport or kerbside recycling. The value-belief-norm theory of environmentalism links pro-environmental behaviour to basic values of altruism; and beliefs about nature, to consequences and personal efficacy, and to norms about obligations (Dietz *et al.*, 2005). Lessons from social marketing theory have also been helpful in understanding how to promote behavioural change, grouping the public into segments, building awareness, and encouraging collective action and reflection. For example, Gilg *et al.* (2005) identified clusters that include committed environmentalists, mainstream environmentalists, occasional environmentalists and non-environmentalists.

Assumptions about individual values and attitudes underlie economic analyses of climate change including Stern's pivotal study on the economics of climate change (Chapter 12; Stern *et al.*, 2006). Although many economists assume that people value the future less than the present, by applying discount rates of 3% or more, Stern took the position that discounting the future is unethical and used a near-zero discount rate, which produced significant future damages that far exceeded the costs of early action on mitigation. Empirical support for individual perceptions of future values, risks and equity concerns showed that most people do value the future, are risk-averse and are concerned about the unequal effects (Chapters 9 and 10) of climate change (Beckerman and Hepburn, 2007; Saelen *et al.*, 2008).

Surveys also provide important insights into how the public views solutions to climate change, especially perceptions of risk that may constrain the acceptability of certain technologies. Some scientists and policymakers are proposing nuclear energy, biofuels, carbon capture and storage (CCS) and geoengineering (such as atmospheric sulfur injection) as the only scalable solutions for reducing levels of greenhouse gases in the atmosphere or limiting the temperature increase. The barriers to rapid global adoption of these technologies include resources and cost, but research suggests that public opposition to CCS and other technologies is also likely (Shackley *et al.*, 2005; Huijts *et al.*, 2007; Singleton *et al.*, 2009; Wong-Parodi and Ray, 2009). For example, research on attitudes to CCS among environmental NGOs finds that although many groups think CCS may be necessary to reach emission reduction goals, they oppose it because of perceived economic and technical uncertainties – including the risk of major accidents, and because they are suspicious of CCS links to oil and coal industries, and that CCS facilities will be located in poor and ethnic communities (Wong-Parodi and Ray, 2009).

In the UK, local and environmentalist opposition to wind farms has grown to the point where very few proposals make it through the planning process. Opposition to renewable energy is becoming a barrier to its implementation in many countries (Barry *et al.*, 2008; Walker *et al.*, 2010). Projects are more likely to be more socially acceptable when they (i) are locally embedded, (ii) provide local benefits, (iii) establish continuity with existing physical, social and cognitive structures, and (iv) apply good communication and participation procedures. The backlash against biofuels has created generic opposition to even sustainable and second-generation biofuel projects, with some Europeans not seeing bioenergy as an environmentally acceptable alternative (Eurobarometer, 2008).

What is the public perception of nuclear power as a solution to climate change? Research in the UK suggests that while higher proportions (55%) of the public are prepared to accept nuclear power if it is seen to contribute to climate change mitigation, very few would actively prefer this as an energy source over renewable sources or compared to energy efficiency, given the choice (Pidgeon *et al.*, 2008). These findings echo more general results on perceptions of nuclear risks, where researchers find continuing concern about the risk of nuclear power to future generations and to public health, and where nuclear risks are seen as unknowable and producing feelings of dread. Classic studies show that the public will perceive technologies as much more risky if they have low-probability but high-magnitude risks; are associated with cancer or risks to future generations; and are difficult to see, complex, involuntary and seen to lack local benefit (Fischhoff *et al.*, 1981; Slovic, 1992).

Governments and private sector actors seeking to develop alternatives to fossil fuels, or projects that counter climate change through sequestration and geoengineering, would be wise to take account of public attitudes as they design and implement projects. A perception of risk or other negative impacts can slow or stop mitigation projects, especially where democratic planning processes allow for public participation. The chances of project approval, or acceptable redesign, are enhanced where the public is involved in decision making in a deliberative fashion (Renn, 2006; Chilvers and Burgess, 2008; Hindmarsh and Matthews, 2008).

16.2 Households and the response to climate change

Households and domestic energy use are a significant source of greenhouse gas emissions in most countries, and changing household behaviour has been identified as an important component of climate policy (Figure 16.4). For example, the IPCC (IPCC, 2007a) estimated that at a carbon price of less than $50 a ton in 2030 greenhouse gases could be reduced by more than 5 Gt CO_2-e per year in the building sector (more than half in

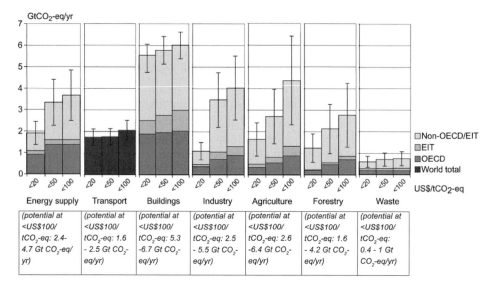

Figure 16.4 Estimated sectoral economic potential for global mitigation for different regions as a function of carbon price in 2030 from bottom-up studies, compared to the respective baselines assumed in the sector assessments. Source: IPCC (2007a).

the developing world), and that emissions from transport and electricity supplies could also be significantly reduced. Similar conclusions derive from the McKinsey global abatement curves, which show relatively low-cost (or even cost-saving) greenhouse gas reductions through building insulation; and more efficient lighting, air conditioning and vehicles (Enkvist *et al.*, 2007). McKinsey estimated that a $40 billion investment in the residential sector could abate many gigatonnes of carbon emissions (40% in the USA and China), with more than 60% of the savings for only $8 billion invested in efficient lighting and appliances (the remainder is for heating and cooling). They highlight barriers to these easy gains that include fuel subsidies, misaligned incentives and lack of information to consumers (Farrell and Remes, 2008).

It is clear that households can help reduce the risks of climate change by choosing low-carbon energy sources, changing their transport options, insulating their homes, buying efficient appliances, reducing their energy demands, and consuming lower carbon foods and materials. However, even well-informed and well-intentioned individuals face considerable barriers to reduce energy use and other emissions, including those associated with cost and lack of available alternatives. In the USA, where households account for almost 40% of carbon emissions, modest changes in energy use and transport could cut household emissions by a third (Gardner and Stern, 2008) – however, change is limited not only by financing, but also by information, time, attention and the inability of renters to alter their energy use. They identified the most effective actions as including buying a more fuel-efficient automobile (saving 13.5% energy for an average individual),

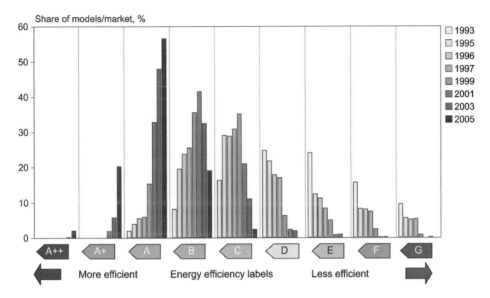

Figure 16.5 The impact of the EU appliance Label (A++ to G, with G being the least efficient) on the market of cold appliances in EU-25. Source: IPCC (2007a).

installing and upgrading insulation and ventilation (7% savings), and replacing light bulbs (4% savings).

Governments and business leadership can encourage households to reduce their emissions through a variety of policies and voluntary programs. European Union and national government programs to foster energy efficiency include grants and tax breaks for home insulation, energy performance standards for housing, and an appliance labelling scheme (Figure 16.5) that resulted in a 25% reduction in appliance energy use (Geller *et al.*, 2006). Private sector initiatives include provision of efficient light bulbs, low-carbon energy options and feed-in tariffs by utilities.

Research in the UK has shown the potential of information and feedback for reducing household energy use. Utility bills that provide information on past energy use, or compare a household to averages for a neighbourhood, can produce significant changes in demand and reinforce energy-saving behaviour of up to 10% (Darby, 2006). Experiments with smart meters that provide home owners with easily visible information on their energy use, cost and emissions can reduce household energy use by 15% through changing habits and stimulating modest investments.

New proposals to label other products – including food – with carbon labels or carbon certification are more problematic, and need to learn from several decades of experience with nutrition labels and organic certification. For example, consumers find it difficult to decode labels, calculating the amount of carbon in a product is fraught with uncertainty, boundaries for what is included vary widely, and costs of labelling and certification can exclude small producers (Boardman *et al.*, 2007; Hogan and Thorpe, 2009).

Individuals in the UK, together with business and government departments, are joining together in efforts to cut UK carbon emissions by 10% in 2010. The 10:10 campaign provides a simple checklist for people to cut their emissions, which includes flying and driving less, turning down thermostats slightly, installing efficient light bulbs, turning products off standby, eating less meat, reducing food waste, recycling and reducing water use. Assuming an average person in the UK is responsible for about 14 t of greenhouse gas emissions annually, each individual is asked to cut emissions by 1.5 t. Table 16.1 shows how each action translates into a reduction in emissions and is a good illustration of the potential of individuals to respond to climate change.

The consumption of food and energy are important sources of greenhouse gas emissions, but asking people to change their lifestyles and aspirations confronts models that see wealth and growth as measures of success. As incomes increase in much of the developing world, consumption of energy tends to increase as does the consumption of meat and dairy products. Finding ways to decarbonise while improving individual livelihoods is a considerable challenge, although there are examples of a decoupling of economic growth and GHG emissions in developed economies (Chapter 11).

16.3 Farmers and climate change adaptation

Farmers are experts at adapting to climate change. Millions of agricultural producers around the world are already adapting to higher temperatures and greater extremes (Chapter 14), often basing their strategies on accumulated local experience in dealing with centuries of climate variability including drought, floods and other hazards (IPCC, 2007b). The decisions of individual farmers will be critical in adapting agriculture and food systems to a warmer world, although their choice and ability to respond is framed by available technology, information and institutions, and by government policies designed to facilitate adaptation.

What choices are farmers making to respond to climate change? Smit and Skinner (2002) developed a typology for Canadian agriculture that focuses on technology options (new crop varieties, early warning systems, irrigation), government programs and insurance (crop insurance, income and production subsidies, disaster relief), farm production practices (crop and livestock diversification, intensification, alternative soil management, timing) and finances (crop insurance and futures, household income diversification). While some of these options are not available to farmers in developing countries – who are often poor and do not have access to government programs – new proposals for adaptation funds and programs in the developing world include options for expanding insurance, implementing information programs and reducing vulnerability to disasters. Agrawal (2008) classified livelihood adaptations to climate risks into mobility (e.g. moving herds or migrating), storage (e.g. of grains), diversification of assets, pooling assets across households and market exchange (e.g. insurance).

Some researchers suggest that, rather than focusing on specific projects, the response to climate change should include increasing the 'adaptive capacity' of individual farmers (the broader ability of farmers to cope with climate change risks through planning and preparing for adaptation) or to build 'resilience' (the ability to absorb shock) within farming systems and livelihoods (Smith *et al.*, 2003; Folke, 2006; Smit and Wandel, 2006).

Dozens of research studies are documenting how farmers are responding to climate change and are identifying barriers to their successful adaptation (Howden *et al.*, 2007; Leary *et al.*, 2007; Schmidhuber and Tubiello, 2007). Box 16.2 provides a case study of farmers in Mexico. The major barriers to successful individual action include (i) uncertainty about how much and how fast climate, especially precipitation, will change into the future, (ii) the ability to change crop and livestock mix given a limited range of culturally, ecologically and economically appropriate options, (iii) cost and access to inputs including affordable water, credit, insurance and fertiliser, and (iv) uncertainty about government programs and markets in both the short and long term.

Box 16.2
Farmers adapting to climate risk in rural Mexico

HALLIE EAKIN

Smallholder farmers in rural Mexico use a variety of adaptation strategies to cope with climate variability illustrating both their creativity in the face of risk and constraints on their ability to respond. A study of three communities in the states of Puebla and Tlaxcala explored how peasant households manage risk through understanding their livelihood strategies in the face of climate and economic change (Eakin, 2005, 2006). Because of recent changes associated with neoliberal policies – including land reform, the North American Free Trade Agreement and the withdrawal of government subsidies – the farmers were trying to adapt to a new production environment as well as to climate: a form of 'double exposure' to globalisation and global environmental change. Short-term coping strategies for drought and other risks included selling livestock, intensifying participation in non-farm activities, seeking financial support through informal loans, replanting their fields, and seeking food assistance and short-term work. Several heads of households migrated to the USA or to the city to try rebuilding incomes.

The most important livelihood strategy in these communities was subsistence maize (corn) production used as insurance to maintain food security despite the vulnerability of maize to drought and frost. Those farmers with very small plots were less able to diversify their crops because of the desire to plant maize, whereas those with larger landholdings were able to respond to government programs that encouraged shifts to less vulnerable oats, barley and livestock (especially dairy cows). An uncertain market for barley made this a risky adaptation, however. Although irrigation can be an important adaptation, the community with access to irrigation had shifted to vegetable growing, exposing them to the volatility of the international market and the risk of losing the considerable investments made in irrigated horticulture. For many farmers, adaptation involved shifting partly out of agriculture, finding off-farm work or migrating to cities and the USA. The proximity of one community to a new factory allowed people to stay in their community and earn higher wages. Over the longer term, community

Box 16.2 (*cont.*)

members saw education as a way to improve the lives of their children, who would no longer work in agriculture. In this region of Mexico, adaptive capacity was associated with a particular combination of key resources that included sufficient land for subsistence, livestock to provide stability and access to the cash economy, and education and social connections to engage in a broader range of work. If they can grow a variety of crops, obtain income from livestock or milk, and access non-farm jobs, then households are more likely to be able to cope not only with climatic impacts, but also to access the technologies, information and knowledge necessary to adapt to the combined effects of socioeconomic and environmental change.

There is growing attention to climate change by major food and agricultural institutions including international organisations, private sector companies and national agricultural departments with organisations such as the Food and Agriculture Organization of the United Nations, the World Bank, the international agricultural research consortium CGIAR, and major national donors all making adaptation to climate change a priority (CGIAR, 2008; FAO, 2009; World Bank, 2010). Individual farmers are seen as key to international initiatives that highlight the importance of local communities, participation, diversity and the value of traditional knowledge (FAO, 2009). In addition to playing a role in adaptation, farmers are also starting to act on mitigation through managing soil carbon, planting trees and reducing emissions from livestock waste.

16.4 Non-state actors and the response to climate change

Mobilising the population to respond to climate change is not just the task of governments and individual citizens. Private sector, non-governmental organisations such as corporations, environmental and humanitarian groups have powerful roles in responding to climate risks, and have growing significance in overall patterns of climate governance (Newell, 2000; Raustiala, 2001; Levy and Newell, 2004; Okereke *et al.*, 2009).

The business sector is important because it produces a large share of greenhouse gas emissions; comprises sectors such as insurance and agriculture, which are vulnerable to climate impacts; and has the financial, technological and organisational resources to respond to climate change. Over the past decade, a significant number of companies, including some of the world's largest GHG emitters, have made voluntary commitments to reduce their emissions. While it is difficult to add up these commitments because of reporting inconsistencies, it is clear that business leadership can and will underlie emission reductions in many regions of the world. For example, 10 major companies participated in the Cement Sustainability Initiative under the World Business Council for Sustainable Development, committing to monitor and disclose CO_2 emissions, and to set targets for emission reductions ranging from 10% to 25% (http://www.wbcsdcement.org) from a 1990 baseline by 2010 to 2015 (Klee and Coles, 2004; Rehan and Nehdi, 2005). Business motivations for

Table 16.1. *The impact of individual action on household greenhouse gas emissions in the UK*

Action	Savings (tonnes of CO_2)
Never fly	1.2
Cut your annual mileage in half	0.7
Sell the second car	0.7
Buy a new car with low emissions then scrap the old one	0.5
Change to an almost entirely vegetarian diet, using mostly unprocessed wholefoods such as grains, seeds and nuts	0.5
Go vegan three days a week	0.5
Install 2-kilowatt solar PV panels	0.4
Major improvement in your home's insulation	0.4
Buy 50% secondhand clothes	0.3
Buy secondhand mobile/cell phones and ensure that three of your electronic devices are recycled	0.3
Cavity wall insulation	0.3
Cycle everywhere	0.3
New boiler if yours is more than 10 years old	0.3
Restrict yourself to one short-haul return flight a year on a carrier with a fuel-efficient fleet	0.3
Share car to work	0.3
Better controls for boiler, hot water tank and radiators	0.2
Buy a wood-burning stove	0.2
But more carefully and never throw food away	0.2
Buy only manmade fibres	0.2
Double glazing if you don't have it	0.2
Go on a day's eco-driving course, fit low-resistance tyres and check air pressure every month	0.2
Increase loft insulation, seal doors and flooring	0.2
Keep your electronic devices (e.g. phones, TVs, computers, DVD players, games machines) one year longer than you would have	0.2
Never buy processed food or ready meals	0.2
Reduce purchases by more than a quarter compared to last year	0.2
Reduce your thermostat temperature by 1 degree	0.2
Solar hot water	0.2
Always use buses instead of the train	0.1
Block direct mail, choose electronic bills and statements, buy secondhand books and share papers	0.1
Buy a new high efficiency A++ refrigerator if yours is more than 4 years old, and only use a small-screen TV	0.1
Buy an automated system to turn off appliances when not in use, get a meter that shows actual energy use	0.1
Buy ultra-low water use cisterns, new water-saving dishwasher, washing machine. Recycle old ones	0.1

Table 16.1 (*cont.*)

Action	Savings (tonnes of CO_2)
Don't ever use a car for shopping. Buy online	0.1
Focus on new fabrics made from bamboo, hemp or other cotton substitutes	0.1
Get rid of the freezer if you can, and replace your small appliances with 'eco' varieties	0.1
Heat one less room	0.1
Install and monitor a water meter. Carefully recycle all waste, compost	0.1
Install a composting toilet	0.1
Never use the tumble dryer	0.1
Only buy newspapers, magazines, books, toilet paper and copier paper made from recycled materials	0.1
Only use your washing machine and dishwasher when full to capacity and at lowest temperature	0.1
Slow-flow showers, not baths	0.1
Switch from a desktop computer to a laptop at home, and recycle the desktop	0.1
Use LED or fluorescent lights where you currently have halogen lights installed	0.1
Use showers, not baths. Install a flow-reducing aerator for the shower head	0.1
Work from home one day a week rather than commuting by car	0.1
Work from home two days a week instead of taking public transport to work	0.1
Total	**11.4**

Source: *The Guardian*, 1 September 2009, http://www.guardian.co.uk/environment/2009/sep/01/how-to-reduce-emissions-10–10. Contributed by Chris Goodall. Copyright Guardian News & Media Ltd 2009.

action on climate change include executive leadership and corporate social responsibility, shareholder pressure, future climate impacts, anticipating and influencing regulation, gaining first mover advantage and experience, and developing new markets in green technologies or carbon trading (Hoffman, 2005; Newell, 2008). A study of the UK FTSE 100 companies (producing 75% of UK emissions) found five factors as reported motivations: profit (e.g. through energy savings), competition for credibility and policy leverage, fiduciary obligation, reducing risk of climate change impacts (e.g. insurance), and ethics (Okereke, 2007).

Environmental non-governmental organisations (ENGOs) often have raising public awareness and mobilising public support for environmental issues at the core of their mission. Over the past two decades, climate change has risen up the agenda of many ENGOs, especially as the risks to ecosystems have become more evident, and more NGOs have participated in international climate negotiations and national climate politics (Corell and Betsill, 2001; Gulbrandsen and Andresen, 2004). Activities include providing information to the public and lobbying, but have recently included undertaking major scientific studies, setting standards, and participating in the implementation of climate mitigation and adaptation actions. The WWF, for example, has a major climate campaign that has included longstanding research into climate impacts on valued ecosystems, the development of standards for carbon offsets and lobbying to include forest protection in the international climate regime (Markham *et al.*, 1993; WWF, 2007; Gold Standard Foundation, 2009; Lenton *et al.*, 2009).

Humanitarian organisations have become involved in climate change because of risks to the poor and the links between disaster relief, vulnerability and the need for adaptation. Oxfam, for example, produced a series of reports on climate impacts and the need for substantial adaptation funds in the run-up to the UNFCCC COP15 negotiation in Copenhagen (Oxfam, 2009). Networks such as the Climate Action Network and 350.org mobilised thousands of protestors in the months prior to, and during, the Copenhagen conference in 2009. Several business networks – such as the World Business Council for Sustainable Development, The Climate Group and the US Climate Action Partnership – also promote action and share strategies to respond to climate change.

16.5 Summary and conclusions

This chapter set out to demonstrate ways in which people matter in the response to climate change, and that there are important insights from social science that can augment the design of local and global solutions to climate change. Important messages are that:

- even when attitudes change they may not translate into behaviour unless there are shifts in habits and norms, and unless behavioural change is enabled by changes in infrastructure and institutions;
- there are considerable emission reductions to be gained from behavioural changes, at the level of both the individual and the firm, and especially in the area of residential energy use;
- some of the large-scale solutions, such as CCS and geoengineering technologies, may face considerable barriers in terms of public acceptability.

People – as citizens or leaders of business, government and civil society organisations – can make choices to reduce emissions and support adaptation that go beyond a direct economic rationale. Culture, ethics, information and perceptions hold some of the keys to a successful response to climate change. We may fail in designing carbon markets, implementing geoengineering and low carbon technologies, and developing international agreements – unless we understand the range of possible human responses to these policies and technologies. Such responses include constraints on individual and organisational decisions, intense criticisms reflecting concerns about social justice and risks, voluntary efforts that go beyond those anticipated by the market, and broad changes in cultural norms and attitudes that promote changes in emissions. People, acting individually and collectively, can inspire and act to change their household, community, region and the world.

References

Agrawal, A. (2008). The role of local institutions in adaptation to climate change. Paper prepared for the *Social Dimensions of Climate Change*, Social Development Department. Washington, D.C.: World Bank.

Barry, J., Ellis, G. and Robinson, C. (2008). Cool rationalities and hot air: A rhetorical approach to understanding debates on renewable energy. *Global Environmental Politics*, **8**, 67–98.

Beckerman, W. and Hepburn, C. (2007). Ethics of the discount rate in the Stern Review on the Economics of Climate Change. *World Economics*, **8**, 187–210.

Boardman, B., Bright, S.-K., Ramm, K. and White, R. (2007). *Carbon Labelling: Report on Roundtable.* 3–4 May, St Anne's College, University of Oxford. Oxford: UK Energy Research Centre.

Bonini, S. M. J., Hintz, G. and Mendonca, L. T. (2008). Addressing consumer concerns about climate change. *The McKinsey Quarterly* (March).

Boykoff, M. T. (2009). We speak for the trees: media reporting on the environment. *Annual Review of Environment and Resources*, **34**, 431–57.

Carvalho, A. (2007). Ideological cultures and media discourses on scientific knowledge: re-reading news on climate change. *Public Understanding of Science*, **16**, 223–43.

CGIAR (2008). *Global Climate Change: Can Agriculture Cope?* http://www.cgiar.org/impact/global/climate.html.

Chilvers, J. and Burgess, J. (2008). Power relations: The politics of risk and procedure in nuclear waste governance. *Environment and Planning A*, **40**, 1881–1900.

Corell, E. and Betsill, M. M. (2001). A comparative look at NGO influence in international environmental negotiations: desertification and climate change. *Global Environmental Politics*, **1**, 86–107.

Darby, S. (2006). *The Effectiveness of Feedback on Energy Consumption. A Review for DEFRA of the Literature on Metering, Billing and Direct Displays.* Oxford: Environmental Change Institute, University of Oxford.

Dietz, T., Fitzgerald, A. and Shwom, R. (2005). Environmental values. *Annual Review of Environmental Resources*, **30**, 335–72.

Eakin, H. (2005). Institutional change, climate risk, and rural vulnerability: Cases from central Mexico. *World Development*, **33**, 1923–38.

Eakin, H. (2006). *Weathering Risk in Rural Mexico. Climatic, Institutional, and Economic Change.* Tucson, AZ: The University of Arizona Press.

Enkvist, P.-A., Nauclér, T. and Rosander, J. (2007). A cost curve for greenhouse gas reduction. *McKinsey Quarterly*, **1**, 34.

Eurobarometer (2008). Europeans' attitudes towards climate change. *Eurobarometer*. Special Eurobarometer 200/Wave 69.2.

FAO (2009). *FAO Profile for Climate Change.* Rome: FAO, http://www.fao.org/climatechange/en/.

Farrell, D. and Remes, J. K. (2008). How the world should invest in energy efficiency. *McKinsey Quarterly* (July).

Fischhoff, B., Lichtenstein, S., Slovic, P., Derby, S. L. and Keeney, R. L. (1981). *Acceptable Risk*. Cambridge, UK: Cambridge University Press.

Folke, C. (2006). Resilience: the emergence of a perspective for social–ecological systems analyses. *Global Environmental Change*, **16**, 253–67.

Gardner, G. T. and Stern, P. C. (2008). The short list: The most effective actions US households can take to curb climate change. *Environment: Science and Policy for Sustainable Development*, **50**, 12–24.

Geller, H., Harrington, P., Rosenfeld, A. H., Tanishima, S. and Unander, F. (2006). Polices for increasing energy efficiency: Thirty years of experience in OECD countries. *Energy Policy*, **34**, 556–73.

Gilg, A., Barr, S. and Ford, N. (2005). Green consumption or sustainable lifestyles? Identifying the sustainable consumer. *Futures*, **37**, 481–504.

Globescan (2009). *Climate Concerns Continue to Increase: Global Poll*. n.p.: Globescan/ BBC World Service.

Gold Standard Foundation (2009). *The Gold Standard: Premium Quality Carbon Credits*. Basel: Gold Standard Foundation, http://assets.panda.org/downloads/gs_overview. pdf.

Gulbrandsen, L. H. and Andresen, S. E. (2004). NGO influence in the implementation of the Kyoto Protocol: compliance, flexibility mechanisms, and sinks. *Global Environmental Politics*, **4**, 54–75.

Hindmarsh, R. and Matthews, C. (2008). Deliberative speak at the turbine face: Community engagement, wind farms, and renewable energy transitions, in Australia. *Journal of Environmental Policy & Planning*, **10**, 217–32.

Hoffman, A. J. (2005). Climate change strategy: The business logic behind voluntary green-house gas reductions. *California Management Review*, **47**, 21–46.

Hogan, L. and Thorpe, S. (2009). *Issues in Food Miles and Carbon Labelling*. ABARE Research Report 09.18. Canberra: Australian Bureau of Agricultural and Resource Economics.

Howden, S., Soussana, J.-F., Tubiello, F. N. *et al.* (2007). Adapting agriculture to climate change. *Proceedings of the National Academy of Sciences (USA)*, **104**, 19 691–96.

Huijts, N. M. A, Midden, C. J. H. and Meijnders, A. L. (2007). Social acceptance of carbon dioxide storage. *Energy Policy*, **35**, 2780–89.

IPCC (2007a). *Climate Change 2007: Mitigation of Climate Change. Contribution of Working Group III to the Fourth Assessment Report of the Intergovernmental Panel on Climate Change*, eds. B. Metz, O. R. Davidson, P. R. Bosch, R. Dave and L. A. Meyer. Cambridge, UK and New York, NY: Cambridge University Press.

IPCC (2007b). *Climate Change 2007: Impacts, Adaptation and Vulnerability. Contribution of Working Group II to the Fourth Assessment Report of the Intergovernmental Panel on Climate Change*, eds. M. Parry, O. Canziani, J. Palutikovf, P. van der Linden and C. Hanson. Cambridge, UK and New York, NY: Cambridge University Press.

Jackson, T. (ed.) (2006). *The Earthscan Reader in Sustainable Consumption*. London: Earthscan.

Klee, H. and Coles, E. (2004). The cement sustainability initiative – implementing change across a global industry. *Corporate Social Responsibility and Environmental Management*, **11**, 114–20.

Leary, N., Adejuwon, J., Barros, V., Burton, I., Kulkarni, J. and Lasco, R. (eds.) (2007). *Climate Change and Adaptation*. London: Earthscan.

Leiserowitz, A. (2007). *Public Perception, Opinion and Understanding of Climate Change – Current Patterns, Trends and Limitations*. Occasional Paper 31. UNDP.

Lenton, T., Footitt, A. and Dlugolecki, A. (2009). *Major Tipping Points in the Earth's Climate System and Consequences for the Insurance Sector*. Berlin: WWF.

Levy, D. L. and Newell, P. J. (2004). *The Business of Global Environmental Governance*. MIT Press.

Markham, A., Dudley, N. and Stolton, S. (1993). *Some Like it Hot: Climate Change, Biodiversity and the Survival of Species*. Gland: WWF International.

McKinsey (2007). How companies think about climate change: a McKinsey global survey. *McKinsey Quarterly* (February).

Newell, P. (2000). *Climate for Change: Non-state Actors and the Global Politics of the Greenhouse*. Cambridge, UK: Cambridge University Press.

Newell, P. (2008). Civil society, corporate accountability and the politics of climate change. *Global Environmental Politics*, **8**, 122–53.

Okereke, C. (2007). An exploration of motivations, drivers and barriers to carbon management: The UK FTSE 100. *European Management Journal*, **25**, 475–86.

Okereke, C., Bulkeley, H. and Schroeder, H. (2009). Conceptualizing climate governance beyond the international regime. *Global Environmental Politics*, **9**, 58–78.

Oxfam (2009). *Suffering the Science: Climate Change, People and Poverty*. Oxford: Oxfam.

Pew Global Attitudes Project (2009). *25-Nation 2009 Pew Global Attitudes Survey*. Washington, D.C.: Pew Research Center.

Pidgeon, N. F., Lorenzoni, I. and Poortinga, W. (2008). Climate change or nuclear power – No thanks! A quantitative study of public perceptions and risk framing in Britain. *Global Environmental Change*, **18**, 69–85.

Raustiala, K. (2001). Nonstate actors in the global climate regime. In *International Relations and Global Climate Change*, eds. U. Luterbacher and D. F. Sprinz. Cambridge, MA: MIT Press, pp. 95–117.

Rehan, R. and Nehdi, M. (2005). Carbon dioxide emissions and climate change: Policy implications for the cement industry. *Environmental Science and Policy*, **8**, 105–14.

Renn, O. (2006). Participatory processes for designing environmental policies. *Land Use Policy*, **23**, 34–43.

Saelen, H., Atkinson, G., Dietz, S., Helgeson, J. and Hepburn, C. J. (2008). *Risk, Inequality and Time in the Welfare Economics of Climate Change: Is the Workhorse Model Underspecified?* Economics Series Working Paper 400. Oxford: University of Oxford, Department of Economics.

Schmidhuber, J. and Tubiello, F. N. (2007). Global food security under climate change. *Proceedings of the National Academy of Sciences (USA)*, **104**, 19703–08.

Shackley, S., McLachlan, C. and Gough, C. (2005). The public perception of carbon dioxide capture and storage in the UK: Results from focus groups and a survey. *Climate Policy*, **4**, 377–98.

Singleton, G., Herzog, H. and Ansolabehere, S. (2009). Public risk perspectives on the geologic storage of carbon dioxide. *International Journal of Greenhouse Gas Control*, **3**, 100–07.

Slovic, P. (1992). Perception of risk: Reflections on the psychometric paradigm. In *Social Theories of Risk*, eds. S. Krimsky and D. Golding. New York, NY: Praeger, pp. 117–52.

Smit, B. and Skinner, M. W. (2002). Adaptation options in agriculture to climate change: A typology. *Mitigation and Adaptation Strategies for Global Change*, **7**, 85–114.

Smit, B. and Wandel, J. (2006). Adaptation, adaptive capacity and vulnerability. *Global Environmental Change*, **16**, 282–92.

Smith, J. B., Klein, R. J. T. and Huq, S. (eds.) (2003). *Climate Change, Adaptive Capacity and Development*. London: Imperial College Press.

Stern, N., Peters, S., Bakhshi, V. *et al.* (2006). *Stern Review: The Economics of Climate Change*. London: HM Treasury.

Stern, P. C. (2000). Toward a coherent theory of environmentally significant behavior. *Journal of Social Issues*, **56**, 407–24.

Walker, G., Devine-Wright, P., Hunter, S., High, H. and Evans, B. (2010). Trust and community: Exploring the meanings, contexts and dynamics of community renewable energy. *Energy Policy* **38** (6), 2655–63.

Wong-Parodi, G. and Ray, I. (2009). Community perceptions of carbon sequestration: Insights from California. *Environmental Research Letters*, **4**, 034002.

World Bank (2010). *World Development Report 2010: Development and Climate Change*. Washington, D.C.: The World Bank.

WWF (2007). *Saving the World's Natural Wonders from Climate Change*. Gland: WWF.

17

The human–Earth relationship:
past, present and future

'Nature takes no account of even the most reasonable of human excuses'[1]

17.1 Introduction

Given that we humans are fundamentally dependent upon Earth and its natural resources
for our very survival, we have always had – and will always have – an intimate relationship
with the planet. The perception that different societies have had of this human–Earth relationship, however, has changed through time. In the first instance, it was addressed entirely
through religion – the belief that (a) superpower(s) control(s) the destiny of the universe.
Thus, it was accepted that it was the god(s) that controlled the seasons, rainfall, availability
of food, and so on.

As scientific understanding has advanced and the number of humans and their activities
have increased, both our perception and the nature of the human–Earth relationship have
changed. A milestone in the evolution of this relationship was reached recently with the
recognition that the Earth functions as a self-regulating system – where not only physical,
chemical and biological, but *also* human processes interact to create the environmental
conditions experienced (Moore *et al.*, 2002). This new understanding of the human role in
the functioning of the Earth System, itself, implies an obligation for active management of
the human–Earth relationship at the global level.

Such management requires the development and use of new methods and approaches,
both with respect to scientific study and the way in which we use science in policy decision-
making. The Copenhagen Accord (Box 13.2) negotiated at the COP15 under the UNFCCC
states a political goal of containing human-induced global warming to within 2 °C. Such
a goal, in fact, recognises the need for new management practices in the human–Earth
relationship. In essence, controlling human emissions of greenhouse gases and, thereby,
human-induced climate change represents the development of mechanisms for sharing global natural resources amongst communities distributed in individual nation-states

---

[1] Joseph Wood Krutch (1893–1970), American author.

Climate Change: Global Risks, Challenges and Decisions, Katherine Richardson, Will Steffen and Diana Liverman *et al.* Published
by Cambridge University Press. © Katherine Richardson, Will Steffen and Diana Liverman 2011.

(Chapter 13). Scientists have quantified the magnitude of a global resource – the atmospheric reservoir for human greenhouse gas 'waste' that can be utilised while still allowing society to limit global temperature increases to under 2 °C (Chapter 8). Now, the political focus is on allocating rights to that resource.

While political and public attention at present focus on climate change and allocating rights to the release of greenhouse gases, current scientific understanding informs us that there are many other aspects of the Earth System in addition to the climate system that are heavily impacted by human activities. Thus, the mechanisms that society develops to deal with the challenge of climate change may be seen as precursors to the kinds of mechanisms that will have to be put in place to manage other aspects of the future human–Earth relationship as well. Currently, the world population is over 6.5 billion. By 2050, it is expected to be around 9 billion. Developing mechanisms to share the Earth's resources among peoples and societies is, perhaps, the greatest challenge humanity faces in the twenty-first century.

17.2 Societal response to advances in scientific understanding

Numerous mythologies and legends bear witness to a wide range of explanations for various natural phenomena embraced by the belief systems our ancestors developed. As a result of advances in scientific understanding of these phenomena, we can now offer an alternative, more evidence-based explanatory framework for the origins and dynamics of the natural world. However, the nexus between religion and science sometimes continues to be an intense battle ground. Galileo and Darwin, were they here today, would certainly attest to the fact that the substitution of scientific understanding for religious beliefs in society can be far from an easy process. The concepts they espoused – that is, that the sun and not the Earth is the centre of the universe, and that human beings are closely related to apes – were not ones that rested well with the dominant religious beliefs of society during their lifetimes (Box 17.1).

Box 17.1
When scientific discoveries threaten human identity

KATHERINE RICHARDSON, HANNE STRAGER AND MINIK ROSING

Seldom do new scientific insights evoke the attention and the emotions of the general public in the manner that climate change has done. Possibly the last time this happened to a similar extent was a century and a half ago, in 1859, when Darwin published his book *On the Origin of Species*. A possible and compelling explanation for why such discoveries provoke an intense and emotive response in society as a whole is that they challenge the prevailing self-identity of humans as a species as well as our perception of the relationship between humans

Box 17.1 (*cont.*)

and the planet. For more than anything else, Darwin's book changed the way we think about humankind's relationship to nature.

Darwin (1859) argued that all species on Earth share a common ancestry. The significance of this is that even humans are 'only' animals. Not only are we related to apes, we *are* apes – apes that share an ancestry with all other living beings down to the most modest bacteria and amoeba. The realisation that humans are a part of nature, and not above or outside of it, dealt a death blow to the common perception of humans having a special God-given position above all other living things. The effect this realisation had on the society of that time was immense and instantaneous. Until Darwin published his ideas about evolution, nature was one thing: wild, untamed, unruly, unorganised and unpredictable. Society, culture, religion and politics were quite another thing. They were two different realms that interacted only in the sense that God had commanded humans to tame, control and dominate nature: 'Be fruitful and increase in number; fill the earth and subdue it. Rule over the fish of the sea and the birds of the air and over every living creature that moves on the ground' (Genesis 1:28).

Many of Darwin's contemporaries found it impossible to accept that a human was just a lowly animal. The church was outraged and claimed that if humans were just animals, then there would be no ethical or moral values; civilisation would fall apart. Gradually, however, Darwin's views were accepted – first by fellow scientists and, later, even by the Church. By the time Darwin died, the Church had embraced him and he is buried in Westminster Abbey, next to Newton.

The new understanding of the relationship between humankind and nature implied by Darwin's theory of evolution also challenged one of the fundamental economic pillars of the time: slavery. In the mid-nineteenth century, many believed that economic stability and growth were not possible without access to slave labour. If Darwin's understanding of the relationship between species was correct, then the supremacy of one race over another, and thus the 'moral justification' for slavery, no longer existed. Clearly, the preservers of the existing economic structure would feel threatened by Darwin's analyses and it is entirely plausible that part of society's resistance to embrace Darwin's theories was based on a subconscious fear they would undermine the existing economic structures. While there is not necessarily a direct link between abolition of slavery and Darwin's studies of the position of humans in nature, conventional history tells us that most societies ceased to embrace slavery as a tool for economic advancement during the last half of the nineteenth century. Ironically, slave labour was, to a large extent, replaced by mechanisation powered through the combustion of fossil fuels.

Today, the burning of fossil fuels has become accepted by many as a prerequisite for economic stability and growth. Now, science is confronting us with the insight that the burning of these fossil fuels is altering the functioning of Earth's climate system. In essence, humans are threatening the viability of their own life support system. The message from science that we impact Earth's climate is provoking a similar kind of emotional response in society as that which followed the publication of Darwin's *On the Origin of Species*. Scientific advances documenting human-induced climate change are, once again,

Figure 17.1 'A Venerable Orang-outang', a caricature from the 22 March 1871 issue of *The Hornet* magazine.

challenging the prevailing understanding of the human–Earth relationship. It is hard for many people to accept that human activity is significant enough to impact the functioning of the Earth as a whole and, once again, the dogmas of what is essential for economic well-being are put under threat.

Religious beliefs cannot be substantiated with concrete observation, experimentation and documentation. Thus, they cannot be 'proven' in the sense that contemporary natural science uses the word. In this sense, 'religion' is intensely personal; that is, a set of beliefs to which one subscribes. These beliefs can be shared with others (e.g. Christianity, Judaism, Islam) or personal. Scientific understanding, on the other hand, develops through the collection of evidence or documentation (usually data). This evidence can be shared and analysed, and conclusions drawn as to the most likely explanation for the collected evidence. In most cases, and certainly that of climate change (Figure 17.2), it takes some time for a new concept or idea to gain support in the scientific community, and for the majority of scientists to become convinced of its validity.

Scientists present their evidence for one another, along with the interpretation they have developed for the mechanism(s) leading to the results they have obtained. Other scientists, then, are invited to support or question the interpretation using the same evidence or by contributing additional evidence. Scientific understanding can only advance through the process whereby scientists question one another and their interpretations of their collected evidence. Thus, in a sense, scientists represent a group of 'organised sceptics'. They are organised in that they are all operating from the same starting point – the advances in understanding made by their predecessors – and they are sceptics in the sense that every new idea or concept introduced must be examined from every conceivable angle before it gains general acceptance and is added to the existing knowledge

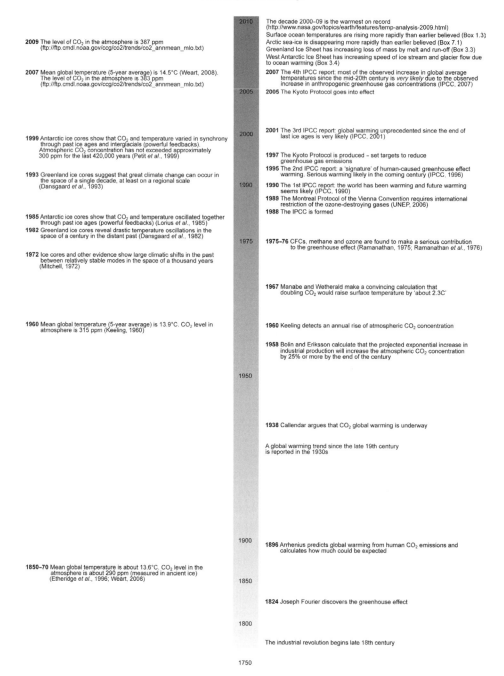

Figure 17.2 Timeline showing the development in scientific understanding and documentation of anthropogenic climate change.

base. The 'debate' that takes place between scientists when new ideas are brought forward is part of any healthy scientific process, but it is often misunderstood by the media and the general public.

Coverage of the scientific 'debate' concerning climate change – that is, the process whereby scientists attempt to interpret the newest data relating to the climate system – has left the impression among many in the general public that there is still great scientific uncertainty with respect to whether or not human activities are influencing the climate. For example, 66% of those interviewed in a survey of 1005 adult Americans (Associated Press/ Stanford University Poll, http://www.pollingreport.com/enviro.htm) conducted between 17 and 29 November 2009, answered 'a lot of disagreement among scientists' when asked the question: 'Do you think most scientists agree with one another about whether or not global warming is happening, or do you think there is a lot of disagreement among scientists on this issue?'. Only 31% replied they thought there was a lot of agreement among scientists.

In fact, studies show that when scientists are asked whether they believe that human activities are influencing the climate, the answer is overwhelmingly 'yes'. The results of one such survey (Kendall Zimmerman, 2008), conducted amongst a large number of scientists who classify themselves as 'earth scientists', are shown in Figure 17.3. In this study, 10 257 scientists were approached by letter and invited to respond; 3146 did so. The responding scientists were subdivided into groups defined by how active they

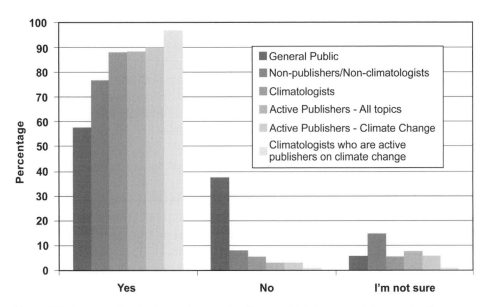

Figure 17.3 Response distribution to the question 'Do you think human activity is a significant contributing factor in changing mean global temperatures?' with 3146 individuals responding. Source: Doran and Kendall Zimmerman (2009).

were as a researcher (i.e. through publication rate) and how closely their research field related to climate change. In the group (77 individuals) defined as those who are actively publishing and where at least half of their recent scientific publications have dealt with climate change, 75 answered 'yes', 1 answered 'no' and 1 answered 'I don't know' to the question 'Do you think human activity is a significant contributing factor in changing mean global temperatures?'.

The study also shows that as the level of research specialisation in climate change decreases, the positive response rate to the question also decreases. Thus, amongst scientists whose research deals directly with climate change, the agreement that human activities are impacting climate is greatest. Taking the entire population of scientists responding to the survey, 82% answered 'yes' to the question. The less specialised the scientists were in climate change research, the more they answered 'don't know' to the question. In all of the groups, however, less than 10% of the respondents answered 'no'. This is in sharp contrast to data from the general public where almost 40% answered 'no' to a similar question. Media coverage (Box 16.1) of climate change 'debate' has quite probably contributed to the general public being less inclined to embrace the conviction that human activities influence the climate system than scientists.

However, for obvious reasons, the balance between religious and scientific input in the formation of personal beliefs differs between the public at large and the subset of the population trained as scientists. It seems unlikely, for example, despite a century and a half having passed since the discovery of evolution, that a survey documenting the percentage of individuals believing in evolution would yield the same results for the general public and the scientific communities. In the same manner, the degree of acceptance of the idea that human activities are influencing the climate system will probably never be the same in the two communities. The very idea that human activities can be of a magnitude that can influence the functioning of global systems is very difficult for many people to accept (Box 17.1). Nevertheless, leaders of the major organised religions agree that human-induced climate change should be combated (http://papalvisit.patriarchate.org/press/articles.php?id=99) and religious groups are lobbying politicians to agree on CO_2 reductions (World Council of Churches, 2009). Religious leaders justify their lobbying for reductions in human greenhouse gas emissions with the argument that humans have a responsibility to take care of 'God's creation'.

Given the nature of the scientific process, where understanding is advanced only when scientists continue to ask questions, it is exceedingly unusual to get 100% agreement among scientists in the interpretation of available evidence or documentation. The lack of 100% agreement among climate scientists is often used in the media and among the general public as an argument for not initiating action to minimise human influence of the climate system. There are several points that should be borne in mind when considering this argument, however.

• Total 100% agreement among scientists is virtually never achievable and, indeed, is actually undesirable in a healthy scientific process.

- There is a very high degree of agreement among active climate change scientists with respect to human activities being a major contributor to global warming (Figure 17.3).
- In no aspect of societal decision-making has a requirement of 100% agreement among experts in the field ever been a prerequisite to the establishment of regulation. There is, for example, certainly not 100% agreement among traffic researchers with respect to 'safe' speed limits. Speed limits are, nonetheless, established.

Ultimately, making societal decisions about responding to the challenge of human-induced climate change is not a question of waiting for 100% agreement among scientists. It is a question of whether or not societal leaders dare to take the chance that the vast majority of climate change scientists are wrong when they conclude that humans are affecting the climate system and that there is a significant risk of a global temperature increase of up to around 6 °C within this century.

17.3 Managing the human–Earth relationship: historical perspectives

Since Plato's classic work, *The Republic*, there has been the tacit understanding in most modern societies that the decisions that will be most beneficial for society are those which are based on the highest level of knowledge ('wisdom' in the words of Plato). Indeed, history demonstrates that all societal regulation of the human–Earth relationship has first been initiated after the knowledge that these activities were causing harm/change in the environment was established. The most basic direct interaction between humans and the Earth relates to food and water supply.

When humans first appeared on Earth, they were dependent upon hunting and gathering for their sustenance. Much later, about 10 000 years ago, they learned to domesticate agricultural animals and cultivate the land; activities that ultimately led to the development of modern agriculture. It seems unlikely these early hunters or farmers ever even considered the possibility that it one day would be necessary by society to regulate these activities. Only as the human population increased and the magnitude of these activities developed did the environmental implications of their uncontrolled development become documented, and controls on these activities were put in place in most regions of the world.

A similar history can be told for waste. In the first instance, humankind dealt with its waste by simple deposition at a location which ensured that the waste did not obviously impact on humans, i.e. through water supply. As humans became more numerous, and the indirect effects of many waste products on humans became understood, regulation of waste disposal was established. The earliest focus here was on the terrestrial environment, as it was either assumed that the atmosphere and the oceans were so large that they could effectively dilute any human-generated waste that they received; or that waste in the ocean or atmosphere could impact human societies. It was only within the past two generations that scientists documented the finite capacity of the oceans and the atmosphere to absorb human-generated waste, and the impact on humans of waste in these components of the

Earth System. This new scientific understanding led to regulation aimed at controlling the use of ocean and atmosphere as waste recipients.

Worth noting here, in relation to the current climate change debate in society, is that the need for control of various waste products from society in the natural environment (pollution) was/is not always readily embraced by the non-scientific community. Resistance, typically, was and is greatest from the segment of society with an economic interest in continuing the polluting practice. In the case of human-generated greenhouse gas emissions and climate change, we are all polluters.

The control systems society currently has in place to manage human impact on the environment are directed, for the most part, towards a specific compartment of the Earth System (e.g. land, sea, air) and are established at the local, national or regional (e.g. the European Union) level. The recognition that human activities can impact the functioning of the Earth System as a whole requires control of aspects of the human–Earth relationship at the planetary scale. Thus, we are entering a new phase in our relationship with the planet. To date, the only example of control at the global – that is, Earth System – level is the Montreal Protocol on Substances That Deplete the Ozone Layer – an international treaty designed to reduce anthropogenic emissions of chlorofluorocarbon (CFC) gases, which degrade the Earth's ozone layer. This treaty came into effect in 1989 and has been highly effective in its mission; most scientists expect the ozone layer to recover. As noted in Chapter 13, however, dealing with climate change on the global political scene is rather more complicated than dealing with the emissions of CFC gases, and new mechanisms for management of the human–Earth relationship at the global level are needed.

17.4 Managing the human–Earth relationship in an Earth System context

What does managing the human–Earth relationship at the global level entail? First, it demands recognition of the fact that human activities influence global processes and Earth System functioning; that is, element cycling and material flows at the global level. Figure 17.4a shows the development of different human activities in the period 1750–2000 (where establishment of McDonalds restaurants, for example, is used as a proxy for globalisation). In all 12 examples, an exponential expansion of the activity occurred during the second half of the twentieth century. Figure 17.4b illustrates on the same timeframe changes in variables representing 12 different aspects of the Earth System. Striking is the fact that the shape of the curves describing change in the Earth System generally mirrors the shape of the curves describing change in the human system.

Demonstrating a correlation between an increase in human activities and changes in the Earth System does not, of course, prove a cause-and-effect relationship between the two. However, when the underlying mechanisms resulting in the changes in the Earth System are examined, causality is shown (for an overview see Steffen *et al.*, 2004). That the human footprint has become such an important factor in controlling Earth System functioning has led some scientists to argue that we have moved out of the geological time period designated as the Holocene (Figure 1.12) and have entered into the 'Anthropocene' (Box 17.2).

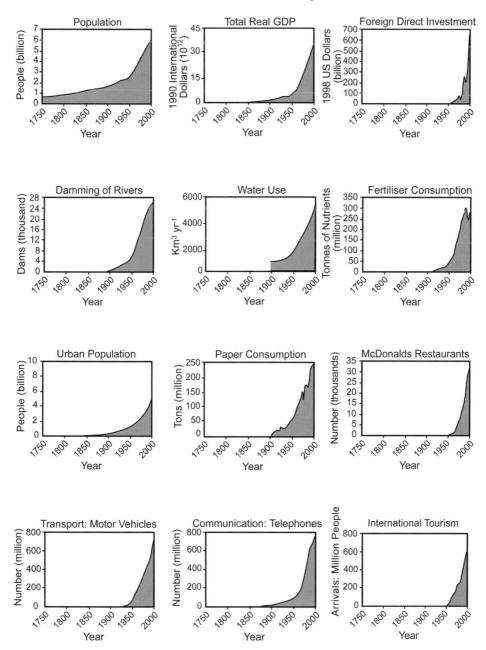

Figure 17.4a The increasing rates of change in human activity since the beginning of the Industrial Revolution. Significant increases in rates of change occur around the1950s in each case and illustrate how the past 50 years have been a period of dramatic and unprecedented change in human history. Source: Steffen *et al.* (2004).

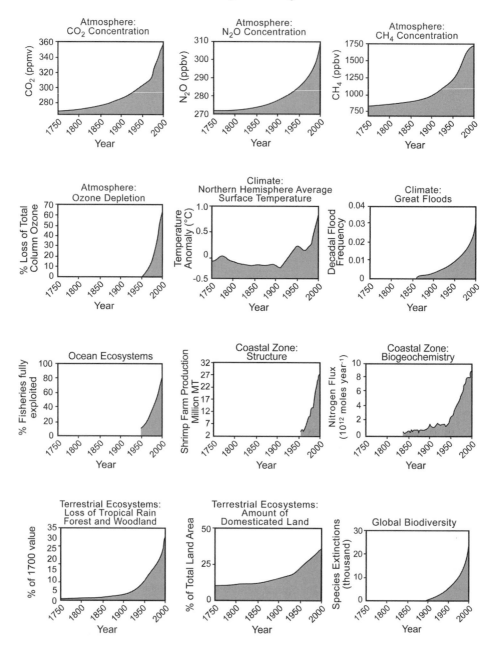

Figure 17.4b Global-scale changes in the Earth System as a result of the dramatic increase in human activity. Source: Steffen *et al.* (2004). With kind permission of Springer Science+Business Media.

Box 17.2

The Anthropocene: are humans creating a new geological era?

WILL STEFFEN, PAUL CRUTZEN, JAN ZALASIEWICZ AND
MARK WILLIAMS

Geologists have long defined major periods of the Earth's history to help them with their work. The definitions are based on major shifts in the Earth's environment, as evidenced by observed changes in stratigraphic records. In many cases, such shifts are associated with massive losses of biodiversity and the emergence of new life forms – shifts which are preserved in the fossil record (Zalasiewicz *et al.*, 2010). A well-known example is the transition from the Cretaceous period to the Cenozoic era about 65 million years ago, a shift that spelled the end of the dinosaurs and ushered in the dominance of mammals.

About a decade ago, a new term came into informal usage in the scientific community to describe an emerging transition in Earth's history, but it did not come from the geological community. Nobel Laureate Paul Crutzen coined the term 'Anthropocene' to describe the rapid and profound changes that are sweeping across the planet today (Crutzen and Stoermer, 2000; Crutzen, 2002). The term was used in juxtaposition to the current formally recognised geological epoch the Holocene, which refers to the most recent, rather long interglacial period that began about 12 000 years ago. In many ways, the Holocene has been a remarkably stable environmental period (certainly more stable than the ice age environments), allowing humanity to develop agriculture, villages and settlements, and eventually more complex societies and civilisations. It has been an accommodating, and generally pleasant, global environmental state for humanity. But we are now rapidly leaving the Holocene conditions behind.

Contemporary climate change, driven primarily by the rapid increase of atmospheric greenhouse gases to levels well beyond their Holocene range and much higher than at any other time during the existence of *Homo sapiens*, is the most well-known piece of evidence for the emergence of the Anthropocene (IPCC, 2007). There are, however, also many other changes afoot, some even more striking (Steffen *et al.* 2004, and references therein). Human activities now fix more nitrogen from the atmosphere than all natural terrestrial processes combined. Most of the world's large rivers are dammed, and about half of the global runoff from the terrestrial surface is intercepted and stored before it can reach the ocean. Since 1950, the world's ecosystems – terrestrial and marine – have been changed more extensively and rapidly than in any other time in human history. And, as for several other transitions in Earth's history, there is a biodiversity signal. The Earth is now in the midst of its sixth great extinction event but the first caused by a biological species. The drivers for these sweeping changes to the global environment are now well understood to arise from the numbers and activities of humans, beginning with the industrial revolution and accelerating through the twentieth century (McNeill, 2000). Together, these changes demonstrate unequivocally that the Earth has moved quickly and decisively out of its accommodating Holocene environmental envelope.

Will the Anthropocene be formally recognised as a new era or epoch in Earth's history? The geological community has an exhaustive and thorough process for evaluating the evidence and determining whether there has indeed been a well-defined transition. That process is underway (Zalasiewicz *et al.*, 2010). An Anthropocene Working Group has been formed as

Box 17.2 (*cont.*)

part of the Subcommission on Quaternary Stratigraphy; the latter is the body that deals with the definition of formal units of the current ice ages. But the Subcommission on Quaternary Stratigraphy is part of the International Commission on Stratigraphy, which then has to report to the International Union of Geological Sciences. All of these bodies will have to agree that the Earth has formally moved into the Anthropocene before the term can become official.

The outcome of the geological deliberations notwithstanding, the term Anthropocene will no doubt continue to be used in the Earth System science community. If Anthropocene is formally recognised by the geological community, will it become a household term for the general public? The terms Pleistocene and Jurassic have achieved that status, thanks to a Hollywood film or two and plans to bring sabre-tooth tigers and woolly mammoths back into existence from fossil DNA. Perhaps when the full implications of the Anthropocene have become clearer as the twenty-first century progresses, humans living near the end of the century may forget about Jurassic Park and look fondly back to the Holocene Heaven that they so rapidly (and officially) left behind.

17.4.1 Climate change is only the tip of the iceberg

Current societal discussions concerning climate change and mechanisms to reduce anthropogenic greenhouse gas emissions (Chapters 11–13) consider only control of the global impact of human activities on one small part of the climate system. Many take as their starting point the fact that a warmer climate would be detrimental to human societies and, therefore, the objective becomes the most cost-effective manner in which to prevent global warming. Here, the idea of a form of geoengineering (Section 7.4, and discussion below) – whereby it is proposed that humans should deliberately alter another part of the climate system (by changing albedo; Box 1.2) so as to counteract the warming created by anthropogenic greenhouse gas emissions – is often invoked as an alternative to greenhouse gas emission reduction.

Using these types of geoengineering schemes as an alternative to CO_2 emission reductions, however, does not respect the fact that, in addition to global warming, there are other detrimental Earth System consequences of increased CO_2 concentrations in the atmosphere; for example, ocean acidification (Section 2.3). Thus, a strategy for combating climate change that relies only on geoengineering to combat global warming does not respect the Earth's functioning as a system. It is well established that the most successful healthcare management requires consideration of the patient as a whole, and not only treatment of the symptoms of the ailment of immediate concern, but also the root cause of the ailment. In the same manner, successful management of the future human–Earth relationship also demands respect and consideration of the Earth System as a whole. Treating symptoms (i.e. global warming) alone will not, in the end, lead to sustainable management of the human–Earth relationship.

Anthropogenic pressures on the Earth System have reached a scale where abrupt environmental change can no longer be excluded (Chapter 7). This is a relatively new insight.

For decades, we have lived in the predominant belief that environmental change occurs in an incremental, linear and predictable fashion. All current control/regulation of the human–Earth relationship takes this premise as its starting point. Take, as an example, fisheries management. An underlying assumption here is that, when the stock size of a fish species falls below a certain level, a cessation of, or more likely a reduction in (given lobbying pressure from the fishers), the fishery will automatically result in the recovery of the stock. Growing evidence indicates that this may be the exception not the rule – not only for ecosystems, but also for the Earth System as a whole.

It now appears that long periods of gradual change can eventually push large subsystems of the Earth System past thresholds, resulting in abrupt and potentially deleterious or even disastrous changes for significant portions of human societies. Future regulation or control of the human–Earth relationship must, therefore, consider the impact of human activities at the global level. As the UN negotiations concerning climate change and a reduction of anthropogenic greenhouse gases have clearly demonstrated, allocation of rights to global resources in the absence of a global governance system is a huge challenge (Chapter 13).

In addition to bringing with it a responsibility for society to monitor and manage the human–Earth relationship at the global level, the growing understanding of the functioning of the Earth System also potentially gives us new possibilities for using science in societal decision-making. Until now, it has only been possible to use science to document that a detrimental change in the environment or Earth System has taken place. Understanding how the Earth System works, the interactions within it, and how human perturbation may give rise to abrupt and dramatic changes in the system, opens the possibility that we may be able to predict when the risk of large-scale change in the environment is high and, thus, act to avoid such situations. One proposal for how understanding of Earth System processes could provide a framework for future management of the human–Earth relationship is described in Box 17.3.

Box 17.3
Planetary boundaries

JOHAN ROCKSTRÖM

Evidence from global environmental change research suggests that humanity no longer can exclude the possibility of crossing hardwired thresholds at the planetary level. This may threaten the self-regulating capacity of the planet to remain in the stable and favourable Holocene state in which human civilisations and societies have developed during the past 10 000 years. Compared with the past 200 000 years or so that we as fully fledged humans have existed on planet Earth, this Holocene state has been extraordinarily stable from an environmental perspective, providing humanity with the precondition for human development as we know it, from the rise of agriculture to the modern industrial societies of today.

Box 17.3 (*cont.*)

Combining these insights from resilience research (the existence of multiple stable states separated by thresholds in ecosystems across scales) with evidence of anthropogenic pressures at the planetary scale, fundamentally changes the agenda on sustainable development – from a focus on minimising environmental impacts of human development, to human development within the safe operating space of a desirable stability domain for planet Earth.

The Planetary Boundaries framework (Rockström *et al.*, 2009a, 2009b) is an attempt to respond to these challenges by providing a new approach to global sustainability. We identify and quantify planetary boundaries for key Earth System processes associated with the risk of undesirable nonlinear change. Within the boundary values we expect that humanity can operate safely in the long term. Transgressing one or more planetary boundaries may be deleterious or even catastrophic due to the risk of crossing thresholds that are associated with a high risk of triggering nonlinear environmental change within continental- to planetary-scale systems. In the first analytical attempt to apply the framework, nine planetary boundaries were identified and, drawing upon current scientific understanding, quantifications were proposed for seven of them:

- climate change (CO_2 concentration in the atmosphere <350 p.p.m. and/or a maximum change of +1 W m^{-2} in radiative forcing),
- ocean acidification (mean surface seawater saturation state with respect to aragonite ≥80% of pre-industrial levels),
- stratospheric ozone (<5% reduction in O_3 concentration from the pre-industrial level of 290 Dobson units),
- biogeochemical nitrogen (N) cycle (limit industrial and agricultural fixation of N_2 to 35 Tg N yr^{-1}) and phosphorus (P) cycle (annual P inflow to oceans not to exceed 10 times the natural background weathering of P),
- global freshwater use (<4000 km^3 yr^{-1} of consumptive use of runoff resources); land system change (<15% of the ice-free land surface under crop land),
- rate of loss of biological diversity (annual rate of <10 extinctions per million species).

The two additional planetary boundaries for which the group of authors was not yet able to determine a boundary level are chemical pollution and atmospheric aerosol loading

- land-use change (limit the terrestrial land area under permanent crop land to a maximum of 15%, compared to the current ~12%).

This first estimation suggests that humanity has already transgressed three planetary boundaries (Figure 17.5): climate change, biodiversity loss and changes to the global nitrogen cycle.

Planetary boundaries are interdependent, because transgressing one may shift the position of other boundaries or cause them to be transgressed. The social impacts of transgressing boundaries will be a function of the social–ecological resilience of the affected societies.

The first planetary boundaries analysis primarily presents the conceptual framework. Improving the boundary definitions and quantifications, as well as the dynamic relations between boundaries, will require major advancements in Earth System and resilience science. The proposed concept of 'planetary boundaries' lays the groundwork for shifting our approach

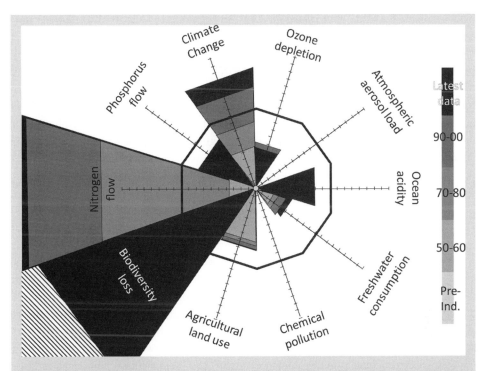

Figure 17.5 Attempt to present the temporal evolution of the proposed planetary boundaries, from a pre-industrial reference point (lightest grey) to the latest accessible data. For several of the boundaries data for early time-steps are lacking. The 'safe' space is defined within the nonagon where the nonagon edge corresponds to the proposed boundary values.

to governance and management – away from the essentially sectoral analyses of limits to growth aimed at minimising negative externalities, and towards the estimation of the safe space for human development. Planetary boundaries define, as it were, the boundaries of the 'planetary playing field' for humanity if we want to be sure of avoiding major human-induced environmental change on a global scale.

17.4.2 Geoengineering

Humankind has unintentionally been altering element cycles and material flows in the Earth System at a global level for at least the past century. Proposals for using geoengineering as a response to the climate change challenge are based on recognition that humans can and do alter global systems and processes, and argue that humankind should *deliberately* alter aspects of the climate system in order to mitigate the cause or reduce the effects of climate change. The idea of deliberately controlling Earth System processes is difficult for many people to accept, and geoengineering is a politically sensitive topic (Chapter 13). Possibly because of its political sensitivity, relatively limited research funding has, until now, been made available for study of the feasibility and the potential side-effects of the various geo-engineering schemes that have been suggested.

Nevertheless, the UK's Royal Society, one of the most respected scientific bodies in the world, issued in 2009 a policy document (The Royal Society, 2009) acknowledging the fact that it may be necessary to employ geoengineering as a supplement to reduction of greenhouse gas emissions in order for society to stabilise the climate system at a level that allows the 2 °C guardrail to be respected. They also called for further research and evaluation of the proposed schemes. The various types of geoengineering proposals (Table 17.1) can be divided into two basic types:

- carbon dioxide removal (CDR) from the atmosphere, whereby human manipulation is on the global carbon cycle,
- solar radiation management (SRM) schemes, whereby human manipulation of albedo reflects some small percentage of the sun's light/heat back into space.

A potential advantage with CDR proposals is that they directly address the root cause of human-induced global warming; that is, a perturbation of the global carbon cycle through the increased concentration of CO_2 in the atmosphere. It often comes as a surprise to people that reforestation schemes can be classified as geoengineering but, in fact, they represent an orchestrated attempt to reduce the human perturbation of the global carbon cycle. As noted in Chapter 4, however, even if it were possible to remove an amount of CO_2 from the atmosphere equivalent to the amount of 'extra' CO_2 humans have introduced into the atmosphere, the distribution of carbon in the various Earth compartments – and, therefore, the global carbon cycle – would not be returned to its pristine state. Thus, reforestation schemes are part of the solution in terms of reducing the human impact on the global carbon cycle but cannot be regarded as a panacea.

Proposed SRM schemes would influence the climate system more quickly that CDR schemes, which is why, in Chapter 7, we have included a discussion of if and how such schemes might potentially be applied in order to avoid the activation of 'tipping points', i.e. the crossing of thresholds in the Earth System. Again, however, it is important to bear in mind that SRM schemes address only a symptom (warming), and not the root cause, of human perturbation of the climate system.

Carbon capture and storage (CCS) schemes represent a subset of the CDR schemes, and have been proposed as a mechanism to allow the continued use of fossil fuels as the primary source of energy for human activities while enabling a decrease in anthropogenic CO_2 emissions. The idea behind such schemes is that CO_2 generated from energy production based on coal is chemically immobilised ('captured') and then stored elsewhere than in the atmosphere (for an overview see IPCC, 2005). Thus, CCS schemes combine several of the geoengineering tools identified in Table 17.1. While the feasibility of such schemes has yet to be demonstrated in large-scale facilities, there is much interest in them, not least of which because some studies indicate that they – in concert with measures aimed at explicitly reducing emissions – potentially could be an important component of regional strategies for meeting emissions reduction targets (Viebahn *et al.*, 2007).

Table 17.1. *Geoengineering schemes under discussion*

Scheme	Description
Carbon burial	Long-term physical storage of atmospheric CO_2 under pressure, confined below the Earth's surface within selected structures such as disused aquifers (Keith and Dowlatabadi, 1992; Stephens and Keith, 2008). Related approaches include deep-ocean carbon sequestration based on hydrate formation. Both have been debated since the early 1990s (Keith and Dowlatabadi, 1992).
Geochemical carbon capture	Chemical transformation of carbon in CO_2 gas to either the dissolved phase (that is, to bicarbonate ions in sea water) or the solid phase (carbonation) (Stephens and Keith, 2008) in the ocean or in brines, analogous to weathering of minerals that occur in nature. The method was first detailed in the early 1990s (Keith and Dowlatabadi, 1992).
Atmospheric carbon capture	Direct capture of CO_2 in air masses by using some form of wind scrubbing with a chemical absorbent. The CO_2 is bound only lightly so that it can subsequently be released and transformed chemically before final storage (Broecker, 2007). Proposed schemes have advocated using medium-sized towers to carry out the wind scrubbing, or using the wind fields around turbines (Broecker, 2007). This idea has been around since the late 1990s (Broecker, 2007).
Ocean fertilisation	Continuous fertilisation, over decades, of ocean waters that have a perennial excess of plant nutrients, in order to boost phytoplankton productivity and consequently increase the sequestration of atmospheric CO_2 into deep water (Boyd, 2008). Similar potential approaches include nitrogen fertilisation of coastal waters (proposed in the late 1990s) or purposeful mixing of deep, nutrient-rich waters into the surface ocean (proposed in 2007) (Lovelock and Rapley, 2007) in the low-latitude ocean. Iron fertilisation has been discussed since the early 1990s (Keith and Dowlatabadi, 1992), enhancing biological uptake of CO_2 from the atmosphere. In terrestrial systems, this can be achieved by increasing the area covered with foliage – especially with forest – or through changes in land-use practice. To stimulate biological uptake of CO_2 in the oceans, 'ocean fertilisation' is proposed.
Stratospheric aerosols	Injection of sulfur particles into the upper stratosphere – using balloons or projectiles – which are there to form aerosols (Crutzen, 2006; Caldeira, 2008). The aerosols alter the Earth's albedo and reflect a proportion of incoming sunlight back into space, mimicking the effect of a volcanic eruption (Crutzen, 2006). This approach has been discussed since the early 1990s (Keith and Dowlatabadi, 1992).

Table 17.1 (*cont.*)

Scheme	Description
Cloud whitening	Spraying of small seawater droplets from many wind-driven vessels into the turbulent boundary layer underlying marine clouds. The scheme is based on observations of the cumulative impact of ship exhausts in busy shipping lanes (Latham *et al.*, 2008). The droplets are thought to increase the reflectance or albedo in existing clouds. This idea was first communicated recently (Latham *et al.*, 2008).
Sunshades in space	Launch of a very large number of sunshades, which will orbit the planet and redirect incoming sunlight in space (Keith and Dowlatabadi, 1992; Keith, 2001). The scheme was first mentioned in the early 1990s (Keith and Dowlatabadi, 1992).

Source: modified from Boyd (2008).

Carbon capture directly from the atmosphere is currently more expensive than capture carried out in conjunction with combustion or production (e.g. cement) facilities. However, this type of capture offers, on the longer term, a method to decouple CCS from energy infrastructure (Keith, 2009) and, ultimately, a mechanism by which atmospheric CO_2 concentrations could be directly reduced. The existing capabilities to capture CO_2 almost certainly give these technologies a role to play in mitigation of anthropogenic CO_2 emissions. How large this role will be remains to be seen. However, as in the case of reforestation, CCS and/or other forms of geoengineering, on their own, will not be enough to address the challenges of human-induced climate change.

17.5 Summary and conclusions

Human beings in their modern form emerged on Earth sometime around 200 000 years ago. Assuming a generation time of about 25 years, this means that until now, there have been approximately 8000 generations of our species. All of those generations have been dependent upon the Earth System for their survival. Approximately six to eight generations ago, humans exchanged animal power with machine power and the industrial revolution was underway. About four generations ago, the motor car was developed. Largely as a result of these two milestones in human innovation, combined with the exponential growth of numbers of our species recorded during the past century, the impact of human activities on the Earth System has become increasingly obvious over the past two generations (Figure 17.4b). This book deals with human impact on the climate system, just one of the many subsystems in the Earth System that are affected by human activities.

After human impact on the Earth's ozone layer, climate change is the second major aspect of Earth System functioning that has been impacted by humans and which is being dealt with on the global political arena. The societal discussions that accompanied the discovery of a human influence on the ozone layer, and the subsequent ratification of the

Montreal Protocol, were not nearly as extensive or emotive as those relating to the scientific documentation of human interference with the climate system. One of the reasons for this is, presumably, the fact that relatively few actors in society were influenced by a ban on the emission of CFC gases. No citizen on Earth would be exempt from the effects of societal control of CO_2 emissions and many would feel the effects of controls put on greenhouse emissions from livestock production.

Only during the present generation have data series of sufficient length been available to demonstrate with reasonable certainty (Chapter 1) that humans are impacting the climate system in a manner that can be expected to result in substantial global warming. Thus, we are the first generation that has the knowledge that human activities are altering the global climate system. This also means that we are the first generation which has the ability – and we would argue the responsibility – to do something about it.

'Doing something about' human-induced climate change is a daunting task, but it also provides opportunities for humanity. As pointed out by Hetherington and Reid (2010), many of the developmental leaps made by human societies in the past have occurred in response to environmental changes. The combination of changing conditions and the concentration of human populations gave rise to revolutionary new ideas that have helped some groups of our species to flourish. Some human societies are more flexible and willing to embrace change than others, and many of the past societies unwilling to embrace change have perished. It seems likely that the capacity for societies to respond to the challenges of climate change will be greatest in societies that are open and willing to accept change. The degree of change that is acceptable to a society depends on the extent of the demands that change will place on society and how well the need for the change is understood. If the change is too large for the society to manage, or if the need for the change is not recognised and understood, or if the society simply is not willing to change, then decline or even extinction ensues. The ability to innovate, adjust rapidly and choose cooperation rather than conflict when facing change are skills that served our successful ancestors well. They are necessary to invoke again if humankind is to successfully redefine its relationship with the planet and meet the challenges posed by human-induced climate change.

References

Arrhenius, S. (1896). On the influence of carbonic acid in the air upon the temperature of the Earth. *London, Edinburgh, and Dublin Philosophical Magazine and Journal of Science*, **41**, 237–75.

Bolin, B. and Eriksson, E. (1958). Changes in the carbon dioxide content of the atmosphere and sea due to fossil fuel combustion. In *The Atmosphere and the Sea in Motion: Scientific Contributions to the Rossby Memorial Volume*, ed. B. Bolin. New York, NY: Rockefeller Institute Press, pp. 130–42.

Boyd, P. W. (2008). Implications of large-scale iron fertilization of the oceans. *Marine Ecology Progress Series*, **364**, 213–18.

Broecker, W. S. (2007). CO_2 arithmetic. *Science*, **315**, 1371.

Caldeira, K. (2008). Can a million tons of sulfur dioxide combat climate change? *Wired*, **16**.07, http://www.wired.com/science/planetearth/magazine/16–07/ff_geoengineering.

Callendar, G. S. (1938). The artificial production of carbon dioxide and its influence on temperature. *Quarterly Journal of the Royal Meteorological Society*, **64**, 223–40.

Crutzen, P. J. (2002). Geology of mankind: The Anthropocene. *Nature*, **415**, 23.

Crutzen, P. J. (2006). Albedo enhancement by stratospheric sulfur injections: a contribution to resolve a policy dilemma? *Climatic Change*, **77**, 211–20.

Crutzen, P. J. and Stoermer, E. F. (2000). The 'Anthropocene'. *Global Change Newsletter*, **41**.

Dansgaard, W., Clausen, H. B., Gundestrup, N. *et al.* (1982). A New Greenland deep ice core. *Science*, **218**, 1273–77.

Dansgaard, W., Johnsen, S. J., Clausen, H. B. *et al.* (1993). Evidence for general instability of past climate from a 250-kyr ice-core record. *Nature*, **364**, 218–20.

Darwin, C. (1859). *On the Origin of Species by Means of Natural Selection, or the Preservation of Favoured Races in the Struggle for Life*. London: John Murray.

Doran, P. T. and Kerdall Zimmerman, M. (2009). Examining the scientific consensus on climate change. *EOS, Transactions of the American Geophysical Union*, **90**, 22–3.

Etheridge, D. M. Steele, L. P., Langenfelds, R. L. *et al.* (1996). Natural and anthropogenic changes in atmospheric CO_2 over the last 1000 years from air in Antarctic ice and firn. *Journal of Geophysical Research*, **101**, 4115–28.

Fourier, J. (1824). Remarques générales sur les températures du globe terrestre et des espaces planétaires. *Annales de Chimie et de Physique*, **27**, 136–67.

Hetherington, R. and Reid, R. G. B. (2010). *The Climate Connection: Climate Change and Modern Human Evolution*. Cambridge, UK: Cambridge University Press.

Intergovernmental Panel on Climate Change (IPCC) (1990). *Climate Change: The IPCC Scientific Assessment. Report Prepared for IPCC by Working Group I*, eds. J. T. Houghton, G. J. Jenkins and J. J. Ephraums. Cambridge, UK: Cambridge University Press.

Intergovernmental Panel on Climate Change (IPCC) (1996). *IPCC Second Assessment: Climate Change 1995. A Report of the Intergovernmental Panel on Climate Change*. Geneva: IPCC.

Intergovernmental Panel on Climate Change (IPCC) (2001). *Climate Change 2001: The Scientific Basis. Contribution of Working Group I to the Third Assessment Report of the IPCC*, eds. J. T. Houghton, Y. Ding, D. J. Griggs, M. Noguer, P. J. van der Linden and D. Xiaosu. Cambridge, UK: Cambridge University Press.

Intergovernmental Panel on Climate Change (IPCC) (2005). *IPCC Special Report on Carbon Dioxide Capture and Storage*, eds. B. Metz, O. Davidson, H. de Coninck, M. Loos and L. Meyer. Cambridge, UK and New York, NY: Cambridge University Press.

Intergovernmental Panel on Climate Change (IPCC) (2007). *Climate Change 2007: The Physical Science Basis. Contribution of Working Group I to the Fourth Assessment Report of the IPCC*, eds. S. Solomon, D. Qin, M. Manning, Z. Chen, M. Marquis, K. B. Averyt, M. Tignor and H. L. Miller. Cambridge, UK and New York, NY: Cambridge University Press.

Keeling, C. D. (1960). The concentration and isotopic abundances of carbon dioxide in the atmosphere. *Tellus*, **12**, 200–03.

Keith, D. W. (2001). Geoengineering. *Nature*, **409**, 420.

Keith, D. W. (2009). Why capture CO_2 from the atmosphere? *Science*, **325**, 1654–55.

Keith, D. W. and Dowlatabadi, H. (1992). A serious look at geoengineering. *Eos, Transactions of the American Geophysical Union*, **73**, 289–93.

Kendall Zimmerman, M. (2008). *The Consensus on the Consensus: An Opinion Survey of Earth Scientists on Global Climate Change*. Chicago: University of Illinois.

Latham, J., Rasch, P., Chen, C. C. *et al.* (2008). Global temperature stabilization via controlled albedo enhancement of low-level maritime clouds. *Philosophical Transactions of the Royal Society of London A*, **366**, 3969–87.

Lorius, C., Jouzel, J., Ritz, C. *et al.* (1985). A 150,000-year climatic record from Antarctic ice. *Nature*, **316**, 591–96.

Lovelock, J. E. and Rapley, C. G. (2007). Ocean pipes could help the Earth to cure itself. *Nature*, **449**, 403.

Manabe, S. and Wetherald, R. T. (1967). Thermal equilibrium of the atmosphere with a given distribution of relative humidity. *Journal of the Atmospheric Sciences*, **24**, 241–59.

McNeill, J. R. (2000). *Something New Under the Sun: An Environmental History of the Twentieth-Century World*. New York, NY: W.W. Norton.

Mitchell, J. M. Jr (1972). The natural breakdown of the present interglacial and its possible intervention by human activities. *Quaternary Research*, **2**, 436–96.

Moore III, B., Underal, A., Lemke, P. and Loreau, M. (2002). The Amsterdan Declaration on global change. *Challenges of a Changing Earth: Proceedings of the Global Change Open Science Conference*, eds. W. Steffen, J. Jäger, D. J. Carson and C. Bradshaw. Amsterdan, The Netherlands, 10–13 July 2001. Springer-Verlag, Berlin, Herdelberg, New York, pp. 207–08.

Petit, J. R., Jouzel, J., Raynaud, D. *et al.* (1999). Climate and atmospheric history of the past 420,000 years from the Vostok ice core, Antarctica. *Nature*, **399**, 429–36.

Ramanathan, V. (1975). Greenhouse effect due to chlorofluorocarbons: Climatic implications. *Science*, **190**, 50–52.

Ramanathan, V., Callis, L. B. and Boughner, R. E. (1976). Sensitivity of surface temperature and atmospheric temperature to perturbations in the stratospheric concentration of ozone and nitrogen dioxide. *Journal of the Atmospheric Sciences*, **33**, 1092–1112.

Rockström, J., Steffen, W., Noone, K. *et al.* (2009a). A safe operating space for humanity. *Nature*, **461**, 472–75.

Rockström, J., Steffen, W., Noone, K. *et al.* (2009b). Planetary boundaries: exploring the safe operating space for humanity. *Ecology and Society*, **14**, 32. (online) URL: http://www.ecologyandsocial.org/vol14/iss2/art32/.

Steffen, W., Sanderson, A., Tyson, P. D. *et al.* (2004). *Global Change and the Earth System: A Planet Under Pressure*. Heidelberg: Springer Verlag.

Stephens, J. C. and Keith, D. W. (2008). Assessing geochemical carbon management. *Climatic Change*, **90**, 217–42.

The Royal Society (2009). *Geoengineering the Climate: Science, Governance and Uncertainty*. London: The Royal Society.

UNEP (2006). *Handbook for the Montreal Protocol on Substances that Deplete the Ozone Layer*, 7th edn. Nairobi, Kenya: United Nations Environment Programme Ozone Secretariat.

Viebahn, P., Nitsch, J., Fischedick, M. *et al.* (2007). Comparison of carbon capture and storage with renewable energy technologies regarding structural, economic, and ecological aspects in Germany. *International Journal of Greenhouse Gas Control*, **1**, 121–33.

Weart, S. R. (2008). *The Discovery of Global Warming*. Cambridge, MA: Harvard University Press.

World Council of Churches (2009). *A Sign of Hope for the Future for People of Good Will*. Statement from the World Council of Churches (WCC) to the High-Level Ministerial Segment of the 15th Session of the Conference of the Parties – COP15 to the UNFCCC 5th Session of the Meeting of the Parties to the Kyoto Protocol – CMP5. 18 December, Copenhagen, Denmark.

Zalasiewicz, J., Williams, M., Steffen, W. and Crutzen, P. (2010). The new world of the Anthropocene. *Environmental Science & Technology*, doi:10.1021/es903118j.

Index

DATE DUE

JAN 0 2 2014	